PRECALCULUS: MODELING OUR WORLD

COMAP'S

PRECALCULUS: MODELING OUR WORLD

DEVELOPED BY

COMAP, Inc.

57 Bedford Street, Suite 210

Lexington, Massachusetts 02420

PROJECT LEADERSHIP

Solomon Garfunkel
COMAP, INC., LEXINGTON, MA

Nancy Crisler
PATTONVILLE SCHOOL DISTRICT, ST. ANN, MO

Gary Froelich
COMAP, INC., LEXINGTON, MA

W. H. Freeman and Company
New York

The Consortium for Mathematics and Its Applications (COMAP)

57 Bedford Street, Suite 210
Lexington, MA 02420

Published and distributed by

W. H. Freeman and Company

41 Madison Avenue, New York, NY 10010

www.whfreeman.com

ISBN 0-7167-4359-0

Printed in the United States of America.
First printing 2001

PROJECT LEADERS: Solomon Garfunkel, COMAP, Inc., Lexington, MA
Nancy Crisler, Pattonville School District, St. Ann, MO
Gary Froelich, COMAP, Inc., Lexington, MA

AUTHORS: Nancy Crisler, Pattonville School District, St. Ann, MO
Gary Froelich, COMAP, Inc., Lexington, MA

CONTRIBUTING AUTHORS: Sharon North, St. Louis Community College, St. Louis, MO
Dale Winter, Harvard University, Cambridge, MA

PUBLISHER: Michelle Julet
ACQUISITIONS EDITOR: Craig Bleyer
MARKETING MANAGER: Mike Saltzman
SUPPLEMENTS AND NEW MEDIA EDITOR: Mark Santee
COVER DESIGN: Salem Krieger
FRONT COVER: (Clockwise from upper left) map: Corbis;
calculator: courtesy of Texas Instruments Incorporated;
traffic: Jim Schafer/eStock Photography/Picture Quest;
jet: Corbis; seismograph: Roger Ressmeyer/Corbis;
temperature sign: Patrick Endres/Alaska Stock; logs: Bill Varie/Corbis
PHOTO RESEARCH: Michele Doherty, Susan Van Etten
ILLUSTRATIONS: Lianne Dunn, Mary Reilly
PRODUCTION DEPARTMENT: Michele Doherty, Daiva Kiliulis,
George Ward, Gail Wessell, Pauline Wright
COMPOSITION: Daiva Kiliulis, Michele Doherty, Mary Reilly
MANUFACTURING: RR Donnelley & Sons Company

Dear Student,

This Precalculus text is a different kind of math book than you may have used, for a different kind of math course than you may have taken. In addition to presenting mathematics for you to learn, we have tried to present mathematics for you to use. We have attempted in this text to demonstrate mathematical concepts in the context of how they are actually used day to day. The word "modeling" is the key. Real problems do not come at the end of chapters in a math book. Real problems don't look like math problems. Real problems ask questions such as: How do we create computer animations? Where should we locate a fire station? How do we effectively control an animal population? Real problems are messy.

Mathematical modeling is the process of looking at a problem, finding a mathematical core, working within that core, and coming back to see what mathematics tells you about the problem with which you started. You will not necessarily know in advance what mathematics to apply. The mathematics you settle on may be a mix of several ideas in geometry, algebra, and data analysis. You may need to use computers or graphing calculators. Because we bring to bear many different mathematical ideas as well as technologies, we call our approach "integrated."

Another very important and very real feature of this course is that frequently you will be working in groups. Many problems will be solved more efficiently by people working in teams. We have done all of this to emphasize our primary goal: Presenting you with mathematical ideas the way you will see them as you go on in school and out into the work force. There is hardly a career that you can think of in which mathematics will not play an important part and in which understanding mathematics will not matter to you.

This course is the gateway to collegiate mathematics. As such you will see a number of essential new concepts and be asked to learn a number of important new skills. But, most of all, we hope you have fun. Mathematics is important. Mathematics may be the most useful subject you will learn. Using mathematics to solve truly interesting problems about how our world works can and should be an enjoyable and rewarding experience.

Solomon Garfunkel
EXECUTIVE DIRECTOR, COMAP

Dear Teacher,

This is a different Precalculus text. It contains all of the topics that you would expect to find in a Precalculus text and more; but it is a text written by COMAP and applications is our middle name. Thus you will see the concepts and skills of Precalculus presented from an applications and modeling point of view. Moreover, this is an activity and problem-solving based book. We want students to experience the mathematics they are learning in a way that will make both the mathematics and the contemporary applications come alive.

We make serious use of graphing calculators and a variety of software, including spreadsheets and geometric utility programs. We introduce and make use of data analysis techniques. All of which permit us to analyze deeper and more meaningful problems. And we have attempted to add material that will in fact, as well as name, better prepare students for their work in calculus—from our introductory work on functions to our final chapter on finite difference equations. We use a spiraling approach, coming back to earlier ideas in later chapters, so that we can keep those ideas, concepts, and skills alive, reinforcing their importance as well as providing additional practice.

This text follows on over twenty years of work by COMAP in curriculum reform. As such there are a great many people whose work past and present should be acknowledged. First and foremost, we wish to recognize the author team led by Gary Froelich and Nancy Crisler. We also want to pay special tribute to the COMAP production staff led by George Ward who worked tirelessly, with unreasonable deadlines, to create this handsome text.

Solomon Garfunkel
EXECUTIVE DIRECTOR, COMAP

COMAP

CONTENTS

PRECALCULUS:
MODELING OUR WORLD

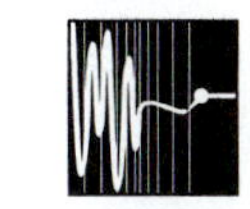

FUNCTIONS IN MODELING

CHAPTER INTRODUCTION

MODELING BEHAVIOR: EXPLANATIONS AND PATTERNS

Providing Explanations

Mathematical modeling is a process by which a real-world situation is replaced with a mathematical one. If the real-world situation and the mathematical setting are well matched, then information obtained in the mathematical setting is meaningful in the real-world setting. Quite often, the mathematical setting includes one or more equations or graphs or other geometric objects that are called the **mathematical model**.

Mathematical models are built to explain why things happen in a certain way. They are also created for the purpose of making predictions about the future. For example, you may want to know why a comet travels in the path that it does and when it will be visible again from the earth.

Finding Patterns

A good mathematical model can help the modeler understand important features of the situation under investigation. The reverse is also true. Knowing features of the situation can help you develop a good model.

When developing a model, modelers often look for the existence of regularity or patterns in the real world. Frequently these patterns involve numbers. Describing these patterns mathematically helps produce information that is useful.

Among the simplest patterns are those that relate one real-world quantity to another. Sometimes these patterns are more evident when viewed visually, that is, with the aid of a graph.

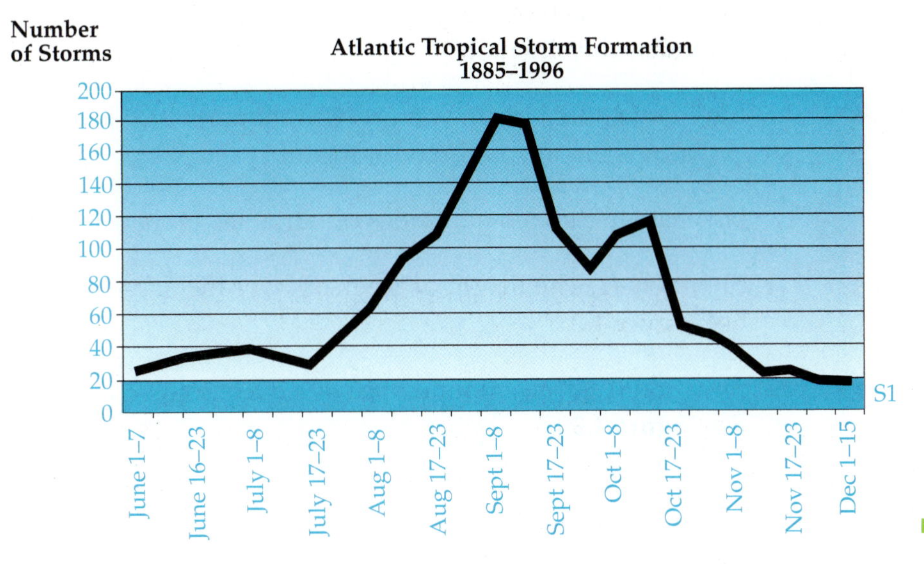

FIGURE 1.1.

Graphs are used so often because they tell a story more easily than it can be told with words or numbers. The story in **Figure 1.1** is apparent at a glance: The number of tropical storms in the Atlantic increases to a maximum around the end of August, then falls off rapidly. There is a moderate increase in early October, after which the decreasing pattern resumes.

The following activity asks you to reflect on several familiar real-world situations in which one quantity is related to a second. You will be asked to think about how graphs describing these situations might appear. Look for patterns and trends as you analyze the possibilities. Focus on features you consider to be important in the graphical models.

Warm-Up Activity

Note that the graphs in the activity (as others you will be asked to draw) are shown without numerical scales. They show qualities that capture the key features of the situation (patterns and trends), but do not show exact quantities.

For each of the following six scenarios, a context and a figure showing several graphs are given. After discussing the context with your partner or group, answer the following questions for each situation.

a) Examine each of the graphs in the figure. Which graph best models the situation?

b) What features made you choose that particular graph? What features made you discount the other graphs?

c) What are the two quantities or variables in the situation?

1. Situation #1: The height of a person over his or her lifetime. See **Figure 1.2.**

2. Situation #2: The circumference of a circle as its radius changes. See **Figure 1.3.**

3. Situation #3: The height of a ball as it is thrown into the air. See **Figure 1.4.**

4. Situation #4: The amount of observable mold on a piece of bread sitting at room temperature from the time it is baked to several months later. See **Figure 1.5.**

5. Situation #5: The daily average low temperature in degrees Fahrenheit in Fairbanks, Alaska from January 1 to December 31. See **Figure 1.6.**

6. Situation #6: The temperature of a cold drink left in a warm room. See **Figure 1.7.**

Discussion/Reflection

1. List some of the important features of the graphs that helped you choose the ones that best represented, or modeled, the given situations.

2. In situations 2 and 6, arrows were drawn on the ends of the graphs to show that the graphs continue indefinitely. Explain why such arrows were not used in the other graphs.

3. You identified the two variables in each of the six situations in Activity 1.1. For which of those situations does it make sense for either of the variables to be negative? Explain.

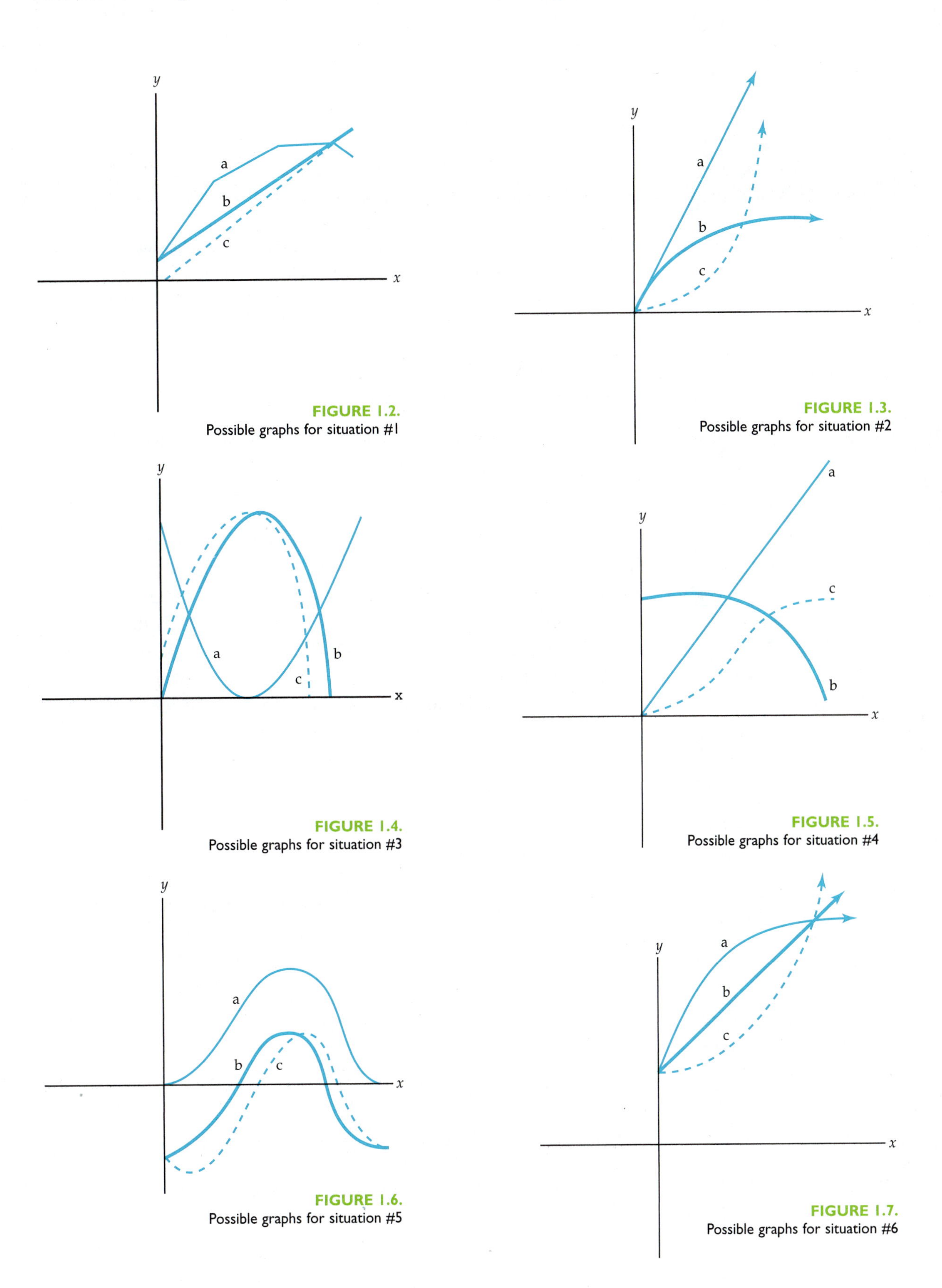

FIGURE 1.2.
Possible graphs for situation #1

FIGURE 1.3.
Possible graphs for situation #2

FIGURE 1.4.
Possible graphs for situation #3

FIGURE 1.5.
Possible graphs for situation #4

FIGURE 1.6.
Possible graphs for situation #5

FIGURE 1.7.
Possible graphs for situation #6

LESSON 1.1

Functions as Models

In the Warm-Up Activity, graphs were used to represent relationships between two quantities. In each case, a graph served as a useful tool in detecting a pattern or general trend. As you will see in this chapter, a graph can also be used to predict a value of one quantity from a value of the other. Often, that is the goal of mathematical model: to predict what will happen to one quantity when another changes. That is the central idea in something mathematicians call a function: to produce exactly one value of one quantity from a known value of another.

In this chapter you will begin a study of mathematical functions and their properties.

FUNCTIONS

For the purposes of this book, the variable G is defined as a **function** of the variable x if each value of x has a unique (one and only one) value of G associated with it. G is called the **dependent** or **response** variable and x is called the **independent** or **explanatory** variable. For example, in situation #2 of Activity 1.1, the circumference of a circle (C) is a function of the radius (r). Thus the dependent variable is C and the independent variable is r. That is, the independent (explanatory) variable "explains" the dependent variable while the dependent (response) variable "responds" to changes in the independent variable.

DOMAINS AND RANGES

The set of all values that make sense for the independent or explanatory variable is called the **domain** of the function. Corresponding to the domain, the set of output values that a function generates from its domain is called the **range** of the function.

EXAMPLE 1

1. Determine a domain for each situation in Activity 1.1.

2. Determine the domain for the function defined by the equation $y = \frac{1}{x}$.

3. Determine the domain of the function that assigns the GNP (Gross National Product) to each year in the 1990's decade.

SOLUTION:

1. The domain for each of the six situations in Activity 1.1 is determined by the context of the problem. The domains in situations 1, 3, 4, and 5 are bounded; that is, there are numbers that are too large or too small to make any sense. Thus, the domains consist of numbers x for which $0 \le x \le n$ where n is some number meaningful to the context.

   The domains in situations 2 and 6 consist of all numbers greater than or equal to zero; there is no largest reasonable value for the independent (explanatory) variable.

2. The domain of $y = \frac{1}{x}$ does not include 0, since division by 0 is impossible. In this case, $x \neq 0$ is a **restriction** on the domain.

3. The domain is the years from 1990 through 1999: 1990, 1991, 1992, 1993, 1994, 1995, 1996, 1997, 1998, 1999.

An easy way to write domains and ranges is to use **interval notation**. A **closed interval** $[a, b]$ indicates all real numbers x for which $a \le x \le b$. Closed intervals include their endpoints. An **open interval** (a, b) indicates all real numbers x for which $a < x < b$. Open intervals exclude their endpoints. **Half-open** or **half-closed** intervals are denoted by $(a, b]$ or $[a, b)$. These notations indicate all real numbers x for which $a < x \le b$ (that is, open at a and closed at b) and $a \le x < b$ (open at b and closed at a) respectively.

EXAMPLE 2

Express each of the following in interval notation.

a) $-2 < x \le 5$ b) $8 \ge x \ge 0$ c) number line: -5, -4, -3, -2, -1, 0

TAKE NOTE

If you want to indicate the unboundedness of x in the positive direction, use the symbol ∞ (infinity). To indicate unboundedness in the negative direction, use $-\infty$. Thus,

$[a, \infty)$ indicates all real numbers x for which $a \le x < \infty$.

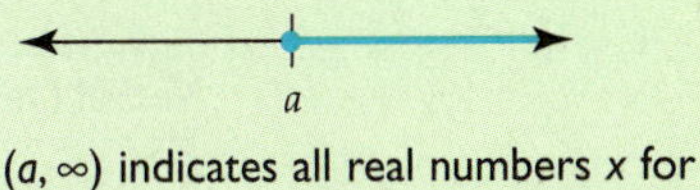

(a, ∞) indicates all real numbers x for which $a < x < \infty$.

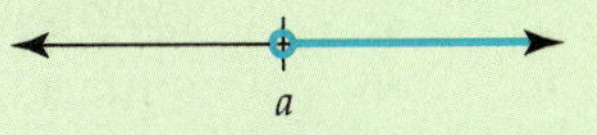

$(-\infty, a]$ indicates all real numbers x for which $-\infty < x \le a$.

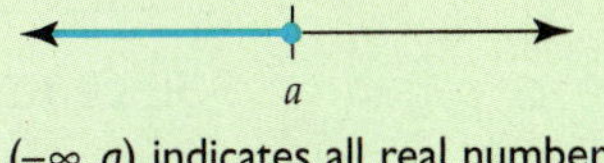

$(-\infty, a)$ indicates all real numbers x for which $-\infty < x < a$.

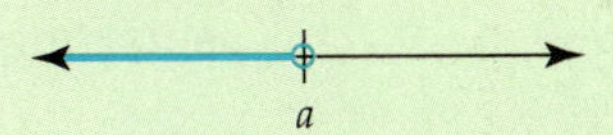

$(-\infty, \infty)$ indicates all real numbers x for which $-\infty < x < \infty$, namely, all real numbers.

SOLUTION:

a) The half-open interval can be expressed as (–2, 5].

b) The closed interval can be expressed as [0, 8].

c) Graphs typically indicate open endpoints with open dots (dots that are not filled in). Closed endpoints are denoted by solid dots. Thus, the graph indicates the half-open interval [–5, –1).

Caution: The notation (3, 4) can be interpreted as the point (3, 4) in the coordinate plane or as the open interval consisting of all numbers strictly between 3 and 4. Context determines the meaning.

REPRESENTATIONS

Functions can be represented by verbal descriptions, symbolic rules, tables, graphs, and visual devices such as function machines and arrow diagrams. A function that represents a given contextual situation is an example of a **mathematical model**.

Verbal Descriptions

For simple functions, a **verbal description** may be all you need. For example, the doubling function doubles each input value, and the squaring function squares each input value.

Symbolic Rules

Since functions involve mathematical operations, **equations** are frequently the most compact representations. Thus, the doubling function can be written as $y = 2x$. The equation $y = x^2$ describes the squaring function. Both of these functions have unrestricted domains.

One way of writing the equation for a function is to use **function notation**. For example, you could write the squaring function as $H(x) = x^2$. The symbol $H(x)$ is read as "the value of H when x is input," or simply "H of x." It means that H is the name of the function, that x is the independent variable, that $H(x)$ is the dependent variable, and that the action is to square the input value.

In function notation, the input variable is named in the parentheses and $H(x)$ is the output variable. However, it is not the letter used that is important when naming a function; it is the rule for the function that counts. For example, $g(x) = 2x + 5$, $f(m) = 2m + 5$ and $h(t) = 2t + 5$ all represent the same function, one that doubles the input value and adds five. Be aware, too, that notations such as $g(x)$ and y are used interchangeably. For example, $y = 2z + 5$ defines the same function as g, f, and h, above. Function notation is especially useful when evaluating functions at specific values.

TAKE NOTE

$g(x)$ is the name of the dependent variable and does not indicate multiplication. So, $g(x)$ does not mean some variable g times some variable x.

EXAMPLE 3

If $g(x) = 4x - 3$, find $g(8)$. Also find the value of x for which $g(x) = -9$.

SOLUTION:

$g(8) = 4(8) - 3 = 32 - 3 = 29.$

If $g(x) = -9$, then $4x - 3 = -9$.

Solve this equation: $4x = -9 + 3 = -6$

$x = -6/4 = -3/2 = -1.5$

Visual Representations

Verbal descriptions usually describe functions as actions, that is, what the functions do to their inputs.

A more visual representation, such as the **function machines** shown in **Figure 1.8**, may convey the function-as-an-action idea more clearly.

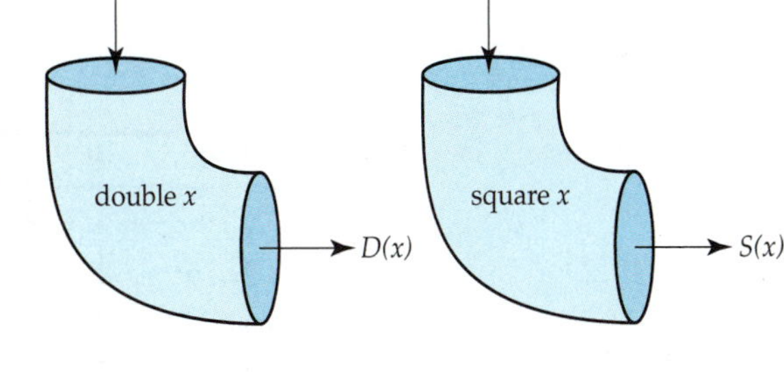

FIGURE 1.8.
Function machines.

The function machines in Figure 1.8 represent $D(x) = 2x$ and $S(x) = x^2$. A value for the independent variable x is selected from the domain. It then goes into the machine and the rule is applied to it. The resulting value of the dependent variable (what comes out of the machine) is a value in the range of the function.

To be a function, each value of the domain must correspond to exactly one value in the range. For example, for the doubling function, $D(3) = 6$

(and only 6). For the squaring function, $S(3) = 9$ (and only 9). Note that $S(-3)$ is also 9, but that is acceptable. Each input needs exactly one output, but it is permissible for an output value to correspond to more than one input value.

Arrow diagrams (**Figure 1.9**) can be thought of as simplified function machines. Again, they represent functions as operations on numbers.

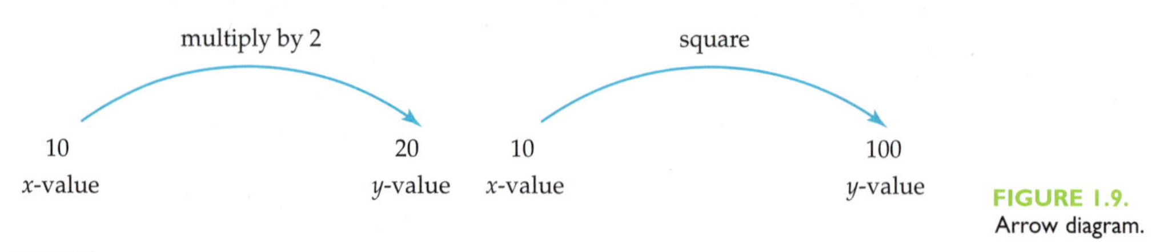

FIGURE 1.9. Arrow diagram.

Tables

Tables are another important representation. In tables, the column headings represent the variables, and the numbers in the columns are the values of the variables. For example, if **Figures 1.10** and **1.11** represent the doubling function D, then D(1) = 2. Exactly two table columns are needed to define a function's input and output, so if a table displays more than two columns, you need to check which two columns represent the function you are studying.

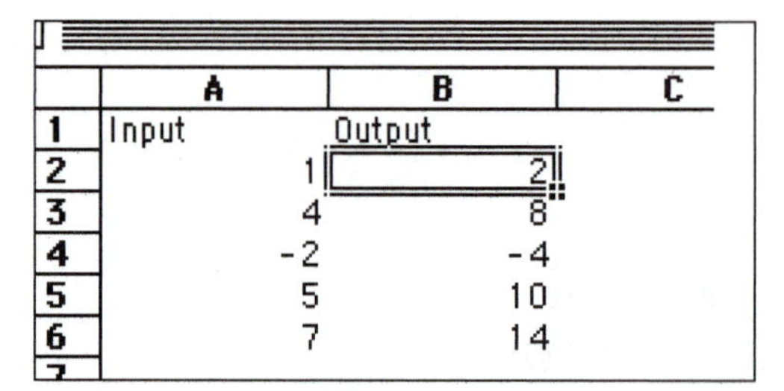

	A	B	C
1	Input	Output	
2	1	2	
3	4	8	
4	-2	-4	
5	5	10	
6	7	14	

FIGURE 1.10. Spreadsheet table for $y = 2x$.

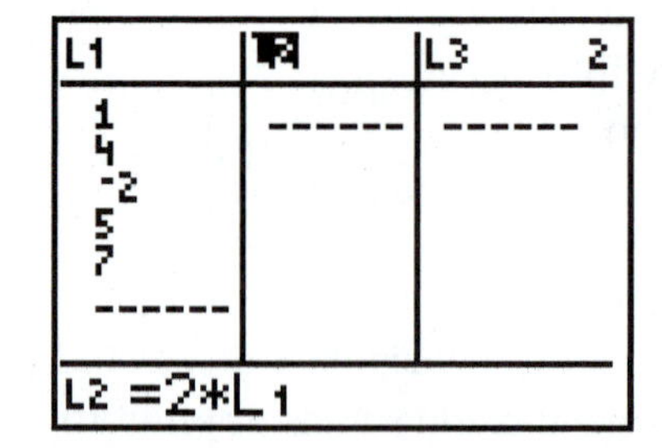

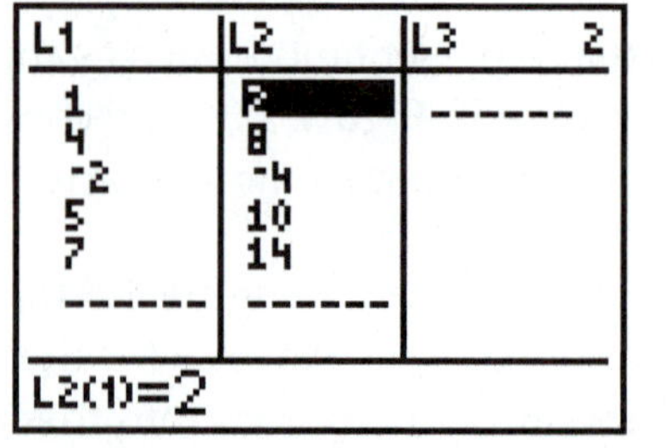

FIGURE 1.11. Calculator lists for $y = 2x$.

It is sometimes helpful to combine representations, especially with tables. Arrow diagrams can show how to get from one column of a table to another. Similar information appears as the formulas for columns of a spreadsheet (top of Figure 1.10) or calculator lists (bottom left of Figure 1.11).

To determine whether a table represents a function, check that each input value has exactly one output value. This is the case for the tables shown in Figures 1.10 and 1.11, so each represents a function. **Table 1.1** does not describe a function since the input value 1 has two values, 7 and 5, associated with it.

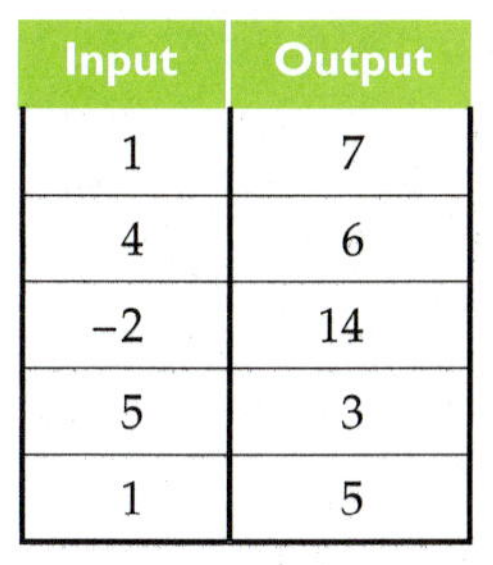

Input	Output
1	7
4	6
−2	14
5	3
1	5

TABLE 1.1.
Input and output values (not a function).

One drawback of a table as a representation of a function is that it may not display all of the values of the function. If that is the case, care must be taken to consider all input values in the domain.

Graphs

Perhaps the most familiar representation of functions is as **graphs**. In graphing functions, it is customary to use the horizontal axis for the domain and the vertical axis for the range. A graph of $y = 2x$ can be constructed from Figure 1.10 by plotting the points and connecting them with a line (**Figure 1.12**).

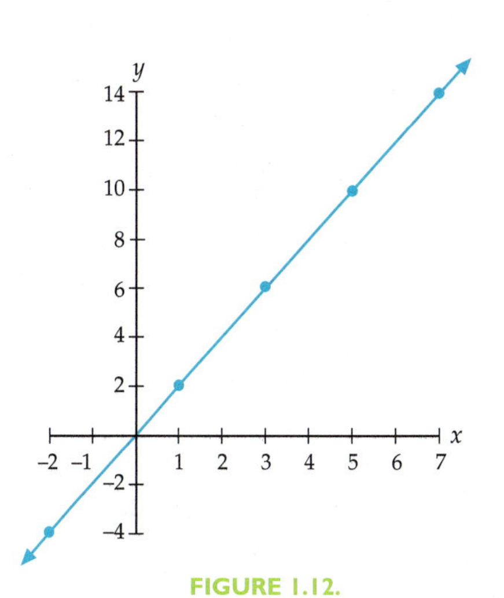

FIGURE 1.12.

In **Figure 1.13**, the dot at the left end of the curve indicates that the graph does not continue further to the left and that $f(-3) = -1$. The arrow at the right indicates that the graph does continue to the right. Thus, the indicated domain of the function is all numbers greater than or equal to –3, or $[-3, \infty)$. Similarly, the range is all numbers greater than or equal to –1, or $[-1, \infty)$.

One advantage of graphs (and tables) over arrow diagrams, function machines, and function notation is that both graphs and tables display many input-output pairs simultaneously, possibly making important properties more visible. However, a graph is not suitable for evaluating a function precisely. Pairs of values that can be plucked from a table or calculated from a formula must be estimated as coordinates of a point on the graph.

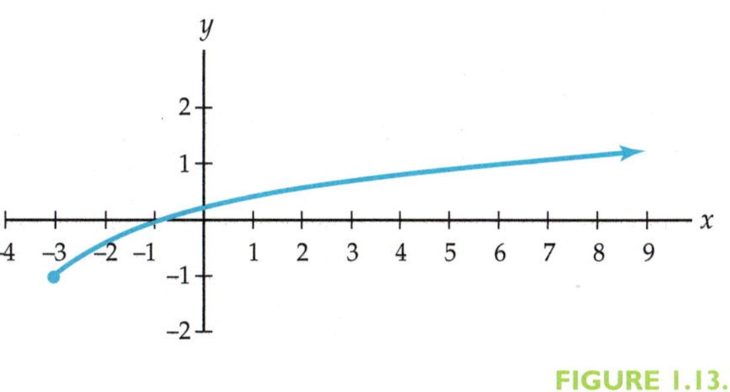

FIGURE 1.13.
Finding the domain and range from a graph.

As you continue to learn more about functions, you will use all of these different representations as tools to help you understand and solve real-world problems.

Exercises 1.1

1. Consider the graphs in **Figures 1.14–1.17**. Use the definition of function to determine whether each graph is the graph of a function. Explain your reasoning.

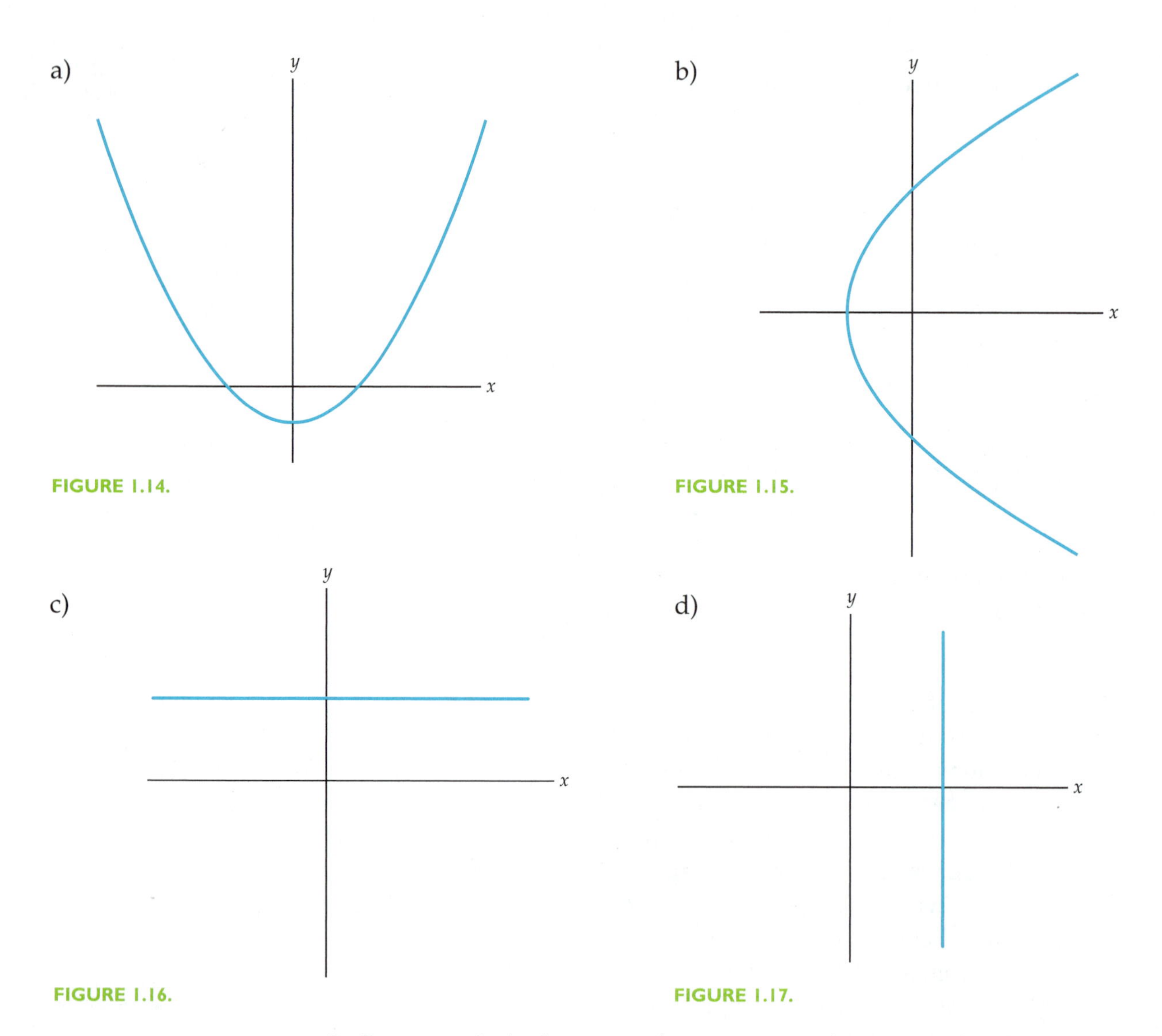

FIGURE 1.14.

FIGURE 1.15.

FIGURE 1.16.

FIGURE 1.17.

2. One test of whether a graph represents a function is the **vertical line test**. This simple test says that if you can find any vertical line that intersects the graph in more than one place, the graph is not a function. If no such line exists, then the graph is a function.

Exercises 1.1

Consider the graphs shown in **Figure 1.18**.

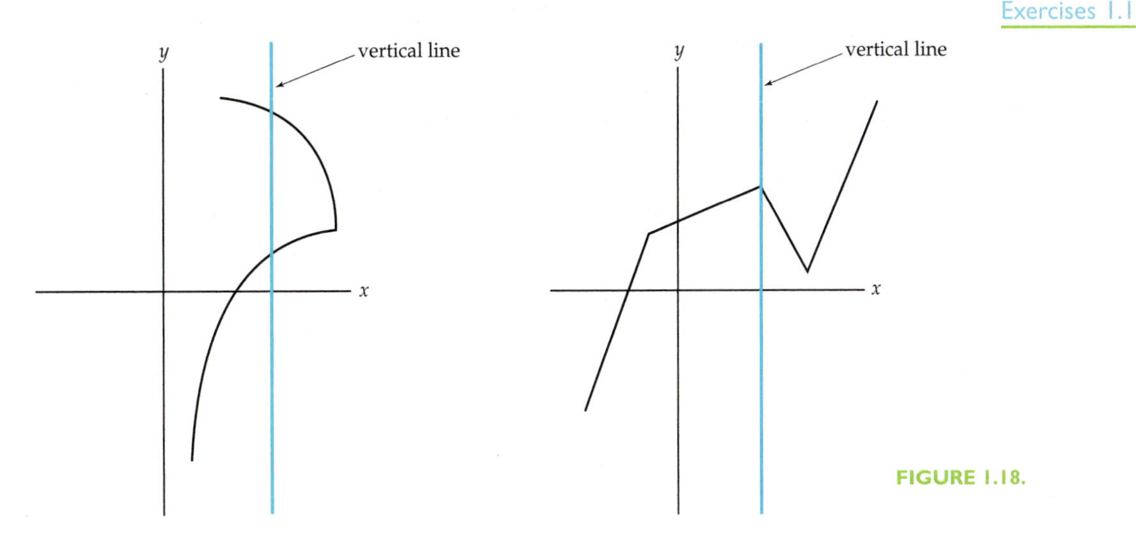

FIGURE 1.18.

Since graph A intersects the vertical line in two places, it does not pass the vertical line test, and therefore, is not the graph of a function. But in graph B, no matter where you move the vertical line, it will never intersect the graph in more than one point, so this is the graph of a function.

a) Use the definition of function to explain why the vertical line test is a valid test.

b) Apply the vertical line test to the graphs in Exercise 1. Which graphs pass the test and are functions, and which do not?

3. In each of **Tables 1.2–1.5**, x is the independent variable and y the dependent variable. Which of the tables represent functions and which do not? Explain.

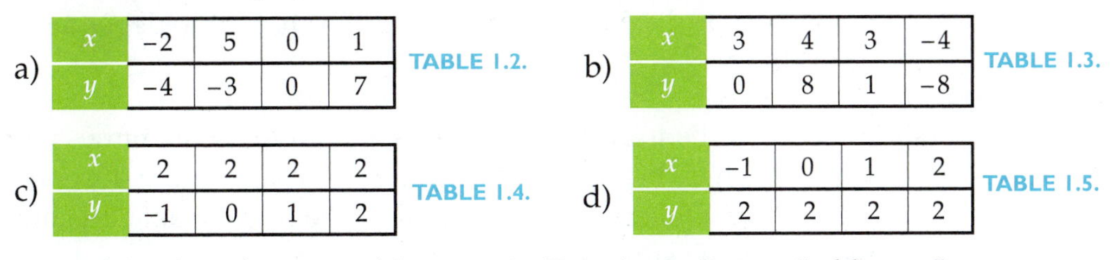

a)

x	−2	5	0	1
y	−4	−3	0	7

TABLE 1.2.

b)

x	3	4	3	−4
y	0	8	1	−8

TABLE 1.3.

c)

x	2	2	2	2
y	−1	0	1	2

TABLE 1.4.

d)

x	−1	0	1	2
y	2	2	2	2

TABLE 1.5.

e) Look back at the vertical line test in Exercise 2. State a "table test" that can be used to identify tables that do not represent functions.

Exercises 1.1

f) Use your table test and the definition of function to explain why $y = x^2$ is a function even though both $x = 3$ and $x = -3$ give the same x-value.

g) Draw arrow diagrams for Table 1.2.

For each function described in Exercises 4–8, determine (a) the value of $f(3)$, and (b) the value(s) of x for which $f(x) = 3$.

4. $f(x) = 5x + 2$

5. $f(x) = |x|$

6. $f(x) = x^2 - 1$

7. See **Figure 1.19**.

8. See **Figure 1.20**.

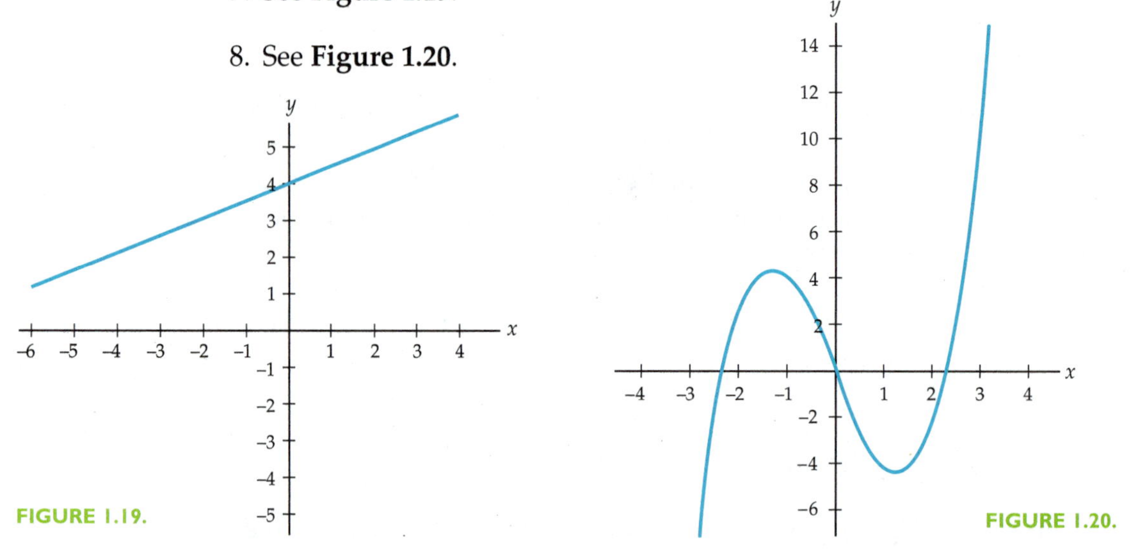

FIGURE 1.19.

FIGURE 1.20.

9. Return to the six situations in the Warm-Up Activity. What is the independent variable in each situation? What is the dependent variable?

10. For each of the following situations, identify two quantities that vary. Which is the independent variable? Which is the dependent variable? Use interval notation to indicate a reasonable domain and range for each situation.

    a) The number of leaves on a tree during the year in New England.

    b) The amount of time spent studying and the grade earned on the test.

c) Depth of the water in a bath tub with a steady stream of water running in.

d) The height of a candle as it burns down.

e) The number of car accidents in a certain city and the amount of alcohol consumed per person.

f) The length of a student's hair over the course of a year.

11. A **qualitative graph** demonstrates the important features of a graph without worrying about exact scales. The graphs in the Warm-Up Activity were qualitative graphs. Sketch a qualitative graph for each of the situations in Exercise 10.

12. Express each of the following in interval notation.

a) $4 < y \leq 10$

b) 

c) $-2 \geq x \geq -3.5$

d) 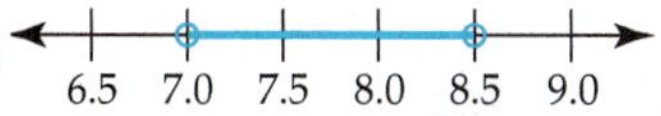

13. Consider the function $f(x) = \dfrac{x+1}{x-2}$.

a) Is the point (5, 2) on the graph of f?

b) Determine the value of $f(-4)$.

c) What is the domain of f?

14. To chemists and others, solubility in water is an important property of a substance. As they investigated this property, they discovered a pattern in the relationship between temperature and solubility, which can be seen by exploring data such as those in **Table 1.6.**

Temperature °C	Grams of potassium chloride (KCl) per 100 Grams of Water
10	30
19	32
30	36
43	40
50	42
59	45

TABLE 1.6.

a) Does the table represent a function? Explain.

b) Which variable is the independent variable and which is the dependent variable?

Exercises 1.1

c) One way to examine data for trends and patterns is to construct a graph of ordered pairs, which is called a **scatter plot**. To construct a scatter plot of the ordered pairs in Table 1.6, plot the data with the independent variable as the x-coordinate and the dependent variable as the y-coordinate. Make a scatter plot of the data in Table 1.6. (The first two ordered pairs are plotted in **Figure 1.21** for you.)

TAKE NOTE

Figure 1.21 shows a scatter plot of the number of grams of KCl that can be dissolved in 100 grams of water **versus** the temperature of the solution. The placement of the word "versus" means that the number of grams of KCl per 100 grams of water is the dependent (response) variable and the temperature is the independent (explanatory) variable.

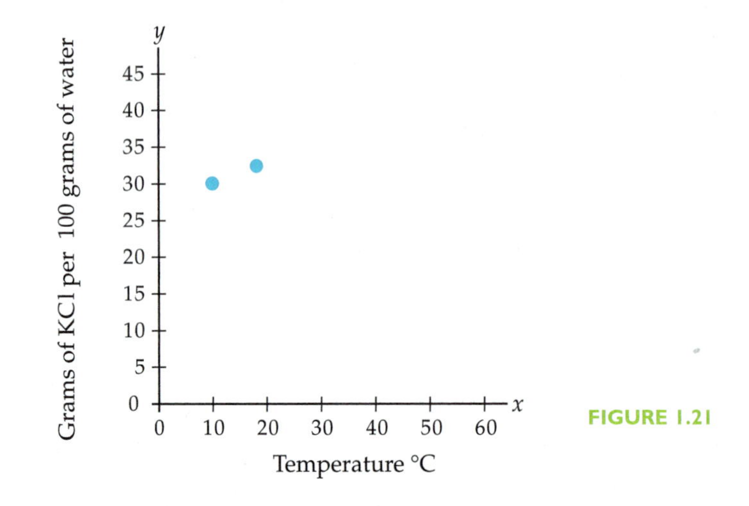

FIGURE 1.21

d) Scatter plots with data that fall along a straight line are said to have a **linear form**. If the data do not fall along a straight line, the scatter plot has a **nonlinear form**. Describe the form of the scatter plot in (c).

15. **Table 1.7** shows data for the temperature of a cup of coffee as it cools down.

TABLE 1.7.

Time (minutes)	0	1	2	3	4	5	6	7	8
Temperature (°C)	80.0	66.3	55.6	47.3	40.8	35.8	31.8	28.8	26.4

a) What is the independent variable? What is the dependent variable?

b) Enter the data from Table 1.7 into your calculator lists and create a scatter plot of the temperature of the coffee versus time.

c) Describe the form of the scatter plot.

d) Describe the pattern you see in the scatter plot.

16. Graphs can be characterized by their curvature. This text uses an informal treatment beginning with this exercise. A graph is concave up if it bends upward and concave down if it bends downward. You might think of a graph that is concave up as being shaped like a cup (or part of a cup) in its upright position, while a graph that is concave down is shaped like an inverted cup. In **Figure 1.22**, graphs (a) and (b) are concave up while (c) and (d) are concave down.

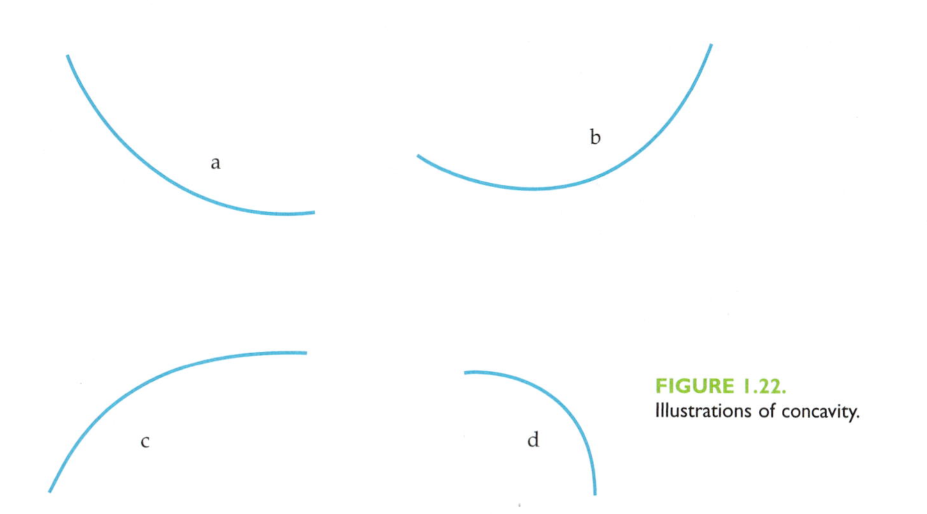

FIGURE 1.22.
Illustrations of concavity.

a) Consider the graph in **Figure 1.23**. When is the graph concave up? When is it concave down?

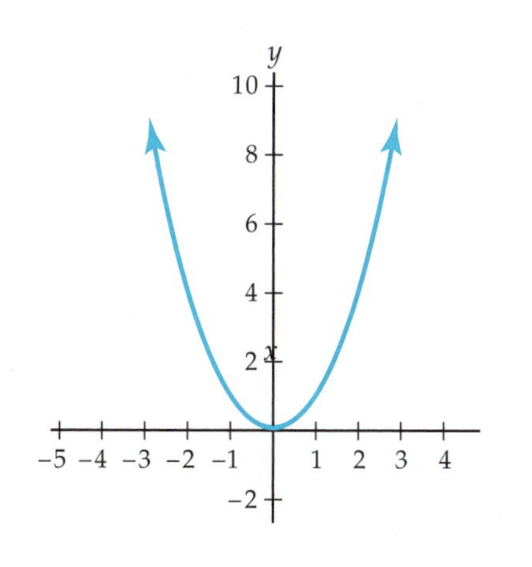

FIGURE 1.23.

Exercises 1.1

b) Consider the graph in **Figure 1.24**. When is the graph concave up? When is it concave down?

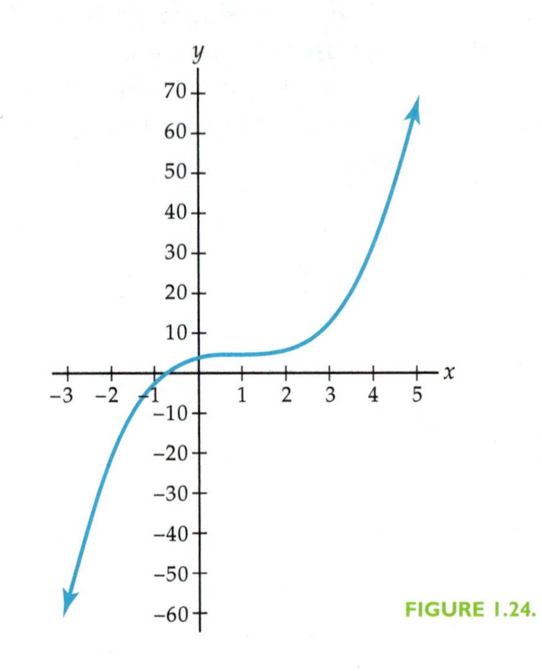

FIGURE 1.24.

LESSON 1.2

Creating A Mathematical Model

In its simplest form, mathematical modeling is the process of beginning with a situation and gaining understanding about that situation through the use of mathematics. It consists of several stages, each of which is vital, although not necessarily mathematical in nature.

THE MODELING PROCESS

The process of modeling can be summarized in the following steps:

Step 1. *Problem Identification*: What is it you would like to do or find out? Make some general observations. Pose a well-defined question asking exactly what you wish to know.

Step 2. *Simplify and Make Assumptions*: Identify the factors that will be used in building a model. Generally you must simplify to get a manageable set of factors.

Step 3. *Build the Model*: Interpret in mathematical terms the features and relationships you have chosen. Your resulting model may be a graph, a table, or an equation. Analyze the model to find answers to the questions originally posed.

Step 4. *Evaluate, Interpret, and Revise as Necessary*: Your conclusions at this point apply to your mathematical model. Verify your conclusions theoretically or by collecting data. Does your model yield results that are meaningful and accurate? If not, refine the model by reexamining your assumptions. Based upon the accuracy of your model, relate the mathematical conclusions to the real world.

Figure 1.25 is useful in describing the modeling process:

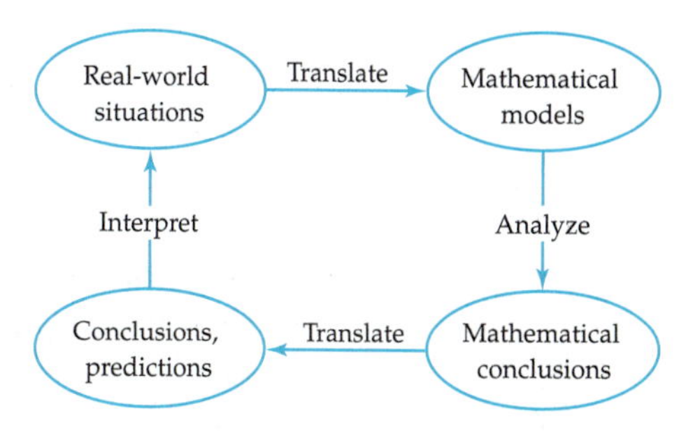

FIGURE 1.25.
A schematic of the modeling process.

USING DATA

In the modeling process data serve at least two distinct purposes. One purpose is to check the accuracy of a proposed model. For example, based on the work of Galileo and Newton, modelers might expect the distance traveled by a falling body to be proportional to the square of the time elasped. In such situations, data are gathered to check the accuracy of the conjecture.

A second use of data is to help the modeler construct a model. In some situations it's impossible to explain a certain behavior by theory, but a prediction is needed anyway. For example, you may not be able to build a model fully explaining the spread of the Asian flu, but must make estimates of the number of cases expected each year to plan for vaccine use. In such cases, data are collected and examined. If a pattern in the data is observed, an attempt is made to capture the trend of the pattern with a model. This resulting model may in turn offer hints on why the variables are related as they are.

Thus, as in the first case, a model may be driven by theory or, as in the second case, driven by data.

Theory ➡ Model ➡ Data

Data ➡ Model ➡ Theory

Both theory-driven and data-driven models are appropriate. Each has its advantages and disadvantages as you will see in this course.

In Activity 1.2, you will participate in the modeling process and build a data-driven model. In the process you will be asked to collect a set of data, search for a pattern, create a graph as a model, and then use your model to make predictions.

Activity 1.1

A first step in developing the desired model is to make observations and formulate a well-defined question asking exactly what you want to know.

When purchasing a man's dress shirt, it is not necessary to specify both the collar and the cuff size. Shirt designers seem to need only the collar size in order to manufacture a shirt that fits both the collar and the cuff of the wearer.

After making this observation, you might ask the following question: Is there a relationship between the collar size and cuff size in men's shirts?

To help answer this question you might decide to collect a sample of data. But before collecting the data, you need to identify the factors that you will use in building your model. For example, for the purposes of this activity, you will assume that the circumference of a person's neck corresponds to collar size and that the circumference of a person's wrist corresponds to cuff size. And since dress shirts are purchased by collar size, neck circumference will be considered the independent (explanatory) variable and wrist size the dependent (response) variable.

1. Collect the data from each group member and complete **Table 1.8.**

2. Prepare a scatter plot of the data set for your group. Clearly label the axes.

4. Create a new data set by combining the data from each group in the class. What patterns do you observe? Draw a function on your scatter plot that provides a good fit to the data.

Student	Neck circumference in cm	Wrist circumference in cm
#1		
#2		
#3		
#4		
#5		
#6		

TABLE 1.8.

5. Use your graph from Item 4 to predict your instructor's wrist circumference based on his or her neck circumference. How good is your prediction?

6. The final step of the modeling process asks you to evaluate your model. Do you think that your model answers your initial question? Is it accurate enough? Is there anything that you might do to improve your model?

USING RESIDUALS TO JUDGE A MODEL

An important tool that mathematicians use to judge a model is residuals. A **residual** is the error between an actual value and the value predicted by a model. For example, if a model predicts a wrist circumference of 7.5 inches for someone whose actual wrist circumference is 7 inches, then the residual is 7 – 7.5 = –0.5. If a model is perfect, all residuals are 0. However, there are few, if any, perfect models in the real world.

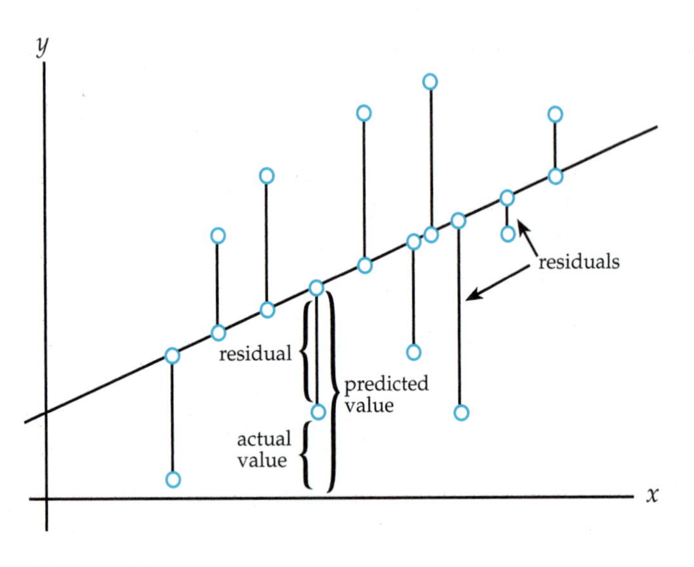

FIGURE 1.26.

Geometrically, a residual is the vertical distance between each data point and the line or curve that is the graph of the prediction equation. See **Figure 1.26**.

It can be difficult to visualize residuals in a scatter plot. Therefore, mathematicians make a special type of scatter plot that shows only the residuals. This residual plot is helpful because it usually has a smaller vertical scale than the original scatter plot. For a good model, the residual plot shows no pattern and the residuals are relatively small.

Example 4

Table 1.9 displays data on neck circumference and wrist circumference in a group of six people.

TABLE 1.9.
Neck circumference and wrist circumference data.

Neck circumference	Wrist circumference
16″	8″
14.5″	7″
15.5″	6.5″
13.5″	6.5″
13″	6.5″
14″	6

One of the people in the group offers the following model: To estimate wrist circumference, subtract 1 inch from neck circumference and divide by 2.

Calculate the residuals for this model and enter them in the table. Show the predicted values and the residuals in the table. Create a scatter plot of the residuals (residual plot).

SOLUTION:

Table 1.10 shows the calculation of the residuals.

Neck circumference	Wrist circumference	Predicted wrist circumference	Residual
16″	8″	$\frac{16-1}{2}=7.5$	8 – 7.5 = 0.5
14.5″	7″	$\frac{14.5-1}{2}=6.75$	7 – 6.75 = 0.25
15.5″	6.5″	$\frac{15.5-1}{2}=7.25$	6.5 – 7.25 = –0.75
13.5″	6.5″	$\frac{13.5-1}{2}=6.25$	6.5 – 6.25 = 0.25
13″	6.5″	$\frac{13-1}{2}=6$	6.5 – 6 = 0.5
14″	6″	$\frac{14-1}{2}=6.5$	6 – 6.5 = –0.5

TABLE 1.10. Residual calculation.

To create a scatter plot of the residuals, plot the residuals against the neck circumference (**Figure 1.27**).

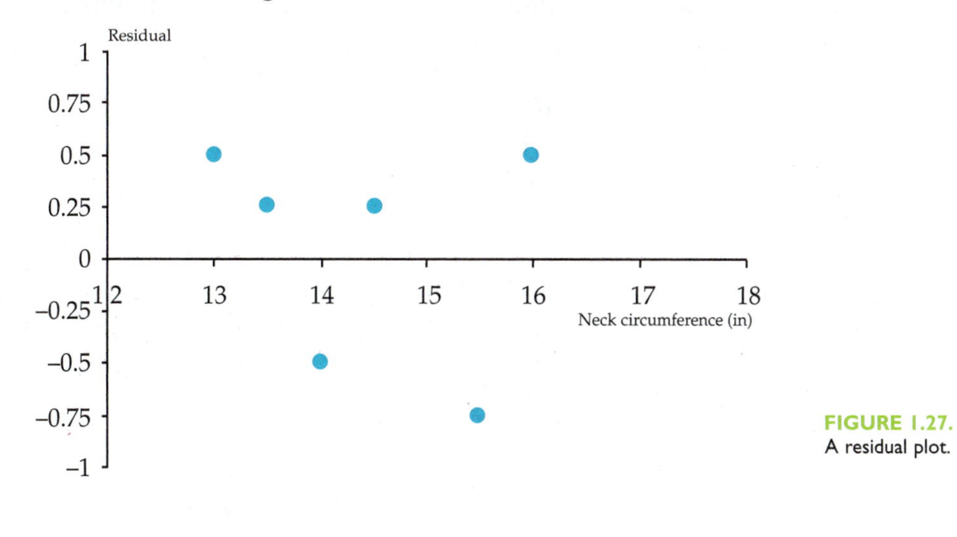

FIGURE 1.27. A residual plot.

The model you created in Acivity 1.1 is an example of a data-driven model. In the following exercises, you create a theory-driven model and use data to evaluate the model.

Exercises 1.2

Consider the following situation:

In order to increase profits, an environmentally conscious timber company wishes to maximize the sustainable yield of lumber per acre over time. To help them do this, they need to develop a model that tells them the best age at which to harvest a stand of trees.

As a first step in developing the desired model, the company needs to find a simplified model that predicts the amount of wood usable for lumber in any given tree. Since time is an important factor to consider, the measurement to determine the volume of the tree must be quick, easy, and accurate.

For example, a lumber cutter can measure the diameter of a tree at its base or some other convenient, specified location. Then the task is to develop a theory-driven model that estimates the volume of a tree as a function of this readily-measurable dimension. In searching for a mathematical interpretation of what is known, the modeler might try to find a formula for the volume of usable wood in a tree in terms of its base diameter.

How can you estimate the volume of a tree given some easily-measurable dimension? The formulation of this well-defined question completes the first step in the process of developing the desired model for the lumber company.

The second step of the modeling process is to make assumptions that help describe the situation mathematically. Answering the following questions may help you simplify the tree-volume situation. Assume that the trees being harvested are Ponderosa Pine.

1. Not all parts of the tree can be cut and made into boards. Which parts do you think are usable? Which are not?

2. Keep in mind that Ponderosa Pine are very tall, thin trees. How might you describe the main body of the tree geometrically?

After considering some assumptions, you are ready to build a model. It may be beneficial to draw shapes, gather additional information about the topic being considered, define variables, and begin thinking about formulas that might apply.

The FYI on this page about the Ponderosa Pine also may be helpful in building the desired mathematical description.

3. Create a mathematical model in the form of an equation that relates the two variables, diameter and volume. Build your theory-driven model on what you know about Ponderosa Pines. Describe how you arrive at your model. Write your equation in a form that permits finding the volume of any Ponderosa Pine tree when given the diameter at its base.

PONDEROSA PINE

The Ponderosa Pine is a large, recognizable species of tree that grows in western North America at altitudes of 1000–4000 feet. The average, mature Ponderosa Pine has a diameter at its base of about 3.5 feet and a height of about 165 feet. It is a much sought-after tree for harvesting because of its abundant clear wood, both near the lower trunk and between its widespread limbs.

Dr. Edward C. Jensen, Director, Forestry Media Center, Oregon State University

TAKE NOTE

You may need one or more volume formulas to develop your model. In case you have forgotten some of them, here are some of the most commonly used volume formulas.

Cube: $V = e^3$, where e is the length of an edge.

Rectangular Solid: $V = lwh$, where l, w, and h are the length, width, and height, respectively.

Sphere: $V = \frac{4}{3}\pi r^3$, where r is the length of the radius.

Prism: $V = Bh$, where B is the area of the base and h is the height.

Pyramid: $V = \frac{1}{3}Bh$, where B is the area of the base and h is the height.

Cylinder: $V = \pi r^2 h$, where r is the length of the radius of the base and h is the height.

Cone: $V = \frac{1}{3}\pi r^2 h$, where r is the length of the radius of the base and h is the height.

Now that you have created your first theory-driven model, you are ready to test it against data that have been gathered from real Ponderosa Pines (see Table 1.9). These data provide the basis for revising your first model. Remember that testing and revising are always part of the modeling process.

The following exercises provide a rough test of your model to see how well it predicts usable volume.

4. For the listed diameters in **Table 1.11**, use your model (equation) to predict the corresponding volume in thousands of cubic inches. Note: You may need to convert units in order to use your equation easily. What is the independent variable? What is the dependent variable?

Exercises 1.2

Diameter in inches	Actual Usable Volume in thousands of cubic inches	Predicted Usable Volume in thousands of cubic inches
36	276	
28	163	
41	423	
19	40	
32	177	
22	73	
38	363	
25	81	
17	23	
31	203	
20	46	
25	124	
19	30	
39	333	
33	269	
17	32	
37	295	
23	83	
39	382	

TABLE 1.11.

5. Are the amounts predicted by your model close to the actual amounts given in the data? Explain.

6. Graph the predictions from your model (predicted volume vs. diameter). Repeat for the actual data from Table 1.9 (actual volume vs. diameter). Then compare the graphs. How are they similar? How are they different?

7. Describe any patterns you see in the graphs.

8. Comparing the overall shapes of your graphs, suggest modifications of your equation that you might try in order to model the actual situation better.

9. Calculate the residuals and prepare a scatter plot of the residuals.

10. Residuals should be relatively small with respect to the data. There should be no pattern. What does the residual plot tell you about your model?

11. To understand why residual plots are important, it can be instructive to intentionally find a wrong model. In this exercise, you first build a table of data with a non-linear function. Then you examine the residuals produced by a linear function.

Exercises 1.2

a) Complete **Table 1.12** of data using the formula $y = (x - 2)^3 + 2$.

x	y
1	$(1 - 2)^3 + 2 = 1$
1.2	
1.4	
1.6	
1.8	
2	
2.2	
2.4	
2.6	
2.8	
3	

TABLE 1.12.

b) The graphing calculator scatter plot in **Figure 1.28** shows the data and the linear regression model $y = 0.712x + 0.576$.

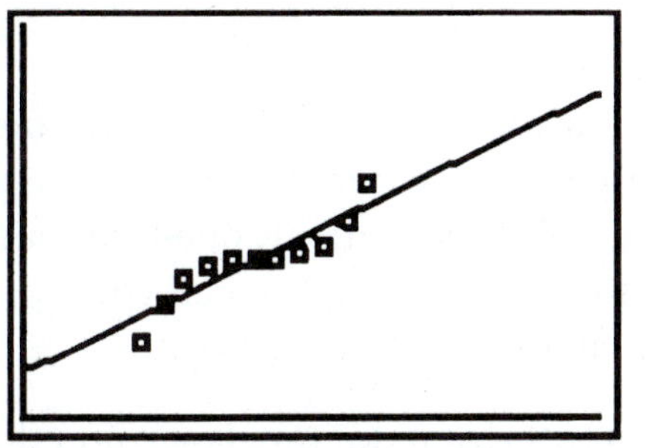

Window: Xmin = 0, Xmax = 5, Ymin = 0, Ymax = 5

FIGURE 1.28.

Is there a pattern in the way the function approximates the data? Explain?

c) Calculate the residuals for the prediction equation $y = 0.712x + 0.576$ and plot them.

d) Describe any patterns visible in the residual plot. What is the reason for the pattern(s)? Why is the residual plot useful?

e) What would you expect to happen if you used a calculator to compute a cubic regression model for these data? If a calculator with a cubic regression feature is available, use it to confirm your answer.

Modeling Linear Patterns: Beginning the Tool Kit of Functions

Because functions are used to build models, knowing various families of functions is important and useful. A family of functions is made up of functions that share some property and differ in other properties. In this lesson, you will begin building a "tool kit" of familiar function families by identifying key features of four functions: the constant, direct variation, linear, and absolute value functions.

You will examine the algebraic forms of their equations, the shapes of their graphs, patterns in their tables of values, and verbal descriptions of their behavior. Initially, one-term parent functions are put in your modeler's tool kit. Later, your investigations will lead to more complex functions.

As you increase your knowledge of these functions, you will also be increasing your skills as a mathematical modeler. The more you know about functions, the easier it will be for you to identify trends in data, explain observed behavior, and improve on your original model.

A FUNCTION TOOL KIT

As you begin to build your function tool kit, you will want to examine carefully the properties of each function. Key features to consider include intervals on which the function is increasing or decreasing, its intercepts (where the graph crosses the x- or y-axis), its peaks and valleys, its behavior as x becomes large, its curvature, and the overall shape of the graph.

Constant Functions

Constant functions are simple and useful tools. The general form of the equation is $f(x) = k$ where k is some constant. For example, when $k = 3$ the constant function can be written as $f(x) = 3$ or $y = 3$.

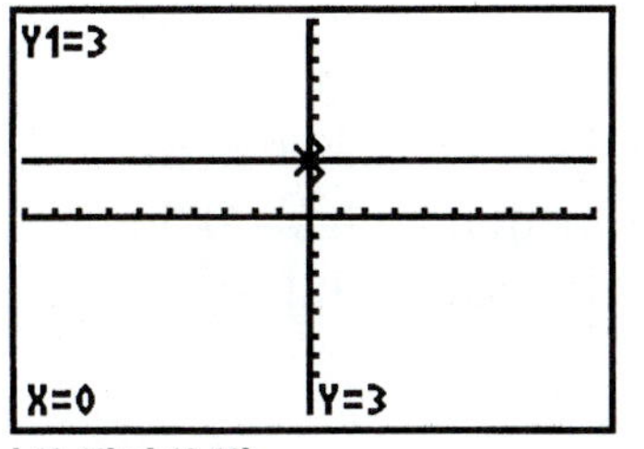

[–10, 10] x [–10, 10]

The calculator screen in **Figure 1.29** shows the graph of $y = 3$ in the window Xmin = –10, Xmax = 10, Ymin = –10, and Ymax = 10.

FIGURE 1.29.
Graph of $f(x) = 3$.

> **TAKE NOTE**
> Shorthand of the form [Xmin, Xmax] x [Ymin, Ymax] is often used to describe a calculator viewing window.

Discussion/Reflection

1. Graph the constant function $y = 3$ and several other functions of your choice, all on the same set of axes. List the features of the graphs that are identical and those that vary or change from function to function.

In Items 2–4, use your functions from Item 1.

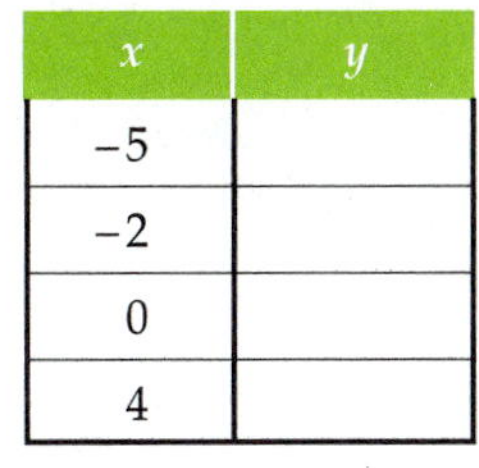

x	y
–5	
–2	
0	
4	

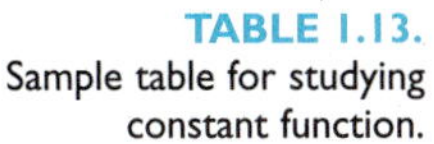

TABLE 1.13.
Sample table for studying constant function.

2. Complete a table similar to **Table 1.13** for each of your constant functions and describe your observations.
3. What are the domain and range of the function $y = 3$?
4. In the equation $y = k$, you might call k a control number because it controls some characteristic of the graph. To investigate the role of a control number, examine graphs of several functions with different values of the control number. What does k control for functions of the type $y = k$?
5. Summarize the key features of constant functions.

Direct Variation Functions

In the real world, the relationship between two quantities can often be expressed in terms of proportionality. For example, the circumference of a circle is proportional to its radius, or the height of a person is proportional to his or her arm length.

If x and y represent two quantities related so that y is a multiple of x, then you can say that y varies directly with x, or y is directly proportional to x. This relationship can be expressed algebraically by the equation $y = kx$, where k is a non-zero constant. The number k is called the constant of proportionality. For example, if $k = 5$, then $\frac{y}{x} = 5$, except if $x = 0$, and the direct variation function can be written as $f(x) = 5x$ or $y = 5x$. See **Figure 1.30** for the graph of $y = 5x$.

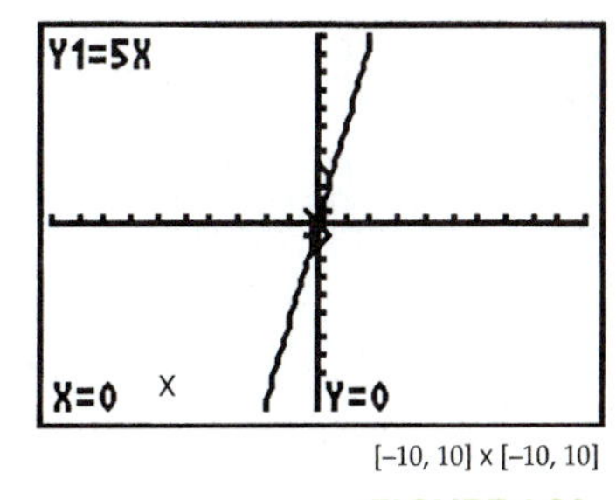

[–10, 10] x [–10, 10]

FIGURE 1.30.
Graph of $f(x) = 5x$.

Discussion/Reflection

1. Graph three different direct variation functions and label them y_1, y_2, and y_3. Be sure to include at least one function with $k < 0$. List the features of your three graphs that are identical and those that vary from function to function.

x	y_1	y_2	y_3
−2			
−1			
0			
1			
2			

TABLE 1.14. Table for studying direct variation.

2. Complete **Table 1.14** for your functions, y_1, y_2, and y_3.

3. What do you observe from your table?

4. In the equation $y = kx$, what does k control in the graphs of direct variation functions?

5. What are the domain and range for each of your functions?

6. Generalize what you observed about your three functions, summarizing the key features of direct variation functions.

7. Explain how you could check data to determine whether they represent a direct variation. If possible, give methods for both graphs and tables.

Linear Functions

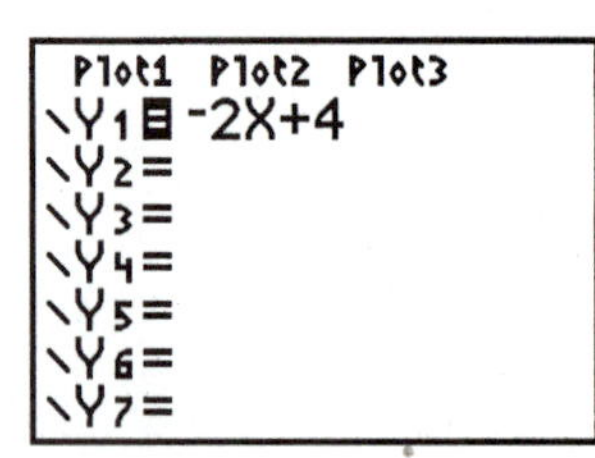

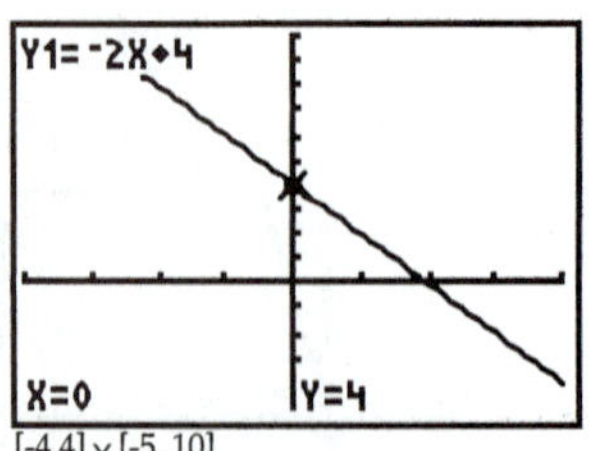

[-4,4] x [-5, 10]

FIGURE 1.31. Calculator screens for graphing $y = -2x + 4$.

The general form for a linear function is $f(x) = mx + b$ where m and b are real numbers. The graph of the function is a nonvertical line with slope m and y-intercept b. Recall that if (x_1, y_1) and (x_2, y_2) are two points on a line, the slope of the line (m) is defined as follows: $m = \frac{y_2 - y_1}{x_2 - x_1}$. The slope is often interpreted as the ratio of the change in y to change in x or as rise over run for two points on the graph of the line.

If the independent and dependent variables increase together, the two variables are **positively related** and the slope of the graph of the linear function is positive. If instead, one variable decreases while the other increases, the variables are **negatively related** and the slope of the graph is negative.

The graph of the linear function in **Figure 1.31** has a slope of −2 and a y-intercept of 4.

Discussion/Reflection

1. Graph at least four linear functions and label them. Make sure that some of the functions have $m < 0$ and some have $b < 0$. List the features of the graphs that are identical and those that vary or change from function to function.

2. Does the point (–60, –38) lie on the graph of $y = \frac{2}{3}x + 2$? Explain.

3. Consider the linear function $y = \frac{2}{3}x + 2$. Interpret the $\frac{2}{3}$ and the 2.

4. The control numbers in $f(x) = mx + b$ are m and b. What does each control?

5. Generalize from your specific cases, and summarize the key features of linear functions of the form $f(x) = mx + b$. In your summary, be sure to include the range and domain of the function as well as what happens when $m = 0$ or $b = 0$.

Example 5

For data that are linear the rate of change, or the ratio of vertical change to horizontal change, is constant for all data values. **Table 1.15** displays the number of minutes talked and monthly bill for a cellular phone service.

1. Choose two pairs of (x, y)-values from the table and calculate $\frac{\text{change in monthly bill}}{\text{change in time talked}}$. Choose several other groups of two pairs and find this ratio. What do you notice?

2. If $\frac{\text{change in monthly bill}}{\text{change in time talked}}$ has the same value for every two pairs of (x, y)-values in the table, the relationship is linear. Is this a linear relationship? If so, determine a linear function that describes this pattern.

3. Interpret the meaning of the slope and y-intercept of your function.

4. Use your function to predict the monthly bill if the total talk time is 43 minutes.

Time talked (minutes) x	Monthly bill (dollars) y
0	15.00
5	15.60
10	16.20
15	16.80
20	17.40
25	18.00
30	18.60

TABLE 1.15.
Phone billing information.

SOLUTION:

1. Choosing (1, 15) and (5, 15.60): $\frac{15.60-15}{5-0}=\frac{0.60}{5}$ = 0.12. This ratio is the same for any two ordered pairs chosen.

2. Yes, the relationship is linear. From Part (1), you know the slope of the line is 0.12. From the table, you know that the y-intercept is 15. Hence, using the slope-intercept form for a linear equation, the linear function is $y = 0.12x + 15$.

3. In this context, the slope represents the cost in dollars per minute of talk-time. That is, every additional minute of talk-time costs an additional 12 cents ($0.12). The y-intercept represents the cost of owning a phone ($15) even is the phone is never used (the monthly service fee).

4. The monthly bill for 43 minutes of talk-time:
$y = 0.12\,(43) + 15 = \$20.16$.

The absolute value function, $f(x) = |x|$ is actually a **piecewise-defined function** (a combination of functions, each defined over specific, but different, parts of its domain).

The function $f(x) = |x|$ can be defined piecewise by $f(x)=\begin{cases} x \text{ if } x \ge 0 \\ -x \text{ if } x < 0 \end{cases}$.

The graph of $f(x) = |x|$ is shown in **Figure 1.32.**

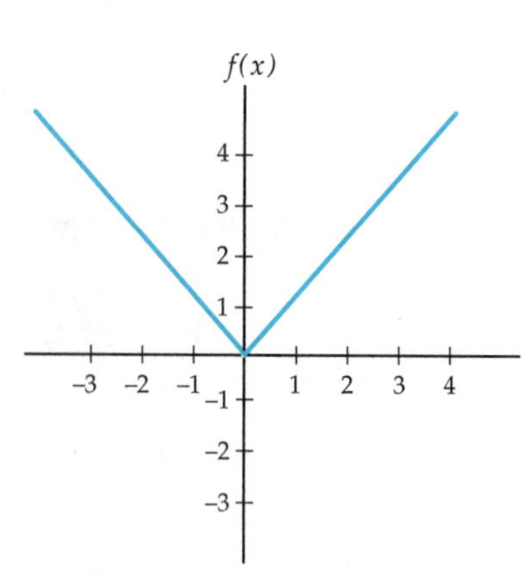

FIGURE 1.32. The graph of $f(x) = |x|$.

EXAMPLE 5

Sketch a graph of the piecewise function

$$f(x)=\begin{cases} x+3 \text{ if } x \le 1 \\ x-3 \text{ if } x > 1 \end{cases}.$$

Where there are breaks in the graph, indicate included points with a shaded dot and excluded points with an open dot.

SOLUTION:

Sketch $y = x + 3$ to the left of 1. Since 1 is included, use a shaded dot at the right endpoint. Sketch $y = x - 3$ to the right of 1. Since 1 is not included, use an open dot at the right endpoint. (See **Figure 1.33.**)

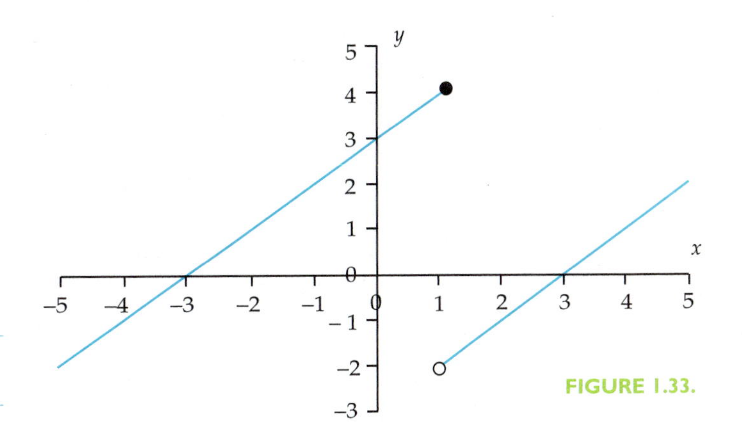

FIGURE 1.33.

DISCUSSION/REFLECTION

1. Complete **Table 1.16** for $f(x) = |x|$.
2. Describe the graph of $f(x) = |x|$ in everyday language.
3. Use mathematical terminology to describe key features of the function.

x	y
−3	
−2.5	
−1	
0	
1.7	
2	

TABLE 1.16.

USING THE TOOL KIT

In many instances, data display a pattern or trend. You now have the ability to recognize these patterns in tables and graphs and decide whether a constant, direct variation, linear, or absolute value function is an appropriate model. In Activity 1.2 you will examine a given set of data, develop criteria for models that describe the data well, determine a model that best describes the data, and use your model to predict for future years.

Activity 1.2

Table 1.17 shows the number of households in the United States with personal computers each year from 1995–2000.

1. Make a scatter plot of these data. Which is the independent variable?

For Items 2–3, use the data in Table 1.17 and your scatter plot.

2. Are the variables positively or negatively related?

Year	Households with PCs (millions)
1995	33.2
1996	38.7
1997	44.0
1998	47.8
1999	51.9
2000	55.1

TABLE 1.17.

3. What features or trends do you notice that would help you choose a function that would best fit the data?

4. What considerations should be made when positioning a line that seems to model the data well, a line of good fit?

5. Develop a linear model that well represents the relationship between year and number of households with a personal computer by first choosing two points that will yield a line of good fit. Draw a line through those two points. Calculate the equation of the line.

(Hint: When you know the slope of a line (m) and one point on the line (x_1, y_1), the equation for the line can be found using the formula $y - y_1 = m(x - x_1)$. This is known at the **Point-Slope Form** of an equation of a line.)

6. Interpret your equation by explaining the contextual meaning of your slope and y-intercept.

7. The graphing calculator uses certain criteria to compute the **least-squares regression** line, LINREG on some calculators, as the line of best fit of a data set. Enter the data into your graphing calculator. Use linear regression to determine the equation of the least-squares line for the data.

8. Use your least-squares line from Item 7 to predict the number of households with a personal computer in 2010.

9. Prepare a residual plot for the least-squares line. Use it to discuss the fit.

TAKE NOTE

Regression options on calculators and spreadsheets vary. As you explore a variety of functions you will become familiar with many of them. These options may include some or all of the following:

LINREG	Linear regression
QUADREG	Quadratic regression
CUBICREG	Cubic regression
QUARTREG	Quartic regression
LNREG	Logarithmic regression
EXPREG	Exponential regression
PWRREG	Power Regression
LOGISTIC	Logistic curve fitting
SINREG	Sine curve fitting

SUMMARY OF PROPERTIES

Constant Functions

The most distinguishing features of the constant function, $y = k$, are the ones that are missing. The constant function has no curvature, no peaks or valleys, and except in the case when $y = 0$, it has no zeros (x-intercepts).

As x increases in both the positive and negative direction, the function remains unchanged. The function is a horizontal line with a y-intercept of k, and a slope of 0. Its domain is all real numbers, and its range is the single number k.

Direct Variation Functions

The graph of any direct variation function, $y = kx$, is a straight line with slope k that passes through the origin, so both the x- and y-intercepts are 0. If $k > 0$, the function is increasing, and if $k < 0$, the function is decreasing. Both the domain and range of the function are $(-\infty, \infty)$.

Linear Functions

The graph of the linear function, $f(x) = mx + b$ is a straight line with slope m and y-intercept b. If $m > 0$, the function is increasing; if $m < 0$, it is decreasing. If $m = 0$, it is the constant function $f(x) = b$. If $b = 0$ the function is a direct variation function $f(x) = mx$ and the graph passes through the origin.

Absolute Value Functions

Since the absolute value function is a combination of two functions, it takes on the characteristics of each of them for different intervals. When $x \geq 0$, the function is defined by the direct variation function $f(x) = x$, an increasing function whose graph is a straight line with a slope of 1. When $x < 0$, the function is defined by the function $f(x) = -x$, a decreasing function whose graph is a straight line with a slope of -1. It has a minimum value of 0 when $x = 0$. The domain of the function is all real numbers. Since the absolute value of any number is always greater than or equal to 0, the range is $(0, \infty)$.

Exercises 1.3

1. For the data in each of **Tables 1.17–1.19** determine whether the data represent a direct variation function. If so, state the constant of proportionality.

a)

x	0.63	1.82	2.45	3.15	4.65	5.87	6.99	7.04	8.93
y	0.07	1.69	4.12	8.75	28.15	56.63	95.63	97.70	199.39

TABLE 1.17.

b)

x	0.500	0.333	0.250	0.200	0.167
y	0.510	0.340	0.255	0.204	0.170

TABLE 1.18.

c)

x	0	1	2	3	4	5	6	7	8
y	0	0.2	0.8	1.8	3.2	5.0	7.2	9.8	12.8

TABLE 1.19.

2. Describe the graph of each of the following:

   a) A linear function, $y = mx + b$, where $m > 0$.

   b) A linear function, $y = mx + b$, where $m < 0$.

   c) A linear function, $y = mx + b$, where $m = 0$.

3. Does **Table 1.20** describe a linear relationship? Explain.

x	−2	−1	0	1	2
y	0	0.5	1	1.5	2

TABLE 1.20.

4. a) Use your calculator to sketch the graph of $y = x$ using the following viewing window: Xmin = −20, Xmax = 20, Xscl = 1, Ymin = −5, Ymax = 5, Yscl = 1. Repeat using the window Xmin = −5, Xmax = 5, Xscl = 1, Ymin = −20, Ymax = 20, Yscl = 1. What do you notice about the appearance of the line in the two different viewing windows?

   b) Care must be taken to get an undistorted view of the slope of a line on graphing utilities because of the rectangular viewing screen. What happens when you graph the function $y = x$ using a window where the Xmax and Xmin are equal to the Ymax and Ymin? For example, try the window Xmin = −5, Xmax = 5, Xscl = 1, Ymin = −5, Ymax = 5, Yscl = 1. Does this window give an undistorted view? Explain.

To avoid this distortion, the window of the calculator must be adjusted so that the screen will be "square." On many calculators a built-in function such as Zsquare automatically squares the screen. If this is not the case with your calculator, consult your manual to find the setting ratio of x to y that will accomplish a square screen.

Exercises 1.3

c) The line $y = x$ bisects the first and third quadrants. Find a window on your calculator that makes it appear that $y = x$ does so.

d) Test your undistorted screen by graphing the following two functions: $y = 2x + 6$ and $y = -0.5x - 2$. First graph them on a standard window of $[-10, 10] \times [-10, 10]$ and then on a square screen. What do you notice about the relationship between the two lines?

TAKE NOTE

Care must be taken when making geometric observations from a calculator screen. Using a square screen can help. As a word of caution: Always take note of the graphing window.

5. Winter Park Resort is the closest ski resort to Denver, 67 miles west of the city via I-70 and US 40. The resort has 2886 skiable acres, four mountains and an average annual snowfall of 370 inches. Statistics describing three of the mountains are given in **Table 1.21**.

	Winter Park	Mary Jane	Vasquez Ridge
Vertical Drop (feet/meters):	2220/676	2610/796	1,214/370
Maximum Run Length (feet)	10,560	23,760	7392
Elevation:			
Top (feet/meters):	11,220/3419	12,060/3676	10,700/3261
Base (feet/meters):	9000/2743	9450/2880	9486/2891

TABLE 1.21.
(Data source: General Information Guide, Winter Park Resort, 1998–1999).

a) Calculate the slope of the descent of the longest run for each mountain.

b) Which mountain has the steepest descent?

c) Create a linear model that gives elevation as a function of horizontal position from the starting point for Vasquez Ridge Mountain.

d) Do you feel this model is realistic? What were some of the assumptions made in designing it?

Exercises 1.3

6. A local cellular company offers two reduced rate plans to college students. The Advantage plan costs $17.00 per month plus $0.27 per minute of use. The Ultimate plan costs $29.00 per month and includes unlimited talk time.

   a) Create a table of values relating talk time and monthly cost for the two plans. Find the monthly cost of each plan paying attention to patterns in the calculations, and then write an equation for the cost of each.

   b) Find the number of talk minutes that leads to the cost of each plan being equal.

   c) Discuss the factors that would lead you to choose the best plan. Which plan would you recommend to a friend on a tight budget?

7. **Table 1.22** shows data for several 1993 automobiles.

Make of car	weight (hundreds of lbs)	City miles per gallon
Acura Integra	27	25
Buick LeSabre	35	19
Cadillac Seville	39	16
Chevrolet Lumina	32	21
Dodge Colt	23	29
Ford Festiva	18	31
Hyundai Scoupe	23	26
Pontiac Firebird	32	19
Volvo 240	30	21

TABLE 1.22. Automobile data from 1993.

   a) Make a scatter plot of these data. What features or trends do you notice that would help you choose a function that would best fit the data?

   b) Use linear regression to determine the equation of the least-squares line for the data. Add this line to your scatter plot. Does it appear to be a good model?

   c) Interpret your equation by explaining the contextual meaning of your slope and y-intercept.

d) What are a reasonable domain and range for your equation considering the given context?

Exercises 1.3

e) Create a residual plot for the least-squares line you found in part (b). Based on your residual plot, does your least-squares line appear to describe the relationship between the weight of the car and its city mileage? Explain.

8. Forensic scientists are often called upon to solve crimes in which only skeletons or parts of skeletons are available as evidence. Sometimes, in attempting to identify the body, these scientists are asked to determine the height of the individual from the remaining bones.

As part of a class project, a group of students decided to see if they could model height as a function of forearm length. After collecting data from individuals in their class, the group decided the best mathematical model for their data was $H = 4r + 65$, where H represents the height of a person and r represents forearm length in centimeters.

PEOPLE AND MATH

The work of a forensic scientist is a lot like that of a detective with math skills including data analysis required. Forensic scientists and technicians work in a variety of environments including medical laboratories in hospitals, research institutions, and crime labs. They conduct various tests on body fluids and tissues, often to determine the cause and time of death. When forensic scientists are trying to solve crimes, they help to gather evidence ranging from finger- and footprints to blood from crime scenes. As part of their job, they determine the specific DNA-genetic makeup of blood, so they can more easily identify possible suspects.

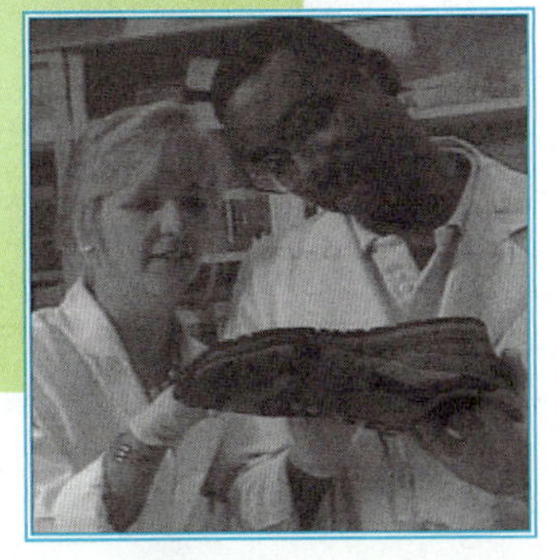

a) Use the students' model to find the missing heights and forearm lengths in **Table 1.23**.

b) Measure the forearm lengths and heights of the members of your class. What seems to be a reasonable domain for this function?

c) Enter into your calculator's lists the data collected by your class. Make a scatter plot of these data. What features do you notice in the data that would help you decide what function would best fit the data set?

Forearm length (cm)	Height (cm)
23.5	159
30.0	185
27.6	175.4
32.8	196
25.8	168

TABLE 1.23.

Exercises 1.3

d) Use linear regression to determine the equation of the least-squares line for your data set. Add the regression line to your scatter plot. Does it appear to be a good model?

e) Create a residual plot for the least-squares line. Does your residual plot support your answer in (d)? Explain.

f) Interpret the slope and y-intercept of your regression equation. Is your interpretation reasonable and consistent with what you know about the world?

g) Is your equation different from that of the class described at the beginning of this exercise? If so, discuss possible explanations.

9. **Table 1.24** displays the sales of recorded compact discs by units shipped (in millions).

Year	CDs shipped (in millions)
1990	286.5
1991	333.3
1992	407.5
1993	495.4
1994	662.1
1995	722.9
1996	778.9
1997	753.1
1998	847.0

TABLE 1.24.
Source: Recording Industry Association of America, Washington D.C.

a) Enter the data from Table 1.24 into your calculator's lists. Make a scatter plot of these data. What features do you notice in the data that would help you decide what function would best fit the data set?

b) Use linear regression to determine the equation of the least-squares line for your data set. Add the regression line to your scatter plot. Does it appear to be a good model?

c) Create a residual plot for the least-squares line. Does your residual plot support your answer in (b)? Explain.

d) Interpret the slope and y-intercept of your linear regression equation.

e) Use your model to predict the number of CDs shipped in 2005? Are you confident with this prediction?

10. According to the Recording Industry Association of America, while the number of CDs shipped has increased, the number of cassettes has decreased since 1990.

Exercises 1.3

a) If 442.2 million cassettes were shipped in 1990 and only 158.5 million cassettes were shipped in 1998, calculate the linear equation that describes the number of cassettes shipped as a function of years since 1990.

b) Explain why a linear function might not be a good model for this situation.

11. The graph of a linear function is shown **Figure 1.34** and the window for the graph in **Figure 1.35**. Estimate the equation of the line.

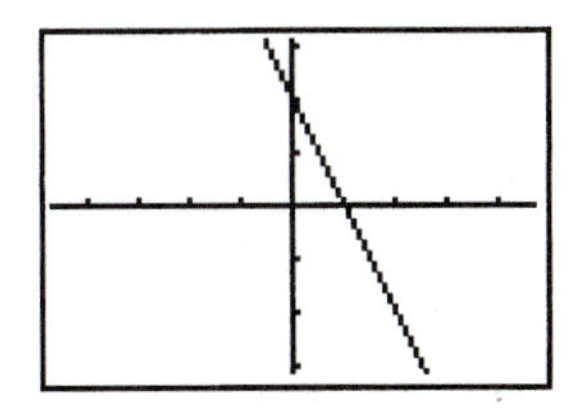

FIGURE 1.34.

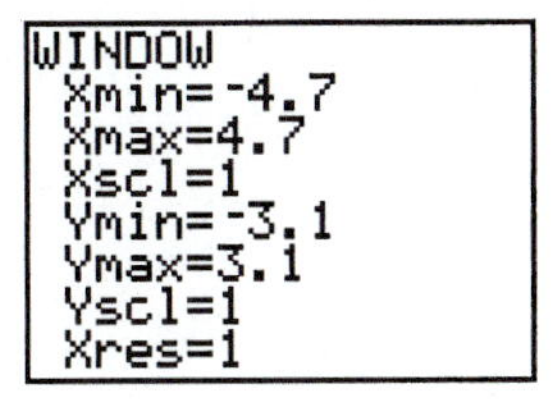

FIGURE 1.35.

12. **Table 1.25** displays the employment rates in the United States from 1975 to 1999.

Year	Percent unemployed
1990	286.5
1991	333.3
1992	407.5
1993	495.4
1994	662.1
1995	722.9
1996	778.9
1997	753.1
1998	847.0

TABLE 1.25.

a) Enter the data from Table 1.25 into your calculator's lists. Make a scatter plot of these data. What features do you notice in the data that would help you decide what function would best fit the data set?

b) Use linear regression to determine the equation of the least-squares line for your data set. Add the regression line to your scatter plot. Does it appear to be a good model?

c) Create a residual plot for the least-squares line. Does your residual plot support your answer in (b)? Explain.

d) Interpret the slope and y-intercept of your regression equation.

e) Use you model to predict the percent employed in the years 1996 and 2008. Are you confident with these predictions?

13. Graph the following piecewise-defined functions:

a) $f(x) = \begin{cases} |x| \text{ if } x < 2 \\ -x+4 \text{ if } x \geq 2 \end{cases}$ b) $f(x) = \begin{cases} |x| \text{ if } x < 2 \\ -x+6 \text{ if } x \geq 2 \end{cases}$

c) $f(x) = \begin{cases} 2-x \text{ if } x < 2 \\ 2 \text{ if } 2 \leq x \leq 4 \\ x-1 \text{ if } x > 4 \end{cases}$ d) $f(x) = \begin{cases} 3-x \text{ if } x < 1 \\ 3x-1 \text{ if } x \geq 1 \end{cases}$

Exercises 1.3

14. Piecewise functions are often used to model real-world behavior. For example, consider a situation in which a rate is changing. A vehicle travels at 30 mph for 1 hour, then enters a freeway and travels at 65 mph for 2 hours. Which of the following piecewise functions describes the distance traveled after t hours. Explain. (For simplicity, consider acceleration instantaneous.)

    a) $d = \begin{cases} 30t \text{ if } 0 \le t \le 1 \\ 65t \text{ if } 1 < t \le 3 \end{cases}$

    b) $d = \begin{cases} 30t \text{ if } 0 \le t \le 1 \\ 30 + 65(t-1) \text{ if } 1 < t \le 3 \end{cases}$

    c) $\begin{cases} 30t \text{ if } 0 \le t \le 1 \\ 30 + 65t \text{ if } 1 < t \le 3 \end{cases}$

15. Graphs and tables are closely related. Equations, arrow diagrams, and function machines are also similar. This exercise looks at the connection between graphs and function machines.

    a) Write the equations for two functions of your own choosing, but not both from the same tool kit family. Then make sample tables and graphs for your functions.

    b) Show how to use your graph to determine the approximate output value for a given input value. (Do so without using either the equation or the table.)

    c) Describe in words how to use any graph as a function machine to obtain the output value corresponding to any given input value.

LESSON 1.4

Expanding the Tool Kit of Functions

After completing Lesson 1.3, it should be clearer to you that increasing your mathematical understanding of certain functions will help in your search for good mathematical models. In this lesson, you will continue to add to your function tool kit by exploring a family of functions known as **power functions**.

Activity 1.3

A **power function** is defined as any function of the form $y = Ax^B$, where $A \neq 0$ and B is a non-zero integer. Initially, only the subset of this family with $A = 1$ will be considered.

1. With the help of your calculator, plot the following functions on the same set of axes: (Hint: Recall that x^{-n} means $\frac{1}{x^n}$; x^{-n} and $\frac{1}{x^n}$ are equivalent expressions.)

   a) $f(x) = x^3$ b) $f(x) = x^2$ c) $f(x) = x^1$

   d) $f(x) = x^{-1}$ e) $f(x) = x^{-2}$ f) $f(x) = x^{-3}$

2. What do your graphs have in common?

3. How do the graphs of these functions differ?

When one half of a graph is the "mirror image" of the other half, the graph is said to be **symmetric** with respect to the line that acts as the mirror. That line is called the **axis of symmetry** for the graph. When a graph or other figure has an axis of symmetry, it can be folded along the axis so that one half coincides with the other. This type of symmetry occurs frequently in the natural world. (**Figure 1.36.**)

FIGURE 1.36.

4. Identify all graphs in Item 1 that are symmetric with respect to some line, and name the corresponding axes of symmetry.

FIGURE 1.37.

A graph is said to be symmetric with respect to the origin if, and only if, for every point (x, y) on the graph, $(-x, -y)$ is also on the graph. This symmetry is a form of point symmetry, which also occurs frequently in the natural world. (**Figure 1.37.**)

5. a) Which of the graphs in Item 1 are symmetric with respect to the origin?

   b) In graphical terms, how do you identify figures that have symmetry with respect to the origin?

6. Which of the functions in Item 1 have restrictions on their domains? Explain.

7. What are the domain and range of each function in Item 1?

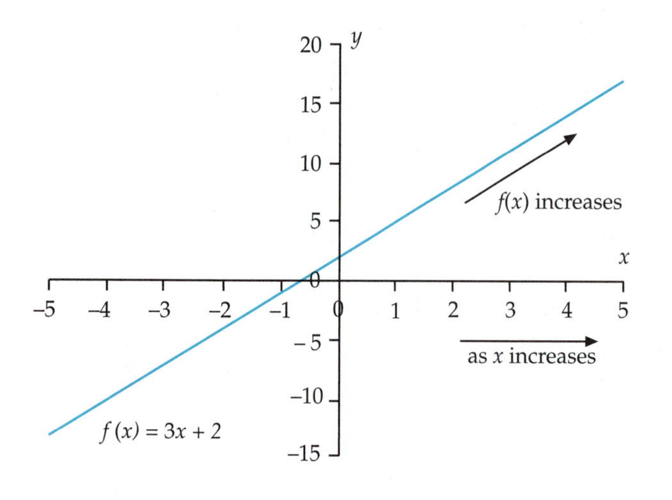

FIGURE 1.38.

When examining properties of functions, it is often helpful to look at the **end behavior** of the function. Examine what happens to values of the function, $f(x)$, as x becomes larger and larger without bound (written $x \to \infty$) or what happens as x decreases through negative values without bound (written $x \to -\infty$). For example, for $f(x) = 3x + 2$, as $x \to \infty$, $f(x) \to \infty$, and as $x \to -\infty$, $f(x) \to -\infty$. (See **Figure 1.38.**) For $g(x) = 3$, as $x \to \infty$, $g(x) \to 3$, and as $x \to -\infty$, $g(x) \to 3$.

8. Consider the graph of the function $f(x) = x^{-1}$. Describe its end behavior.

9. Without graphing the functions, predict the shapes of $f(x) = x^4$, $f(x) = x^5$, and $f(x) = x^{-4}$. Consider such things as symmetry, concavity, intercepts, end behavior, domains and ranges.

10. From your explorations, generalize the key features of power functions.

THE LADDER OF POWERS

The Ladder of Powers, a simple and useful tool in mathematical modeling, is nothing more than a list of power functions. The order in which functions are placed on the Ladder of Powers depends on how rapidly the functions increase or decrease. Above $y = x$ on the ladder of powers, the functions are increasing more rapidly as you move up the ladder. Below $y = x$ on the ladder, the functions are decreasing more rapidly as you move down the ladder.

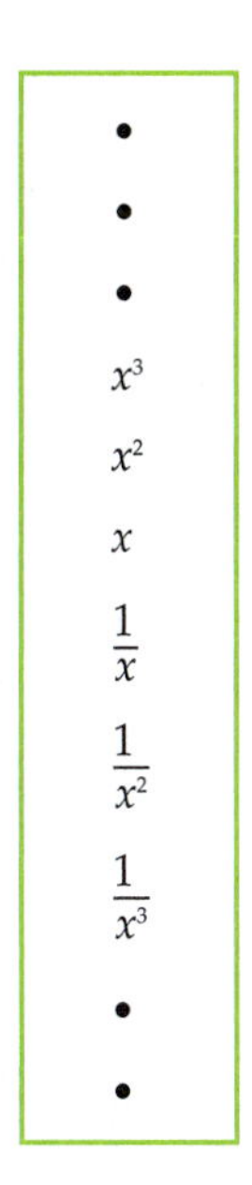

FIGURE 1.39. The Ladder of Powers.

Figure 1.39 shows the ladder placement of the power functions you explored in Activity 1.3. As additional functions are encountered, you will be asked to add more rungs to your ladder.

To get a sense of the Ladder of Power's usefulness in identifying trends as you model data, consider the data in **Table 1.26**.

x	1	2	2.5	4	5	6
y	1	5.66	9.88	32	55.9	88.2

TABLE 1.26.

Figure 1.40 shows a scatter plot of the data. Note that the graph appears to be increasing and concave up. These key features suggest that a power function, $y = x^B$, where $B > 1$ might be used to model the data.

Figure 1.41 shows the graph of $y = x^2$ added to the scatter plot. Notice that this curve does not appear to be a good fit as the data appear to bend upward more and increase more rapidly than $y = x^2$. However, the function does increase and bend in the right direction.

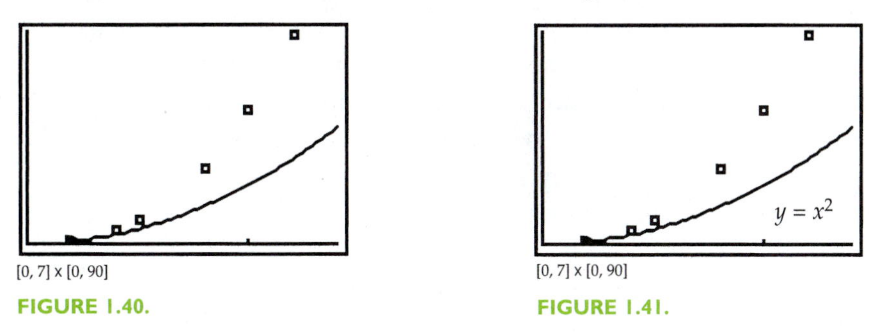

FIGURE 1.40. **FIGURE 1.41.**

This is where The Ladder of Powers can help you. As you noticed, $y = x^2$ does not bend up enough for the data. In other words, to fit the given data, you need a function that has more upward curvature.

When this happens, find the function you just used on your Ladder of Powers, then move up the ladder to find a function with more curvature. In this case, the function above $y = x^2$ on the ladder is $y = x^3$.

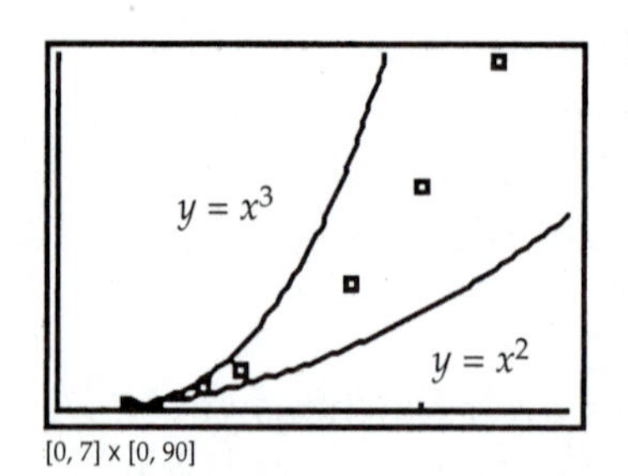

[0, 7] x [0, 90]

FIGURE 1.42.

Figure 1.42 shows the graph of $y = x^3$ added to Figure 1.41.

Notice that the graph of $y = x^3$ bends upward more and increases more rapidly than the data. Now that you have found a power function that increases too rapidly for the data and one that doesn't increase rapidly enough, what do you do?

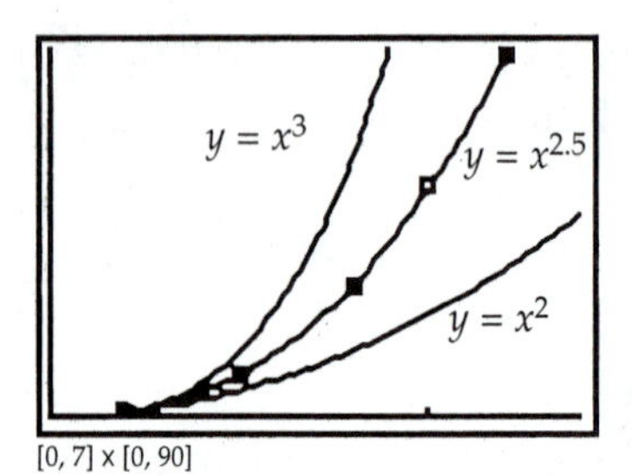

[0, 7] x [0, 90]

FIGURE 1.43.

One way to find a curve with a better fit is to try to fit a power function with a non-integer exponent to the data, for example, $y = x^{2.5}$. Another way to solve the problem is to use the calculator to try a power regression on the data. Either option shows that $y = x^{2.5}$ provides a good fit to the data. See **Figure 1.43**.

RATIONAL POWER FUNCTIONS

The function that fits the data in Table 1.26 had a non-integer exponent. Functions of the type $f(x) = Ax^{\frac{m}{n}}$ where $A \neq 0$ and m and n are non-zero integers are called rational power functions. Recall that the expression $x^{\frac{m}{n}}$ is equivalent to $\sqrt[n]{x^m}$ or $\left(\sqrt[n]{x}\right)^m$ when $x > 0$ and m and n are integers. This leads to questions about the domains of functions of the form $f(x) = x^{\frac{m}{n}}$, where m and n are non zero integers and $\frac{m}{n}$ is not an integer. In this chapter, the domains of these functions will be restricted to non-negative real numbers, and only the portions of these graphs that are in the first quadrant will be considered.

Exercises 1.4

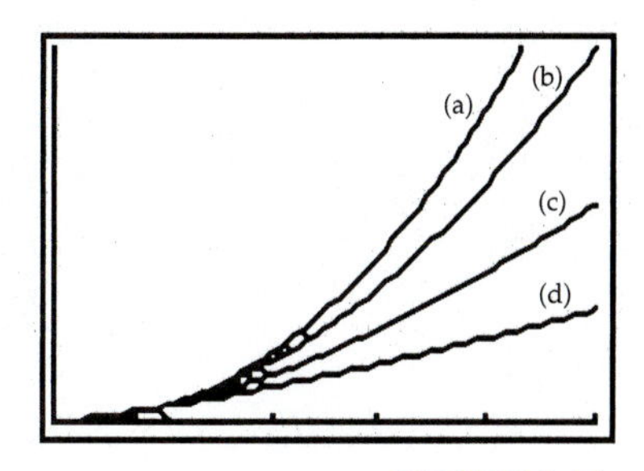

FIGURE 1.44.
Four rational power functions.

1. The calculator screen in **Figure 1.44** shows the graphs of $y = x^2$, $y = x^{5/4}$, $y = x^{5/3}$, $y = x^{11/5}$, in the window $[0, 5] \times [0, 25]$. Without using your calculator, match the graphs to the equations. Explain how you arrived at your conclusions. Use your calculator to verify your answer.

2. Discuss the symmetry of the graph of each of the following.

   a) $f(x) = \frac{1}{x}$ b) $f(x) = |x|$ c) $y = x^{3.1}$ d) $y = x^3$

   e) $y = x^{-2}$ f) $y = x$ g) $x = y^2$

3. Graph all the functions shown in the Ladder of Powers (Figure 1.39) on a single set of axes using the domain $0 < x \leq 5$. From your observations, explain the order of these functions on the ladder.

4. Return to the data in Table 1.26.

   a) Graph the function $y = x^2$ along with the data. Then calculate and graph the residuals. Describe the residual plot, and explain what it tells you about $y = x^2$ as a model for these data.

   b) Repeat Exercise 4(a) using the function $y = x^3$.

5 a) Describe the key features of functions of the form $y = x^n$ where n is a positive odd integer.

   b) Describe the key features of functions of the form $y = x^n$ where n is a positive even integer.

6. When fitting a curve to the data in Table 1.26, you found that for the function $y = x^B$, B does not have to be an integer. For example, $y = x^{2.5}$ is permissible. Graph the rational power functions (a)–(e), and describe key features. Restrict your domain by using a window of $[0, 10] \times [0, 10]$.

   a) $y = x^{\frac{1}{2}}$ b) $y = x^{\frac{2}{3}}$ c) $y = x^{1.4}$

   d) $y = x^{\frac{7}{10}}$ e) $y = x^{\frac{1}{4}}$

7. If you were to put the functions in Exercise 6 on your Ladder of Powers, where and in what order would you place them? Why?

Exercises 1.4

8. Use a window of [0, 5] x [0, 5] to graph the following rational power functions. Describe key features.

   a) $y = x^{-\frac{1}{2}}$ b) $y = x^{-\frac{2}{3}}$ c) $y = x^{-1.4}$

   d) $y = x^{-\frac{9}{10}}$ e) $y = x^{-\frac{1}{4}}$

9. If you were to put the functions in Exercise 8 on your Ladder of Powers, where would you place them and in what order? Why?

Up to this point, your explorations of power functions, $f(x) = Ax^B$, $A \neq 0$ have been limited to those functions with $A = 1$. In Exercises 10–12, you will explore the effects of the control number A.

10. a) Complete a copy of **Table 1.27** for the functions $y_1 = 2x^2$, $y_2 = 3x^2$, $y_3 = 0.5x^2$, $y_4 = -2x^2$, $y_5 = -3x^2$, and $y_6 = -0.5x^2$.

TABLE 1.27.

x	y_1	y_2	y_3	y_4	y_5	y_6
−5						
3						
1						
0						
1						
3						
5						

   b) On a single set of axes, sketch each of the functions in part (a).

   c) Each of the functions in part (a) involves x^2. Think of $y = x^2$ as the parent function for the family $y = Ax^2$. Add a column for $y = x^2$ to your table in part (a). Compare the columns for the six functions in part (a) to that of $y = x^2$. How are the x- and y-values for corresponding points related?

   d) Include the graph of $y = x^2$ in your graph from part (b). Compare the graphs of the six functions in parts (a) and (b) to that of $y = x^2$. Describe the similarities and differences that you observe.

11. a) Complete a copy of **Table 1.28** for the functions $y_1 = \frac{2}{x}$, $y_2 = \frac{0.5}{x}$, $y_3 = -\frac{2}{x}$.

   b) Graph all three functions on a single set of axes.

x	y_1	y_2	y_3
−5			
3			
−1			
0			
1			
3			
5			

TABLE 1.28.

c) Each of the functions in part (a) involves $\frac{1}{x}$. Think of $y = \frac{1}{x}$ as the parent function for the family $y = A\left(\frac{1}{x}\right)$. Add the table and graph for $y = \frac{1}{x}$ to those of parts (a) and (b). How are the x- and y-values for corresponding points related? Describe the similarities and differences that you observe.

Exercises 1.4

d) Use a spreadsheet or arrow diagrams to explain the relationships between the parent function $y = \frac{1}{x}$ and its relatives $y = A\left(\frac{1}{x}\right)$ that you described in part (c).

12. Summarize the effect on the graph of a power function $y = Ax^B$ when $A > 1$. What if $0 < A < 1$? $A < -1$? $-1 < A < 0$?

13. The calculators screens shown in **Figure 1.45** (a–c) show values for three functions. Write a function represented by each table. (Hint: one function is of the form $y = ax$, another of the form $y = ax^2$, and a third of the form $y = ax^3$.)

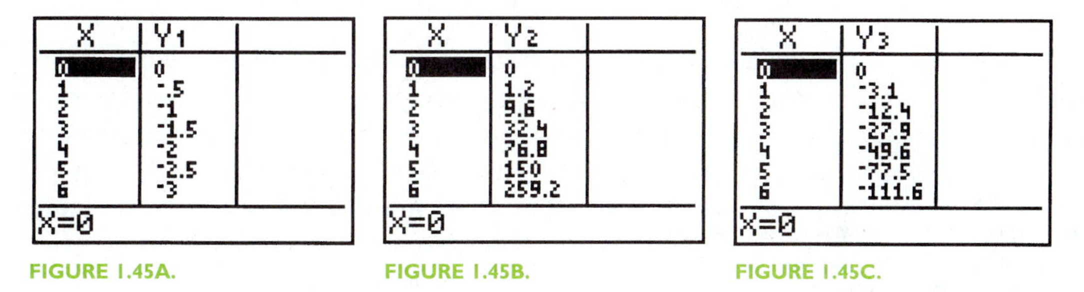

FIGURE 1.45A. FIGURE 1.45B. FIGURE 1.45C.

14. You are given the power function, $f(x) = Ax^{2.1}$, and a point, (8, 71), on its graph. Use substitution to find A.

15. Investigation. Now that you are more familiar with power functions, revisit the Ponderosa Pine data from Lesson 1.2.

a) Fit a power function of the form $y = Ax^2$ to the actual data in Table 1.11. (Hint: Before graphing the function, use the method suggested in Exercise 14 to find a reasonable value for A.) Does your function appear to be a good fit? Explain.

Exercises 1.4

b) From your observations, what might you do to find a curve with a better fit?

c) Fit a power function of the form $y = Ax^3$ to the data. Is it a better fit than your $y = Ax^2$ model?

d) Use your calculator to carry out a power regression on the data. What function did the calculator return?

e) What does the value of A in the regression equation tell you about Ponderosa Pines?

f) Graph the resulting power function and a scatter plot of the data. Does the model appear to be a good fit?

g) To get additional information about the fit, examine a residual plot. Describe the residuals. Do they indicate a good fit?

h) Your theory-driven model in Lesson 1.2 was probably a power function of the form $y = Ax^3$. The data-driven power function from Exercise 15(f) probably fits much better. Return to your thinking that led to that first model and discuss the relative merits of the following two models: $y_1 = 0.0039x^{3.137}$ and $y^2 = 0.0061x^3$.

Some ecologists have criticized the power function model saying that it only describes what is happening rather than explaining it, while others view this basic relationship as one of ecology's few genuine laws.

© Susan Van Etten

16. For over two hundred years, ecologists have hypothesized that as the area of a region increases, so does the number of different species inhabiting the region. The species-area relationship seems to hold true for large and small areas as well as for plants and animals. **Table 1.29** shows native plant species data for coastal areas of California at or above 33-degree latitude.

Location	Area	Species
Tiburon Peninsula	5.9	370
San Francisco	45	640
Santa Barbara area	110	680
Santa Monica Mountains	320	640
Marin County	539	1060
Santa Cruz Mountains	1386	1200
Monterey County	3324	1400
San Diego County	4260	1450
California Coast	24,520	2525

TABLE 1.29. Ecology data.

Source: *UMAP Module 768*, "The Species-Area Relations," Kevin Mitchell and James Ryan.

Exercises 1.4

a) Use your calculator to find the power function that best fits the plant species data. State the function.

b) Consider the context of this problem and interpret the values of A and B in your regression equation, $y = Ax^B$. (Hint: What key features of the graph do A and B control?)

c) Use your equation to predict the number of species for a 20,000 mi^2 region.

17. **Investigation**. Distant lights appear less bright than those of comparable size that are close by. This happens because as the distance from the light source increases, the light emitted by the source covers a larger area and becomes more spread out. Handout 1.4 provides directions for investigating what happens to light intensity as the distance from the light source varies.

18. The data in **Table 1.30** were collected in an experiment conducted to investigate the relationship between a light bulb's intensity and the distance from the light source. A 40-watt bulb's intensity is measured in an otherwise dark room.

Distance (meters)	Intensity
1.0	0.246
1.1	0.196
1.2	0.166
1.3	0.142
1.4	0.124
1.5	0.106
1.6	0.095
1.7	0.084
1.8	0.078

TABLE 1.30.

a) Plot intensity as a function of distance.

b) Find a power function that provides a good fit of the data. Graph this function on the scatter plot.

c) Estimate the intensity at a distance of 2 meters.

d) The theoretical model describing the relationship between intensity I and distance from the source x is given by the formula $I = \frac{a}{x^2}$.

Is the model you constructed consistent with this theoretical model? Explain.

19. a) A Honda S2000 sports car traveling at 60 mph can stop in 117 ft. Assuming that stopping distance varies directly as the square of velocity, calculate the stopping distance of an S2000 traveling at 95 mph. *(Source: Consumer Reports, August 2000)*

b) On wet pavement the stopping distance of the S2000 traveling at 60 mph increases to 147 ft. Calculate the stopping distance of an S2000 traveling at 75 mph on wet pavement.

Transformations of Functions

You now have a tool kit that contains several functions: $f(x) = k$, $f(x) = kx$, $f(x) = mx + b$, $f(x) = |x|$, and $f(x) = Ax^B$. At this point, you should be familiar with the general shape and important features of the graph of each of these functions. Your exploration of the coefficients of power functions in Lesson 1.4 showed one way to alter a parent function to create related functions with similar but slightly different properties. Such changes are called transformations. In this lesson, you will extend your knowledge of functions in general while exploring what happens to the graphs of functions when you apply transformations.

Activity 1.4

In Lesson 1.4, you found that when you multiplied the function $f(x) = x^B$ by some number A, the graph of the resulting function was either stretched or compressed vertically depending on the value of A.

But what about other functions in your tool kit? If you multiply one of them by some constant, will the geometric effect of the transformation be the same as when you multiplied the power function by A, or will something different happen? Part I of this activity helps you answer these questions.

Part I: Multiplying a function by a constant

1. Consider the absolute value function $f(x) = |x|$. Explore this parent function's family by graphing $y = f(x)$ and $y = cf(x)$ on the same set of axes. That is, look at $y = |x|$ and $y = c|x|$. Think of c as a control number. Vary the value of c, including large and small, and positive and negative values, until you feel certain of the effect of multiplying this function by a constant. Make careful sketches of the graphs so that similarities and differences are evident. Write a short but detailed description of your findings. Include both qualitative and quantitative descriptions.

2. Repeat this exploration for the parent functions $f(x) = x^2$, $f(x) = x^3$, and $f(x) = x$.

3. **Figure 1.46** shows a sample spreadsheet for two functions you may have examined in Item 1.

   a) Identify the two functions of x. (Note: Columns A and B define one function. Columns A and C define the other.)

   b) Displayed above the body of the spreadsheet is the formula that was used to create Column C. Explain how this formula helps explain your answers in Items 1 and 2.

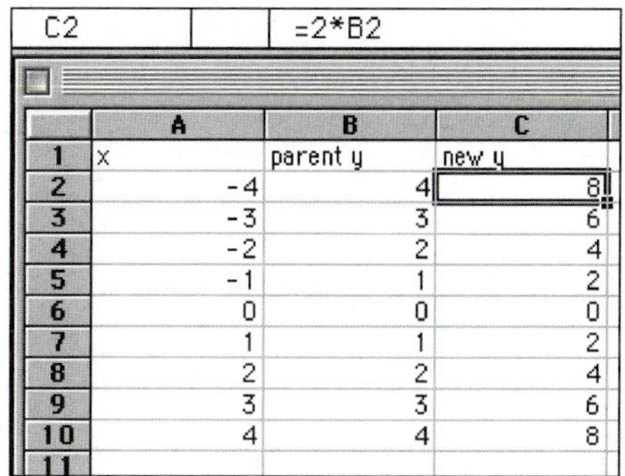
C2 =2*B2

	A	B	C
1	x	parent y	new y
2	-4	4	8
3	-3	3	6
4	-2	2	4
5	-1	1	2
6	0	0	0
7	1	1	2
8	2	2	4
9	3	3	6
10	4	4	8
11			

FIGURE 1.46. Spreadsheet for transformation.

Part II: Adding a constant to a function

In Part I, you examined the effect of multiplying a function by some constant c. In this part of the activity, you will examine the effect of adding a constant to a function.

1. Once again, consider the absolute value function $f(x) = |x|$. Explore what happens to this parent function when a constant is added to it. Make careful graphs of $y = f(x)$ and $y = f(x) + c$ on a single set of axes. Think of c as a control number. Vary the value of c, including large and small, and positive and negative values, until you feel certain of the effect of adding a constant to the absolute value function. Write a short but detailed description of your findings. Include both qualitative and quantitative descriptions.

2. Repeat this exploration for the parent functions $f(x) = x^2$, $f(x) = x^3$, and $f(x) = x$.

3. **Figure 1.47** shows a sample spreadsheet for two functions you may have examined in Item 1.

   a) Identify the two functions.

   b) Notice the formula that was used in creating Column C. Explain how the spreadsheet formula helps explain your answers in Items 1 and 2.

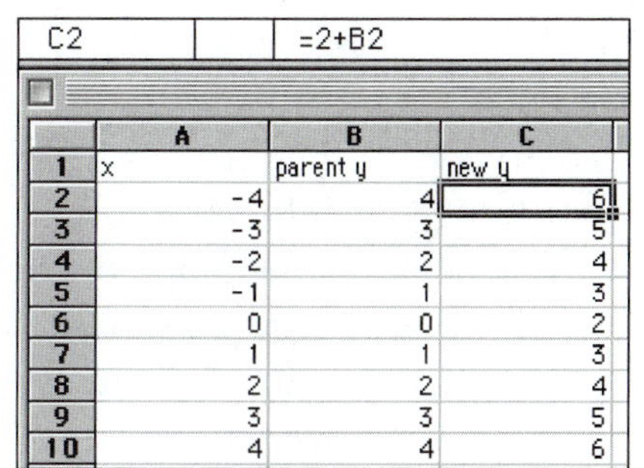
C2 =2+B2

	A	B	C
1	x	parent y	new y
2	-4	4	6
3	-3	3	5
4	-2	2	4
5	-1	1	3
6	0	0	2
7	1	1	3
8	2	2	4
9	3	3	5
10	4	4	6

FIGURE 1.47. Spreadsheet for transformation.

EFFECTS OF TRANSFORMATIONS

A **transformation** associates each point of a graph with a new point. The most common types of transformations cause a movement of the graph or a stretching (or compression) of the graph. Some transformations leave one or more points of the graph unaffected.

Of the two transformations you examined, $y = cf(x)$ and $y = f(x) + c$, both affect the graph vertically. The first, $y = cf(x)$ either compresses or stretches the graph by a factor of $|c|$, and if $c < 0$, the graph is also reflected across the x-axis. The transformation $y = f(x) + c$ translates (or shifts) the graph $|c|$ units upward when c is positive and downward when c is negative.

A spreadsheet is helpful in explaining the effects of these transformations. Figures 1.46 and 1.47 show several values in the domain of a function, the values of the function at each value of x, and the corresponding values after the transformations. Since columns A and B define the graph of the parent function and columns A and C define the graph of the transformed function, it is clear that the x-value is not affected by the transformation. Thus the graphical effect of each of these transformations is vertical.

Arrow diagrams can be used to convey the same relationships (see **Figure 1.48**).

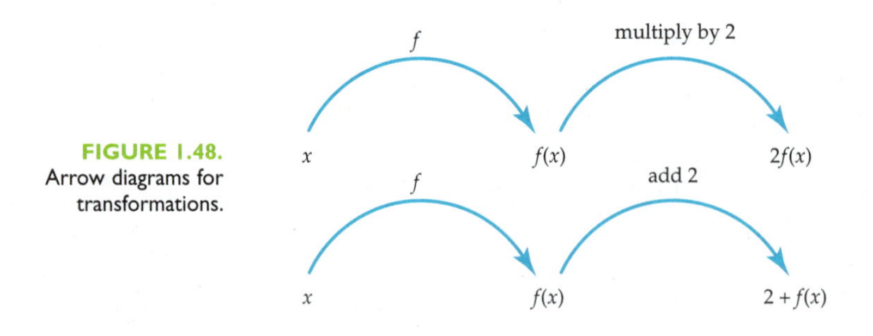

FIGURE 1.48. Arrow diagrams for transformations.

To use arrow diagrams to understand the effects of the transformations of the graph of f, it is necessary to identify the two quantities that define each graph. Here, x and $f(x)$ define the graph of the parent function, and x and $2f(x)$ or x and $2 + f(x)$ define the transformed graph. Again, the x-values are unchanged and the effects are vertical only.

EXAMPLE 6

If $f(x) = 2x$, explain the transformations $f(x) + 3$ and $-f(x)$.

SOLUTION:

The graph of $f(x) + 3 = 2x + 3$ is the graph of $f(x) = 2x$ translated 3 units upward. The graph of $-f(x) = -2x$ is the graph of $f(x) = 2x$ reflected in the x-axis. See **Figure 1.49**.

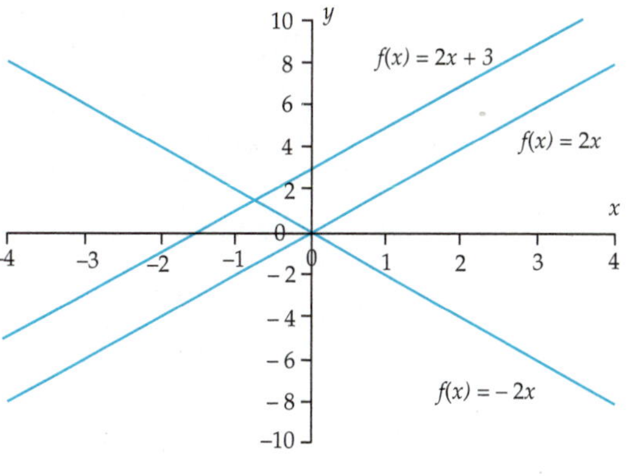

FIGURE 1.49.

COMBINATIONS OF TRANSFORMATIONS

It is possible to combine transformations. Consider the parent function graphed in **Figure 1.50**, and the related transformation $y = 2f(x) + 3$, which involves both multiplication and addition.

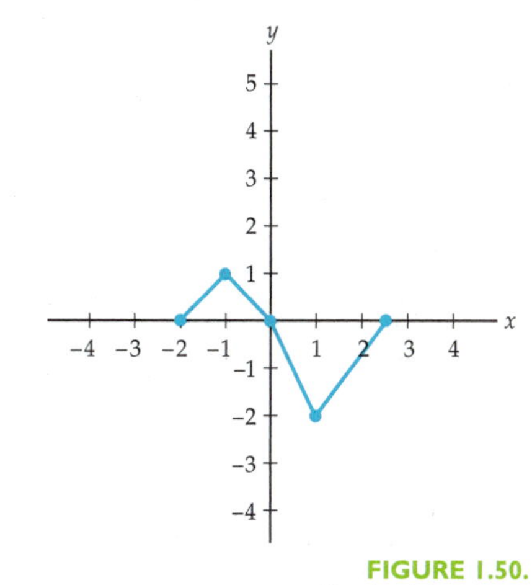

FIGURE 1.50. Graph of the parent function $y = f(x)$.

The graph of this new function can be sketched quickly by stretching the graph of $y = f(x)$ vertically by a factor of 2, then shifting the resulting graph vertically 3 units upward. This combined transformation is illustrated in an arrow diagram in **Figure 1.51**.

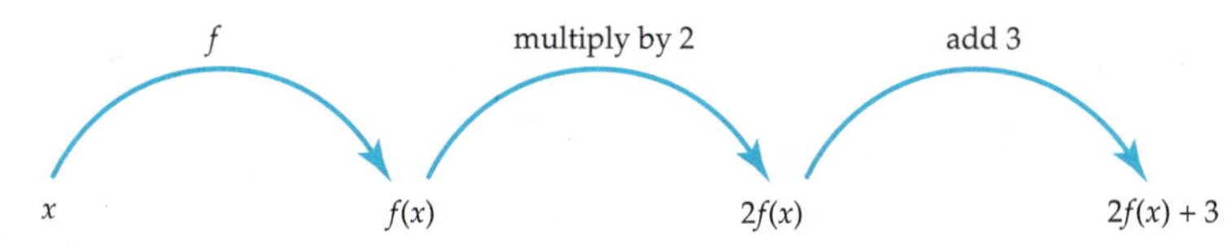

FIGURE 1.51. Arrow diagram of combined transformations.

Similar information is shown in the spreadsheet in **Figure 1.52**.

Column C is defined as 2*B, and column D is defined as 3 + C. Thus the formula defining column C involves only the multiplication by 2, and the formula for column D involves only the addition of 3. That is, the columns are completed by following the arrow diagram shown in Figure 1.51.

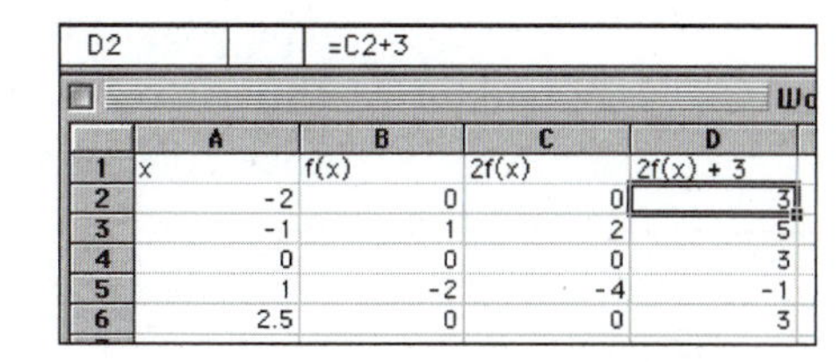
D2 =C2+3

	A	B	C	D
1	x	f(x)	2f(x)	2f(x) + 3
2	-2	0	0	3
3	-1	1	2	5
4	0	0	0	3
5	1	-2	-4	-1
6	2.5	0	0	3

FIGURE 1.52. Spreadsheet for combined transformations.

For graphing, columns A and B define the graph of $y = f(x)$, and columns A and D define the graph of $y = 2f(x) + 3$. Thus, the same x-values (column A) are used for both graphs, so the transformations are only vertical, with the stretch taking place before the vertical shift.

When beginning to sketch the new graph, it is often helpful to focus on special points of the function. For example, the point (1, –2) is on $y = f(x)$ in Figure 1.50. Applying the arrow diagram of Figure 1.51 to the point (1, –2) produces **Figure 1.53**. Compare this diagram to the $x = 1$ row of the spreadsheet in Figure 1.52.

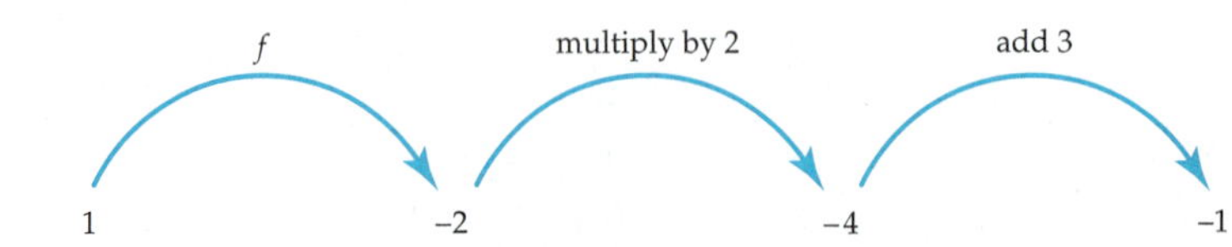

FIGURE 1.53.
Tracking the point (1, –2) using an arrow diagram.

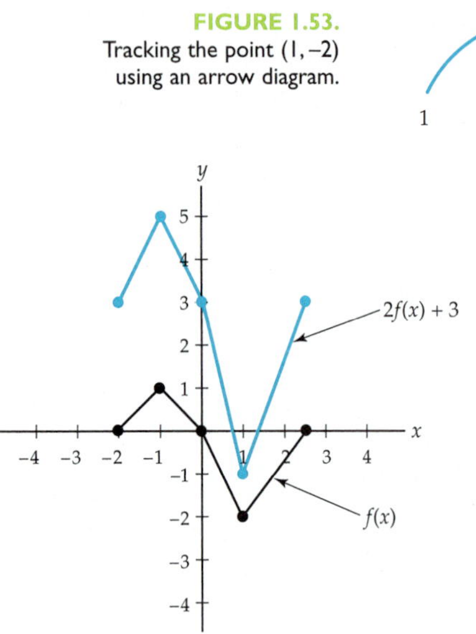

FIGURE 1.54.
A parent graph and its transformed relative.

Notice carefully how a point's two coordinates appear here. The numbers before and after the "f arrow" define the point on the parent graph. The first and last numbers in the entire sequence define the corresponding transformed point. Thus, (1, –2) becomes (1, –1). The spreadsheet verifies this result.

It is not necessary to transform each point on the graph in order to sketch the new graph, since the same reasoning is followed for every point on the graph. The result is shown in **Figure 1.54**. The spreadsheet in Figure 1.52 shows the corresponding information for each of the "corner" points.

EXAMPLE 7

Use transformations to sketch the graph of $y = -2|x| + 1$.

SOLUTION:

The parent function is $y = |x|$. The order of operations is summarized in the arrow diagram of **Figure 1.55**, which, in turn, can be used to define four columns of a spreadsheet or table. See **Figure 1.56**.

FIGURE 1.55.
Arrow diagram for $y = -2|x| + 1$.

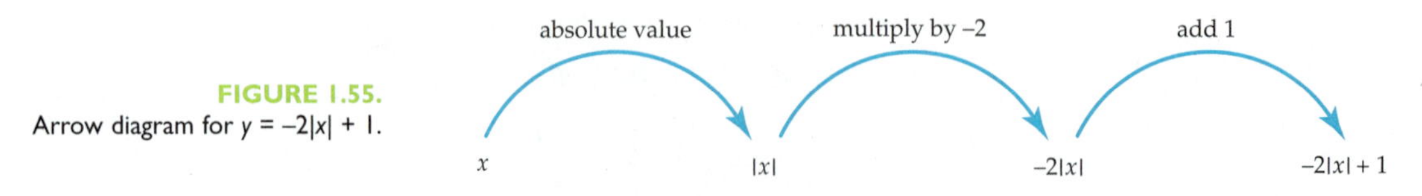

Rows 8 and 6 of the spreadsheet show that (2, 2) on the parent function becomes (2, –3), and the point (0, 0) becomes (0, 1).

D2	=C2+1			
	A	B	C	D
1	x	f(x)	-2f(x)	-2f(x) + 1
2	-4	4	-8	-7
3	-3	3	-6	-5
4	-2	2	-4	-3
5	-1	1	-2	-1
6	0	0	0	1
7	1	1	-2	-1
8	2	2	-4	-3
9	3	3	-6	-5
10	4	4	-8	-7

FIGURE 1.56.
Spreadsheet for $y = -2|x| + 1$.

The output of the parent function has been multiplied by a factor of –2, and 1 has been added to that result. Thus the graph's heights are doubled, the entire graph is reflected across the x-axis and then it is translated (shifted) up 1 unit. **Figure 1.57** shows the sequence of steps.

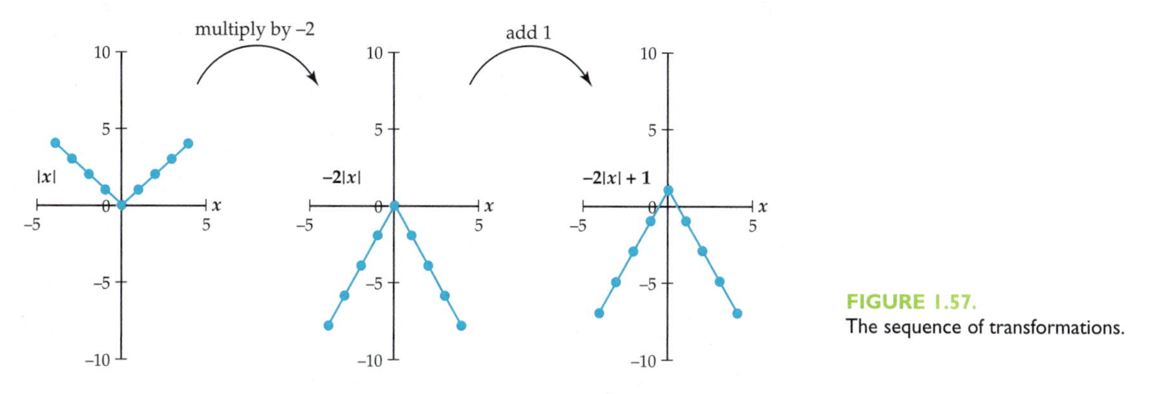

FIGURE 1.57.
The sequence of transformations.

Two additional transformations—multiplying the independent variable by a constant and adding or subtracting a constant to the independent variable—are explored in the following exercises. A good understanding of the geometry of all of the transformations in **Table 1.31**, along with your general knowledge of the tool kit functions, will help you to recognize graphs as well as to sketch them accurately and efficiently.

Form	Process	Effect
$y = cf(x)$	Multiply the output of a by a constant	Vertical stretch or function compression
$y = f(x) + c$	Add a constant to or subtract a constant from the output of a function	Vertical shift up or down
$y = f(cx)$	Multiply the independent variable by a constant	Horizontal stretch or compression
$y = f(x + c)$	Add a constant to or subtract a constant from the independent variable	Horizontal shift left or right
$y = -f(x)$	Multiply output by –1	Reflect the graph of f about the x-axis
$y = f(-x)$	Multiply input by –1	Reflect the graph of f about the y-axis

TABLE 1.31.
Basic transformations.

Exercises 1.5

1. **Figure 1.58** shows the graphs of two functions. Use these graphs to answer the following questions.

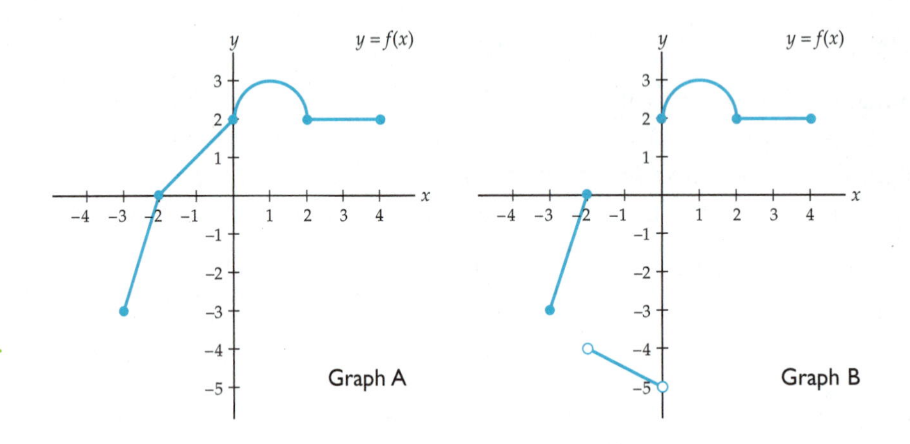

FIGURE 1.58. Graphs for Exercise 1.

a) What are the domain and range of each function in Figure 1.58?

b) For each graph in Figure 1.58, determine $f(-1)$, $f(0)$, and $f(3)$.

c) For each graph in Figure 1.58, determine all values of x for which $f(x) = 1$.

d) Graph $y = f(x) - 1$ for each graph. What are the domain and range of each new function?

e) Graph $y = 0.5f(x)$ for each graph in Figure 1.58. What are the domain and range of each new function?

f) What effect does adding a constant to a function have on the domain and range of the function? What effect does multiplying a function by a constant have?

2. Use your knowledge of transformations to sketch the graph of $y = 3x^2 - 1$. Briefly explain the process.

3. a) Describe how the graph of $y = x$ can be transformed to give the graph $y = 4x - 2$.

   b) The point (0, 0) on $y = x$ becomes what point on $y = 4x - 2$?

Exercises 1.5

4. **Figure 1.59** is a copy of Figure 1.50. **Figure 1.60** is an altered version of Figure 1.51.

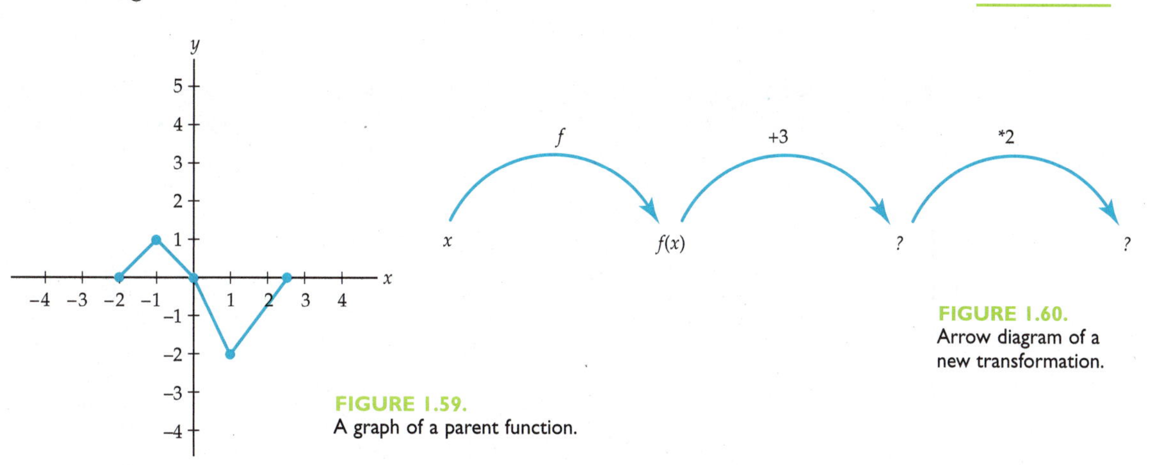

FIGURE 1.59. A graph of a parent function.

FIGURE 1.60. Arrow diagram of a new transformation.

a) Note that the final two labels for the results of the transformations have been omitted from the arrow diagram. Draw a copy of Figure 1.60 on your paper and add proper labels below the last two arrowheads. Then write a single equation for this combination of transformations.

b) Describe the geometric effects of the transformation defined by this arrow diagram. You might wish to use a spreadsheet or table to help organize your thinking about the variables that define the graphs.

c) Graph the result of this transformation. Is this transformation equivalent to the one you read about in Figure 1.51? Explain.

5. **Figure 1.61** shows the graphs of the parent function $y = |x|$ and a transformed relative. Write an equation for the relative and explain. Can you find a second equation that is also correct?

6. If $f(x) = 2x^2 + 1$, write an equation for $3f(x)$. Write an equation for $f(3x)$.

7. If $f(x) = 2x^2 + 1$, write an equation for $f(x) - 3$. Write an equation for $f(x - 3)$.

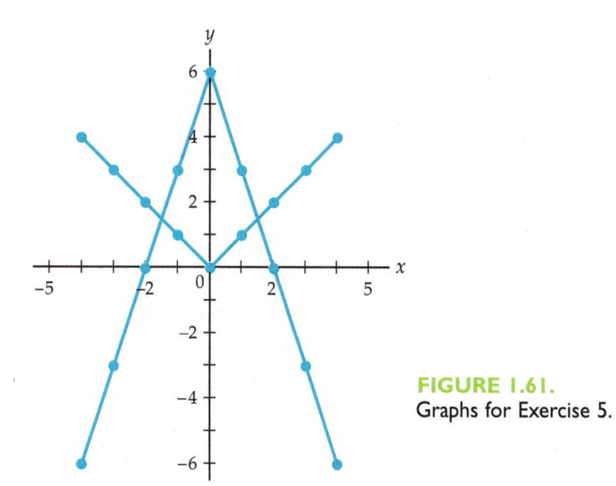

FIGURE 1.61. Graphs for Exercise 5.

Exercises 1.5

8. **Table 1.32** shows the normal high temperature in Salt Lake City Utah for each month. The temperatures are displayed in degrees Celsius, degrees Kelvin, and in degrees Fahrenheit.

Month	Degrees Celsius	Degrees Kelvin	Degrees Fahrenheit
Jan.	2.2	275.35	36
Feb.	6.7	279.85	44
Mar.	11.1	284.25	52
Apr.	16.1	289.25	61
May	22.2	295.35	72
June	28.3	301.45	83
July	33.3	306.45	92
Aug.	31.7	304.85	89
Sept.	26.1	299.25	79
Oct.	18.9	292.05	66
Nov.	10.6	283.75	51
Dec.	3.3	276.45	38

TABLE 1.32. Normal high temperatures in Salt Lake City.

Source: U.S. Department of Commerce, National Oceanic and Atmospheric Administration, Climatic Averages and Extremes for U.S. Cities.

a) Graph both Celsius and Kelvin temperatures in the same viewing window. How are the two graphs related?

b) What transformation could be used to convert temperatures measured in degrees Celsius to degrees Kelvin?

c) Graph both Celsius and Fahrenheit temperatures in the same viewing window. How are the graphs related?

d) What transformations could be used to convert temperatures measured in degrees Celsius to degrees Fahrenheit?

9. Of the basic transformations listed in Table 1.31, you are now familiar with the first two, namely the vertical transformations $y = cf(x)$ and $y = f(x) + c$. This exercise examines $y = f(cx)$ and $y = f(x + c)$.

a) Sketch $f(x) = x^2$. Predict what the graph of $y = 2f(x)$, that is, $y = 2x^2$, looks like. Confirm your prediction by graphing the equation.

b) Predict the graph of $y = f(2x)$, that is, $y = (2x)^2$. Then on a single set of axes, use your calculator to help you sketch both $y = x^2$ and $y = (2x)^2$. How does the graph compare to your prediction?

c) Describe the differences between the graphs of $y = 2f(x) = 2x^2$ and $y = f(2x) = (2x)^2$. Pay particular attention to the effect of the constant 2 in each case.

d) Parts (b) and (c) indicate that multiplying the independent variable by 2 causes a horizontal compression by a factor of 1/2, but what happens when you multiply by –2? Use your calculator to graph $f(x) = (2x)^3$ and $f(x) = (-2x)^3$ on a single set of axes. Describe the apparent effect. To confirm your conjecture, try another pair of related functions such as $f(x) = (2x)^2$ and $f(x) = (-2x)^2$. You may also wish to try tracking points.

10. Use your calculator to graph $f(x) = \left(\frac{x}{2}\right)^2$. What is the apparent effect of multiplying x in the equation by 1/2? Explain.

11 a) Graph $y = x^3$, $y = (3x)^3$, and $y = (-3x)^3$. Describe your observations. Include the effects on the graph of $y = x^3$ caused by multiplying the input by 3 or –3 before cubing.

b) Construct a pencil-and-paper table, arrow diagram, or spreadsheet for each of $y = (3x)^3$ and $y = (-3x)^3$. Show a separate column for each step of the calculation. Then use your table to explain the geometric effects you observed in part (a).

12. Let $f(x) = |x|$, $g(x) = |x - 3|$, and $h(x) = |x + 3|$.

a) Evaluate $f(1)$, $g(4)$, and $h(-2)$.

b) On a single set of axes, sketch the graphs of f, g, and h. Describe the similarities and differences in your three graphs.

c) Summarize the effect on the graph of a function $y = f(x)$ when a constant is subtracted from the independent variable. If you are not sure, try other functions, such as x^2 and x^3 before drawing conclusions. Be sure to observe cases in which $c < 0$ and those in which $c > 0$.

d) Assuming $c > 0$, explain why subtracting c units from the independent variable causes the graph to shift c units horizontally to the right. (You may wish to use a table or arrow diagram to help with your explanation.)

Exercises 1.5

13. **Figure 1.62** shows the entire graph of a particular parent function, $y = g(x)$.

a) Make a table of the three key points in this graph. Label your columns "input for g" and "output of g." State the domain and range of the function g.

b) Evaluate $g(-2)$ and $g(0)$. Determine x so that $g(x) = 3$.

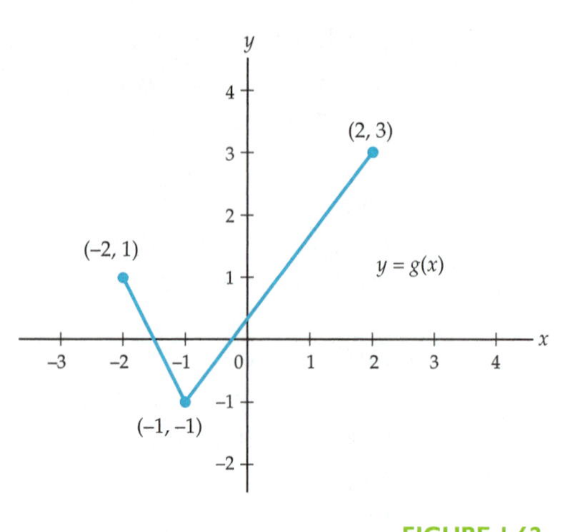

FIGURE 1.62. The graph of $y = g(x)$.

x Input for $g(x-2)$	$x-2$	y Output for $g(x-2)$
	–2	1
	–1	–1
	2	3

TABLE 1.33.

c) Complete **Table 1.33** to find the key points for the graph of $y = g(x - 2)$. Then sketch the graph. What are the domain and range of the new function? How do the "input for $g(x - 2)$" and "output for $g(x - 2)$" entries in the new table compare to those from part (a)?

d) Make a table for $y = g(x + 1)$ and sketch its graph. What are the domain and range of the new function? How do the entries in the new table compare to those from part (a)?

e) Make tables for $y = g(x/3)$ and $y = g(2x)$ and sketch their graphs. State the domain and range of each of the new functions. How do the entries in the new tables compare to those from part (a)?

f) What effect does multiplying the independent variable by a constant have on the domain and range of the function? What effect does adding a constant have?

14. **Table 1.34** lists 8 functions, each a transformation of the parent function. Predict what each graph should look like, then use your calculator to sketch it. Complete Table 1.34 based on your observations. (Item (a) has been done for you.)

15. The function $y = 2(x - 1)^2 - 3$ is a transformation of which of your tool kit functions? Without the help of your calculator, use transformations to sketch the graph. Explain the steps you take in graphing the function.

Function	Domain	Range	Transformation (computation)	Geometric effect of the transformation
$(y=\sqrt{x})$ (parent)	$x \geq 0$	$y \geq 0$	———	———
a) $y=3\sqrt{x}$	$x \geq 0$		Multiply the output of the function by 3	Vertical stretch by a factor of 3
b) $y=\sqrt{3x}$				
c) $y=\sqrt{x}+1$				
d) $y=\sqrt{x+1}$				
e) $y=\frac{1}{2}\sqrt{x}$				
f) $y=\sqrt{\frac{1}{2}x}$				
g) $y=-\sqrt{x}$				
h) $y=\sqrt{-x}$				

Exercises 1.5

TABLE 1.34. Transformations for Exercise 14.

16. Write an equation for each graph in **Figures 1.63–1.68**. The parent functions are $y = x^2$, $y = |x|$, and $y = \sqrt{x}$.

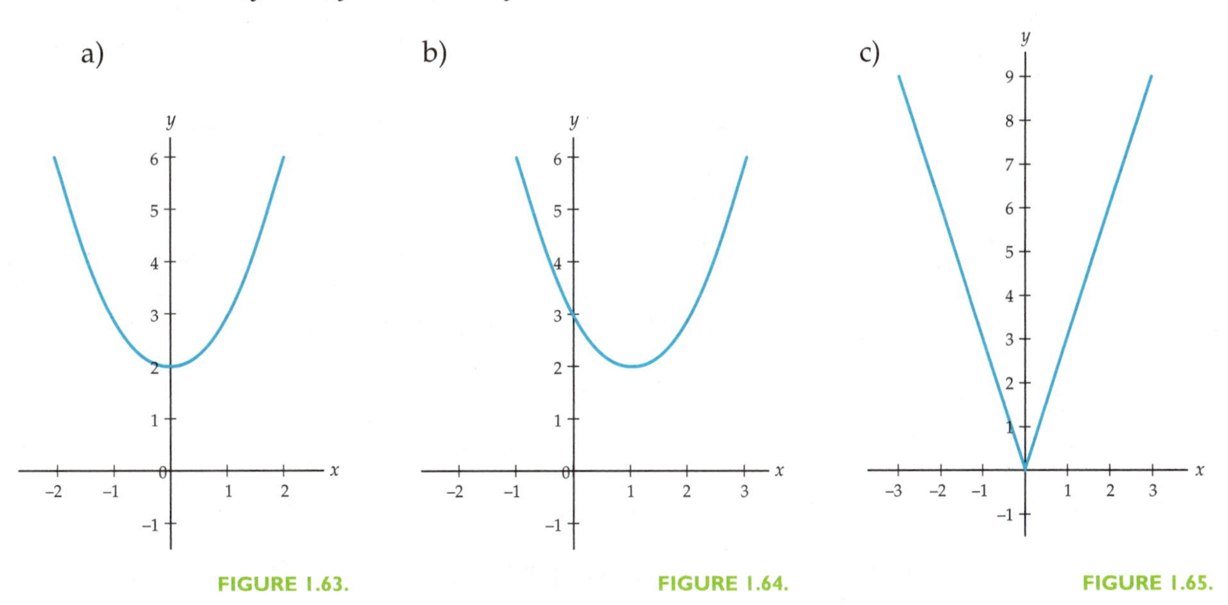

FIGURE 1.63. **FIGURE 1.64.** **FIGURE 1.65.**

Exercises 1.5

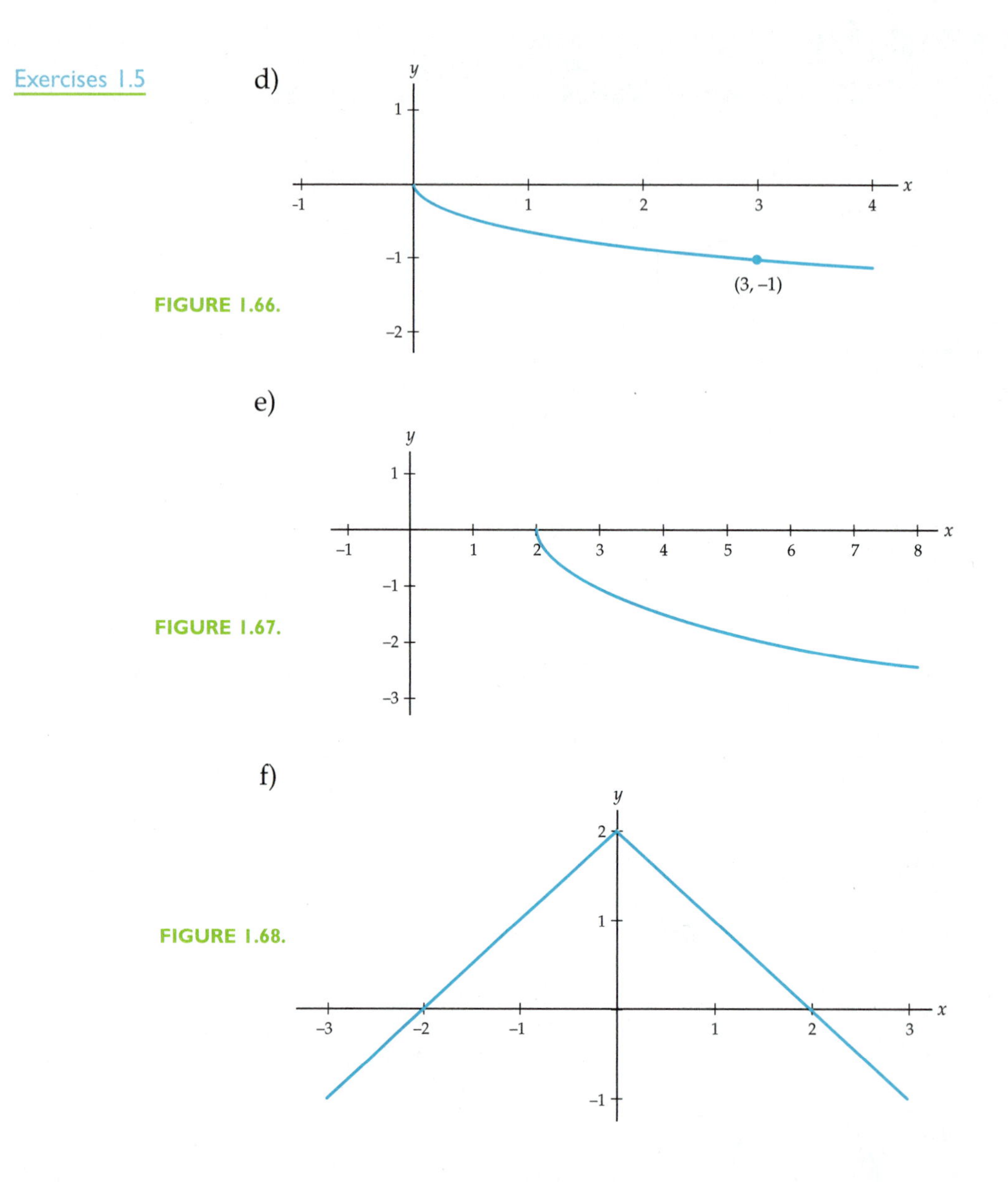

FIGURE 1.66.

FIGURE 1.67.

FIGURE 1.68.

17. Write an equation for $y_1 = 2f(x)$ and for $y_2 = f(0.5x)$ if $f(x) = 4|x| - 1$. Which of the functions, y_1 or y_2, transforms the graph of f horizontally? In what way? Be specific.

18. a) Use your calculator to graph $y = x^2$ and $y = (-x)^2$. Describe this transformation. Explain.

    b) Let $f(x) = 3x + 4$ be the parent function. Define the related function $y = f(x - 2) + 6$. Graph both functions on a single set of axes. Describe the effect of the transformation. Explain what you see when you compare the graphs.

19. In Exercise 18, you saw two cases in which transformations did not change the graphs of the original functions. It is also possible for two different transformations to produce identical final graphs.

   a) Let $f(x) = -2x + 3$ be the parent function. Define two related functions by $y_1 = f(x-2)$ and $y_2 = f(x) + 4$. Sketch the graphs of y_1 and y_2 on a single set of axes. Describe the results and explain.

   b) Let $h(x) = x^2$ be the parent function. Find a horizontal stretch $y = h(cx)$ that produces the same graph as the vertical stretch $y = 2h(x)$.

20. The graph of $f(x) = (x-2)^3 + 1$ is a transformation of the parent function $y = x^3$.

   a) Without graphing it, describe the symmetry of the graph of the parent function.

   b) Sketch the graph of $f(x) = (x-2)^3 + 1$.

   c) Does this graph have symmetry with respect to the x-axis? The y-axis? The origin? Describe any symmetry it does have.

   d) Geometrically, a graph is said to have **symmetry with respect to a point** if, when the graph is rotated 180° about the point, it is superimposed on itself. Does the graph in part (b) have symmetry with respect to any point? If so, what point?

   e) In Activity 1.3 of Lesson 1.4, Item 5 on page 44 gave an algebraic definition for symmetry with respect to the origin. A graph is said to be symmetric with respect to the origin if, and only if, for every point (x, y) on the graph, $(-x, -y)$ is also on the graph. Keeping that definition in mind and using your knowledge of transformations, write an algebraic definition for symmetry with respect to a point (h, k). Be sure to check your definition with several points on your graph.

21. **Investigation**. Considering the functions $y_1 = x^2$ and $y_2 = Ax^2$, y_2 varies directly with y_1, and the constant of proportionality is A. That is, $y_2 = Ay_1$, and the graph of y_2 vs y_1 is part of the line through (0, 0) with slope A. A similar result applies for any power function and its corresponding parent: the output of $y_2 = Ax^B$ varies directly with the output of $y_1 = x^B$. That fact provides a powerful tool for modeling data when you can guess the tool kit family to which the model belongs.

Exercises 1.5

a) **Table 1.35** shows values for $y_1 = x^2$ and $y_2 = 2x^2$. Verify that the graph of y_2 vs y_1 is part of the line through (0, 0) with slope 2.

b) **Table 1.36** shows data from Table 1.26 of Lesson 1.4.

x	y_1	y_2
–5	25	50
3	9	18
1	1	2
0	0	0
1	1	2
3	9	18
5	25	50

TABLE 1.35.
Data for $y_1 = x^2$ and $y_2 = 2x^2$.

x	y
1	1
2	5.66
2.5	9.88
4	32
5	55.9
6	88.2

TABLE 1.36.
Data from Lesson 1.4.

Add a column for x^2 and plot y versus x^2. Is y proportional to x^2? Is $y = Ax^2$ a good model for these data?

c) Repeat part (b) for y versus x^3.

d) Repeat part (b) for y versus $x^{2.5}$.

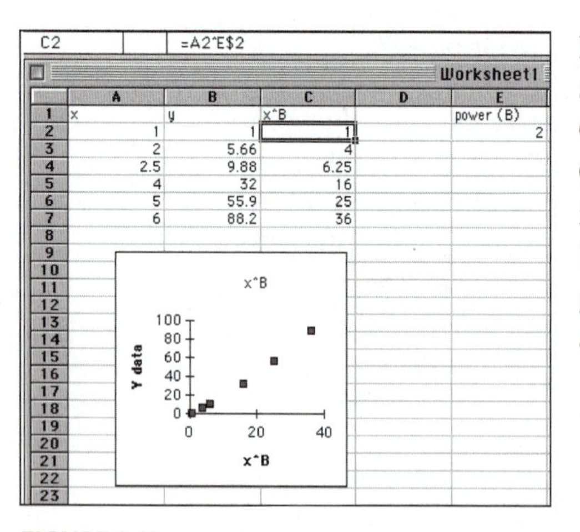

FIGURE 1.69.
Spreadsheet for testing power models.

Figure 1.69 shows a sample spreadsheet for automating this kind of power search. Columns A and B are the original data. Column C is the transformed x-values (data from the suspected parent function—in Figure 1.69, from x^2). Cell E2 is the power used to generate column C, and the graph is defined as column B versus C.

e) Explain how the user, just by changing E2, can find the best power model for a given set of data. That is, what should the user look for as different values are tried in cell E2?

f) Build a spreadsheet like the one in Figure 1.69. In column D add a formula to check for direct variation using the ratio definition of proportionality. Then use your spreadsheet on the Ponderosa Pine data. (Note: If your calculator allows list operations, try doing the same thing on it, too.)

LESSON 1.6

Operations on Functions

As you proceeded through the lessons in this chapter, you reviewed familiar functions and examined the effects of transformations on these functions. In this lesson, you will explore geometrically and algebraically the effects of adding, subtracting, multiplying, and dividing two or more of the functions in your tool kit.

Activity 1.5

Use your calculator to examine the data in **Table 1.37** that shows 11 years of private college costs from 1975 to 1985.

Year	Total Cost
1 (1975)	$ 4205
2	4460
3	4680
4	4960
5	5510
6 (1980)	6060
7	6845
8	7600
9	8435
10	9000
11 (1985)	9659

Corbis

TABLE 1.37.
Private College Costs (1975–1985).

Source: Fall 1985 issue of *The College Board News*

1. Describe the scatter plot of your data.
2. Find a linear regression equation for the data.
3. Plot both the data and your line in the same window. Describe the quality of the fit for this model.
4. To quantify the fit you described in Item 3, examine the residuals and graph them. Is there a pattern? If so, describe it.

5. From your observations of the data, the linear regression line, and the patterns in the residuals, try to find a model that provides a better fit. Check your new model by examining its residuals. Explain why you think your new model provides a better fit than the linear one.

   Hint: Since Residual = Data y-value – Model y-value, then Data y-value = Model y-value + Residual.

ADDITION OF FUNCTIONS

In Activity 1.5, you may have found that by adding two functions together, you were able to create a new function that provided a model that more closely fit the data. If you have an initial model and can fit an equation to the related residuals, then the sum of the two equations should be an improvement.

TAKE NOTE

If a definite trend or pattern is apparent in the residuals, it is sometimes possible to refine the original model by modeling the residuals, and this new model often provides a better fit to the data. However, even though this appears to be a better model for the data given, it may not be an improvement for the given situation. If given a choice, a modeler always prefers a model that explains the situation that is under investigation to one that just fits the data collected. You will deal more with this kind of consideration in Chapter 3.

Functions, like numbers, can be added together. In symbols, the sum of two functions f and g can be expressed as $f + g$. For example, if $f(x) = 3x + 5$ and $g(x) = x^2 - 2x + 2$, then the output of the sum of the two functions, $f + g$, for a particular input, x, can be found by adding $f(x)$ and $g(x)$. That is, $(f + g)(x)$ is defined as $f(x) + g(x)$:

$$(f + g)(x) = f(x) + g(x) = (3x + 5) + (x^2 - 2x + 2) = x^2 + x + 7.$$

The domain of the resulting function is the **intersection** of the domains of the original two functions. That is, it consists of all values of x that are common to the domains of both f and g.

A warning about notation: $(f + g)(x)$ means "the value of the function whose name is $f + g$, evaluated when x is the input." It does *not* indicate multiplication of the quantity $f + g$ and the quantity x.

EXAMPLE 8

Given: $m(x) = 2x^{-1} + 5$ and $n(x) = \sqrt{x+4} + 3$. Write an equation for $m + n$ and state its domain.

SOLUTION:

$$(m + n)(x) = (2x^{-1} + 5) + (\sqrt{x+4} + 3) = 2x^{-1} + \sqrt{x+4} + 8.$$

The domain of m is all numbers other than 0. The domain of n is all numbers greater than or equal to –4. Thus the domain of $m + n$ is the intersection of the domains of m and n, which is all real numbers in either of (–4, 0) or (0, ∞).

Representing Addition of Functions

As with transformations, adding functions can be represented in many ways. Tables and graphs are perhaps the two most useful representations. **Figure 1.70** shows a portion of a spreadsheet for adding two functions. Note that the addition is done exactly as you might expect; the number in the $f(x)$ column is added to the number in the $g(x)$ column to form the $(f + g)(x)$ column.

D2 =C2+B2

	A	B	C	D
1	x	f(x)	g(x)	(f+g)(x)
2	-1	-1	2	1
3	0	0	3	3
4	1	1	4	5
5	2	2	5	7
6	3	3	6	9
7	4	4	7	11
8	5	5	8	13

FIGURE 1.70. Adding functions f and g using a table.

Information for graphing the sum can be read directly off the table. However, it is possible, and frequently useful, to build the graph of the sum of two functions from the original graphs—without actually adding any numbers. Since $f(x)$ means "the height of f above the input number x," the sum $f(x) + g(x)$ is exactly the combined heights of the two functions at a particular x-value.

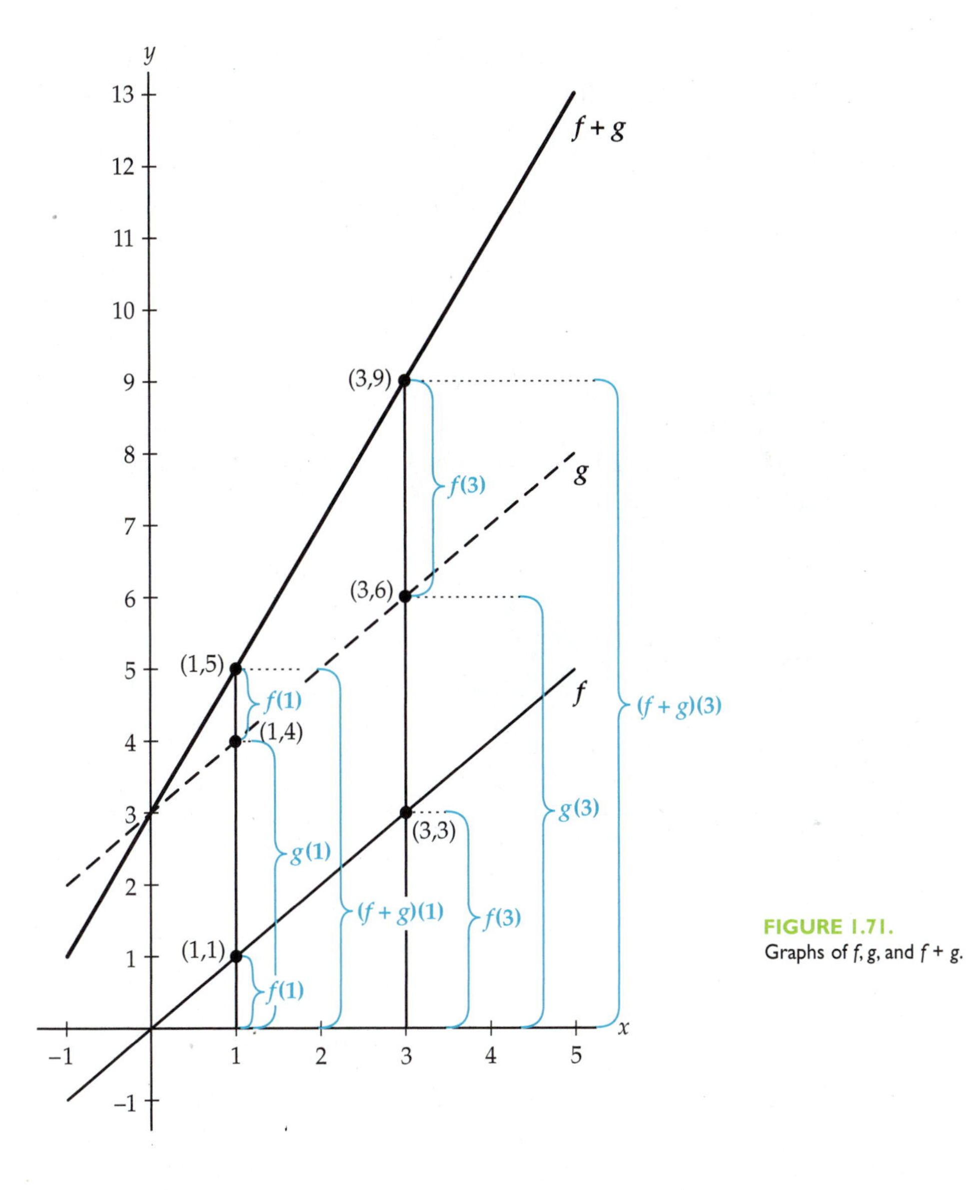

FIGURE 1.71.
Graphs of f, g, and $f + g$.

For example, in **Figure 1.71,** when $x = 3$, the height of f is 3, so $f(3) = 3$. At the same x, the height of g is 6, so $g(3) = 6$. Thus $(f + g)(3) = 9$. However, distances (heights) can be added geometrically. A correct graph of the sum of two functions can be obtained by transferring heights. Thus, the scales on the axes here are not needed.

OTHER OPERATIONS ON FUNCTIONS

Not only can functions be added, but they can be subtracted, multiplied, and divided as well.

$$(f-g)(x)=f(x)-g(x)$$

$$(f\cdot g)(x)=f(x)\cdot g(x)$$

$$\left(\frac{f}{g}\right)(x)=\frac{f(x)}{g(x)}$$

In each case, the domain of the resulting function is the intersection of the domains (x-values common to both domains) of the functions f and g. When dividing functions, e.g. $\frac{f}{g}$, it is also necessary to exclude those values of x for which $g(x) = 0$.

TAKE NOTE

While distances can be added and subtracted directly on graphs without reference to scales, multiplication and division are scale-dependent. These operations must be done numerically, not graphically.

Exercises 1.6

1. In parts (a)–(c), find equations for $f + g, f - g, f \cdot g$, and $\frac{f}{g}$, and state the domains of the resulting sum, difference, product and quotient.

   a) $f(x) = 3x^2 - 1;\ g(x) = 2x - 4$

   b) $f(x) = \sqrt{x};\ g(x) = 3x - 1$

   c) $f(x) = 1 + \frac{1}{x};\ g(x) = \frac{1}{x}$

2. Given $f(x) = 4x - 1$ and $(f + g)(x) = 7 - \frac{1}{2}x$, write an equation for the function g.

3. Given $f(x) = 3x^2 - 6$, $g(x) = 5 - 4x^2$, and $h(x) = 3x + 1$, find the sum of the three functions.

4. **Figure 1.72** shows the graphs of two functions.

   a) Find the sum of m and the piecewise function n graphically.

   b) Determine the (piecewise-defined) equations for both m and n. Then write the (piecewise-defined) equation for $m + n$ for the domain $0 \le x \le 3$.

   c) Use your answer to (b) to check your graphical answer to (a).

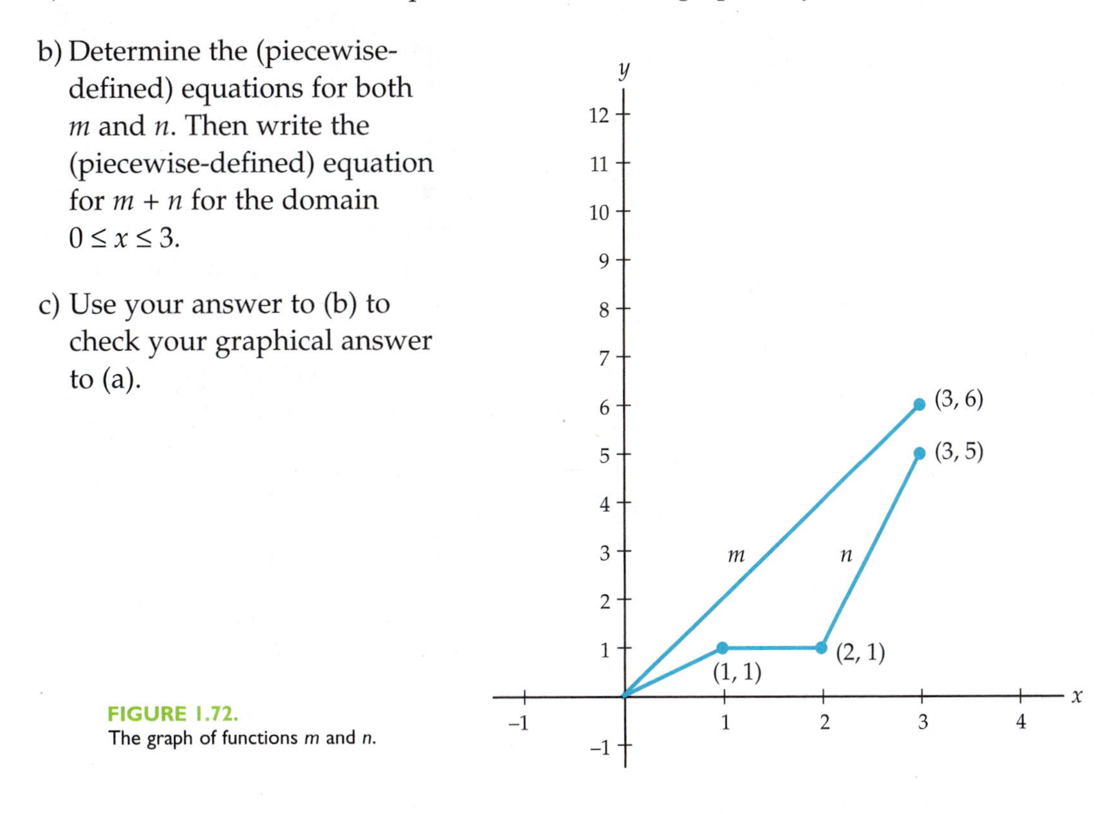

FIGURE 1.72.
The graph of functions m and n.

Exercises 1.6

5. Graph the sum of the two functions whose graphs are given in **Figure 1.73.**

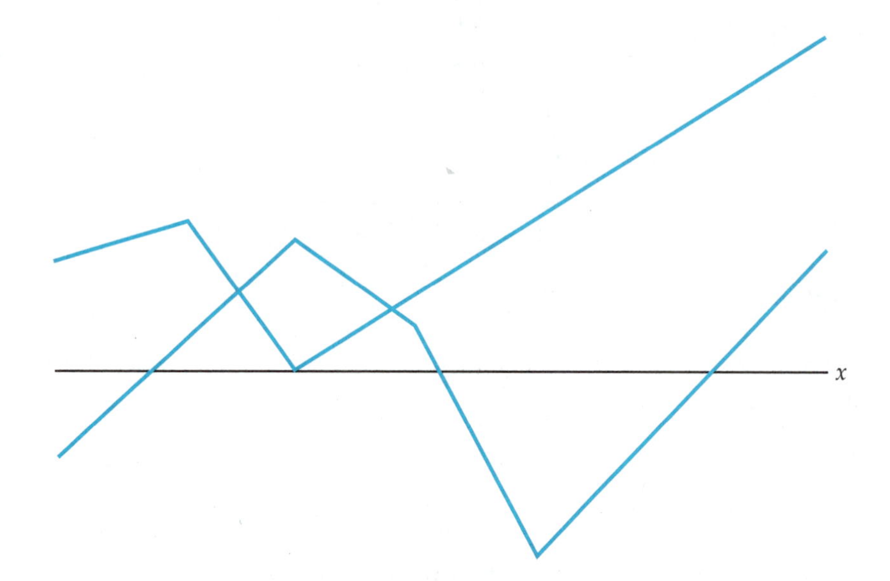

FIGURE 1.73.
Graphs of two functions.

6. a) Use your calculator to graph $f(x) = 3x^2$ and $g(x) = 2x^3$ on the same set of axes using a window of $[-5, 5] \times [-5, 5]$. Graph $f + g$. Do the characteristics of this new graph more closely resemble the graph of f or the graph of g?

   b) Choose any second degree power function you wish for f and any third degree power function for g. Repeat (a). Again, which power function does the graph of $f + g$ most closely resemble? (Hint: Make sure you choose calculator windows that allow you to see the characteristic shape of each function.)

   c) Graph $f(x) = 2x^4$ and $g(x) = 3x^3$. Once again, graph $f + g$ and tell which function the graph of $f + g$ most closely resembles.

   d) What you've observed in (a)–(c) illustrates the dominance of higher powers of x over lower powers of x when $x > 0$. Explain why this is true.

7. Graph the functions f and g where $f(x) = 3x + 4$ and $g(x) = x + 1$. Find $(f - g)(x)$ graphically and algebraically. Do your two answers support each other? Explain.

Exercises 1.6

8. Let $f(x) = |x|$ and $g(x) = x$.

   a) Without using a graphing utility, sketch the graphs of $f + g$, $f - g$, and $f \cdot g$ on the domain [–3, 3].

   b) Explain the features of the graph of $f \cdot g$.

9. The heat index is a measure of apparent temperature and is based on air temperature and relative humidity. When the relative humidity is high, the body is unable to perspire efficiently resulting in higher internal body temperature. During summer months meteorologists provide the heat index and the air temperature to aid the community in making plans. The following formula is used to calculate the heat index:

   $HI = -42.3799 + 2.04901523T + 10.1433127R - 0.22475541TR - 6.83783 \times 10^{-3}T^2 - 5.481717 \times 10^{-2} R^2 + 1.22874 \times 10^{-3}T^2R + 8.5282 \times 10^{-4}TR^2 - 1.99 \times 10^{-8}T^2R^2$

   where

   T = ambient dry bulb temperature degrees Fahrenheit

   R = relative humidity in percentage

   The equation is only useful for temperatures 80 degrees or higher, and relative humidities 40% or greater.

   Table 1.38 gives heat advisory information based on the heat index.

Category	Heat Index	Possible Heat Disorders for People in High Risk Groups
Extreme Danger	130°F or higher	Heatstroke or sunstroke likely
Danger	105°F–129°F	Sunstroke, muscle cramps, and/or heat exhaustion likely with prolonged exposure
Extreme Caution	90°F–104°F	Sunstroke, muscle cramps, and/or heat exhaustion possible with prolonged exposure
Caution	80°F–90°F	Fatigue possible with prolonged exposure

TABLE 1.38.

Source: National Weather Service, Birmingham AL Homepage.

   Calculate the heat index for a 94-degree F day with relative humidity of 80% and classify these conditions using the table above.

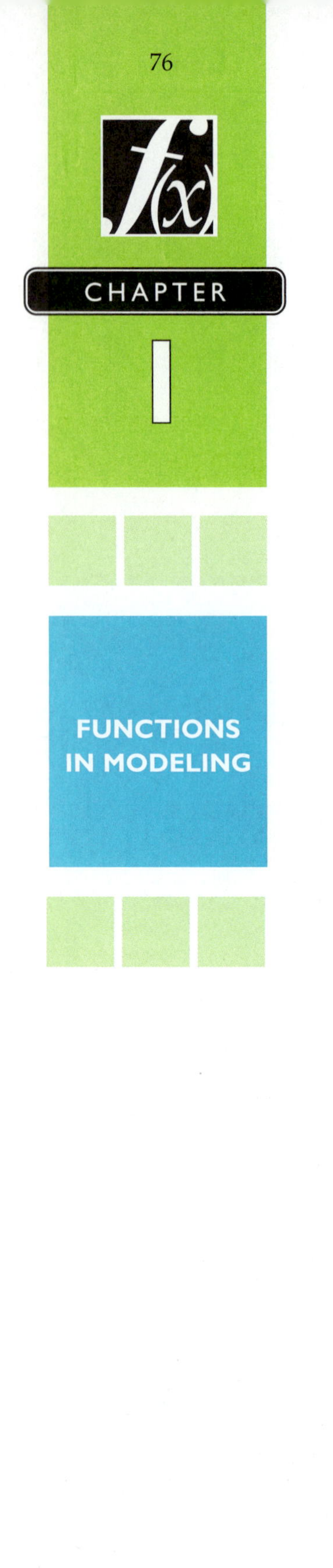

Chapter 1 Review

1. Write a summary of the important mathematical ideas found in Chapter 1.

2. Do the graphs in Figures **1.74–1.76** show functions? Explain.

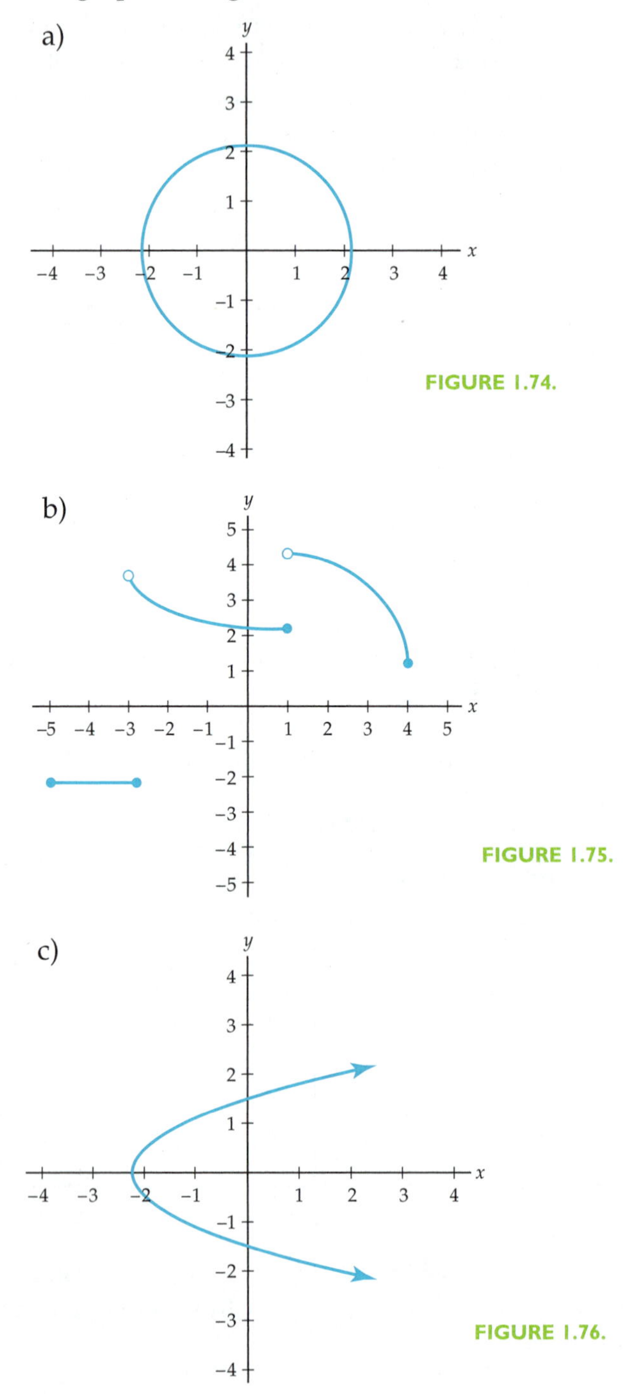

FIGURE 1.74.

FIGURE 1.75.

FIGURE 1.76.

3. Give the domain and range for the functions in (a)–(d):

   a) $f(x) = x^{\frac{1}{2}}$

   b) $y = 2x + 6$

   c) See **Figure 1.77**.

   d) See **Figure 1.78**.

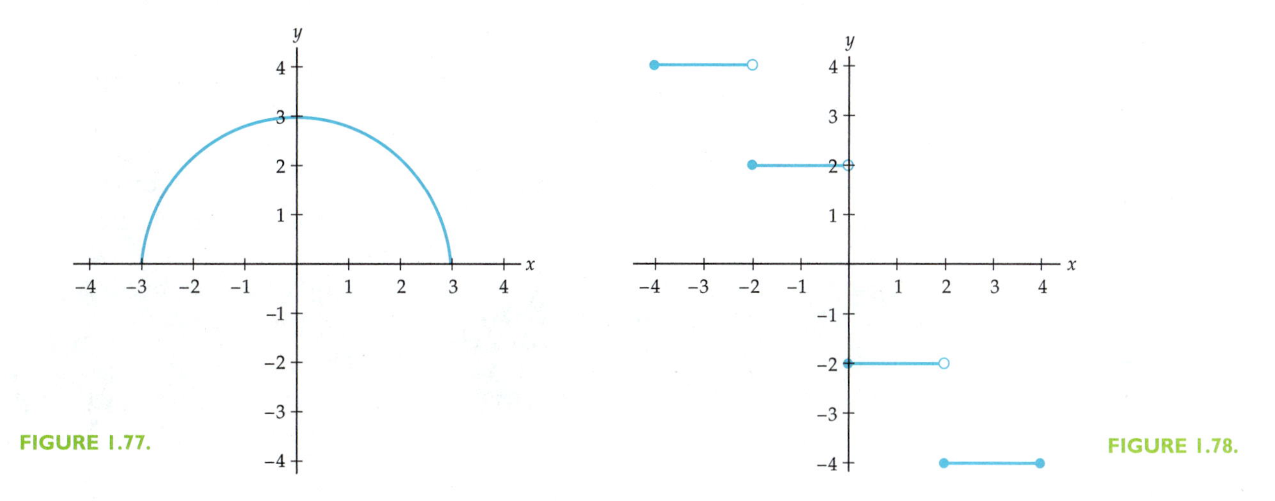

FIGURE 1.77.

FIGURE 1.78.

4. Sketch a qualitative graph for each of the following situations.

   a) The temperature of a glass of ice water sitting in a warm room.

   b) The temperature of a hot cup of coffee sitting in a cool room.

5. During the month of November, grocery stores throughout the United States have sales on turkeys. **Figure 1.79** shows an advertisement for one of those sales.

Turkey Sale!	
6-8 pounds	**$ 1.09** per pound
over 8 pounds but less than 16 pounds	**.89** per pound
16 pounds or over	**.49** per pound

FIGURE 1.79.
An ad for turkeys.

a) For this situation, you are interested in how much your turkey will cost. What is the independent variable? the dependent variable?

b) Does this situation describe a function? Explain why or why not.

c) Give a reasonable domain for the function.

d) Sketch a graph of the function. To what tool kit family does it belong?

e) Suppose you have $8.00 to spend for your Thanksgiving turkey. What size turkey can you purchase?

6. Without the use of your calculator, complete **Table 1.39** for the indicated functions on the domain [–2, 2] and sketch each graph.

TABLE 1.39.

Function	Value of function when x = –2	Value of function when x = 2	Description of function for the interval [–2, 2] (increase/decrease, concavity, symmetry, etc.)
$f(x) = -x + 1$			
$g(x) = 4$			
$h(x) = \lvert x \rvert$			
$j(x) = x^2$			
$k(x) = x^3$			

7. From the graph in **Figure 1.80**, analyze the function H. Identify intervals on which the function is increasing. On which it is decreasing? Concave up? Concave down? Is there any symmetry exhibited? What appears to be the domain of the function? The range?

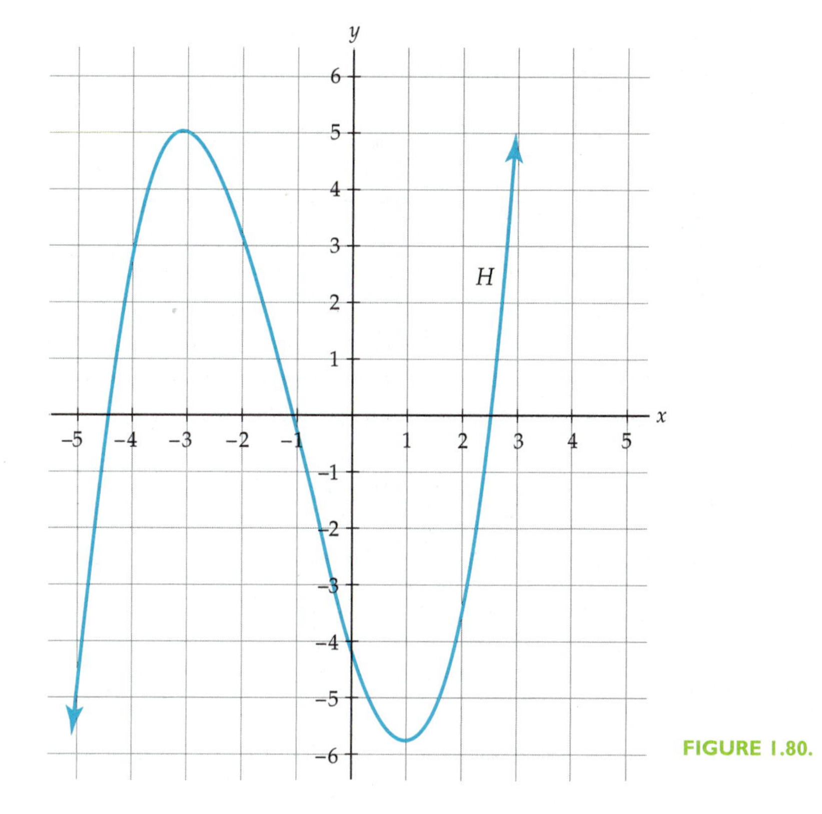

FIGURE 1.80.

8. When Sidney examined a set of data given to him by his teacher, he decided to fit the one-term model $y = x^{-1}$ to the data. His calculator screen is shown in **Figure 1.81**.

   After examining the graph and scatter plot, he decided that since his function didn't decrease as rapidly as his data, he would move down his Ladder of Powers and try $y = x^{-2}$. His new screen is shown in **Figure 1.82**.

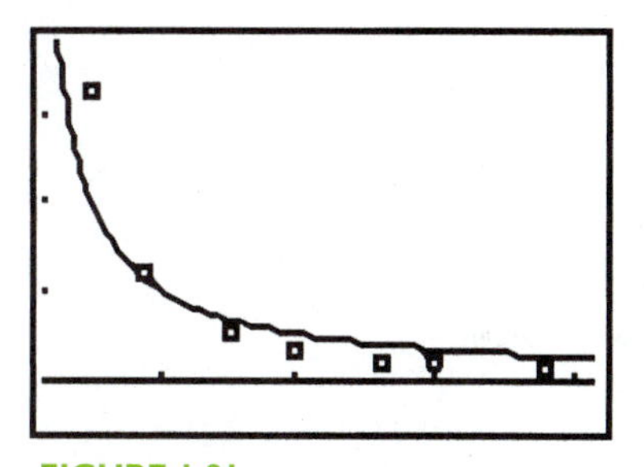

FIGURE 1.81.
Scatter plot of data and $y = x^{-1}$.

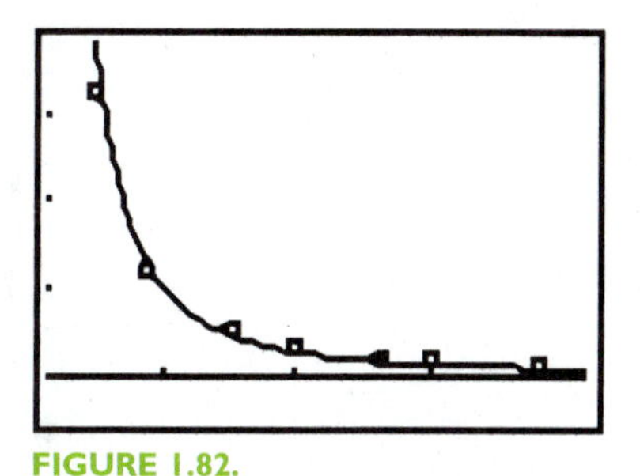

FIGURE 1.82.
Scatter plot of data and $y = x^{-2}$.

a) Give Sidney a suggestion that might help him find a better one-term function to fit his data.

b) Sidney's original data are shown in **Table 1.40**.

TABLE 1.40.
Sidney's data.

x	0.5	0.9	1.5	2	2.6	3	3.8
y	3.25	1.2	0.5	0.3	0.2	0.15	0.1

Make a scatter plot of the data and fit your suggested model from (a) to the data. Is your suggested model better than the ones used by Sidney? If not, try to find one that is better. Based on your graphs and a residual plot, explain why your model is better than Sidney's two attempts.

9. Let $f(x) = 3x - 5$, $g(x) = 2x^2$ and $h(x) = 4 - x$.

   a) Find $f(x) + g(x) - h(x)$.

   b) Find $3 \cdot f(x) + h(x)$.

   c) Evaluate the expression $g(2) - f(2)$ and $f(2) - g(2)$. Does the order in which you subtract functions make a difference?

   d) Evaluate the expression $g(3) \cdot h(2)$. Does the order in which you multiply make a difference?

   e) What is the domain of the function I if $I(x) = g(x)/h(x)$?

10. The following items refer to **Figure 1.83**.

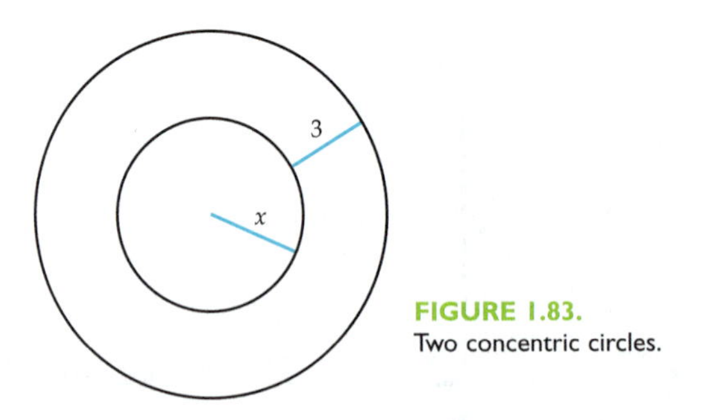

FIGURE 1.83.
Two concentric circles.

   a) Write a function for L, the area of the large circle.

   b) Write a function for S, the area of the small circle.

c) Let $A(x) = L(x) - S(x)$. What is the meaning of the function A?

d) Find A when x is 5 cm.

11. **Figure 1.84** shows the graphs of two functions.

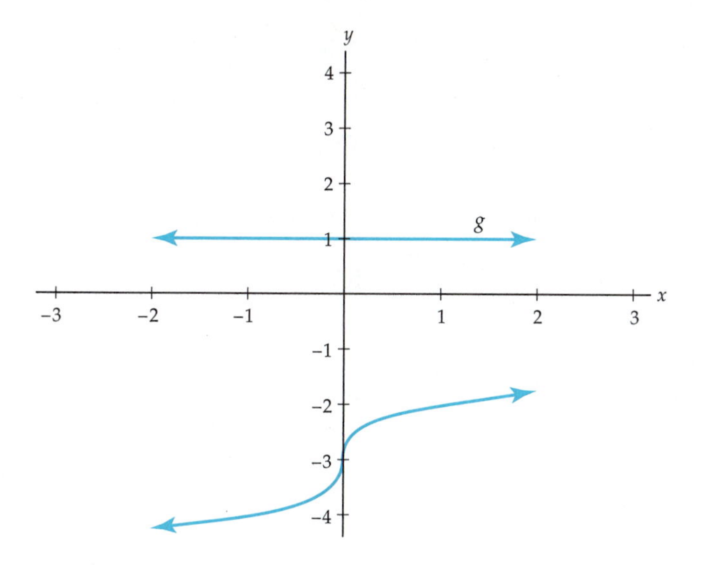

FIGURE 1.84.
Two functions.

a) Sketch the graph of $y = f(x) + g(x)$.

b) Sketch the graph of $y = g(x) - f(x)$.

12. The area of the square in **Figure 1.85** can be represented by the function $A(x) = x^2$.

Sketch a figure to represent the following:

a) $4(A(x))$

b) $A(4x)$

FIGURE 1.85.

13. Let $f(x) = x^4$. Find a function whose graph is the graph of f but is:

a) shifted three units downward.

b) compressed vertically by a factor of 0.5.

c) reflected across the x-axis.

d) stretched horizontally by a factor of 6.

e) reflected across the y-axis.

f) shifted to the left 2 units.

g) shifted downward 1 unit and to the right 3 units.

14. Let $f(x) = (x)^{1/2}$. Graph and write an equation for each of the following related functions.

    a) The function $g(x)$ is a reflection of $f(x)$ in the y-axis.

    b) The function $h(x)$ is a reflection of $f(x)$ in the x-axis.

    c) The function $k(x)$ is $f(x)$ shifted to the right 3 units, then reflected in the x-axis.

15. a) $g(x) = (-x)^{1/2}$

    b) $h(x) = -(x)^{1/2}$

    c) $k(x) = -(x - 3)^{1/2}$

16. State the domain and range of $f(x)$, $g(x)$, $h(x)$ and $k(x)$ in Exercise 15.

17. Write a function for each of the following descriptions.

    a) The change in y-values per unit change in x-values is zero. It contains the point (1, 6).

    b) The rate of change of the function is constant and the function passes through the points (1, 3) and (–2, 4).

18. Nicole was asked to formulate a model that fit the data in **Table 1.41**.

x	y
5	0.65
7	1.38
8.5	2.15
10	3.15
12	4.85
14.5	7.64

TABLE 1.41.
Nicole's data.

She tried fitting the function $y = 0.009x^{2.5}$ to the data. Her residual plot is shown in **Figure 1.86**.

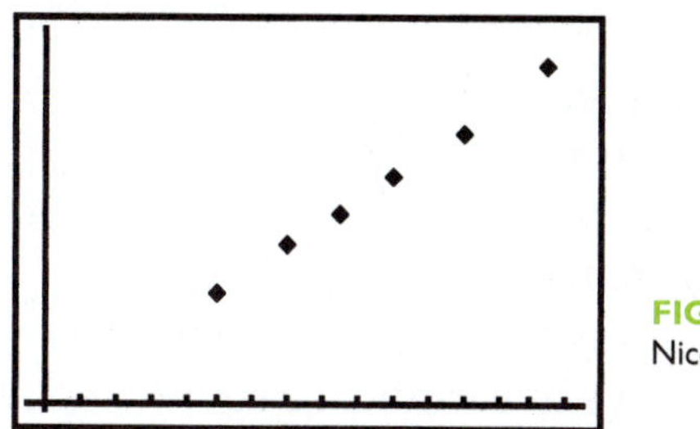

FIGURE 1.86.
Nicole's residual plot.

a) What does the scatter plot of the residuals tell you about Nicole's model?

b) Explain to Nicole how she might refine her model.

c) Use your suggestion from (b) to find a better model.

d) Convince Nicole that your model provides a better fit to the data than her model does.

Chapter 2

THE EXPONENTIAL AND LOGARITHMIC FUNCTIONS

Corbis

CHAPTER INTRODUCTION

Earthquakes are natural events that occur thousands of times a day throughout the world. People who live in earthquake country go about their lives as if the phenomenon doesn't exist, but are very aware of the potential for a disaster. In a matter of seconds, the shaking of a strong earthquake can cause extensive damage and destruction.

Scientists who study earthquakes might pose questions such as:

- What causes the shaking?
- What factors determine how much damage occurs?
- How are earthquakes measured?
- How can earthquakes be classified in a way that ordinary people can understand?
- Can earthquakes be predicted?

The shaking during an earthquake is caused by the passage of seismic (or sound) waves through the earth, beginning where a sudden movement takes place along a fault. The energy contained in those waves can cause a rolling action along the earth's surface, like a ripple moving on a still pond, or crush solid rock by compression. The earth moves surprisingly slowly during an earthquake; the velocity at which the two sides of a fault move is only about 0.1 miles per hour. However, the seismic waves that result from that movement travel along the fault at a velocity of nearly 5000 miles per hour.

Earthquake damage is due primarily to the generated seismic wave, and intensity is a measure of the severity of its shaking. Earthquakes vary immensely in their intensities, from small tremors that can be detected only with seismographs to great earthquakes that can cause extensive damage over widespread areas. Factors that affect the amount of destruction include distance from the epicenter, depth of the originating earth movement, type of soil, building practices, and population density. When a great earthquake occurs near a highly populated region, tremendous destruction can take place within a few seconds. In 1976, a single earthquake in Tangshan, China killed 600,000 people. In 1755, the city of Lisbon, a major capital at the time, was destroyed with high loss of life. In more recent times, earthquakes have caused fires originating in the gas and electrical lines that interweave modern cities.

One way to report earthquake size is to describe how much energy is released; the 1960 Chile earthquake gave off 1.25×10^{26} ergs, or roughly the equivalent of 32 billion tons of dynamite.

Most people use a magnitude scale, such as the Richter scale, in referring to earthquake size. A seismograph detects and records the amount of movement, and a magnitude number is assigned. Richter's original scale went up to 10, and was designed to communicate the relative size of one quake compared to another. The Chile earthquake had a Richter magnitude of 9.0; a shake caused by a mine explosion might only have a magnitude of 2.0. These numbers tell scientists familiar with the scale that the Chile earthquake released 10,000,000 times the energy as the mine explosion.

Thousands of minor earthquakes occur each day, but a major seismic event happens only once every 2–3 years. Since large earthquakes are preceded by a pattern of foreshocks and followed by aftershocks, seismologists study the sizes and locations of earthquakes to determine patterns that assist them in predicting major earthquakes. The data in **Table 2.1** and the chart in **Figure 2.1** represent the 783 earthquakes, strong enough to be felt, that occurred on the Pacific coast between January 1, 1990 and July 11, 1996.

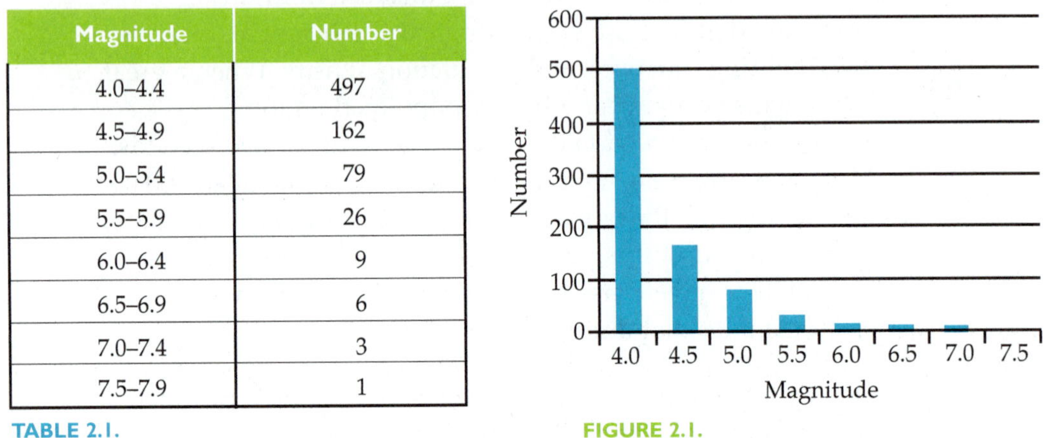

Magnitude	Number
4.0–4.4	497
4.5–4.9	162
5.0–5.4	79
5.5–5.9	26
6.0–6.4	9
6.5–6.9	6
7.0–7.4	3
7.5–7.9	1

TABLE 2.1.

Source: Martin Van Bonsangue, *Humanistic Mathematics Network Journal* #17, page 19.

FIGURE 2.1.

Given that a quake has been felt, what is the likelihood that it is a major earthquake? How do you think the magnitudes of earthquakes greater than 4.0 are distributed? In this chapter, you will use a new mathematical tool to classify and compare earthquakes that differ by orders of magnitude, and to understand a variety of models better.

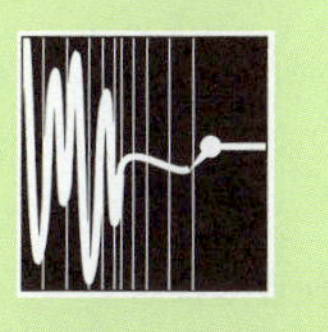

LESSON 2.1

Exponential Functions

The mathematical tool kit begun in Chapter 1 needs to be expanded to include the exponentials. This family of functions, describing patterns of repetitive multiplication, requires two parameters (control numbers)—a starting value, denoted by a, and a multiplier, called the base and denoted by b. This chapter begins by reviewing properties of exponential functions and their graphs, ideas you may have studied in previous courses.

Activity 2.1

1. One way to generate an exponential pattern of growth is by having each successive change be a percentage of the amount currently present.

   a) Use a graphing calculator or spreadsheet to build a table that starts with 200, and goes up by 25% each time, like the one started in **Table 2.2**. Record at least 6 terms in your table.

Term No.	Value
0	200
1	250

TABLE 2.2. Start of sample table for Activity 2.1.

   b) Complete a table like the one in **Table 2.3** by adding a column to display **first differences**—the differences between successive pairs of values. First differences indicate the amounts of increase in the value column. Describe any patterns you notice in these numbers.

Term No.	Value	Difference
0	200	250 – 200 = 50
1	250	

TABLE 2.3.

   c) Complete a table like the one in **Table 2.4** by adding a fourth column to display ratios between successive values in the table (next ÷ current). Describe any patterns you notice in these numbers.

Term No.	Value	Difference	Ratio
0	200	50	250 ÷ 200 = 2.5
1	250		

TABLE 2.4.

d) Graph value versus term number. Describe the graph.

e) A graph of the pairs ($\text{value}_{\text{current}}$, $\text{value}_{\text{next}}$) is called a **recursive graph**. Make such a graph and describe it.

f) Graph the ordered pairs (value, first difference). Describe the graph. What can you deduce about the relationship between current value and its change?

2. In general, any function with an equation of the form $f(x) = ab^x$, $b > 0$, $b \neq 1$, $a \neq 0$, is called an **exponential** function. Using your calculator, graph several different examples of equations of this form, and sketch their graphs. Then answer the following questions.

a) Describe the general shape of the graph of an exponential function.

b) How does the value for the base, b, affect the exponential graph?

c) How does varying the constant a affect the graphs?

3. Contrast the way in which exponential functions change with the way in which linear functions change.

4. In Chapter 1, you used several properties of functions to characterize graph behaviors and to classify the various members of the tool kit. Identify each of the following properties that are relevant to exponential functions, and describe how they relate to their graphs:

a) domain/range

b) symmetry

c) increasing/decreasing

d concavity

e) end behavior

EXPONENTIAL FUNCTIONS

You may recall seeing exponential patterns of growth in other mathematics courses. The basic mathematical calculation in each case involves repetitive multiplication by a specific number, called the **growth factor**, or an increase (decrease) by a specific percent of the amount present, called the **relative rate**. A calculation like 10(1.5)(1.5)(1.5)(1.5) … can be rewritten as $10(1.5)^x$, where x represents the number of multiplications in the calculation. Exponential functions have the same form, namely $y = a \cdot b^x$. The name, exponential, comes from having the variable be the exponent.

The ideas reviewed in Activity 2.1 are characteristic of patterns in tables and graphs of exponential functions. For example, **Table 2.5** starts with a value of 10 and shows both the increasing pattern that comes from repetitive multiplication by 1.5 and the decreasing pattern that comes from several successive reductions by 20%. In the first case, 1.5 is the growth factor and 0.5 (50%) is the relative rate, whereas in the second case, 0.8 is the decay factor and 0.2 (20%) is the relative rate.

Value	Next Value / Current Value	Increase	Increase / Current Value	Value	Next Value / Current Value	Decrease	Decrease / Current Value
10	1.5	5	0.5	10	0.8	2	0.2
15	1.5	7.5	0.5	8	0.8	1.6	0.2
22.5	1.5	11.25	0.5	6.4	0.8	1.28	0.2
33.75	1.5	16.875	0.5	5.12	0.8	1.024	0.2
50.625	1.5	23.3125	0.5	4.096	0.8	0.8192	0.2
75.9375				3.2768			

TABLE 2.5. Exponential Growth/Decay Patterns.

The values in the growth pattern are increasing, but the amount of increase is also increasing. Similarly, the values in the decay pattern are decreasing, but the amount of decrease is also getting smaller. In both situations, the amount of change (first difference) is proportional to the value, and the pattern in the amount of increase (decrease) explains why the graphs either grow without bound or slowly approach zero. The ratio between successive terms remains constant, and the recursive graph of $\text{Value}_{\text{next}}$ vs. $\text{Value}_{\text{current}}$ lies on a straight line through the origin (**Figure 2.2**). Either of these tests is a convenient way to identify a table of exponential values.

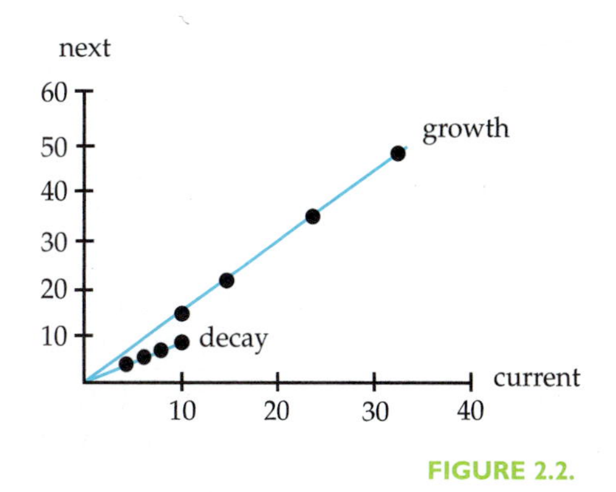

FIGURE 2.2.

Graphs of exponential functions have distinctive features as well. They have a domain that is defined

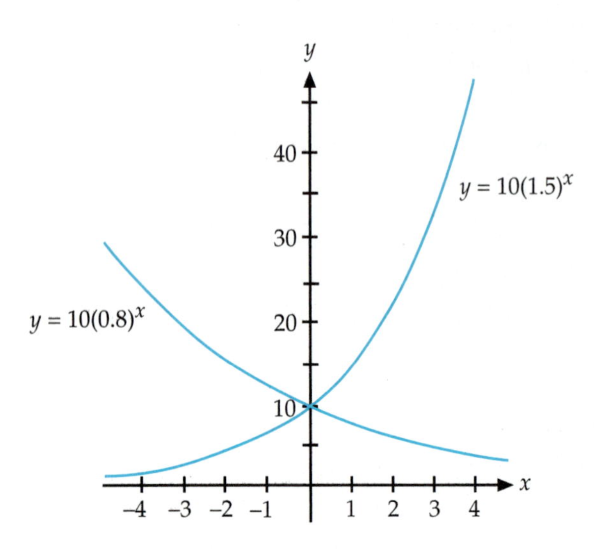

FIGURE 2.3.
Graphs of $f(x) = 10(1.5)^x$ and $f(x) = 10(0.8)^x$.

for all real numbers, but the range is restricted to being either $y > 0$ or $y < 0$. The graphs of $f(x) = 10(1.5)^x$ and $f(x) = 10(0.8)^x$ are shown in **Figure 2.3** as examples of the general behavior of exponential functions of the form $f(x) = a \cdot b^x$.

Both graphs are concave up everywhere, since they both have $a > 0$. The graph either increases everywhere (when $b > 1$) or decreases everywhere (when $b < 1$). Both graphs cross the y-axis at 10 because that is the value of a in the equations. Neither graph shows a symmetry pattern. The end behavior exhibits values either approaching 0 on the right and $+\infty$ on the left (when $b < 1$) or approaching 0 on the left and $+\infty$ on the right (when $b > 1$). Changing a to a negative number reflects the graph across the x-axis (**Figure 2.4**).

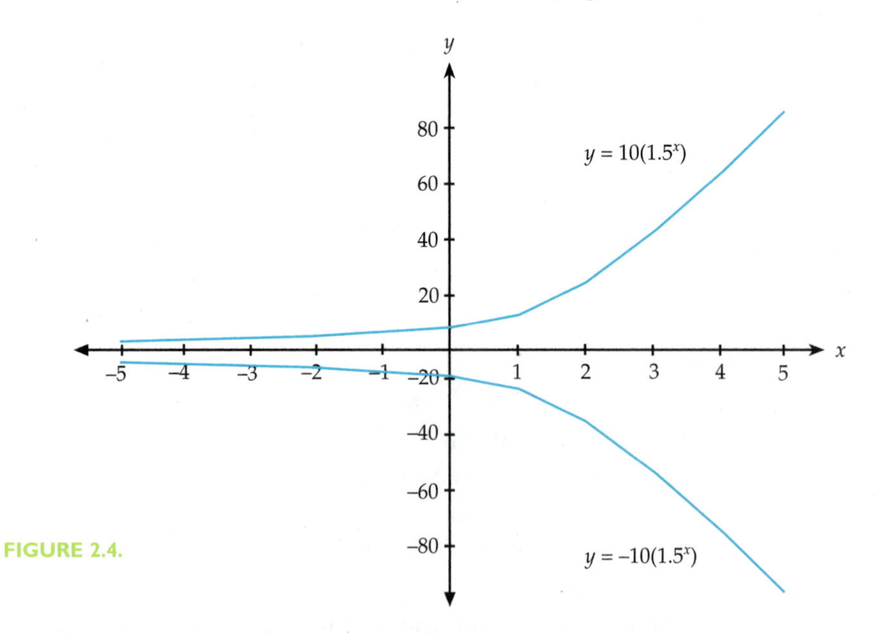

FIGURE 2.4.

Examining how the amount of change relates to current amounts reveals a powerful feature of all exponential functions. Values don't simply go up or down, but increase/decrease by amounts that are proportional to the current value, or in other words, by amounts that are a given percentage of the current value. **Table 2.6** lists some familiar situations in which change is proportional to the amount present.

Behavior	Nature of Change
Savings	The amount earned on a compound interest savings account is proportional to the amount of money present in the account.
Population Growth	The number of births and deaths is proportional to the size of the population.
Radioactive Decay	Radioactive substances decay (change into other substances) over time. The amount of decay is proportional to the amount of the substance present.
Spread of a Disease	The number of new cases of a disease (in early stages) is proportional to the number of people who have the disease.

TABLE 2.6. Examples of exponential change.

In the next activity, you will explore the first situation in Table 2.5. In particular, you will investigate how compounding interest (earning interest on previously earned interest) affects your savings account balance. In the process, you will learn about a new base that is frequently used to describe growth patterns.

Activity 2.2

Banks provide savings accounts that pay compound interest at time intervals specified when the account is opened. The interest earned during one period is redeposited, and more money is available to earn interest in the next time period. This allows slightly more money to be earned during the next time period, then the process repeats.

Suppose you wish to invest \$100 in a bank that pays 8% interest per year, but allows you to choose the number of compounding intervals, say n, in the year. Of course, you'd like to choose n so that you earn as much as possible, and you'd like to know just how much you could earn that way. Looking at a few cases might help.

$n = 1$: If interest is calculated only once, then after one year (\$100)(0.08) = \$8 is the interest earned, so the balance is \$108. As a one-step calculation, $108 = 100(1 + 0.08/1)^1$.

$n = 2$: If interest is calculated semi-annually (twice a year) in the first 6 months the interest earned is (\$100)(0.08/2) = \$4, giving a balance of \$104. During the last 6 months, (\$104)(0.08/2) = \$4.16 is earned, so the final balance is \$108.16. Note that $(104)(0.08/2) = 100(1 + 0.08/2)^2$ since $104 = 100(1 + 0.08/2)^1$.

1. **Table 2.7** summarizes the $n = 1$ and $n = 2$ cases. Complete a copy of Table 2.7. Explore additional cases to predict the maximum balance possible at the end of one year.

Compounding Period	n	Expression to Calculate	Balance after 1 Year
Annual	1	$100(1 + 0.08/1)^1$	$108.00
Semiannual	2	$100(1 + 0.08/2)^2$	$108.16
Quarterly	4		
Monthly	12		
Daily	365		
Hourly	8760		

TABLE 2.7.
Compound interest computations.

2. Item 1 should convince you that there is a ceiling to the amount you can earn by compounding. The expression controlling the amount of growth from compounding is $(1 + r/n)^n$, where r is the interest rate. Above, the 100 represents the starting amount. In order to understand the growth better, consider a starting amount of 1, with $r = 1$, so $(1 + 1/n)^n$, examining each part of the calculation in turn.

n	$y = 1/n$	$y = 1 + 1/n$
1		
2		
5		
10		
50		
100		

TABLE 2.8.
Values for Item 2.

a) Start with the fraction inside the parentheses. Copy **Table 2.8** and compute values of $y = 1/n$ to complete column 2. Then graph y versus n. What is its long-term behavior?

b) Now examine the entire expression inside the parentheses. Compute values of $y = 1 + 1/n$ to complete column 3 of Table 2.7, and graph y versus n. What is the long-term behavior this time? Is your result consistent with what you observed in part (a)?

c) Now consider the entire expression $y = (1 + 1/n)^n$. Using only your results in (b), predict its long-term behavior.

d) Check your prediction by computing values of $y = (1 + 1/n)^n$ to complete a copy of **Table 2.9**. Graph y versus n, and describe the long-term behavior.

n	$y = \left(1 + \frac{1}{n}\right)^n$
1	
2	
5	
10	
50	
100	
500	

TABLE 2.9.
Values for Item 2(d).

3. a) How might you adjust your work in Item 2 to reflect the original problem's condition of a $100 initial deposit?

b) How might you adjust your work in Item 2 to reflect the original problem's condition of an interest rate of only 8%?

BASE *e*

Your results in Activity 2.2 may have been a bit surprising. You know that 1 raised to any power is always exactly 1, and you have observed in this lesson that an exponential with base larger than 1 grows without

© Susan Van Etten

bound. Intuition might suggest that $(1 + 1/n)^n$ should either get closer and closer to 1, or become infinitely large as n grows.

Of course, neither intuitive model is correct. The base is not *exactly* 1, and the expression is not *really* exponential (the base is not constant). In fact, as n grows large, a kind of dynamic tug-of-war develops between the shrinking base and the ever-increasing exponent. The *real* surprise comes in finding that the expression approaches a constant.

The number 2.718281828..., representing the limiting value of the expression $(1 + 1/n)^n$, is defined formally by calculations similar to those you did in Activity 2.2 and is given the name 'e'. The number e, also called **Euler's number**, is an irrational number (like π) and can be computed precisely to many decimal places. Euler's number has many applications. Here is its formal definition:

$$\lim_{n\to\infty}\left(1+\frac{1}{n}\right)^n = e.$$

The definition is read as "the limit of the expression $\left(1+\frac{1}{n}\right)^n$, as n grows without bound, equals the number e." In other words, the value of the expression $\left(1+\frac{1}{n}\right)^n$ gets closer and closer to the value e as the value of n gets larger and larger.

In Activity 2.2, you found that a deposit of $1 in a bank account paying simple interest at a rate of 100% produces a balance of $2 after one year. If, instead, the bank compounded the interest continuously you would end up with e dollars, or approximately $2.71 after one year.

It has been said that many problems in today's society are caused by the failure of people to understand the nature of exponential growth. As an example, if you live in an area that is experiencing population growth, the failure of local government to plan for increases in support services such as transportation, utilities, and schools could result in unnecessary hardships. As someone who will have to deal with everchanging situations in the future, adding exponential functions to your tool kit will make you a more powerful modeler!

Exercises 2.1

1. a) In Activity 2.1, any function with an equation of the form $f(x) = ab^x$, $b > 0$, $b \neq 1$, $a \neq 0$ was defined as an exponential function. Why are restrictions placed on b?

   b) Explain why $a \neq 0$.

   c) Why can't equations of the form $y = a \cdot b^x$ have roots? (The term root is another name for a zero of an equation; that is, a value of x that produces a value of 0 for y.)

2. In 1990, wildlife biologists in a regional park estimated that there were 140 deer in the park. In 1995, their estimate changed to 365 deer. Assume that the size of the population grows exponentially.

   a) What is the growth factor for this single 5-year time period?

   b) Write an equation describing the deer population P in terms of n, the number of 5-year periods since 1990.

   c) What is the annual growth factor? Explain how you found it.

   d) What is the annual relative rate? How did you determine that value?

   e) Write an equation describing the deer population P in terms of t, the number of 1-year periods since 1990.

   f) Assuming that the population continues to grow exponentially, explain how to use each of your equations to predict the deer population in the park in 2010.

   g) How could you modify your equation in (b) so that it still describes the population in 5-year intervals, but uses the annual growth factor?

3. A bank account paying 8% annual interest compounded quarterly actually pays 2% interest each quarter. The **annual yield** is slightly higher than 8% due to the compounding.

   a) If $1500 is deposited when the account is opened, how much interest is earned during the first year?

   b) What is the annual yield?

c) If the money is invested for a 5-year period, what will the balance be at the end of that interval?

Exercises 2.1

4. As previously noted, if you deposit $1 in a bank account paying interest at an annual rate of 100% compounded continuously, you would end up with e dollars after one year.

   a) With continuous compounding, how much would be in the bank after two years?

   b) With continuous compounding, how much would be in the bank after five years? After t years?

   c) Use your calculator to find 100 * e^(0.08). How does that answer compare to the work done in Item 1 of Activity 2.2?

   d) Review your answer to Item 3 of Activity 2.2 and this exercise. Then generalize that work to write an expression for the balance after depositing A dollars at $100r\%$ compounded continuously for t years. Use numbers to check your expression for a specific case.

5. Repeat Exercise 3 using an 8% annual interest rate compounded continuously. Compare your answers to those in Exercise 3.

6. Explain how to solve the equation: $20(2)^x = 640$ by using:

   a) tables or contextual reasoning,

   b) algebra and basic properties of exponents (see Appendix A), and

   c) a graphing calculator.

7. Radioactive substances decay exponentially. **Half-life** is a convenient way to describe the rate at which such a substance breaks down; it's the time it takes for exactly half the substance to decay. Suppose a scientist starts with 500.0 grams of a substance with a half-life of 9 hours.

   a) Develop an equation describing the amount $A(n)$ of the substance remaining after n half-lives, and use it to determine how much of the substance will be remaining after 24 hours.

   b) A different researcher has a 200.0-gram sample of a radioactive substance that decays at the rate of 6% per hour. Use the graphing method developed in Exercise 6 to determine the half-life of this substance.

Exercises 2.1

c) Describe the limitations for solving 7(b) using the other two methods (tables and algebra) that were developed in Exercise 6.

8. a) Sketch graphs of each of the following exponential equations on a single set of axes: $y = 5(1.2)^x$, $y = 5(1.2)^{x+2}$, $y = 5(1.2)^x + 2$, $y = 7.2(1.2)^x$, and $y = (5 \cdot 1.2)^x$.

   b) Which two equations are algebraically equivalent?

   c) Which equations are translations of the first equation?

   d) How can you tell from the graphs that $a \cdot b^x \neq (a \cdot b)^x$?

9. Sketch the graphs of $f(x) = 3^x$ and $f(x) = \left(\frac{1}{3}\right)^x$ on a single set of axes. Describe the similarities and differences that you observe.

10. In Activity 2.1, you explored the pattern generated by tabulating values of first differences, $y_{\text{next}} - y_{\text{current}}$ (also written as Δy, the change in the y-values). This calculation also has a geometric interpretation. Consider **Table 2.10**, based on the exponential equation $y = 8(1.5)^x$, and its graph (see **Figure 2.3**).

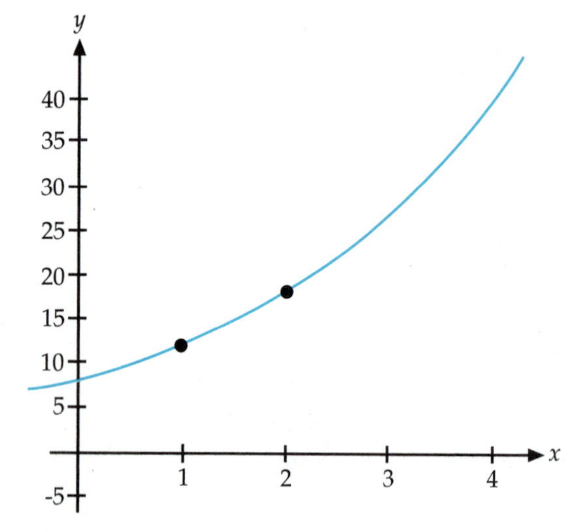

FIGURE 2.5.
Graph of $y = 8(1.5)^x$.

x	y	Δy	Δx	$\Delta y/\Delta x$
0	8	4	1	4
1	**12**	**6**	**1**	**6**
2	**18**	9	1	9
3	27	13.5	1	13.5
4	40.5			

TABLE 2.10.
Calculations using $y = 8(1.5)^x$.

a) What do $\Delta y = 6$, $\Delta x = 1$ and $\Delta y/\Delta x = 6$ tell you about the two points shown on the graph? (See bold entries in Table 2.10.)

b) How does the fact that the values for $\Delta y/\Delta x$ increase relate to the properties of the graph?

c) What kind of function describes the values in the $\Delta y/\Delta x$ column?

11. **Investigation**. Build a spreadsheet with columns labeled x value, y value, Change in y, and (Change in y) ÷ (y value).

    a) Complete the first column with the numbers 0, 1, 2, 3, 4, etc., and use an exponential function of your choice to fill in the values for y. Then enter formulas for calculating the other two columns. Describe the values in the last column.

    b) Explore various patterns of numbers as x values. Describe your observations.

    c) Try different exponential equations to fill the column of y values. Describe the behavior of the last column.

    d) Try using a different function from the tool kit to fill the column of y values. Describe the behavior of the last column.

12. A line segment that connects two points on a curve is called a **secant** (**Figure 2.6**). Using a spreadsheet, explore the ideas developed in Exercise 10(c) again. What kind of pattern is formed by the slopes of secant lines when the original y values come from a:

    a) constant function?

    b) linear function?

    c) power function?

    d) sinusoidal function? (Hint: Examine the pattern for lots of function values, and use a small value for Δx.)

    e) another exponential function?

FIGURE 2.6.
A secant line.

13. Secant lines are useful tools for determining when a curve is concave up or down.

    a) Sketch the graph of $f(x) = 2^x$, along with secant lines to your graph for the intervals [–2, –1], [0, 1], and [3, 4]. Describe the positions of the secant lines.

    b) What is the relationship between the concavity of a function (concave up or down) and the relative position of secant lines to the curve? Test your conjecture by sketching a few qualitative graphs and drawing secant lines on the curves in several places.

Exercises 2.1

14. Does an exponential function describe y in terms of x for the values in **Table 2.11**? Explain.

x	3	5	7	9	11
y	3.38	7.59	17.09	38.44	86.50

TABLE 2.11. Sample data for Exercise 14.

15. The population of Vallejo, California grew from 100,628 in 1980 to 116,952 in 1990.

   a) If you assume that the population grew exponentially during that time, what was the percent of increase each year?

   b) If Vallejo continues to grow at this rate, what would be the projected population for the year 2000 census? Did you need your answer from part (a) to make this prediction?

16. Derive a relationship between the base of an exponential function and the constant defined by the ratio $\Delta y/y_n$.

17. Exercise 7 asserted that radioactive decay occurs exponentially. Suppose that the data shown in **Table 2.12** represent the decay of a radioactive dye used in a certain medical procedure. Determine whether the decay really is reasonably modeled by an exponential function. Justify your conclusion.

Time (minutes)	Activity (counts/minute)
0	10023
1	8174
2	6693
3	5500
4	4489
5	3683
6	3061
7	2479
8	2045
9	1645
10	1326

TABLE 2.12. Decay data.

18. a) Consider the linear function $y = 3x - 2$. Investigate the rate at which y changes in a linear function by completing **Table 2.13** and looking at first differences.

   b) Describe any patterns you notice in your table from (a). Explain why this is true.

   c) Use first differences to determine whether a linear function describes y in terms of x for the values in **Table 2.14**.

x	y
0	
1	
2	
3	
4	
5	

TABLE 2.13.

x	0	1	2	3	4	5
y	2.71	1.368	0.026	–1.316	–2.658	–4.000

TABLE 2.14.

Exercises 2.1

19. Use the Laws of Exponents to explain the following properties of the exponential function $f(x) = b^x$:

 a) $f(m) \cdot f(n) = f(m + n)$ b) $f(m) \,/\, f(n) = f(m - n)$ c) $[f(m)]^n = f(m \cdot n)$

20. Using transformations, derive the general equation for an exponential function that goes through the points $P_1(x_1, y_1)$ and $P_2(x_2, y_2)$.

21. You developed a Ladder of Powers in Chapter 1. This exercise looks at how you might place exponential functions on that ladder. Consider two functions that grow at very different rates. For example, use a slow-growing exponential function like $y_1 = 5(1.02)^x$ and a fairly fast-growing power function, like $y_2 = 5x^4$.

 a) Complete a copy of **Table 2.15**. Use the results to compare (qualitatively) the two functions' values and their respective rates of growth.

x	y_1	y_2
0		
1		
2		
3		
5		

x	y_1	y_2
10		
20		
40		
60		
100		

x	y_1	y_2
200		
500		
1000		
2000		
3000		

TABLE 2.15. Blank table for Exercise 21.

 b) Explain from the actual computations involved why the functions exhibit the behavior you described in (a).

 c) What implication do your findings have for classifying exponential functions on the Ladder of Powers developed in Chapter 1?

22. **Table 2.16** displays data, reported by the National Soft Drink Association, on the average number of 12-ounce soft drinks consumed per person in the United States each year from 1945–1984:

 a) Construct a model for average consumption versus time (since 1945). How did you decide which tool kit function to use?

Exercises 2.1

TABLE 2.16.
Average soft drink consumption per person data.

Year	Number	Year	Number	Year	Number	Year	Number
1945	88	1955	125	1965	172	1975	294
1946	87	1956	126	1966	188	1976	328
1947	102	1957	126	1967	199	1977	349
1948	107	1958	125	1968	221	1978	370
1949	108	1959	132	1969	228	1979	389
1950	106	1960	128	1970	241	1980	401
1951	108	1961	131	1971	259	1981	415
1952	117	1962	144	1972	271	1982	420
1953	119	1963	151	1973	286	1983	438
1954	117	1964	165	1974	288	1984	470

b) Use your model to predict the average number of six-packs of soft drink an average person drank in 1999. Comment on your prediction.

23. **Table 2.17** shows the population of the United States, as determined by the U. S. Census Bureau every ten years.

TABLE 2.17.
U. S. Census data.

Year	Population	Year	Population	Year	Population
1790	3,894,000	1860	31,184,000	1930	123,077,000
1800	5,085,000	1870	38,156,000	1940	132,122,000
1810	6,808,000	1880	49,371,000	1950	152,271,000
1820	10,037,000	1890	63,000,000	1960	180,671,000
1830	12,786,000	1900	76,094,000	1970	205,052,000
1840	16,988,000	1910	92,407,000	1980	227,225,000
1850	23,054,000	1920	106,461,000	1990	249,440,000

a) Construct a model that describes the population growth over the time period 1790–1990.

b) The graph of the data looks like a classic exponential graph. However, there are some places in which the growth pattern changed slightly. At which dates does this occur? What are some plausible factors that could explain these changes in the pattern of growth?

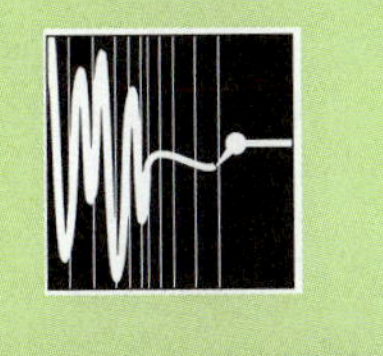

LESSON 2.2

Logarithmic Scale

The chapter introduction raised the issue of geologists' classification of earthquakes. There are many factors that go into creating an effective scale for measuring the size of an earthquake. In the following activity, you will explore some of these factors, and how mathematics can be applied to this measurement problem.

Activity 2.3

Table 2.18 shows the computed intensities for a portion of the earthquake activity for Central America and California on 1 May 1997; other earth movements took place on that day, but were not sizeable events. (Note: For the purposes of this chapter, no units are used with intensity values; however, they are directly proportional to energy values discussed in the Introduction.)

1. According to an ancient legend, really *big* earthquakes happen at sunset. A geologist investigating the scientific merit of that claim decides to graph seismic energy versus time of day for the data in Table 2.18.

Time	Latitude	Longitude	Intensity
3:30 A.M.	32.2°N	115.8°W	2.143×10^{13}
5:08 A.M.	32.0°N	115.8°W	2.244×10^{13}
6:22 A.M.	34.4°N	118.6°W	1.233×10^{14}
6:22 A.M.	34.4°N	118.7°W	2.000×10^{13}
8:02 A.M.	32.2°N	115.8°W	2.891×10^{13}
9:45 A.M.	37.5°N	121.7°W	3.991×10^{13}
11:37 A.M.	19.0°N	107.4°W	2.518×10^{18}
12:52 P.M.	39.4°N	119.8°W	1.262×10^{14}
4:25 P.M.	18.7°N	107.1°W	3.991×10^{15}

Time	Latitude	Longitude	Intensity
4:41 P.M.	18.7°N	106.9°W	2.518×10^{15}
5:00 P.M.	33.5°N	118.0°W	2.000×10^{13}
5:17 P.M.	34.1°N	116.4°W	2.698×10^{13}
8:12 P.M.	34.0°N	116.8°W	2.576×10^{13}
8:41 P.M.	39.6°N	122.0°W	1.517×10^{14}
9:22 P.M.	35.8°N	117.6°W	2.193×10^{13}
10:07 P.M.	36.0°N	120.6°W	1.262×10^{14}
10:36 P.M.	38.9°N	122.8°W	2.000×10^{13}

TABLE 2.18.
Earthquake Activity—5/1/97 (adapted from Council of the National Seismic System data).

a) Describe how to scale the horizontal axis to represent the earthquake times.

b) Describe how to scale the vertical axis to represent the seismic energy intensity for the earthquakes.

c) Use a graphing calculator to make a scatter plot of the intensity versus time data in the table. Describe your graph. How can you adjust the window or data to make the graph more informative?

d) According to these data, do really big earthquakes occur only at sunset? Explain.

Charles Richter

The geologist Charles Richter was faced with the problem of constructing a scale that would handle a large range of values, such as the earthquake data for 1 May 1997. The intensity of an earthquake is proportional to its energy output. Richter used a description called **magnitude**, calculating $M = \log(I/I_0)$. Here I_0 is the intensity of the smallest detectable earthquake at that time (intensity 2.0 x 10^{11}). In answering Items 2 and 3, use a spreadsheet or statistical lists on a calculator to investigate how the magnitude equation handles the problem of scaling intensities.

2. First, perform only the calculation inside the parentheses (I/I_0) where I_0 = 2.0 x 10^{11}. Record each answer in a copy of **Table 2.19**. This ratio, I/I_0, is the **relative intensity** of the quake.

   a) How does this first step of the magnitude calculation affect the values of intensity given in the table? What kind of transformation is created?

I	2.143 x 10^{13}	2.244 x 10^{13}	1.233 x 10^{14}	2 x 10^{13}	2.891 x 10^{13}	3.991 x 10^{13}	2.518 x 10^{18}	1.262 x 10^{14}
I/I_0								

I	3.991 x 10^{15}	2.518 x 10^{15}	2 x 10^{13}	2.698 x 10^{13}	2.576 x 10^{13}	1.517 x 10^{14}	2.193 x 10^{13}	1.266 x 10^{14} 2 x 10^{13}
I/I_0								

TABLE 2.19. Calculation of relative intensities.

   b) What is the contextual meaning of a computed ratio of 100 in part (a)? In general, what is the meaning of the calculation (I/I_0)?

   c) Graph relative intensity vs. time. Does calculating relative intensity solve Richter's problem of creating a useful scale for displaying earthquake data? Explain.

3. Now, using a scientific calculator capable of finding logs, carry out the second step of the magnitude calculation by calculating the log of each I/I_0 value found in Table 2.19. Record each answer in a table like **Table 2.20.**

   a) What does the log step of the calculation do to the relative intensities?

I/I_0	107.2	112.2							
Magn.	2.03								

TABLE 2.20.
Calculating magnitude from relative intensity.

   b) Graph magnitude vs. time. Does the use of logs in the magnitude calculation solve Richter's problem of creating a useful scale for displaying earthquake data? Explain.

   c) Based on the pattern in your scatter plot, do you think there were any pre- or after-shocks? If so, identify them.

4. Understanding the individual steps in the magnitude calculation provides insight into what Richter was trying to accomplish. Examine some of the details in greater depth to become more familiar with log calculations.

   a) An earthquake with a relative intensity of 100 (=1.00 x 10^2) has a magnitude value of 2.00. Where did the 2 come from? What does it really mean? Under what circumstances would an earthquake magnitude be between 2 and 3? That is, when would it be 2 and a decimal, like 2.xx?

   b) Use trial and error to find relative intensities that produce Richter numbers of exactly 3, 4, 5, ... , 10. Describe any pattern you see.

   c) 1 May 1997 was chosen because an earthquake measuring 7.1 on the Richter scale took place. Look back over the calculations, and explain where the 7 came from.

   d) What is the meaning of the integer part of a magnitude number?

5. Draw a number line and number its scale from 0 to 10 by ones. Label the line Richter number.

   a) Indicate on the number line where each earthquake from 1 May 1997 belongs, according to its Richter number.

b) Re-scale the number line (write the new scale markings below your Richter number scale markings) so that it indicates the relative intensities as well as the Richter numbers. Describe the relationship between the two scales.

RELATIVE SIZES AND ORDERS OF MAGNITUDE

The calculation of relative intensity in Activity 2.3 illustrates a pretty standard technique in mathematics for moving from the question "how big is it" to "how much bigger is it than . . ?". Establishing relative size requires some kind of standard for comparison, and Richter used the smallest possible earthquake detectable *at that time*. This eliminated decimal values smaller than one. In addition, using ratios allowed him to compare two measurements directly, not just to compare to a standard. For example, if one earthquake had a relative intensity of 100,000 and another 1000, the first would be 100 times as large as the second, since 100,000 ÷ 1000 = 100.

The data reported by the Council of the National Seismic System included only earthquakes of magnitude 2.0 or larger. There were many more earthquakes with less energy that were detected but not reported. Even then, there were 17 seismic events in that region of the continent in one day. This represents a typical day, except for the "big one" (M = 7.1). Modern apparatus is sensitive enough to detect earth movements as small as a rock breaking on a lab table (M = –1.5).

The Richter scale for describing earthquake intensities is an example of a **magnitude scale**, named after the fact that the scale expresses the order of magnitude (or power of 10) for the number, not the actual number. It is used whenever the quantities being represented have values over a large range of sizes. The problem of graphing the data in Activity 2.3 revealed that a *uniform* scale does not adequately depict numbers over several orders of magnitude. A small scale interval shows all the values correctly, but requires a very large grid, while a large scale interval reduces small values to essentially zero. Instead, using a scale that represents the various powers of ten retains relative sizes, and places all numbers on a more useful scale. (**See Figure 2.7**.)

The mathematical tool for changing the scale in such a way is the **logarithm** function, or "log" for short. The logarithm converts *any* positive number into its "power-of-ten exponent". That's why the Richter scale is more completely referred to as the Richter magnitude scale.

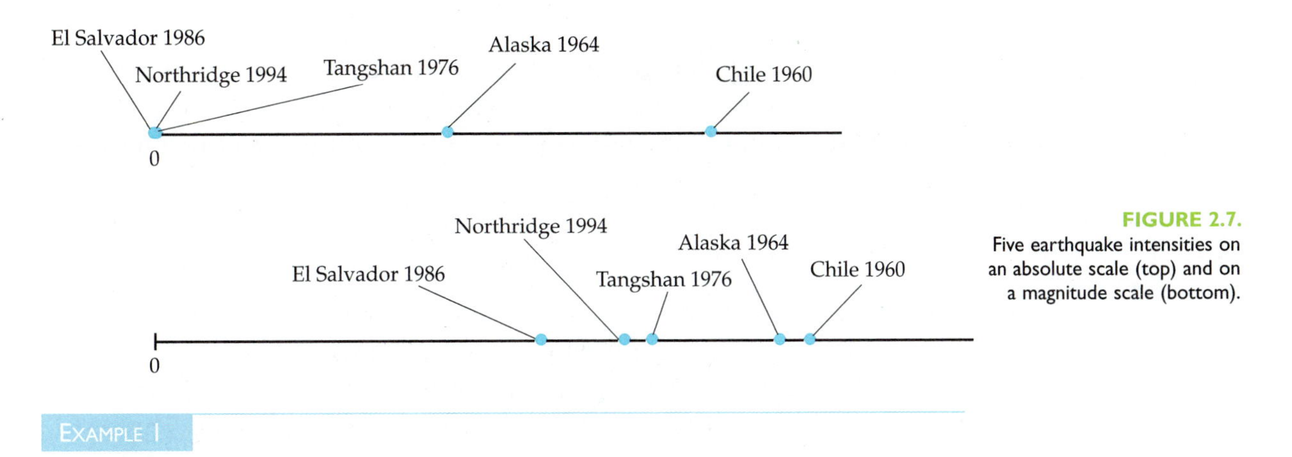

FIGURE 2.7. Five earthquake intensities on an absolute scale (top) and on a magnitude scale (bottom).

EXAMPLE 1

Without using your calculator, evaluate the following logarithms:

a) log(100)

b) log(1/10)

c) log(50)

SOLUTION:

Since $10^2 = 100$, log(100) = 2. Likewise, $10^{-1} = 1/10$, so log(1/10) = –1. 10 < 50 < 100, so log(50) is between 1 and 2. The problem of obtaining a better approximation will be revisited.

The key to constructing a magnitude scale is to use equally-spaced scale markings to represent increasingly large intervals. For example, the Richter magnitude interval [0, 1] represents relative intensities from 1–10, while the Richter magnitude interval [1, 2] represents relative intensities from 10–100. A magnitude scale accomplishes two things. First, the resulting numbers are conveniently small and convey information about the order of magnitude. Second, relative intensity calculations reduce to subtraction.

EXAMPLE 2

Consider two earthquakes with Richter numbers 7 and 4. Determine how much more powerful the stronger is than the weaker.

SOLUTION.

The corresponding relative intensities are 1×10^7 and 1×10^4, so the first is 10^3 times as powerful as the second. Note that $7 - 4 = 3$, so the calculation of relative strength reduces to the subtraction of Richter numbers.

Sound is another familiar physical phenomenon that is measured (and described) in a manner similar to that of earthquakes. Sounds are produced when vibrating objects cause pressure waves in some medium, like air or water, that then carries the energy of the vibrations. These pressure variations are detected as sound when a receiver, such as a human eardrum, is caused to vibrate in a similar fashion. The loudest sounds the human ear can detect are more than a trillion (10^{12}) times louder than the softest sounds. A sound's **intensity** is measured as power per unit area. The **sound level** (loudness) of a sound wave is commonly specified with a value B, measured in **decibels** (abbreviated dB), which is defined as:

$$B = 10 \cdot \log(I/I_0).$$

I_0 is the threshold for hearing, defined to be $I_0 = 1 \times 10^{-12}$ watts/m^2. Calculations of sound levels and relative intensities are very similar to those of earthquakes.

EXAMPLE 3

How many times more powerful is the sound of a jet plane taking off ($B = 150$ dB) than the noise generated by a car traveling at 60 mph (sound intensity at that speed is 10^{-4} watts/m^2).

SOLUTION:

Let I_2 represent the intensity of the plane, and I_1 the intensity of the car.

Step 1: Find the relative intensity for the sound level of the car:
$I_1/I_0 = 10^8$. (Thus the sound level is $(10)(8) = 80$ dB.)

Step 2: Find the relative intensity for the sound level of the plane:
$150 = 10 \cdot \log(I_2/I_0)$, so $\log(I_2/I_0) = 15$, or $I_2/I_0 = 10^{15}$.

Step 3: Compare the relative intensities of the plane and car: $(I_2/I_0)/(I_1/I_0) = 10^7$. Therefore, $I_2/I_1 = 10^7$, and the jet engine is 10 million times louder.

Note: Subtracting the two decibel levels to compare relative intensities again helps with the comparison, but you must also divide by 10: $150 - 80 = 70$, and $70/10 = 7$, so $I_2/I_1 = 10^7$, as before.

Other contexts in which magnitude scales are used include light intensity and nuclear radiation.

Activity 2.4

The Richter scale is a nice example of a scale based upon relative size. In this activity, you will examine more closely how each interval is subdivided, and explore some interesting properties of this kind of scale.

1. In Item 5 of Activity 2.3, you constructed a number line with two different (but equivalent) scales, describing relative intensity and magnitude. With the earthquake context removed, the result of re-scaling the number line is shown in **Figure 2.8**. Remember, the key to constructing a magnitude scale is to use equally-spaced scale markings to represent increasingly large intervals.

Power of 10	0.01	0.1	1	10	100	1000	10,000	100,000	Number
Exponent	–2	–1	0	1	2	3	4	5	Logarithm

FIGURE 2.8. Re-scaling between numbers and their logarithms.

a) Explain how to use Figure 2.8 to find the exponent needed to write 10,000 as a power of 10. How would you find the exponent without the figure?

b) Again, use the Power of 10 scale to explain how to estimate (to the nearest integer) the exponent needed to write 200 as a power of 10. Then use a calculator and the definition of logarithm to get the actual value to verify your result.

c) Without using a calculator, determine the number whose logarithm is 8; that is, solve $\log(x) = 8$. Explain your method.

d) Without using a calculator, evaluate log 0.001. Explain.

2. **Figure 2.9** shows a more detailed view of a portion of the Powers of Ten scale from Figure 2.8.

Power of 10	1	2	3	10	20	30	100
Exponent	0			1			2

FIGURE 2.9.
Powers of Ten Scale.

a) Carefully measure and redraw Figure 2.9 sideways on a clean sheet of paper. Label the scale with the powers of ten given in Figure 2.9, then label the exponent values as was done in Figure 2.8.

b) Extend your number line as far in each direction as your paper will permit. Again, measure accurately and label powers of ten and exponents.

> **TAKE NOTE** Throughout the remainder of this chapter, paired scales such as these will be called the power scale and the exponent scale to indicate which one is being used. The exponent scale displays the logarithms of the power-scale values.

c) The interval [0, 1] on the exponent scale corresponds to the interval [1, 10] on the power scale. Find at least four other intervals on your number line that have exponent-scale intervals exactly 1 unit long. Include at least one interval having non-integer exponent-scale endpoints. What property is shared by all the corresponding power-scale intervals?

d) Repeat part (c) using exponent-intervals of length 2.

e) With a ruler, carefully re-measure the lengths of the power-scale interval [1, 2] and the exponent-scale interval [0, 1]. Use these two lengths to compute an approximate value for log(2). (Hint: What is the length of the power-interval [1, 2] in exponent-scale units?)

f) Repeat part (c) using the power-scale interval you measured in (e).

g) Generalize any patterns you have observed about interval distances on the Powers of Ten scale.

h) Apply your result in (g) to find x in terms of a and b so that the exponent-scale lengths of the power-scale intervals $[1, x]$ and $[a, b]$ are equal.

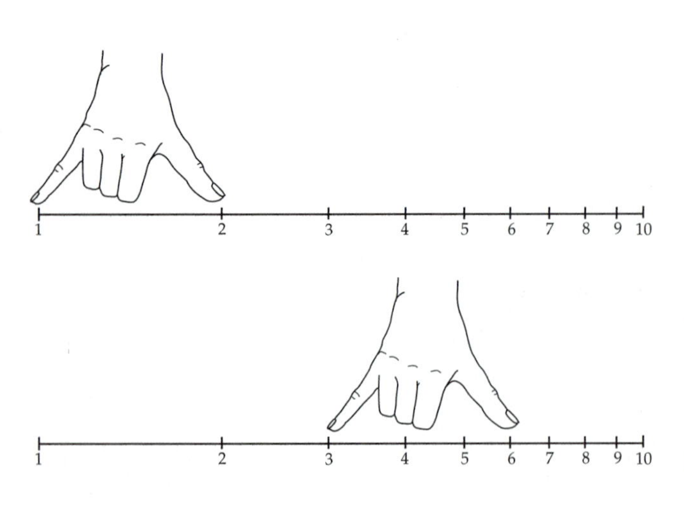

FIGURE 2.10.
Adding logs to multiply.

3. Emily claims that moving to the right one unit on the exponent scale is like multiplying by 10 on the power scale. In fact, she actually believes that she can multiply any two power-scale numbers just by adding their corresponding logarithms. She "adds logs" by starting at the first power-scale number on the exponent scale and then moving to the right on the exponent scale by the distance from zero to the second exponent scale number, then finding the power-scale value corresponding to that final position. (See **Figure 2.10**.)

   a) Based on your work in Item 2, do you agree with Emily? Explain.

   b) Use a ruler to measure carefully the power-scale interval [1, 20]. Then use your measurements from 2(e) to check Emily's claim for (2)(10) = 20.

   c) Explain what sliding to the left on the exponent scale should mean. Check your conjecture with an example, or prove it using definitions and properties you already know.

"The exponent-scale length of the power-scale interval $[1, x]$" is exactly the number $\log(x)$ since the corresponding exponent-scale interval begins at 0. Thus $\log(x)$ can be thought of as a number on a scale or as a distance between two scale locations, whichever is most helpful in a given situation.

4. a) The exponent-scale length of the power-scale interval $[a, b]$ is exactly $\log(b/a)$. Explain.

   b) If you know the exponent-scale lengths of the power-scale intervals $[1, a]$ and $[1, b]$, explain how you could determine $\log(b/a)$.

   c) If you know $\log(a)$, explain how you could calculate $\log(a^5)$. (Think distance.)

5. **Challenge**. Carefully measure and redraw **Figure 2.11** on a clean sheet of paper.

1 2 3 10

FIGURE 2.11. Powers of Ten scale for Item 5.

   a) Without using a calculator, accurately determine where the remaining integers 4, 5, . . . , 10 fit on your scale, and draw them in. Explain how you determined where to place each number.

   b) The number 8/3 can be thought of as $2 \cdot 2 \cdot 2 \div 3$. Explain how to determine where 8/3 lies on your Powers of Ten scale from part (a).

   c) Mark the midpoint of the interval [1, 10] on your drawing. What number represents the location for that point?

LOGARITHMIC SCALE

A Power of Ten scale is created when you graduate the scale markings of a number line so that successive powers of ten are equal distances apart. By renumbering the scale marks to indicate the exponents needed to produce the powers of ten, you produce a logarithmic scale; logarithms are exponents. For example, $10^? = 53$ is solved (by calculator, of course) by evaluating log(53).

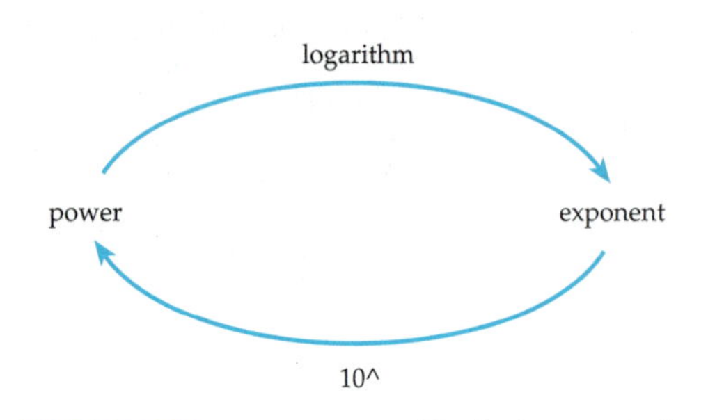

FIGURE 2.12. Arrow diagram relating exponents and powers in a powers of ten diagram.

The powers of ten diagrams you have used show two number lines, one for the powers, scaled as described above, and one for the exponents, scaled at equal distances. The relationship between the power and exponent scales of such a diagram are summarized in the arrow diagram shown in **Figure 2.12**.

Each increase by 1 on the logarithm scale represents a ten-fold increase on the power of ten scale, regardless of the starting point. Thus, moving to the right by one log unit (adding 1) corresponds to multiplying by 10 on the power scale. Moving to the left (subtracting) represents division.

Sometimes you may wish to visualize the paired scales not as scales at all, but as columns in a table or spreadsheet. Values in the exponent column increase by adding a constant (1) while values in the power column increase by multiplying by a constant (10). Other magnitude scales could be developed by using different constants.

Actual distances on a powers of ten scale are determined by the exponent scale, since that is the scale that is uniformly spaced. As such, lengths of power-scale intervals are counter-intuitive. Rather than the usual interpretation as *an amount of change* (implying a *difference* calculation), lengths are interpreted as *relative change* (implying a *ratio* calculation). For example, on a powers of ten scale the length of the interval [60, 90] is the same as that for [4, 6], since they both represent intervals whose right endpoints are 1.5 times their left endpoints.

Since 0 on the log scale is aligned with 1 on the power scale, $\log(x)$ represents both the location on the log scale of the number x on the power scale and the distance from 1 to x on the power scale. Thus, log(1.5) is the length of the power interval [1, 1.5], and (continuing the example above) is also the length of the intervals [60, 90] and [4, 6], since each of these three intervals represents growth by a factor of 1.5.

Adding logarithms is moving to the right on a powers of ten scale. Since equal distances represent growth by a common factor, adding logs corresponds to multiplying powers. Subtracting logs is equivalent to dividing powers. In a spreadsheet representation, then, differences of 1/2 in the exponent column correspond to factors of $\sqrt{10}$ since it takes two of them to make a factor of 10.

Example 4

Given that log(2) ≈ 0.301 and log(3) ≈ 0.477, interpret each of the following arithmetic problems using distances, logarithms, and powers.

a) $2 \times 3 = 6$

b) $3/2 = 1.5$

c) $2^3 = 8$

Solution:

(For simplicity, the notation [*a*, *b*] denotes both the power-scale interval [*a*, *b*] and its exponent-scale length.)

First, using the arrow diagram in Figure 2.12, note that log(2) ≈ 0.301 means that $10^{0.301} \approx 2$, and log(3) ≈ 0.477 means that $10^{0.477} \approx 3$.

a) [1, 2] + [2, 6] = [1, 6] from the geometry of the number line. But [2, 6] has the same length as [1, 3] since each represents growth by a factor of 3. Thus [1, 2] + [1, 3] = [1, 6], or log(2) + log(3) = log(6): 0.301 + 0.477 ≈ 0.778. Also $2 \times 3 = (10^{0.301}) \times (10^{0.477}) = 10^{0.778} = 6$.

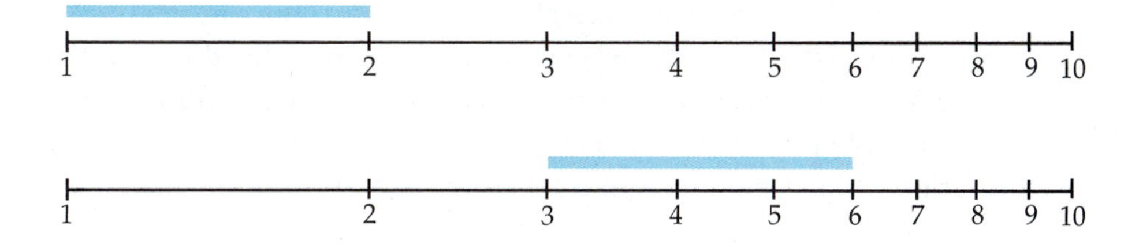

Table 2.21 summarizes this and the remaining calculations.

Arithmetic Problem	Distance Interpretation	Power Interpretation	Logarithm Interpretation
$2 \times 3 = 6$	$[1, 2] + [1, 3] = [1, 6]$	$2 \times 3 = (10^{0.301}) \times (10^{0.477}) = 10^{0.778} = 6$	$\log 2 + \log 3 = \log 6 = 0.778$
$\frac{3}{2} = 1.5$	$[2, 3] = [1, 3] - [1, 2] = [1, 1.5]$	$\frac{(10^{0.477})}{(10^{0.301})} = 10^{0.176} = 1.5$	$\log 3 - \log 2 = \log 1.5 = 0.176$
$2^3 = 8$	$[1, 2] + [2, 4] + [4, 8] = [1, 8]$	$(10^{0.301})^3 = (10^{0.903}) = 8$	$\log 2^3 = 3\log 2 = 0.903$

TABLE 2.21. Interpreting powers and logarithms.

The power interpretations in Example 4 are based on three Laws of Exponents:

i. $(b^m)(b^n) = b^{m+n}$, ii. $(b^m)/(b^n) = b^{m-n}$ and iii. $(b^m)^n = b^{m \cdot n}$.

The logarithm interpretations are equivalent, tracking the arithmetic of the exponents, since logs *are* exponents.

i. When you multiply two numbers together, add their exponents (logarithms); $\log(c \times d) = \log(c) + \log(d)$.

ii. When you divide a number by another, subtract their logs; $\log(c/d) = \log(c) - \log(d)$.

iii. When you raise a power to another exponent, multiply the new exponent by the log of the first number; $\log(c^A) = A \times \log(c)$.

The logarithm interpretations are closely related to the physical properties of distance on an exponent scale. Historically, this geometric interpretation of Powers of Ten as a distance along a logarithmic scale enabled people for several hundred years to do sophisticated calculations, like establishing models for planetary motion, using slide rules.

Exercises 2.2

1. Without using your calculators, find the logarithms of the following numbers:

a) 100,000
b) 0.001
c) 1/10000
d) 10^{-11}
e) 1,000,000
f) 10^{16}
g) 1/100
h) 0
i) –100
j) 10
k) 1
l) 10^{35}

2. Each of the following numbers is a logarithm value. Without using your calculator, what is the power of ten associated with it?

a) 5
b) –2
c) 0
d) 1

3. Estimate the log of each of the following numbers; then use your calculator to find the value exactly.

a) 1956
b) 0.117
c) 245.67
d) 0
e) –0.335

4. In constructing a Powers of Ten number line, the scale chosen determines the actual interval lengths.

a) If the scale is [1, 10] = 5.0 cm, calculate the interval length [1, 43].

Exercises 2.2

b) If the scale for a power of ten line is [1, 10] = 3 cm, calculate the interval length [1, 7].

c) If the scale for a power of ten line is [1, 2] = 2 cm, calculate the interval length [1, 10].

5. Use only the fact that log 2 ≈ 0.3010 (and answers to earlier parts of this exercise) to calculate each of the following logarithms.

   a) log(5)

   b) log(20)

   c) log(25)

   d) log(32)

   e) log(12.5)

   f) log(1000) (Using only log(2) and log(5))

6. a) If log(2) = 0.3010 and log(3) = 0.4771, what is the exponent-length of the power interval [60, 90]?

   b) Using only the portion of the Powers of Ten scale from 1–10, explain how to determine the length for [10, 12].

7. Use the properties of the Powers of Ten scale to complete each of the following.

   a) Estimate log(2), knowing only that $2^{10} = 1024$.

   b) Estimate log(3), using your answer from part (a) and the fact that 81 ≈ 80.

   c) Estimate log(7), using your answer from part (a) and the fact that 49 ≈ 50.

8. a) How many times more energy is released in an earthquake with relative intensity 10,000 than by an earthquake with relative intensity 1000? How did you determine your answer?

   b) How does a scientist know that the magnitude 9 Chile earthquake mentioned in the introductory reading was 10,000,000 times larger than a magnitude 2 mine explosion?

c) How many times larger is an earthquake with magnitude 5 than an earthquake with magnitude 2? How did you determine your answer?

Exercises 2.2

9. The Northridge, California earthquake in January of 1994 had a magnitude of $R = 6.9$.

a) How much more powerful was the Northridge shaker than the threshold level for earthquakes?

b) When reporting a magnitude 6.9 earthquake, the measure of R is recorded to one decimal place. The value of R is actually in the interval $6.85 \le R < 6.95$. Recording the relative intensity to 9 significant figures implies too much accuracy for the precision of the original measurement. Use $R = 6.85$ and $R = 6.95$ to show that the Northridge earthquake was between 7 and 9 million times more powerful than the threshold level for earthquakes.

c) Would it make a difference if the value of R were stated as $R = 6.90$? Explain.

10. An earthquake is generally felt when its Richter scale rating is 4.0 or greater. The data and graph in **Figure 2.13** represent the 783 felt earthquakes occurring on the West Coast between 1 January 1990 and 11 July 1996.

Magnitude	Number
4.0–4.4	497
4.5–4.9	162
5.0–5.4	79
5.5–5.9	26
6.0–6.4	9
6.5–6.9	6
7.0–7.4	3
7.5–7.9	1

Source: Martin Van Bonsangue, *Humanistic Mathematics Network Journal* #17, page 19.

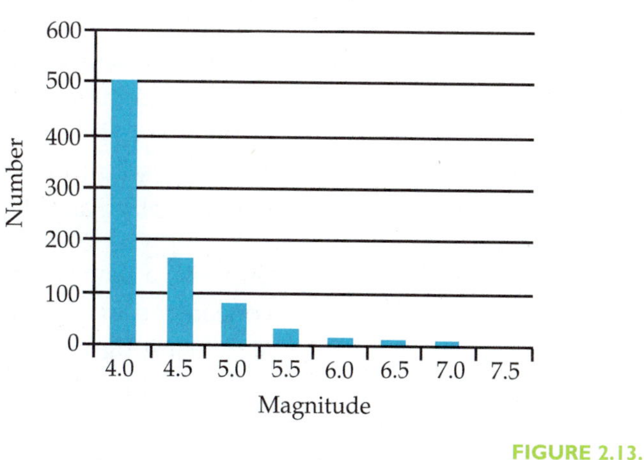

FIGURE 2.13. Earthquakes felt on the West Coast, 1/1/90–7/11/96.

a) Given that a felt earthquake will occur, how likely is it that it will be a great earthquake (magnitude at least 7.0)?

b) How does the distribution of earthquake magnitudes in the number line you created in Item 5 of Activity 2.3 for 1 May 1997, compare to that displayed in Figure 2.13 for a 6-year period?

Exercises 2.2

11. Find the difference in sound levels (measured in decibels) for the following situations:

 a) Pneumatic drill (I_2) versus street with heavy traffic (I_1); relative ratio $I_2 / I_1 = 100$.

 b) Jet engine at 150 feet (I_2) versus music from a 10W stereo at 12 feet (I_1); relative ratio $I_2 / I_1 = 1000$.

 c) Background noise after installation of acoustic insulation versus before installation of acoustic insulation; relative ratio $I_2 / I_1 = 0.5$.

12. One way to determine the acidity of a solution is to measure the concentration of hydronium ions. Chemists use the formula $pH = -\log[H_3O^+]$, where the symbol $[H_3O^+]$ represents the concentration of hydronium ions (in moles per liter) in the solution. A solution is called **acidic** if $pH < 7$ and **basic** if $pH > 7$.

 a) Find the concentration of hydronium ions of a solution that is neither basic nor acidic—that is, a solution with $pH = 7$.

 b) Find the pH level of a solution with a hydronium ion concentration of 7.0×10^{-6}. Classify the solution as acidic or basic.

 c) Find the pH level of a solution with an hydronium ion concentration of 5.0×10^{-8}. Classify the solution as acidic or basic.

 d) Suppose a laboratory uses a process for measuring the concentration of hydronium ions that is precise to two significant figures. Find a lower estimate and an upper estimate for the pH level of a solution with a concentration measured accurately to 1 decimal place as 6.4×10^{-6}.

 e) Suppose a laboratory uses a process for measuring the concentration of hydronium ions that is precise to three significant figures. Find a lower estimate and an upper estimate for the pH level of a solution with a concentration measured as 6.40×10^{-6}.

13. **Investigation.** Stellar magnitude is used to describe the brightness of a star or other celestial body. Smaller numbers indicate brighter objects, and a scale difference of 5 magnitudes indicates that one object is 100 times as bright as the other.

Exercises 2.2

a) Construct a pair of scales, one for relative brightness and one for stellar magnitude, that shows the above-defined relation between them.

b) What relative brightness corresponds to a difference of 1 magnitude?

NASA

Since smaller magnitudes indicate brighter objects, a number of relatively nearby objects actually have negative stellar magnitudes. **Table 2.22** provides information about the stellar magnitudes of some familiar objects in the sky.

Object	Stellar Magnitude
Sun	–26.7
Full Moon	–13
Venus (at brightest point)	–4.4
Jupiter	–2.5
Sirius (brightest star)	–1.42
Polaris (North Pole star)	2.0

TABLE 2.22.
Stellar magnitudes of familiar objects.

c) Use your scales to estimate to the nearest order of magnitude how many times brighter the full moon is than Venus.

LESSON 2.3

Changing Bases

The previous lesson introduced the concept of a logarithmic scale. This kind of scale has the geometric property of preserving ratios (instead of differences) between scale markings. The numbers used for the Richter magnitude scale are the exponents of powers of ten, or logarithms, associated with given ratios. Since you based your scale on powers of ten, these logarithms are known as base-10 logs. In this lesson, you will explore logarithms created from bases other than ten, and examine standard mathematical operations that can be applied to these new numbers.

Activity 2.5

1. **Figure 2.14** shows a powers of ten scale like that used in Lesson 2.2. Points representing powers of 10 are indicated with dots. Below the scale is a number line showing the corresponding base-10 logarithms.

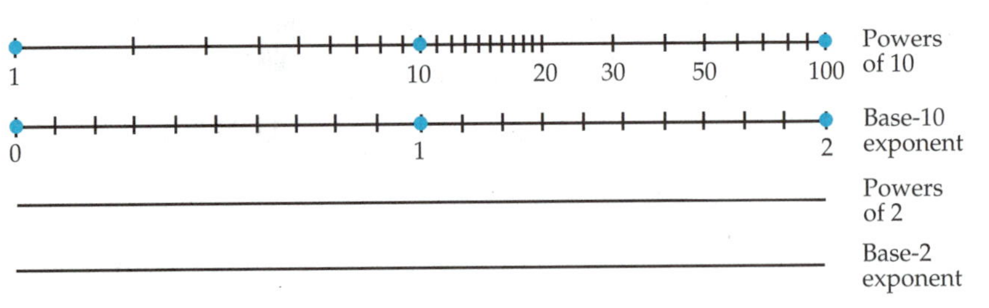

FIGURE 2.14.
A powers of ten scale and its associated base-10 log scale.

Logarithms describe the relationship between the scales. For example, the *logarithm to the base 10 of 100 is 2*, since $10^2 = 100$; you write $\log_{10}(100) = 2$, or $\log_{10}100 = 2$. **Figure 2.15** shows the same relationship more generally.

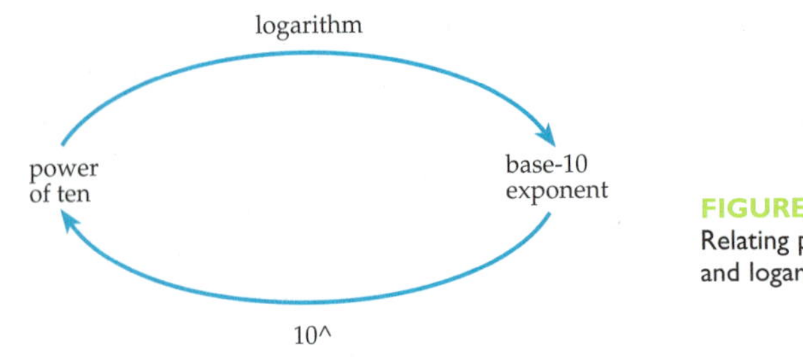

FIGURE 2.15.
Relating powers and logarithms.

a) Use a copy of the Figure 2.10 to build a Powers of Two scale on the indicated number line by marking six dots to represent the integer powers of 2 between 1 and 100. Explain how you can use the logarithm property that equal distances maintain equal ratios (see Question 2(e) of Activity 2.4) to locate 32 and 64 on that scale.

b) Complete your copy of Figure 2.14 by using your work from part (a) to construct a $\log_2$ number line on the bottom line. Be sure to align the left sides of your lines and to scale consistently with the other lines.

c) Sketch an arrow diagram similar to that in Figure 2.15 to describe the relationship between the base-2 power and log scales.

2. In answering Item 1, you used the property that intervals defined by successive powers of 2 have equal length on the logarithm scale. This question checks other properties of base-10 logarithms, which you developed in Activity 2.4, for your new logarithm scale.

   a) According to your scales, $\log_2 64 = 6$. Explain how to see that fact directly on your scales, then write an equation that justifies it algebraically.

   b) On the Powers of Two scale, the interval [1, 16] can be formed by "adding" the intervals [1, 8] and [8, 16]. Rewrite that fact as an equation using exponents of base 2. Write an equivalent equation that uses the $\log_2$ scale.

   c) Consider forming the interval [1, 8] by "subtracting" [8, 32] from [1, 32]. Rewrite that process as an equation using base-2 exponents. Write an equivalent equation that uses the $\log_2$ scale.

   d) Write an equation using base-2 exponents that explains how to determine the powers-of-two interval [1, 16] just from the powers-of-two interval [1, 4]. Write an equivalent equation using the $\log_2$ scale.

   e) What value does the point halfway between 0 and 1 on the $\log_2 x$ scale represent on the powers-of-two scale? Explain.

3. You now have 4 scales—the original lines labeled "powers of 10" and "base-10 exponent", and the new ones you drew that are labeled "powers of 2" and "base-2 exponent." Remember, logarithms are exponents.

a) The exponent scales are both number lines. That is, the scale on each of these lines is uniform—equally-spaced! What does one "unit" of distance represent in each of these cases?

b) Since the two log scales are both uniformly spaced, distances on one scale must be directly proportional to distances on the other scale. Compute by direct measurement the ratio: (unit distance on a $\log_2$ scale) ÷ (unit distance on a $\log_{10}$ scale). Interpret the significance of that ratio in terms of logs.

c) Compute the ratio: (unit distance on a $\log_{10}$ scale) ÷ (unit distance on a $\log_2$ scale). Interpret its significance in terms of logs.

In looking at the log scales you constructed, it should appear that units on the base-2 log scale are about 1/3 the size of those on the base-10 log scale. So, just as a rough rule of thumb, you should expect $\log_2 x$ to be about 3 times as large as $\log_{10} x$ no matter what x is. (It takes about 3 times as many little units to go as far as some number of big units.) But that's just an approximation.

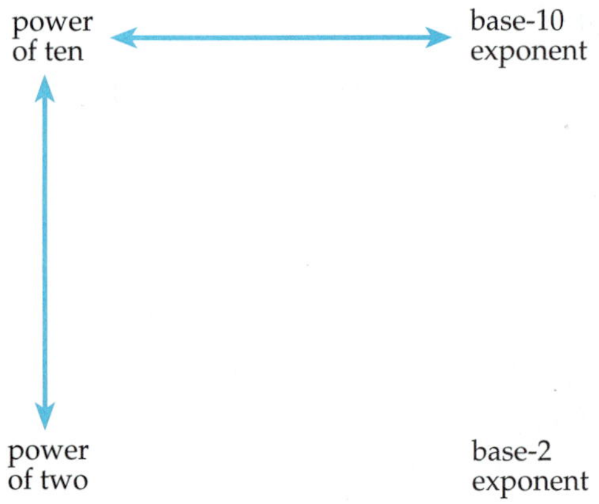

power of ten ⟷ base-10 exponent

power of two   base-2 exponent

FIGURE 2.16. Arrow diagram for base-10 and base-2 logs and powers.

4. Consider the partially-completed arrow diagram in **Figure 2.16**. Each item in the figure represents one of the four scales you drew in Item 1. Moving between the two power scales is simple; they're identical, so there is nothing to do. Your calculator provides a way to get from base-10 powers to base-10 logs and back (see Figure 2.15).

   a) Draw a copy of Figure 2.16 on your paper. Label the top arrows with the calculator commands (keys) that define them.

   b) Use your work in Item 3 to explain how a calculator can compute $\log_2 x$ exactly from $\log_{10} x$. Or, construct a spreadsheet to do the computation. Illustrate your reasoning with $x = 5.41$. Compare your result to predictions from your scales in Item 1. Then add your rule and an appropriate arrow to your copy of Figure 2.16.

   c) Repeat part (b) for the reverse problem of computing $\log_{10} x$ exactly from $\log_2 x$. Again, illustrate using $x = 5.41$. Add an appropriate arrow and rule to your diagram.

d) Complete the diagram by adding arrows between the bottom two labels. Be sure to write calculator-appropriate rules defining those arrows. (It's not fair just to repeat your answer to 1(c); the calculator doesn't have a LOG_2 button.)

LOGARITHMS AND BASES

On a log scale, log(x) is directly aligned with the number x on the corresponding power scale, and 0 on the log scale is matched with 1 on the power scale. Thus log(x) represents both the log-length of a power interval (having a given ratio, namely x) and the log-scale location of the power x. The location interpretation provides the usual definition of logarithms: the logarithm of a number x to the base 10 is "the power y to which one must raise 10 in order to get a value of x." More formally, $y = \log_{10}x$ (read "log to the base 10 of x") means that $10^y = x$, and vice versa. For example, $\log_{10}1000 = 3$ since $10^3 = 1000$.

For any positive value of b not equal to 1, the base-b logarithm is defined in a similar manner. The logarithm of a number x to the base b is "the power y to which one must raise b, in order to get a value of x". In other words, $y = \log_b x$ (logarithmic form) means that $b^y = x$ (exponential form), and vice versa. Look at **Figure 2.17**. The value of x must be positive, but the value for y can be either positive, negative, or 0. As before, powers of b form the integer values on the base-b logarithm scale.

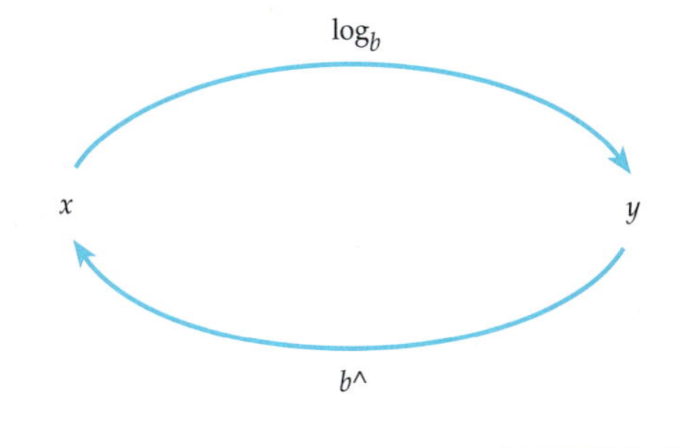

FIGURE 2.17. Relating logarithmic and exponential forms.

EXAMPLE 5

Check that $\log_3(7) = 1.771$ (approximately).

SOLUTION:

$\text{Log}_3(7) = 1.771$ means $3^{1.771} = 7$. Using a calculator, $3^{1.771}$ is just over 6.998, or about 7 except for rounding.

Example 6

Write 3 as a power of 10.

SOLUTION:

$\text{Log}_{10}(3) = y$ means $10^y = 3$. Therefore, $10^{\log(3)} = 3$, so $10^{0.4771} = 3$.

Base-10 logarithms are used often enough to merit a special name, **common logarithms**, or just plain logs. By agreement, when one is working with base-10 logarithms, the base number does not have to be written. Thus, log 1000 means $\log_{10}1000$, and is 3. This notation is consistent with typical calculator keyboards; the **LOG** button returns the base-10 logarithm of a number.

Logarithm scales in all bases have properties—the Laws of Logarithms—that follow from the exponential definition of logarithms (see also Lesson 2.2):

1. $\log_b 1 = 0$ since $b^0 = 1$,
2. $\log_b b = 1$ since $b^1 = b$,
3. $\log_b(m \cdot n) = \log_b m + \log_b n$,
4. $\log_b(m/n) = \log_b m - \log_b n$, and
5. $\log_b(m^p) = p \cdot \log_b m$.

Example 7

Use the Laws of Logarithms to write $3\log_5 2 - \log_5 x$ as a single term.

SOLUTION:

By Law 5, $3\log_5 2 = \log_5 2^3 = \log_5 8$.

By Law 4, $\log_5 8 - \log_5 x = \log_5 \left(\frac{8}{x}\right)$.

EXAMPLE 8

Solve $400(1.02)^t = 600$ by using the Laws of Logarithms.

SOLUTION:

Since $400(1.02)^t = 600$, then $\log 400(1.02)^t = \log 600$.

But $\log 400(1.02)^t = \log 400 + \log 1.02^t = \log 400 + t \cdot \log 1.02$, so $t \cdot \log 1.02 = \log 600 - \log 400$. Thus, $t = (\log 600 - \log 400) / \log 1.02 \approx 20.48$.

CHANGING BASES

Remember that logarithms may also be thought of as units of distance for measuring ratios of numbers. For example, using base 10, $\log(b/a) = 1$ if and only if the interval $[a, b]$ represents a multiplication by 10; that is, if $b = 10 \cdot a$. Thus the interval from 1–10 is 1 unit long on the log scale. The same is true for the interval from 10–100 or from 2–20. Likewise, $\log_2(b/a) = 1$ if and only if $b = 2 \cdot a$. In general, the base determines the ratio that defines one unit on the log scale.

What may be surprising is the fact that on any particular power scale defined by a particular base, successive powers of *any* base form a uniform scale. So powers of 2 are uniformly spaced on a base-10 power scale! Thus the log scale for one base is necessarily proportional to the log scale for any other base. The relationship between the bases establishes a size transformation between the log scales.

EXAMPLE 9

a) How does the unit distance on a $\log_3$-scale compare to that on a $\log_{10}$-scale?

b) Use that relationship and your calculator to evaluate $\log_3(7)$.

SOLUTION:

a) Since the scales are proportional, find the ratio of the two unit distances, each measured in the same scale.

$\text{Log}_{10}(10)$ represents one unit on the base-10 log scale. $\text{Log}_3(3)$ is one unit on the base-3 log scale, corresponding to the interval [1, 3] on the power scale. This same distance on the base-10 scale is $\log_{10}(3/1) = \log_{10}(3)$. See **Figure 2.18**.

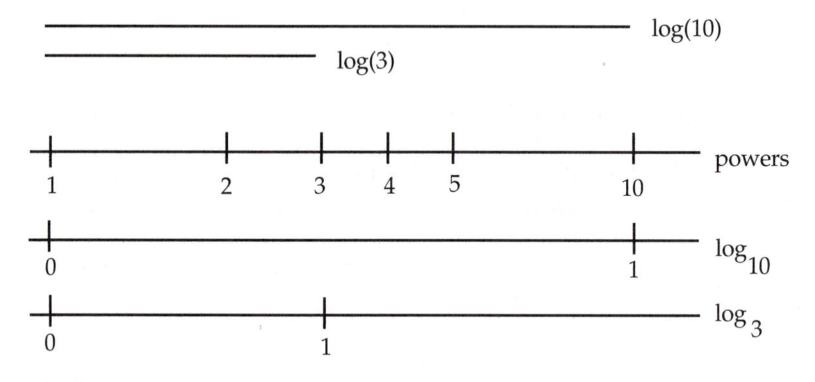

FIGURE 2.18. Comparing scales for $\log_{10}$ and $\log_3$.

Therefore one unit on the $\log_3$-scale is exactly $\log_{10}(3)/\log_{10}(10)$ times the length of one unit on the $\log_{10}$-scale. Since $\log_{10}(10) = 1$ (remember, it's the unit), the constant of proportionality is just $\log_{10}(3)$, or about 0.4771 (from a calculator), a little less than half as long.

Alternately, in base-3 units, the $\log_{10}$-scale unit is exactly $\log_3(10)/\log_3(3)$, or $\log_3(10)$, times the $\log_3$-scale unit. Since calculators don't have a $\log_3$ key, this value cannot be calculated directly. However, is should be clear from the above proportionality that the ratio must be about 1/0.4771, or 2.096; that is, exactly $1/\log_{10}(3)$. This suggests that $\log_3(10) = 1/\log_{10}(3)$.

b) $\text{Log}_{10}(7) = 0.8451$, accurate to four places. Since the $\log_3$-scale unit is about half the length of the $\log_{10}$-scale unit, base-3 lengths of power intervals will be proportionately larger—about twice as large. More exactly, $\log_3(7) = \log_{10}(7)/\log_{10}(3)$ or about $(0.8451) \cdot (2.096) = 1.771$.

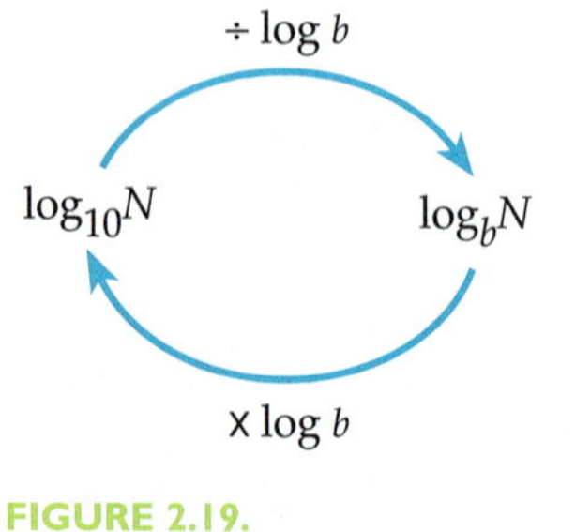

FIGURE 2.19. Changing bases.

The process of changing bases involves stretching or shrinking the number line, changing the ratio that represents one unit. **Figure 2.19** is an arrow diagram representing base changes as descibed above. Dividing log(N) by log(b) converts from base-10 to base-b units. Multiplying by log(b) reverses the process. This change-of-base process can be generalized to go from *any* base to any other base. However, most calculators have base-10 logs built in, so 10 is a frequently-used base.

Changing bases provides a powerful tool in solving exponential equations.

EXAMPLE 10

Suppose that an initial investment of \$400 is put into a bank account that earns 8% interest per year compounded each quarter. How long will it take for the investment to grow to \$600 from accumulating interest only?

SOLUTION A (LOGARITHMIC FORM):

Using the given information, you must solve $400(1.02)^t = 600$, where t is the number of quarters that have elapsed since the initial deposit. (You solved similar problems in Lesson 2.1 by graphing.) Change the equation to base 1.02. See **Figure 2.20**.

Divide both sides by 400 , so that: $1.02^t = 1.5$
Take logs to the base 1.02: $t = \log_{1.02}(1.5)$

Evaluate by converting from base 10:

$t = \log 1.5 \,/\, \log 1.02$, or $t \approx \dfrac{0.1761}{0.0086}$.

Thus, it takes about 20.48 quarters, or just over 5 years.

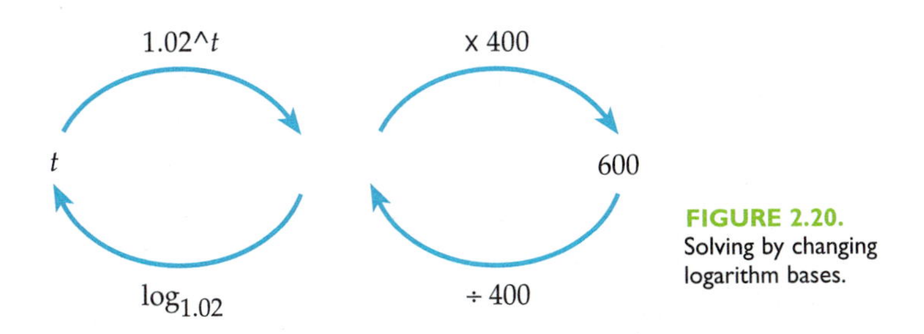

FIGURE 2.20. Solving by changing logarithm bases.

TAKE NOTE

Exponential and power equations differ, so their solutions use different operations. Example 9 involves an exponential equation; the variable is the exponent. Contrast that with the problem of trying to determine an interest rate that would give you \$600 after only 3 years. Now you must solve: $400 \cdot (b)^{12} = 600$. The variable is the *base*; this is a power equation. **Figure 2.21** diagrams its solution. You may find it useful to think of logs as *undoing* exponentiation, and vice versa, much in the same way that roots *undo* powers. This idea will be developed more fully later.

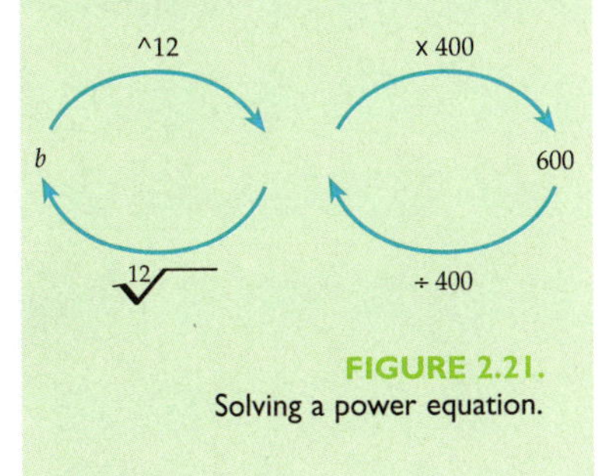

FIGURE 2.21. Solving a power equation.

SOLUTION B (EXPONENTIAL FORM):

Divide both sides by 400, so that: $1.02^t = 1.5$

Rewrite both numbers as powers of 10: $(10^{\log 1.02})^t = 10^{\log 1.5}$

Since both sides have the same base, equate exponents
$t \cdot \log 1.02 = \log 1.5$

So, $t = \log 1.5/\log 1.02 \approx 20.48$ quarters.

Compare this result to that in Example 8.

TAKE NOTE In several of the previous examples, the solutions used the fact that $m = n$ if and only if $b^m = b^n$. This fact follows intuitively from the graph of the related exponential function, $y = b^x$, in which two points have the same y-value only in case they come from the same x-value. Since logarithms are exponents, this property permits altering equations both by *exponentiating both sides to the base* b and by *logging both sides to the base* b.

Logarithms are used to describe a variety of events, and the choice of base frequently can reflect the underlying nature of the event. Base-10 logarithms provide a convenient scale with which to describe quantities like earthquake magnitudes, since the Richter numbers convey something about the relative powers of ten. As you observed in Activity 2.2, base-e exponential functions can be used to describe continuous growth at a constant percentage. Base-e logarithms can be used to solve a variety of problems related to this type of growth.

Logarithms to the base e have many desirable properties that you will learn about as you study calculus or science. These properties make e the base of choice in so many scientific applications. The $\log_e x$ is called the **natural logarithm**, and is traditionally written in shorthand as "$\ln x$" (read L.N. of x). Its formal definition is similar to that for logs to other bases: $\ln x = y$ means $e^y = x$. Finally, since it defines a logarithm scale, all properties of logarithms hold for natural logs as well:

(i) $\ln 1 = 0$

(ii) $\ln e = 1$

(iii) $\ln (a \cdot b) = \ln a + \ln b$

(iv) $\ln (a/b) = \ln a - \ln b$

(v) $\ln a^n = n \cdot \ln a$

Exercises 2.3

1. A scientific calculator has both e^x (or EXP) and LN keys. Use your calculator to answer these questions.

   a) Evaluate e^(1). What number should be displayed?

   b) Now evaluate the LN of the previous answer. (Depending on your calculator, you may have to type in the number after pressing the LN button, or simply press the LN button.) What number is displayed now?

   c) Use words and an arrow diagram to summarize your observations about the relationship between the e^x and $\ln x$ operations.

2. Rewrite expressions so that both sides involve exponentials to the same base, then solve the equation.

   a) $10^x = 10^{3.86}$

   b) $5 \times 10^x = 3865$

   c) $4 \times 2^x = 836$

3. a) Complete a copy of **Table 2.23**. How are the final two columns of values related?

x	3^x	$10^{0.47712x}$
1		
5		
10		
15		
20		

TABLE 2.23.
For Exercise 3.

   b) Explain your observation from part (a) by using the properties of exponents or logarithms.

   c) Explain any discrepancies between the observed pattern and that predicted in part (b).

Exercises 2.3

4. Follow the steps taken in Activity 2.5 to construct a $\log_5$ number line that is to scale with the $\log_{10}$ and $\log_2$ number lines previously made in the activity.

   a) Where is the number 2.7 on your new number line aligned with the power scale?

   b) From your work in Activity 2.5, describe at least two different ways to mark the unit length for the $\log_5$ number line without direct measurement (i.e., no ruler, etc.).

   c) Describe at least two different ways to calculate the unit length for the $\log_5$ number line in order to mark it using a ruler.

5. Rewrite the following logarithmic equations in exponential form ($b^y = x$).

   a) $\log_5 25 = 2$

   b) $\log_4 64 = 3$

   c) $\log_3 81 = 4$

   d) $\log_e 1 = 0$

6. Rewrite the following exponential equations so that they are in base 10.

   a) $y = 5^x$

   b) $y = 3^x$

   c) $y = e^x$

7. Rewrite the following exponential form equations so that they are in base e:

   a) $y = 2^x$

   b) $y = 3^x$

   c) $y = 10^x$

8. Using the properties of logarithms, write the following expressions as a single term. Assume that $x > 0$.

Exercises 2.3

a) $\log_4 x + \log_4 3$

b) $2\log_4 x - \log_4 3$

c) $3\ln x + 2\ln 3$

d) $3\ln 10 - \ln x$

9. Rewrite as the sum, difference, or product of log(x) and log(y); assume that $x > 0$ and $y > 0$.

a) $\log xy^2$

b) $\log \frac{x^2}{y}$

c) $\ln\left(\frac{x^3}{y^4}\right)^{\frac{1}{5}}$

10. Figure 2.19 summarized moving between common logs (base 10) and an arbitrary base. Make an arrow diagram summarizing the relationships between natural logs (base e) and an arbitrary base.

11. Rewrite each indicated log as an expression involving only natural logarithms.

a) $\log_2 10$

b) $\log_5 26$

c) $\log_7 50$

12. Rewrite each indicated log as an expression involving only common logarithms.

a) $\log_2 10$

b) $\log_5 26$

c) $\log_7 50$

13. Solve each of the following exponential equations.

a) $3^{x+2} = 243$

b) $5^{2x+2} = 125$

c) $4^{\frac{x}{3}+2} = 64$

Exercises 2.3

d) $4^{x^2+1} = 256$

e) $10^{2x+1} = 1000$

14. Solve each of the following logarithmic equations.

a) $\log_a x = 2\log_a 8 - \log_a 4$

b) $\log_a x = 3\log_a 2 + \log_a 4$

c) $\log_a x = \frac{1}{2}\log_a 4 + \log_a 3$

15. Carbon-14 (C_{14}) is a type of carbon that exists naturally in all living objects. It is **radioactive**, which means that it gradually decays into other elements. In the case of C_{14}, approximately 0.012% of it decays each year, and scientists can use that fact as a kind of clock for determining how long ago an object lived.

a) Suppose you begin with 500 grams of C_{14}. Write an exponential equation that describes the amount of C_{14} *remaining* after t years of decay.

b) Rather than referring to the fraction that decays in one year, it is more convenient to describe the rate of decay in terms of its **half-life**—the time it takes for 50% of the substance to break down. How long does it take for a 500-gram sample of C_{14} to decay to 250 grams?

c) How long does it take for a 2500-gram sample of C_{14} to decay to 1250 grams?

d) Why does the answer for half-life *not* depend on how much was originally in the sample?

16. For quantities that grow exponentially over time, such as populations or finances, a convenient way to describe the growth rate is with **doubling time**—the time it takes for the quantity to double.

a) The population of the United States was 266 million people at the beginning of 1997. At that time, it was increasing at an average growth rate of 0.7% per year. Calculate the doubling time.

b) Write an exponential function that describes a population growing at $100r\%$ per year, beginning with population P_0 at time t_0. Use your function to write an equation describing the time at which the population will have doubled. Then solve that equation.

c) In finance, the **Rule of 72** is sometimes used to estimate doubling times. Here's how it works: If the annual interest rate is $r\%$, then the doubling time is approximately $72/r$. Use the Rule of 72 to estimate how many years it will take for an investment to double, using several different reasonable annual interest rates. Record your estimates in a table.

d) Now use the formula you derived in part (b) to calculate the exact doubling time for each rate you tried. How accurate is the Rule of 72? Why might such a rule be used by people involved in financial calculations?

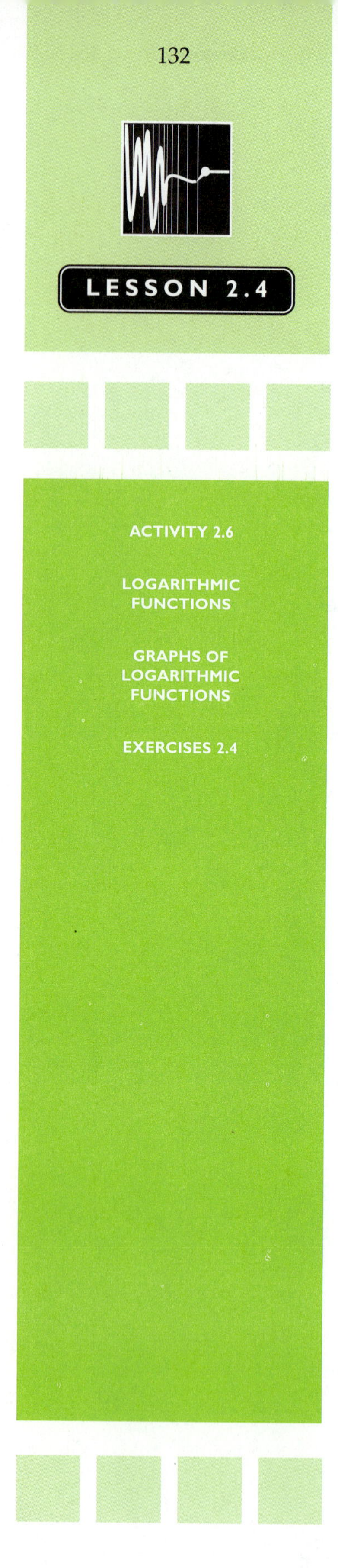

Logarithmic Functions

So far you have explored logarithms as a new type of number, and have found and applied some of their properties. Now the tool kit development expands to include logarithmic functions.

Activity 2.6

1. a) Without using a calculator, make a table of values for x and y that satisfies the equation $y = \log_3 x$. Include x-values above and below 1. Then plot the graph of your table.

   b) Explain why x cannot be negative.

   c) Describe how you found the values in your table.

   d) Examine pairs of points on your graph whose x-coordinates differ by 1 unit. Is there a pattern among their y-values? Repeat for pairs of points whose y-coordinates differ by 1 unit.

   e) Use part (c) and the definition of logarithms to explain the shape of your graph in part (a).

2. Consider the functions $y = \log_{10} x$, $y = \ln x$, and $y = \log_2 x$.

   a) What is an effective calculator WINDOW setting for viewing graphs of all these functions at once?

   b) Explain how to enter $y = \log_2 x$ into the calculator to display its graph.

   c) Sketch the graphs in your selected window. Be sure to include behavior near the y-axis.

   d) What do the three graphs have in common? How can you tell them apart?

   e) Describe the characteristics (domain/range, end behavior, roots and concavity) of these graphs.

3. Remember that logarithms are defined in terms of an exponential equation.

   a) Make an arrow diagram relating 2^x and $\log_2 x$. Then make two tables of values, one for the equation $y = 2^x$, and the other for the equation $y = \log_2 x$.

   b) Using a graphing calculator, graph the equations of $y = 2^x$ and $y = \log_2 x$ together. Be sure to use square axis scaling. Sketch the graph.

   c) How are the domain and range of the exponential function $y = 2^x$ related to the domain and range of the logarithmic function $y = \log_2 x$?

   d) How is the graph of the exponential function $y = 2^x$ related to the graph of the logarithmic function $y = \log_2 x$? How is it different?

   e) How does the formal definition of logarithm relate the graphs?

Review your work on transformations in Lesson 1.4 of Chapter 1.

4. Consider the transformed logarithmic function $y = 3\log_2(x - 4) + 1$.

   a) Describe how the function $y = \log_2 x$ was altered to produce this new function. (Note: Describe the computation of numbers for y, not the graph. An arrow diagram or spreadsheet may be helpful.)

   b) Without using a calculator, sketch your prediction for the graph of this new function. Explain how you determined your prediction.

   c) Use a graphing calculator to check your prediction. What equation did you enter into the calculator?

LOGARITHMIC FUNCTIONS

Lessons 2.2 and 2.3 focus on two representations of logarithms. Both representations communicate similar ideas and provide powerful ways to view the nature of logarithmic functions.

The first representation is a scale, representing successive powers of a base as equally-spaced along a number line. Thus, exponents become scale marks.

The second representation is the algebraic relationship among bases, exponents, and logarithms, both in exponential and in logarithmic form. That is, $y = \log_b x$ if and only if $b^y = x$.

To illustrate these ideas, consider the information provided in **Table 2.24**.

TABLE 2.24. Exponential/Logarithmic relationship in base 2.

A	0	1	2	3	4	5
B	1	2	4	8	16	32

One of the variables lists powers of 2, while the other lists the exponents responsible for generating those powers. One of the variables exhibits a constant ratio between successive terms, while the other shows a constant difference between terms. Plotting (A, B) pairs yields a different graph than plotting (B, A) pairs, yet the graphs have the same shape. The pairs satisfy the relationship $B = 2^A$; they also satisfy the relationship $A = \log_2 B$. In both situations, B represents powers of 2 and A represents the exponents. The distinction between the graphs comes from identifying domain and range variables.

Look carefully at the *relationship* described by the equation $y = \log_2 x$. **Figure 2.22** shows three visual representations of that function.

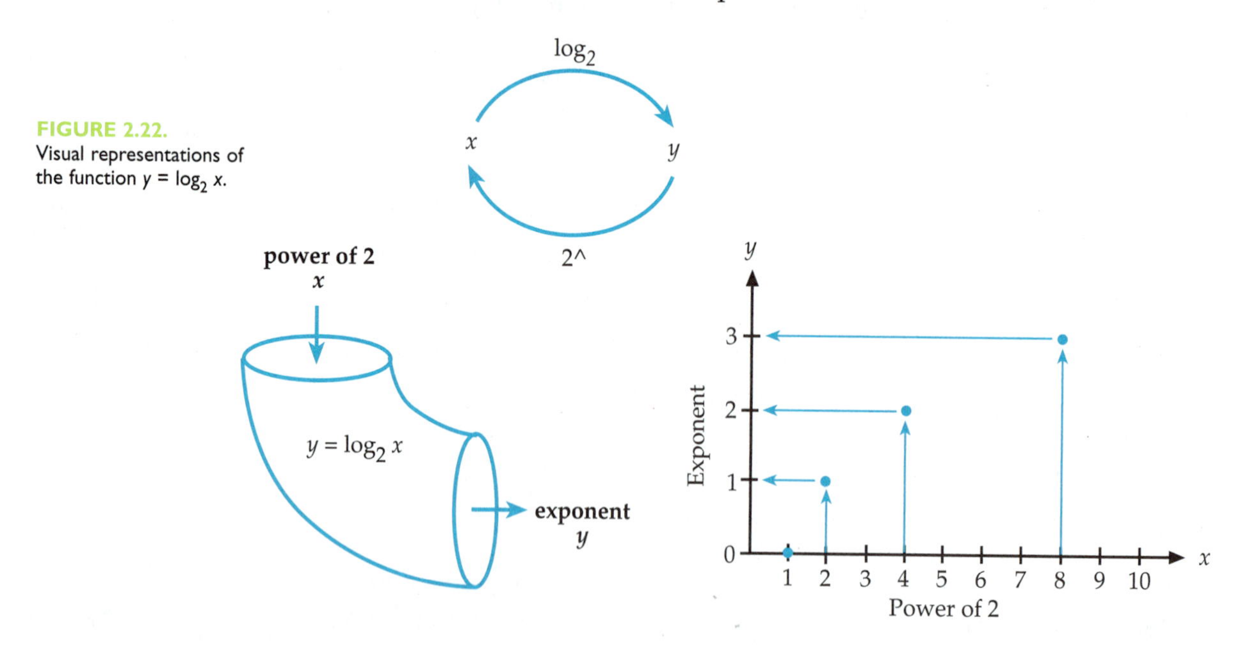

FIGURE 2.22. Visual representations of the function $y = \log_2 x$.

This function takes as input the variable x, which represents any positive number x, thought of as a power of 2. It returns as output the exponent needed for that power. In other words, $2^y = x$, which is just how logarithms are defined. The graph shows successive numbers having constant *ratios* (powers of 2) being compressed into a scale in which those same numbers have constant *differences* (exponents). In either case, reversing the process starts with an exponent and ends with some number (power of 2), which is what an exponential function does.

GRAPHS OF LOGARITHMIC FUNCTIONS

The graph of each logarithmic function is directly related to the graph of the exponential function having the same base. Remember, $y = \log_b x$ if and only if $b^y = x$. The coordinate pair (A, B) on a logarithmic graph corresponds to the coordinate pair (B, A) on an exponential graph. The domain of the logarithmic function is the same as the range of the exponential function; the range of the logarithm is the same as the domain of the exponential function. The end behavior for exponential graphs along the negative x-axis mimics the behavior for logarithmic graphs along the negative y-axis. The y-intercept for the exponential function corresponds to the x-intercept (root) of the logarithmic function. These behaviors are a consequence of the definition of logarithm in terms of an exponential relationship.

Graphs of all equations of the form $y = \log_b x$ exhibit the same characteristic shape—a curve that is quite steep for x near 0, but which flattens out as x increases. Their domains are restricted to positive values for x, but the ranges include all numbers. They have only one x-intercept, at $x = 1$, and are concave down everywhere. The end behavior for the graphs is for $y \to -\infty$ as $x \to 0$ ($x > 0$ only), and for $y \to +\infty$ as $x \to +\infty$. **Figure 2.23** shows typical logarithmic graphs.

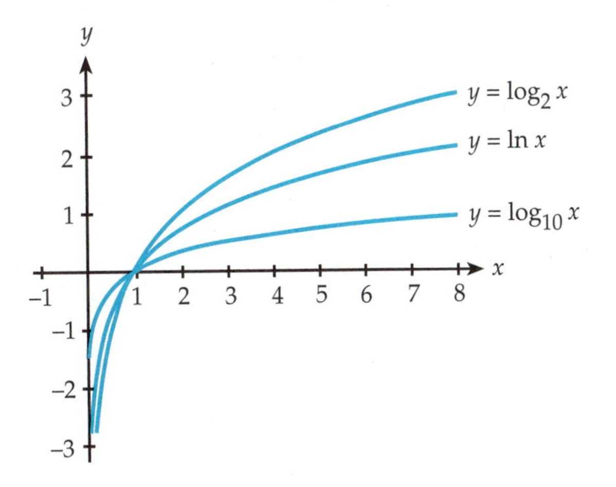

FIGURE 2.23. Graphs of logarithms to the bases 2, e, and 10.

The base determines the particular shape of the curve, with graphs for small bases bending more gradually, while those for larger bases show a dramatic change from near-vertical to near-horizontal near the root. The larger the base, the closer the graph is to the x-axis for any particular value of x other than 1.

As with any tool kit function, graphs of logarithmic functions may be transformed by translations or stretches via addition and multiplication, respectively. The procedures you learned in Chapter 1 remain effective.

Exercises 2.4

1. Why is the graph of $y = \ln x$ between the graphs of $y = \log_2 x$ and $y = \log_{10} x$?

2. Using numbers, explain why, for the graph of $y = \log_2 x$, $y \to -\infty$ as $x \to 0$.

3. An **identity** is an equation that is true for all values of its variable. Verify by graphing that the following equations are identities.

   a) $2^x = 10^{x \cdot \log 2}$

   b) $\ln x = \log x / \log e$

   c) $\log x^3 = 3\log x$

   d) $\log_5 x^4 = 4\log_5 x$, for $x > 0$

4. a) Describe the following features of the graph of $y = \log_5 x$: domain/range, roots, y-intercept, concavity, and end behavior.

   b) How does this graph compare to that of the function $y = \log x$?

5. The graphs of the three logarithmic functions $y = \log_{1.5} x$, $y = \log_2 x$, and $y = \log_5 x$ are shown in **Figure 2.24**. Match each of the graphs with its equation, and use that information to determine the equation of the exponential function that is also shown.

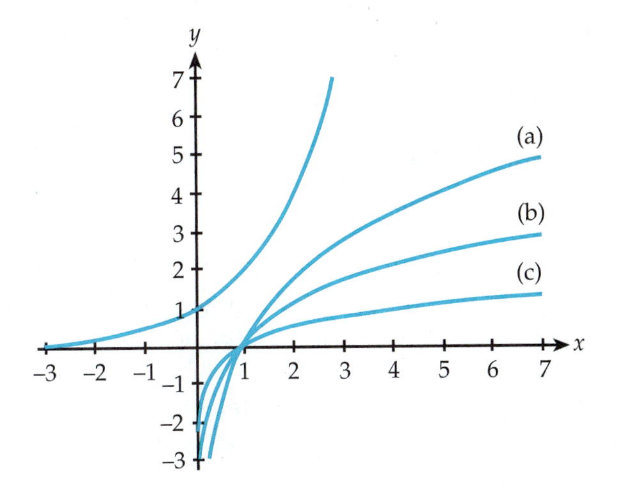

FIGURE 2.24. Graphs for Exercise 5.

Exercises 2.4

6. Looking at Figure 2.24, Ramon did not see the labels and thought the graphs were just vertical stretches of the graph of the base-10 log function. Use what you know about changing bases and about transformations to prove that his point of view is also correct.

7. Use a graphing calculator to find an approximate solution to the equation $\ln x = 3 - x$. Show your method clearly.

8. Consider the equation $y = 2 \cdot \log_4(x + 3) - 1$ as a transformation of $y = \log_4(x)$.

   a) Make an arrow diagram for this transformation, then predict the effects of the transformation on the graph.

   b) Use statistical lists (or a spreadsheet) to show the implementation of your arrow diagram.

   c) Sketch the graph, and explain how the transformations affect the location and shape of the graph. Which two columns of your spreadsheet (or lists) define the input and output of the final, transformed graph. Which two columns define the input and output of the parent function?

9. Use your knowledge of translations and your understanding of the graphs and properties of logarithms to predict the graph of the function $y = 2 \cdot \ln(x - 3) + 5$. Then, check your prediction using a graphing calculator.

10. On a single set of axes, graph the functions $y = x^{1/2}$, $y = x^{1/3}$, and $y = \log x$.

    a) Which graph grows most slowly as x increases?

    b) Consider functions of the form $x^{\left(\frac{1}{n}\right)}$, where n is an integer. Are there values of n for which such curves grow more slowly than $y = \log(x)$?

    c) Based on the work in this problem, where would you position the function $y = \log x$ on the Ladder of Powers?

11. In Lesson 2.1, slopes were used to detect exponential patterns in data. The same analysis can be applied to the newest tool kit function—the logarithm.

a) Make a copy of **Table 2.25**, fill in the column of function values for $y = \log_3 x$, then complete the table.

Exercises 2.4

x	$y = \log_3 x$	Δx	Δy	$\Delta y/\Delta x$
1				
3				
9				
27				
81				

TABLE 2.25.
Difference table for $y = \log_3 x$.

b) Describe any patterns you observe in the table's columns. Explain.

c) Recall that $\Delta y/\Delta x$ can be interpreted as the slope of a secant line. How does the pattern in the slope column help explain the shape of the graph of $y = \log_3 x$?

d) Make a copy of **Table 2.26**, completing it using the exponential function $y = 3^x$. Compare patterns you see for $y = 3^x$ to those generated by $y = \log_3 x$.

x	$y = 3^x$	Δx	Δy	$\Delta y/\Delta x$
0				
1				
2				
3				
4				

TABLE 2.26.
Difference table for $y = 3^x$.

e) Use your understanding of the definition of logarithms to explain your observations in part (d).

12. **Investigation.** Exercise 11(b) hints at yet another function related to logarithms. Here's the idea: You may recall similar work in earlier courses. When you zoom in on a smooth curve, it looks very much like a line. The more you zoom in, the more natural it is to think about the slope of the graph. Take successively closer looks at the curve near some fixed x-value, and examine the slopes of the corresponding secant lines.

Exercises 2.4

In this investigation, work with the natural log, $y = \ln x$. Use **Table 2.27** to zoom in at $x = 2$. That is, use $x = 2$ as the left endpoint for each secant, but choose right endpoints nearer and nearer to 2. Observe the ratio $\Delta y/\Delta x$.

a) Complete a copy of Table 2.27.

TABLE 2.27. Investigating secants of the natural log.

x_1	$y_1 = \ln x_1$	x_2	$y_2 = \ln x_2$	Δy $(y_2 - y_1)$	Δx $(x_2 - x_1)$	$\frac{\Delta y}{\Delta x}$
2	0.6931	10				
2	0.6931	5				
2	0.6931	3				
2	0.6931	2.5				
2	0.6931	2.1				
2	0.6931	2.01				

b) Describe any patterns you observe in the various columns.

c) Using a graph of $y = \ln x$, show how the slope calculations relate to the graph; interpret geometrically your observations for the table of $\Delta y/\Delta x$ values.

d) In (a)–(c), you were told to use $x = 2$ as the left endpoint. Repeat the process several more times, each time varying the choice for the left endpoint. (For each left endpoint, move the right endpoint close to it to calculate zoomed secant slopes.)

Record your choices for left endpoint and the corresponding zoomed slope in a data table like **Table 2.28**. How do the left endpoint value and the ratio $\Delta y/\Delta x$ appear to be related for the function $y = \ln x$? (Data analysis may help establish the nature of the relationship precisely.)

TABLE 2.28. Slopes of secants for the natural log.

Left Endpoint x						
Zoomed Slope: $\Delta y/\Delta x$						

Exercises 2.4

13. **Investigation.** You live in an information age. Whether it's CDs, digital cameras, or computers, information is stored as collections of 0s and 1s in some kind of binary code that is interpreted by the specific application. But what *is* information? How much binary code is needed to convey information? The basic measure of information is a "bit." For the purposes of this investigation, you may think of one bit as the information contained in the answer to one yes-no question.

Part 1: Exploration

Corbis

a) Consider a simple situation in which you must select one particular CD from 2 that are gift-wrapped. You know that one of them is the latest recording by your favorite artist, and you don't have it yet. What's the smallest number of questions you can ask in order to identify the CD you want? What question(s) would you ask? How many bits of information are needed?

Let f denote the information function, where $f(n)$ represents the number of bits of information needed to identify 1 item from among n items. Thus, you just computed $f(2)$.

b) Consider an even simpler situation. Suppose there is only 1 CD from which to select, and you know it's the latest recording by your favorite artist. How many questions do you have to ask now? How many bits of information are needed here? That's $f(1)$.

c) Henry needs to identify a particular CD from among 4 CDs. He saw your solution to the 2-CD problem and reasons as follows, "Let me group the 4 CDs into groups of 2 CDs per group. That leaves me with 2 groups, and I just saw how to identify one item (group) from a pair. After that, I'll have one group (2 CDs), but that's just the 2-item problem again. So I should need exactly $f(2) + f(2)$ questions (bits)." Comment on Henry's reasoning, then find $f(4)$.

d) How many questions do you have to ask if there are 8 CDs? That is, find $f(8)$. Explain your reasoning carefully. If possible, provide more than one explanation of your computation.

Exercises 2.4

e) Maria has to select from among 32 CDs. She heard Henry's explanation for $f(4)$ and reasons that she could arrange the 32 CDs into 4 groups of 8 CDs each. Then $f(4)$ questions would identify the correct group and an additional $f(8)$ questions would find the right item. Compute $f(32)$ using this approach, then check it directly.

Part 2: Modeling

a) Based on its contextual meaning, identify a reasonable domain for the information function, f, used in Part 1.

b) Should the information function be increasing, decreasing, or neither? Explain based on the contextual meaning.

c) Generalize the observations made by Henry and Maria to write $f(MN)$ as the sum of two values of f. Explain your reasoning.

d) Use the values of f that you computed in Part 1 to begin its graph. Label your axes carefully. Compute more values, either directly or by using properties you identified above, to extend the graph until you can identify it as a member of your tool kit.

e) Through your work in Parts 1 and 2 (a)–(d), the following properties have been attributed to the information function, f.

- domain: $N > 0$
- initial condition: $f(1) = 0$
- additional observed value: $f(2) = 1$
- increasing: $f(N) > f(M)$ whenever $N > M$
- addition property: $f(MN) = f(M) + f(N)$

Verify that the function you named in Part 2 (d) satisfies all these properties.

LESSON 2.5

Modeling with Exponential and Logarithm Functions

In the earlier lessons of this chapter, the tool kit of functions was expanded to include both exponential and logarithmic functions. This lesson will explore modeling real-world phenomena with these new functions.

Activity 2.7

Guess the Power

In Chapter 1, the proportionality of $y = kx^n$ and $y = x^n$ was used to check data against a given power function. For data following a power equation of the form $y = kx^n$, the graph of y_{data} versus x^n lies along a line through (0, 0), and the ratios y_{data}/x^n are essentially constant. These factors allowed you to determine the power, n, without having to rely on power regression.

1. a) When examining data of the form (x, y), explain why both a straight-line graph of y versus x through the origin and having a constant ratio of y/x are valid tests for y varying directly with x.

   b) Although it was not discussed in Chapter 1, some people prefer a slight variation on the procedure outlined above when guessing fractional powers. For example, if you think that the power is 3/2, the method above graphs y_{data} versus $x^{3/2}$ or divides y_{data} by $x^{3/2}$. However, you might instead graph $(y_{\text{data}})^2$ versus x^3 or divide $(y_{\text{data}})^2$ by x^3.

   Verify that the two methods are equivalent.

Table 2.29 contains data from a physics lab in which students mounted steel rings of various diameters, d, on a hacksaw blade, and measured the time, t, it took for each ring to swing back and forth 25 times.

TABLE 2.29.
Data for physics lab on period of motion.

Ring Diameter (cm)	3.51	7.26	13.7	28.5	38.7
Time for 25 Swings (sec)	9.35	13.3	19.2	27.0	32.9

The goal of the remainder of this activity is to develop a model that predicts the time for 25 swings in terms of the diameter of a ring.

2. a) Graph the data and describe the scatter plot.

   b) Using statistical lists or spreadsheets, apply the *graphical* proportionality test for power functions to the data in Table 2.29. Graph several combinations—t vs. d^2, t vs. d^3, t^2 vs. d, t^3 vs. d, and others if necessary. According to your graph, what equation best describes the variation between diameter and time?

   c) Verify the results from part (b) by examining *ratios* between powers of t and powers of d.

   d) Verify the equation(s) determined in parts (b) and (c) by using LinReg on the re-expressed data in part (b), and PwrReg on the original data. Use residuals to determine whether the models describe the relationship between d and t reasonably well.

Calculate the Power

Here is a way to straighten the scatter plot without any trial-and-error work.

3. a) Again use statistical lists or spreadsheets. Calculate the quantities $\log d$ and $\log t$, and graph $\log t$ vs. $\log d$. Describe the resulting scatter plot.

   b) Write an equation that describes the relationship between $\log d$ and $\log t$. Interpret the meaning of the slope.

   c) Verify apparent linearity in part (a) by calculating slopes between successive points in that graph. Are the slopes roughly constant?

   d) Using your equation relating $\log t$ and $\log d$, explain how to write an equation for the relationship between t and d. Use your method to write an equation describing t in terms of d. (Hint: Laws of Logarithms).

Exponential Models

4. The data in **Table 2.30** come from the U.S. Department of Energy, and represent the average cost of natural gas (in dollars/1000 ft^3) to consumers over the period from 1973–1983.

Time (t)	0	1	2	3	4	5	6	7	8	9	10
Ave. Cost (C)	1.29	1.43	1.71	1.98	2.35	2.56	2.98	3.68	4.29	5.17	6.06

TABLE 2.30.
Average cost of natural gas from 1973–1983 (U.S. Dept. of Energy).

a) Describe the scatter plot of C versus t.

b) Explain why you can't apply the "log-log" analysis from Item 3 to these data.

c) Remove the first data point to allow you to apply a log-log re-expression. Describe the graph of log C versus log t.

d) In general, the effect of taking the logarithm of numbers is to pull large numbers down a lot. In trying to straighten the scatter plot, predict and check the effect of taking logarithms of only one of the variables; that is, for C versus $\log(t)$, and for $\log(C)$ versus t.

e) Write an equation that describes the straight-line graph from (d). Then re-express it to write C in terms of t. How well does this new equation describe the original data?

5. Describe a procedure (re-expression) that might straighten logarithmic data.

MODELING POWER FUNCTIONS

In Activity 2.7, you were presented a set of data (Table 2.29). A scatter plot of these data indicated that a power function or an exponential function might describe the relationship between the two variables, t and d. One way to check this assumption is to re-express the data as log t and log d and then observe whether a scatter plot of log t versus log d or log t versus d is linear.

Using the Laws of Logarithms, the equation $y = a \cdot x^p$ is equivalent to $\log(y) = \log(a) + p \log(x)$ for $x > 0$. Since x and y are the variables in the original equation, $\log(x)$ and $\log(y)$ are variable quantities in the second equation. Likewise, $\log(a)$ and p are constants. Thus, $\log(y)$ is linear in terms of $\log(x)$, with slope p and vertical intercept $\log(a)$ if and only if

$y = a \cdot x^p$, so knowing the slope and intercept allows you to calculate the power, p, and coefficient, a, immediately.

Any power function $y = a \cdot x^p$ will lead to a linear graph of log(y) versus log(x). The reverse is also true: *Any* linear graph of log(y) versus log(x) means that y is a power function of x. So in modeling data, a log-log linearity test determines not only whether the data follow a power function form but also the exponent in that form, and both without guessing.

Example 11

Use a log-log re-expression to check the data in **Table 2.31** for a power model.

Solution:

Table 2.31 shows the computations. The transformed scatter plot in **Figure 2.25** is the graph of log(y) versus log(x). Slope is calculated as $\Delta\log(y)/\Delta\log(x)$.

x	y	$\log x$	$\log y$	$\frac{\Delta \log y}{\Delta \log x}$
1	2	0	0.301	1.5
4	16	0.602	1.204	1.5
9	54	0.954	1.732	1.5
16	128	1.204	2.107	1.5
25	250	1.398	2.398	

TABLE 2.31.
Power Function with Log-Log Linearity Test.

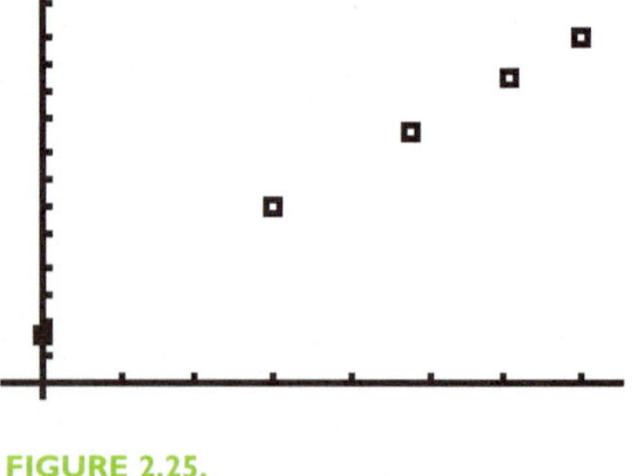

FIGURE 2.25.
Log-log scatter plot. [–0.1, 1.5] x [–0.1, 2.8]

The equation of the line (re-expressed) can be read from row 1 (the intercept) and column 5 (the slope), or calculated from the graph using slope-intercept or point-slope calculations. The result is $\log y = 0.301 + 1.5 \cdot \log x$. To transform this into a model for the original data, $p = 1.5$ and $\log(a) = 0.301$, so $a = 10^{0.301} = 2$. Thus, $y = 2 \cdot x^{3/2}$.

MODELING EXPONENTIAL & LOGARITHMIC FUNCTIONS

Much as is the case for power functions, the Laws of Logarithms show how to transform exponential data. The equation $y = a \cdot b^x$ is equivalent to $\log(y) = \log(a) + x \log(b)$. Now, x and $\log(y)$ are the variable quantities in the second equation, and $\log(a)$ and $\log(b)$ are constants. Thus, $\log(y)$ is linear in terms of x, with slope $\log(b)$ and vertical intercept $\log(a)$, if and only if $y = a \cdot b^x$. Again, knowing the slope and intercept allows you to calculate the base, b, and coefficient, a, for the original data.

Any exponential function $y = a \cdot b^x$ will lead to a linear graph of $\log y$ versus x, and vice versa. So in modeling data, a "semi-log" linearity test determines not only whether the data follow an exponential function form but also the base for that function, and both without guessing. For example, if $\log y = mx + k$, then $y = 10^{mx+k} = 10^k \cdot 10^{mx} = A_0 \cdot 10^{mx}$, where $A_0 = 10^k$.

The situation is even easier for logarithmic data. When a linear relationship is established between y and $\log x$, the equation $y = m \cdot (\log x) + b$ is the final (logarithmic) model.

EXAMPLE 12

Suppose that doctors are monitoring the growth of tumor cells in a cancer patient in order to initiate a remedial treatment before the cell count reaches 60,000. The treatment they prescribe will vary depending upon the estimate of how quickly the cells are growing. They have collected the data in **Table 2.32** and have asked you to estimate when the cell count will reach 60,000.

Time in days	Cell count
0	597
2	893
4	1339
6	1995
8	2976
10	4433
12	6612
14	9865
16	14719
18	21956
20	32763

TABLE 2.32.
Tumor cell counts.

SOLUTION:

First, graph the data. Refer to **Figure 2.26**.

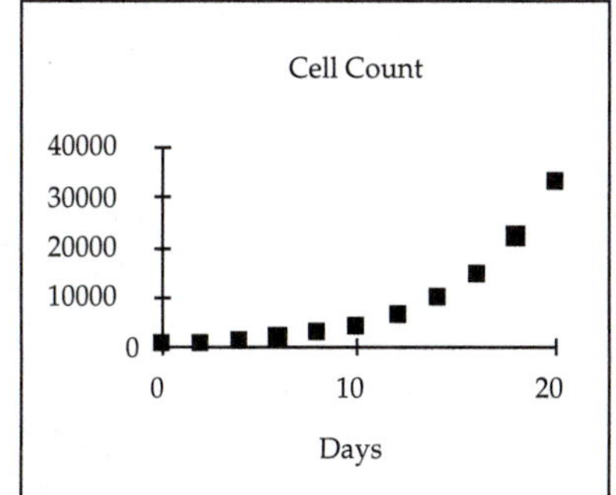

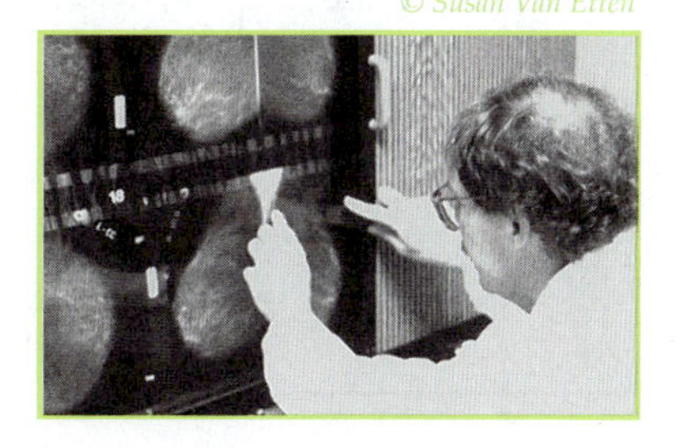

FIGURE 2.26.
Scatter plot of cell growth data.

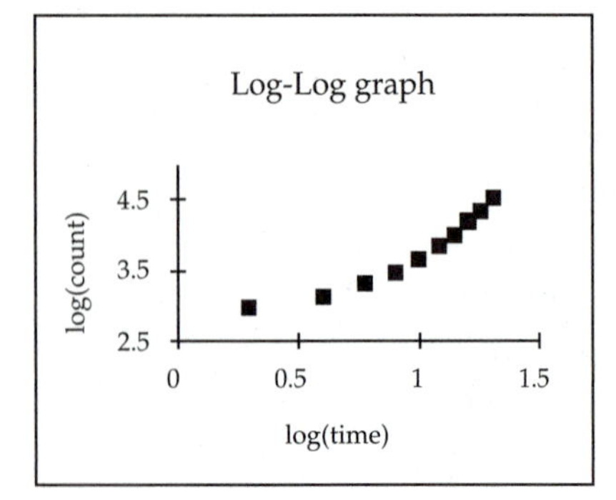

FIGURE 2.27.
Log-log re-expression.

The graph is curved. It might be a power function or an exponential function. (Since it does not contain (0, 0), it can't be exactly a power function, but data rarely follow their models exactly!) Check to see whether the data fit a power function by applying a log-log test. See **Figure 2.27**.

The re-expressed scatter plot is definitely *not* linear, so this is not a power relationship. Next, graph $\log(y)$ versus x to see whether the data are exponential. See **Figure 2.28**.

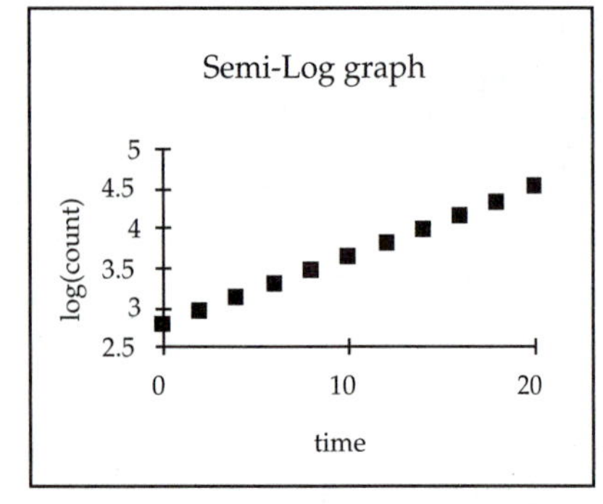

FIGURE 2.28.
Semi-log re-expression.

This looks very linear. The corresponding table is shown as **Table 2.33**. (Note that spreadsheet-calculated values display more precision than may reasonably be believed. Read values to only 3 or 4 decimal places.)

A reasonable equation for the re-expressed line is $\log(y) = 0.087x + 2.78$. This may be solved for y (the transformed equation is $y = 603(1.22)^x$), but since you really want to know the time at which the cell count is 60,000, you can use $y = 60{,}000$ in this equation and find x: $\log(60{,}000) = 0.087x + 2.78$ gives $x = 23.0$ days.

Time in days	Cell count	log count	semi-log slope
0	597	2.77597433	0.08743856
2	893	2.95085146	0.08796456
4	1339	3.12678058	0.08658116
6	1995	3.2999429	0.08684501
8	2976	3.47363293	0.0865324
10	4433	3.64669773	0.08681756
12	6612	3.82033284	0.08688212
14	9865	3.99409709	0.08689061
16	14719	4.16787831	0.08683746
18	21956	4.34155322	0.08691522
20	32763	4.51538366	

TABLE 2.33.
Calculations for semi-log re-expression.

In general, always consider whether you wish to transform a model back to the original quantities or use the re-expressed model directly. In many cases, it is better to leave the model in terms of the re-expressed data, since any errors that may have been introduced (precision, round-off, etc.) in the process will be magnified in the the original data.

Exercises 2.5

1. Revisit your placement of the logarithm function on the Ladder of Powers (see Exercise 10 of Lesson 2.4).

   a) Graph $y = \log x$ and $y = x^{0.1}$ using a WINDOW of [0, 1000000] x [0, 10]. Describe what you see on the graphic screen.

   b) Now change the WINDOW to [0, 1 x 10^{11}] x [0, 30]. Describe the new graph.

   c) What does this imply about the placement of the function $y = \log x$ on the Ladder of Powers?

   d) Repeat this experiment several more times comparing $y = \log(x)$ to other functions of the form $y = x^{1/n}$ for various values of $n > 10$. Describe your graphs. Where would you put $\log(x)$ on the Ladder of Powers now?

2. Refer back to Example 12, modeling tumor growth. Use regression analysis on your calculator to find an exponential function that fits the cell count data. If the model is not $y = 603(1.22)^x$, graph both models on the same set of axes. Are the two graphs significantly different?

3. Suppose data such as those in **Table 2.34** are related logarithmically.

x	10	20	50	80	150	290	450
y	0.0	0.60	1.34	1.73	2.24	2.78	3.13

TABLE 2.34. Logarithmic data for Exercise 3.

   a) Find a model for these data by regression or re-expression. How well does your model fit the data?

   b) Noticing that all the x-values are multiples of 10, you decide to graph y versus t, where $t = x \div 10$. Find a model for the new data. How well does your model fit the new data?

   c) Show that the equations from parts (a) and (b) are equivalent.

Exercises 2.5

4. Complete **Table 2.35** as indicated. Use the results to graph log y versus log x and determine an equation relating x and y.

TABLE 2.35.
Data for Exercise 4.

x	2.5	4.0	5.8	7.9	10.2	12.7
log x						
y	3.0	4.0	5.0	6.0	7.0	8.0
log y						

5. For each of the data sets in **Tables 2.36–2.38**, determine whether y is a power, exponential, or logarithmic function of x. Find the model, and justify your answer.

TABLE 2.36.
Data for Exercise 5(a).

a)

x	3	12	27	48	75
y	1	2	3	4	5

TABLE 2.37.
Data for Exercise 5(b).

b)

x	1	2	3	4	5
y	1	1.26	1.44	1.59	1.71

TABLE 2.38.
Data for Exercise 5(c).

c)

x	0.5	0.25	0.125	0.0625	0.03125
y	1	2	3	4	5

6. Graph the following functions together on one piece of log-log graph paper on the domain [0.1, 10]: $y = x^3$, $y = x^2$, $y = x^{1/2}$, $y = x^{-1/2}$, $y = x^{-1}$ and $y = x^{-2}$. Describe your graphs.

7. You place $10,000 in a savings certificate that compounds interest monthly at 0.5% per month. You dutifully record the amount A in the account for a few months, then decide to construct a model to predict the balance over time.

   a) Write an equation that describes the value of the CD account in terms of the number of months since your initial deposit.

   b) Calculate data for the first 5 months, then verify that semi-log re-expression of these data leads to the same equation that you wrote in part (a).

   c) How long it will take your money to quadruple?

Exercises 2.5

8. The ability to predict weather and natural disasters or design pharmaceuticals, among other things, depends partly on implementing sophisticated models on supercomputers. Such models are limited by the speed of the computing device. Supercomputer speed has grown steadily since 1991. **Table 2.39** shows nine recent speed records. Speed is recorded in gigaflops (10^9 floating point operations per second.) Time is measured in months since January, 1991.

   a) Construct a model that predicts how long it takes for computer scientists to record a particular computation speed.

   b) Predict when computer speed will reach 500 gigaflops.

Month	Speed Record
0	5
4	8
6	16
14	24
27	60
33	125
41	145
44	170
48	280

TABLE 2.39. Recent supercomputer speed records.

9. A radioactive dye is injected into a patient's veins to facilitate an x-ray procedure. **Table 2.40** shows the radioactivity levels each minute for a 10-minute period.

   a) Use re-expression to linearize these data, and fit a linear equation to them.

   b) Rewrite your model to relate radioactivity levels to elapsed time.

   c) You actually considered these data using other techniques in Exercise 17 of Lesson 2.1. Do your conclusions from that exercise still seem valid?

   d) When will the radioactivity level drop to 500 counts per minute?

Time (minutes)	Acitvity (counts/min.)
0	10023
1	8174
2	6693
3	5500
4	4489
5	3683
6	3061
7	2479
8	2045
9	1645
10	1326

TABLE 2.40. Radioactivity levels.

10. All living things absorb carbon dioxide (CO_2) that contains Carbon-14 (C_{14}), a radioactive substance. After death, the absorption of carbon dioxide stops and the radioactivity level of C_{14} decays exponentially. Scientists can use measurements of that activity to estimate how long ago the organism died. **Table 2.41** shows hypothetical data for a tree of unknown age.

Exercises 2.5

Year	Acitvity Level (decays per minute per gram)
1950	15.010
1960	14.991
1970	14.973
1975	14.964
1980	14.954
1985	14.945
1990	14.936
1995	14.926

TABLE 2.41.
C_{14} decay data.

a) Knowing that the decay is exponential, use re-expression to find an equation relating the amount of activity to the number of years since 1950.

b) Living trees maintain a C_{14} activity rate of 15.3 decays per minute per gram. Use your model to estimate when the tree died.

c) The accepted decay rate of 15.3 decays per minute per gram is rounded to three significant figures. Repeat part (b) to find the range of estimates implied by this level of precision.

11. An earthquake is generally felt when its Richter scale rating is 4.0 or greater. The data and graph in Table 2.1 and Figure 2.1 in the introduction to this chapter represent the 783 felt earthquakes occurring on the West Coast between 1 January 1990 and 11 July 1996. Fit an equation to the number of quakes (y) versus magnitude (x) for these data. (Decide what value of x you will use for each magnitude range.)

Year	Population	Year	Population
1790	3,894,000	1890	63,000,000
1800	5,085,000	1900	76,094,000
1810	6,808,000	1910	92,407,000
1820	10,037,000	1920	106,461,000
1830	12,786,000	1930	123,077,000
1840	16,988,000	1940	132,122,000
1850	23,054,000	1950	152,271,000
1860	31,184,000	1960	180,671,000
1870	38,156,000	1970	205,052,000
1880	49,371,000	1980	227,225,000
		1990	249,440,000

TABLE 2.42.
U. S. Population data.

12. In Exercise 22 of Lesson 2.1, you examined data on U. S. population. Those data are repeated in **Table 2.42**. It is customary, in the absence of other information, to assume that human populations grow roughly exponentially. Perhaps a better assumption is that populations grow roughly exponentially, but that the rate of growth (the base) changes every now and then to reflect differences in environment and culture.

Under semi-log re-expression, *any* exponential pattern becomes linear. Different bases produce different slopes, but they all produce lines. Carry out a semi-log re-expression on the U. S. population data and look for line *segments*—intervals over which growth had roughly the same rate. If possible, explain any deviations from observed patterns and/or changes in growth rates.

Exercises 2.5

13. **Table 2.43** provides data on Ponderosa Pine trees introduced in Chapter 1. Graph log y versus log x to determine a power model predicting the usable volume of wood from the diameter of the tree.

Diameter (in)	36	28	28	41	19	32	22	38	25	17
Volume (in^3)	276	163	127	423	40	177	73	363	81	23

Diameter (in)	31	20	25	19	39	33	17	37	23	39
Volume (in^3)	203	46	124	30	333	269	32	295	83	382

TABLE 2.43.
Volume and diameter data for Ponderosa Pines.

14. **Investigation:** This exercise is a follow–up to Exercise 12 of Lesson 2.4, examining slopes of secants to the graph of $y = \ln x$. Recall that as the right endpoint moved along the curve closer to the left endpoint, two things happened. First, the line segment became shorter and changed direction until it matched the direction of the curve at the left endpoint, approximating a tangent line to the function at that left endpoint. Second, the sequence of slopes of the secant lines converged to a particular value approximating the rate of change for the *function* at that left endpoint.

a) To review your work in Lesson 2.4, make a copy of **Table 2.44**, and fill in the missing values. What is the approximate rate of change for the function $y = \ln x$ at the point where $x = 3$?

x_1	$y_1 = \ln x_1$	x_2	$y_2 = \ln x_2$	$\Delta y = (y_2 - y_1)$	$\Delta x = (x_2 - x_1)$	$\frac{\Delta y}{\Delta x}$
3	1.098612	5				
3	1.098612	4				
3	1.098612	3.5				
3	1.098612	3.1				
3	1.098612	3.01				
3	1.098612	3.001				

TABLE 2.44.
The rate of change of $y = \ln(x)$ at $x = 3$.

Exercises 2.5

b) In Lesson 2.4, you completed analysis similar to that in part (a), but for a variety of x-values. Use your data from that exercise, or repeat the process for a variety of different left endpoints. Record the selected x-values (left endpoints) and the corresponding rate of change values in a table like **Table 2.45** (the results from Exercise 12 are included):

Left Endpoint x–coordinate (x)	2						
"Rate of change" value (y')	0.5						

TABLE 2.45. Rates of change for $y = \ln(x)$ for various x-values.

c) Use the values in your table and techniques from this lesson to develop a model that relates the rate of change values (for the function $y = \ln x$) to the x–coordinates.

d) Repeat the work done in parts (a) and (b) with other functions of the form $y = \log_b(x)$, one function at a time. Summarize your work in another table, keeping track of the base for the logarithmic function and the model relating its rate of change to its x-values.

e) Describe the common features of the various rate of change models for logarithmic functions.

f) Develop a relationship between the bases for the various logarithmic functions and the constants of proportionality in the corresponding rate of change models. (Hint: Plot proportionality constants versus base and fit a model.)

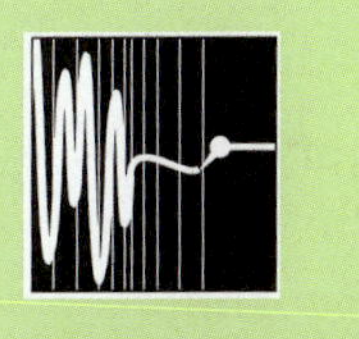

LESSON 2.6

Composition and Inverses of Functions

Exponential and logarithmic functions are related in a special way, due to how the logarithmic function is defined: $y = \log_b x$ means $b^y = x$. In this lesson, this relationship is examined more formally and is extended to other function pairs.

Activity 2.8

1. a) Complete a copy of **Table 2.46**, evaluating the first equation for each value of x provided and recording those y_1-values in the second row. Then evaluate the second equation using row 2 as inputs, and record the new answers in row 3. Find an equation that describes the values in row 3 in terms of the original values for x.

x	0	1	2	3	4	5
$y_1 = 2x + 1$						
$y_2 = 3y_1 - 7$						

TABLE 2.46.
For Item 1(a).

b) Now perform a similar combination using exponential and logarithmic functions. Use statistical lists or a spreadsheet to show the results of each step of the calculation in evaluating $y = \ln(e^x)$ for $x = \{0, 1, 2, 3, 4, 5\}$. Record the values in a table like **Table 2.47**. How are x and y related?

x	e^x	$\ln(e^x)$
0		
1		
2		
3		
4		
5		

TABLE 2.47.
For Item 1(b).

c) In completing part (b), you did the exponentiation first and the logarithm second. Repeat part (b), this time reversing the order of operations so that $y = e^{\ln(x)}$. Again, describe the relationship between x and y.

x	y
0	
0.5	
1	
2	
3	
4	

TABLE 2.48.
For Item 2(a).

2. a) Complete **Table 2.48** using the equation $y = e^x$, and make a scatter plot of the values.

   b) Form new ordered pairs by reversing the two columns, and graph the results.

   c) Compare your graphs from parts (a) and (b).

   d) If your calculator was already set up to display the scatter plot for part (a), explain a quick way to display the scatter plot for part (b).

   e) One way to think of interchanging the two variables is to imagine folding the y-axis down until it becomes the x-axis, and vice versa. Where would the crease for such a fold be?

   Rather than using just a scatter plot, graphing an equation gives more information.

   f) Interchange the variables in the original equation in 2(a). Then solve that new equation for y.

   g) Using the Zoom-Square feature on your graphing calculator, graph on a single set of axes both the function $y = e^x$ and the one you found in part (f). Describe any symmetries you observe.

3. Consider the following cubic function: $y = 2(x - 3)^3 - 5$.

L_1 (= x)	L_2	L_3	L_4	L_5 (= y)
1				
2				
3				
4				
5				
6				
7				

TABLE 2.49.
For Item 3(b).

   a) Sketch an arrow diagram showing the operations that must be done (in order) to evaluate the equation for a particular value of x.

   b) Use statistical lists or a spreadsheet to implement your arrow diagram. Record your results in a copy of **Table 2.49**, then graph L_5 versus L_1; that is, graph y versus x.

c) For each pair of adjacent columns, write an equation that describes that step. That is, what equation takes each column into the next column?

d) Use substitutions to rewrite the expressions for L_3, L_4 and L_5 in terms of x only. Check that your equation for L_5 matches the original equation.

In Item 2 you built a new function by reversing x and y in a given function. Item 4 repeats that process for the cubic function of Item 3.

4. a) Return to your table from Item 3(b) and graph column L_1 versus L_5.

b) Think of reversing each step you identified in Item 3(a). Complete an arrow diagram "reversing" 3(a). This defines a new function.

c) Now write a single equation representing your new function. (For calculator purposes, use x as its input and y as its output.)

d) Graph both the original cubic $y = 2(x - 3)^3 - 5$ and your new function on the same set of axes. Be sure to Zoom-Square. Compare the two graphs; do they have the same symmetry that you observed in Item 2?

COMPOSITION OF FUNCTIONS

Functions are frequently used in conjunction with other functions. For example, Chapter 1 introduced adding functions. **Composition** is an even more useful way of combining functions. To compose two functions, the output of one becomes the input of the other. You used statistical lists or spreadsheets to carry out compositions in Activity 2.8. **Figure 2.29** illustrates the composition of two tool kit functions to form the new function $y = 2x^{3/2}$:

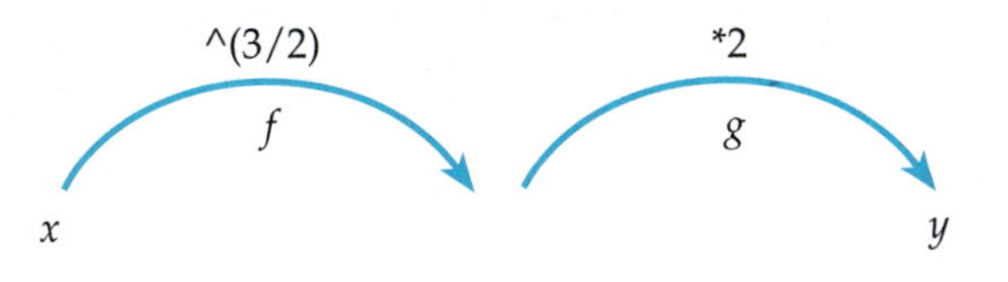

FIGURE 2.29. Arrow diagram of composition creating $2x^{3/2}$.

Defining $f(x) = x^{3/2}$ and $g(x) = 2x$, Figure 2.29 identifies these as the components of $y = 2x^{3/2}$. For example, g doubles its input. But in this composition, the inputs to g are output values of the function f. So f has input x and output $f(x)$, and g has input $f(x)$ and output $g(f(x))$, read as "g of f of x". Substituting the given formulas, $g(f(x)) = g(x^{3/2}) = 2(x^{3/2})$.

This process defines a new function, written $g \circ f$, having input x and output $g(f(x))$. Read this new function name as "g composed with f," "g following f," or just "g of f." So $g \circ f(x) = 2x^{3/2}$. Similarly, $f \circ g$ represents f following g. However, as is clear from arrow diagrams, the order in which a composition is formed is important. In general, $f \circ g$ and $g \circ f$ are different functions.

Example 13

Let $f(x) = x^2$ and $g(x) = 2x - 5$. Write formulas for the compositions $f \circ g$ and $g \circ f$.

Solution:

$f \circ g(x) = f(2x - 5) = (2x - 5)^2$.
On the other hand, $g \circ f(x) = g(x^2) = 2(x^2) - 5 = 2x^2 - 5$.

One can extend compositions of functions to three or more functions.

Example 14

Let $f(x) = x - 3$, $g(x) = x^2$, and $h(x) = 2x + 5$.
Then $h \circ g \circ f\,(x) = h\,(g(f(x))) = h\,(g(x - 3)) = h\,((x - 3)^2) = 2\,(x - 3)^2 + 5$.

The beauty of the approach used in Examples 13 and 14 is that all transformations are just special cases of compositions.

INVERSE FUNCTIONS

Some function pairs have the property that one function cancels the effect of the other when they are composed. In that case, a composition (either $f \circ g$, or $g \circ f$) returns the original input. In other words, $g(f(x)) = f(g(x)) = x$, which means that the functions f and g "undo" each other. Such function pairs are called **inverse functions**. You have long used inverse functions in solving equations. **Figure 2.30** illustrates $g \circ f$ for $f(x) = 2^x$ and $g(x) = \log_2 x$.

x	$f(x)$	$g(f(x))$
−3		
−2		
−1		
0	1	0
1	2	1
2	4	2
3	8	3

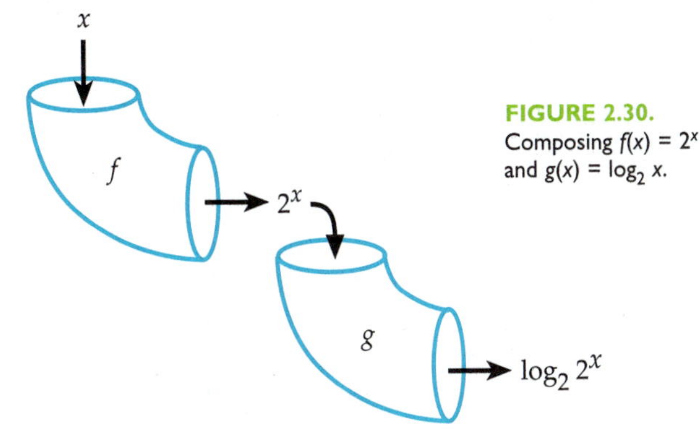

FIGURE 2.30. Composing $f(x) = 2^x$ and $g(x) = \log_2 x$.

For each x, $g(f(x)) = \log_2 2^x = x$. The composition $g \circ f$ returns the original input. Similarly, the composition $f(g(x))$ again returns x. Note, however, that its domain is only $x > 0$. The composition of f and g (in either order) results in the **identity function**, a function that assigns each number to itself, on an appropriate domain. Thus f is the inverse function of g and vice versa. The symbol for the inverse of f under composition is f^{-1}, read "f inverse". Note that the −1 in f^{-1} is not an exponent: f^{-1} does not mean $\frac{1}{f}$.

FINDING INVERSES

Throughout your last several math courses you have used operations to undo other operations when solving equations. Most recently this inverse relation allowed algebraic solution of exponential equations. For example, if $2^x = 3$, then $x \ln 2 = \ln 3$, and $x = \frac{\ln 3}{\ln 2} \approx 1.47$.

In general, the functions $f(x) = b^x$, $b > 1$, and $g(x) = \log_b x$ are inverses. The compositions $f \circ g$ and $g \circ f$ have already been applied in the chapter before any mention of the word composition. **Table 2.50** summarizes key identities.

The Inverse Properties for Base b Logarithms

$f \circ g$: $b^{\log_b x} = x, b > 0, b \neq 1, x > 0$

$g \circ f$: $\log_b b^x = x, b > 0, b \neq 0$

The Inverse Properties for Natural Logarithms

$f \circ g$: $e^{\ln x} = x, x > 0$

$g \circ f$: $\ln e^x = x$

TABLE 2.50. Inverse properties for logarithms and exponentials.

In general, inverse functions are powerful problem-solving tools because they allow you to "undo" functions, simplify expressions, and solve equations. The process used to determine an inverse depends on the representation of the function.

Energy Used	Cost
100kwh	\$20
200kwh	\$40
300kwh	\$60
400kwh	\$75
500kwh	\$90

Cost	Energy Used
\$20	100kwh
\$40	200kwh
\$60	300kwh
\$75	400kwh
\$90	500kwh

TABLE 2.51.
Table representation of a function and its inverse.

Tables

For tables, find the inverse by interchanging columns. That is, reverse the roles of input and output values. **Table 2.51** shows an example.

Electric bills are computed by a function that assigns a cost to energy used. The inverse function determines the energy used from a given cost:

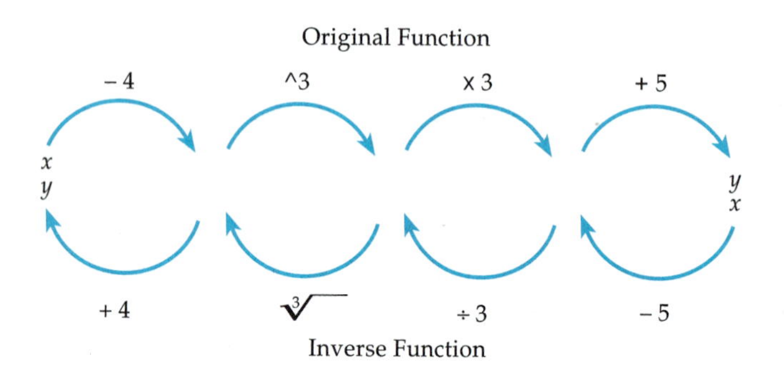

FIGURE 2.31.
Arrow diagram of a function and its inverse.

Arrow Diagrams

A composition defining the function $y = 3(x - 4)^3 + 5$ is illustrated in the top half of **Figure 2.31**. Find the inverse by reversing the entire sequence, doing the opposite steps in the opposite order, as shown in the lower portion of Figure 2.33. In the case illustrated, the inverse function is given by $\sqrt[3]{\frac{x-5}{3}} + 4$.

Graphs

A graph of $y = f(x)$ represents taking an input and changing it into an output value. In **Figure 2.32**, point A, with coordinates (a, b), represents taking input a and returning output b. By reversing the arrows, the same graph can be used to evaluate the inverse function, as represented by point B, with coordinates (c, d). Thus, $f^{-1}(d) = c$.

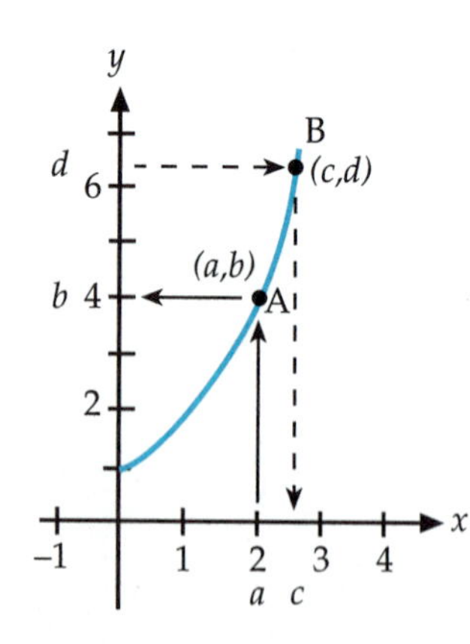

FIGURE 2.32.
Using the graph of a function to evaluate it and its inverse.

Although you can use the graph of f to find values of f^{-1}, this is not the same as graphing f^{-1}. Remember, by convention the horizontal axis represents input values and the vertical axis represents output values.

If (a, b) is on the graph of f, then (b, a) must be on the graph of f^{-1} and vice versa. Two graphs that have the property that point (a, b) is on one graph if and only if point (b, a) is on the second graph are said to be **symmetrical with respect to the line $y = x$.** Thus, to graph the inverse of a function, reflect the graph of the original function in the graph of $y = x$. **Figure 2.33** illustrates this symmetry for the graphs of $y = \log_2 x$ and $y = 2^x$.

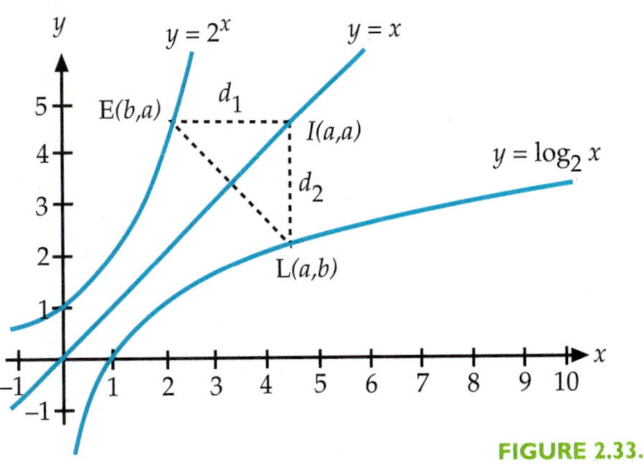

FIGURE 2.33. Symmetry between inverse functions.

Equations

To find the equation for the inverse of a function, f, expressed as an algebraic equation, reverse the roles of the variables. Convention writes the output variable alone, so solve the equation of f for its input variable (which is the output variable of the inverse function).

EXAMPLE 15

1. Find an equation for the inverse of $y = 2(x - 4)^{3/2} + 3$, where $x \geq 4$.

SOLUTION 1:

Subtract 3 from both sides: $y - 3 = 2(x - 4)^{3/2}$

Divide by 2: $\frac{y-3}{2} = (x - 4)^{3/2}$

Raise to the 2/3 power: $\sqrt[3]{\left(\frac{y-3}{2}\right)^2} = x - 4$

Add 4: $\sqrt[3]{\left(\frac{y-3}{2}\right)^2} + 4 = x$

Relabel variables: $\sqrt[3]{\left(\frac{x-3}{2}\right)^2} + 4 = y$

2. Find an equation for the inverse of $y = 3(2^x)$.

SOLUTION 2:

Apply log to each side: $\log y = \log(3(2^x))$

Apply properties of logs: $\log y = \log 3 + x\log 2$

Subtract log3: $\log y - \log 3 = x\log 2$

Apply properties of logs: $\log\left(\frac{y}{3}\right) = x\log 2$

Divide by log2: $\dfrac{\log\left(\frac{y}{3}\right)}{\log 2} = x$

Relabel variables: $\dfrac{\log\left(\frac{x}{3}\right)}{\log 2} = y$

A note about notation. In dealing with "mathematical" functions, it is customary to use x as the input variable and y as the output variable, both for the original function and its inverse. Thus, in Example 15, the final step is interchanging the *labels* for the variables.

For "applied" functions, though, it is best to use variable names that indicate their meanings. Thus, in the power-cost example (see Table 2.49), the original function might use P as input and C as output, with the inverse function having input C and output P. In other words, do *not* "swap letters" in finding inverses in applications.

Function Check

Recall the definition of a function and the corresponding Vertical Line Test for checking whether something really *is* a function.

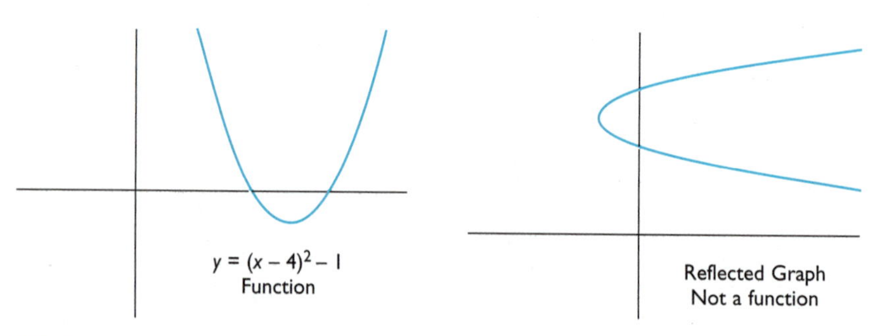

FIGURE 2.34.
Graph of a function and its reflection about the line $y = x$.

Given the graph of any function, you can reflect it about the line $y = x$ to get a graph of its inverse. However, as you can see from **Figure 2.34**, the reflected graph does not always represent a function.

Similarly, given a data table for a function, you can interchange the columns of data to produce a table representing its inverse. However, as you can see from **Table 2.52**, the resulting table might not describe a function.

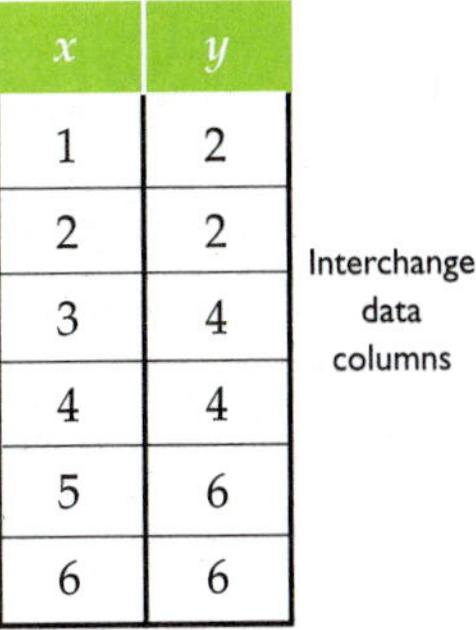

x	y
1	2
2	2
3	4
4	4
5	6
6	6

Interchange data columns

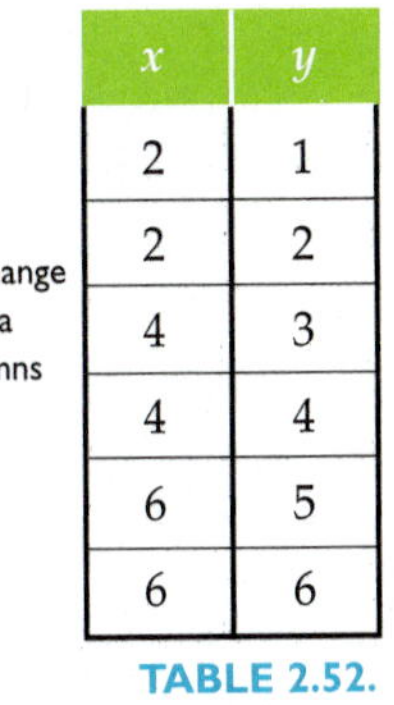

x	y
2	1
2	2
4	3
4	4
6	5
6	6

TABLE 2.52.
A table and its inverse.

In order to be a function, the inverse must take each input to exactly one output. But inputs for the inverse are outputs from the original function. Thus, in order that the inverse be a function, the original function must have only one input for a given output. That is, the original function cannot return the same value for different x-values. A function whose inverse is also a function is called **one-to-one**.

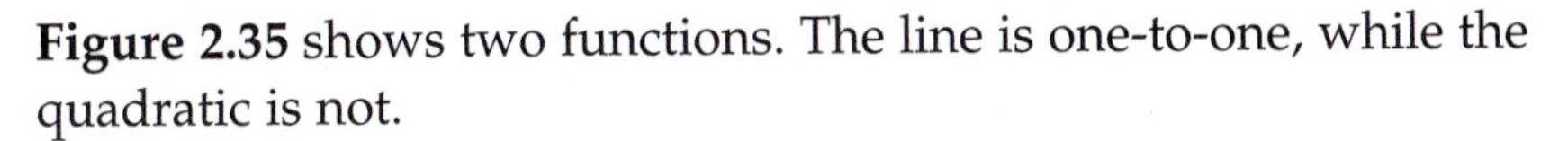

Figure 2.35 shows two functions. The line is one-to-one, while the quadratic is not.

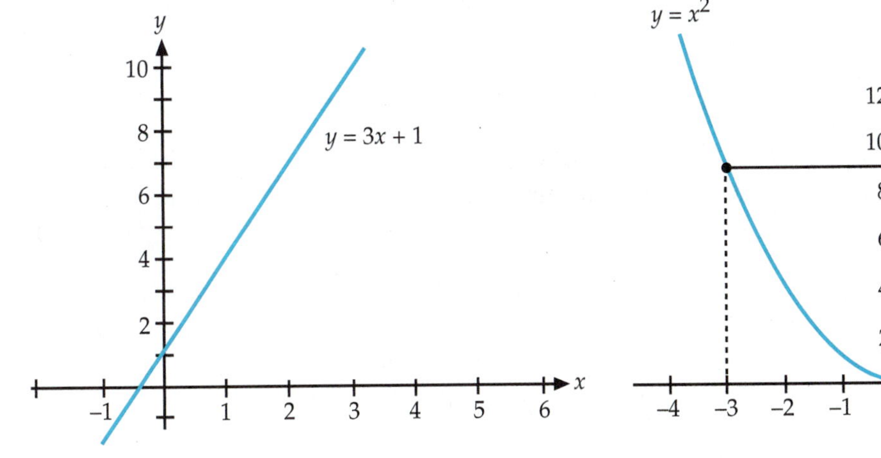

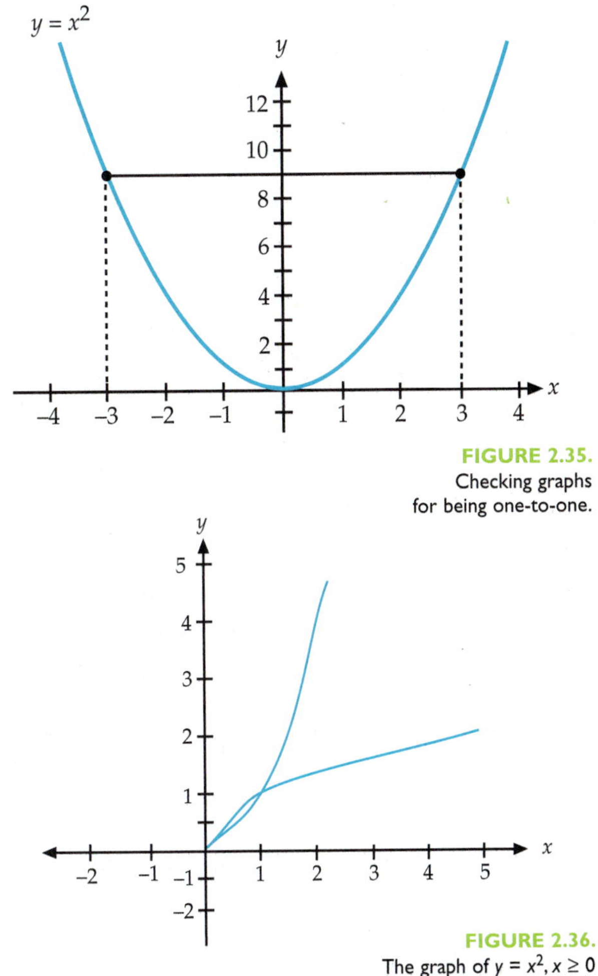

FIGURE 2.35.
Checking graphs for being one-to-one.

Notice that $y = x^2$ returns the number 9 for both $x = 3$ and $x = -3$. In reversing the squaring process, it would be ambiguous to start with 9. However, if you *know* a function is one-to-one (or restrict its domain to force it to be one-to-one), then the inverse is a function. See **Figure 2.36**.

In checking whether a function is one-to-one it should be no surprise that if the graph intersects a horizontal line more than once, the function is not one-to-one. This Horizontal Line Test follows directly from the Vertical Line Test for functions.

FIGURE 2.36.
The graph of $y = x^2, x \geq 0$ and the graph of $y = \sqrt{x}$

Exercises 2.6

1. a) Simplify $\ln(e^x)$, $e^{\ln x}$, $10^{\log(x)}$, and $\log(10^x)$. Identify the domain of each expression.

   b) Solve each of the equations $x = e^y$ and $x = \ln y$ for y. Identify your algebraic steps and explain.

2. Determine whether each of the following sets of ordered pairs defines a function. If not, explain. If yes, is it one-to-one? If so, list the ordered pairs for the inverse function. If not, explain.

   a) {(1, 3), (2, 4), (3, 5), (4, 3)}

   b) {(3, 5), (4, 6), (6, 8), (8, 10)}

   c) {(1, 1), (1, 3), (3, 5), (4, 6)}

   d) {(35, 48), (48, 35), (20, 20), (15, 15)}

3. For each of the following families of tool kit functions, indicate any restriction on its domain necessary to make it one-to-one.

   a) $y = mx + b$

   b) $y = |x - 3|$

   c) $y = a \cdot x^b$, $a > 0$, where b is a positive integer.

   d) $y = a \cdot b^x$, $a > 0$, $b > 0$, $b \neq 1$

   e) $y = \log_b x$, b > 1

4. Write equations for $g(f(x))$ and $f(g(x))$ in terms of x for each of the following. (An arrow diagram may be helpful if you have difficulty.)

   a) $f(x) = x + 8$, $g(x) = 3x$

   b) $f(x) = x + 2$, $g(x) = x^2$

   c) $f(x) = e^{2x}$, $g(x) = \frac{1}{2}\ln x$

Exercises 2.6

5. An *iterative function* is a special kind of composition that applies the same function rule repeatedly at each step of the process. Using the function $f(x) = x^2 - 2$, write an equation in terms of x for each of the compositions indicated in parts (a)–(c). Evaluate each equation for $x = -1$. Explain why the results are easier to understand using the idea of composition than by using the equations you found.

   a) $f(x)$

   b) $f(f(x))$

   c) $f(f(f(x)))$

   d) Suppose $g(x) = 1.005x$. Write equations for the iteration of g, and interpret your results.

6. Write each function h as a composition of two (or more) functions. Be sure to define the component functions and their order clearly.

   a) $h(x) = \sqrt[3]{x^2 - 4}$

   b) $h(x) = (x + 4)^2 + 2(x + 4)$

   c) $M(x) = \log \frac{x}{x_0}$. (Converts earthquake intensity x into a Richter magnitude M)

   d) $M(x) = 63360x$ (Converts miles into inches)

7. A pebble is dropped into a calm pond, causing ripples in the form of concentric circles. The radius (in feet) of the outer ripple is described by the equation $r = 0.5t$, where t is time (measured in seconds after the pebble strikes the water). The area of the circle is given by the function $A(r) = \pi r^2$. Write and interpret an equation for $A \circ r(t)$.

8. Find the inverses. Graph f and f^{-1} on the same set of axes. (Use "square" axis scales.)

   a) $f(x) = 3x + 5$

   b) $f(x) = x^2, x \geq 0$

   c) $f(x) = x^2, x \leq 0$

Exercises 2.6

d) $f(x) = 0.4x^{\frac{3}{5}} + 2, x \geq 0$

e) $f(x) = 1/x, x \neq 0$

f) $f(x) = -x$

9. Explain the Horizontal Line Test in your own words. Be sure to say both what it is used for and how it is used.

10. a) Graph the equation $y = 2(x - 4)^2 - 1$. How can you tell from your graph that the function is not one-to-one.

 b) Make an arrow diagram showing the operations that must be done (in order) to evaluate the equation for a particular x.

 c) If possible, reverse each step in your arrow diagram from (b). Then write a single equation representing your new function.

 d) Did you encounter any obstacles in (c)? Explain.

 e) Restrict the domain of the function $y = 2(x - 4)^2 - 1$ so that the graph of the restricted function is one-to-one. Now write an equation describing its inverse.

11. This lesson claims that the graphs of a function and its inverse are symmetric with respect to the line $y = x$. Formally, two graphs are symmetric with respect to a given line if and only if that line serves as the perpendicular bisector of all line segments joining corresponding points in the two graphs. Show that inverses really are symmetric with respect to $y = x$. (Refer back to Figure 2.33 if necessary.)

12. In the context of coding, a letter is assigned a position number p and then the number is transformed in some way, like multiplying or adding. Suppose on a particular day an encoder at Station A uses the rule $C(x) = 2x + 14$ to code a message. A decoder at Station B thought that the decoding rule for that day was $D(x) = x/2 - 7$.

 a) Decoding rule D does not have the opposite steps in the opposite order from coding rule C. By examining $(D \circ C)(x)$, show that $D(x)$ will decode a message written using the rule $C(x)$ anyway. Explain the kind of relationship necessary to have an effective coding and decoding process.

b) For the coding process, x represents the position number of a letter of the alphabet. For the decoding process, x is some number that is a code value. How is it possible to take the composition of C and D, when the variable x means two different things?

Exercises 2.6

c) If you know that $D \circ C$ forms an identity function, is it necessary to examine $C \circ D$ as well? Or is knowing that one composition forms an identity function sufficient to guarantee that the two functions are inverses of each other?

13. The Fahrenheit scale assigns 32° to the freezing point of water and 212° to the boiling point, while the Celsius scale assigns 0° and 100° respectively (see **Table 2.53**).

Freezing Point	32°F	0°C
Boiling Point	212°F	100°C

TABLE 2.53. Key values for temperature scales.

a) Find a function f that converts Celsius to Fahrenheit.

b) Find a function g that converts Fahrenheit to Celsius.

c) Are the two functions f and g from parts (a) and (b) inverses of one another? Justify.

d) Graph the functions f and g. Are the graphs symmetrical about the line $y = x$?

e) Is there a temperature at which the numerical values measured using the Fahrenheit and Celsius scales are equal?

14. In Lesson 2.5 Exercise 7, you wrote an equation modeling the growth of a certificate of deposit having an initial deposit of $10,000 with interest compounded monthly at 0.5% per month.

a) That model was exponential, predicting the balance as a function of time. Use that work to write a model predicting the number of months as a function of the balance.

b) Use your equation to verify your "time to quadruple" from Lesson 2.5.

c) When will the certificate be worth $15,000?

Exercises 2.6

PEOPLE AND MATH

Although they are not doctors, **Physicians' Assistants** can now also prescribe medications for patients. They and many other professionals, including the pharmacist who fills the prescription and the nurses who administer medications in hospitals or other settings must be vigilant about drug dosage (and possible interaction with other drugs). Safe and effective dosage depends on knowledge of both the drug and the patient. For example, the strength of drug dosage must vary according to several factors, including the body weight of the patient. In this case, understanding the math involved in keeping an effective level of a drug in the bloodstream can be the difference between a cure and a relapse. Depending on the drug and its effect on the body, dosage calculations can also be the difference between life and death.

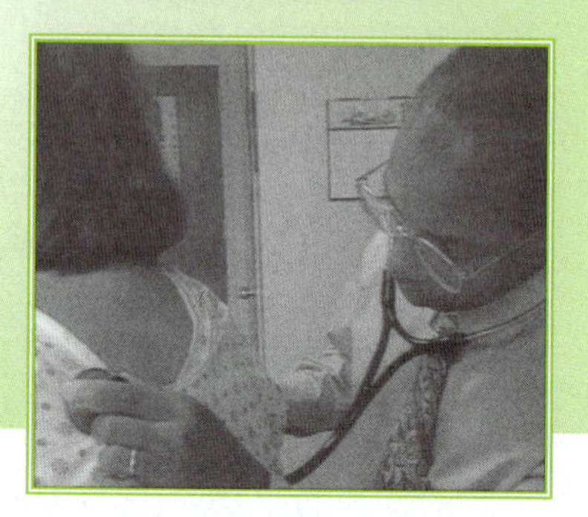

15. Accidentally, a patient has taken an overdose of a drug known to follow the model $y(t) = y(0)e^{-rt}$, where $y(t)$ is the amount of the drug remaining in the bloodstream t hours after taking the amount $y(0)$. The constant r is the decay rate, which varies among individual patients and is important to determine as accurately as possible. The attending physician wishes to prescribe treatment, and knows that the drug was taken 3 hours ago, but because the patient is unconscious the physician knows neither the amount taken, $y(0)$, nor the elimination constant r for this particular patient.

    The drug concentration in the blood can be measured directly and is proportional to the amount of drug present in the bloodstream. Therefore, the doctor has two blood samples taken, and finds that $y(3) = 2.27$ mg, and $y(3.5) = 1.52$ mg. In modeling the situation, she also assumes that the drug dissolved almost instantaneously, so $y(0)$ represents the amount taken, which is what needs to be determined.

    a) Use the two observed amounts to determine the amount of drug taken, $y(0)$.

    b) When will the amount of drug in the bloodstream drop below the critical level of 0.5 mg?

Exercises 2.6

16. How is the amount of time it takes for an investment to double affected by the rate of return for the investment? Does the doubling time depend on the initial amount invested? In each case, assume continuous compounding.

17. An engineer is designing an x-ray laboratory. Various materials are being considered for the shielding used to reduce the exposure to harmful radiation. One particular material being considered is known to reduce the radiation by approximately 20% of the radiation level for each millimeter used.

   a) Develop a model to predict the radiation level y present behind the shielding as a function of the thickness x of the material in millimeters, where y_0 is the radiation level at the outside face of the shielding (see **Figure 2.37**).

   b) How many millimeters of this material are required to reduce the radiation by 50%?

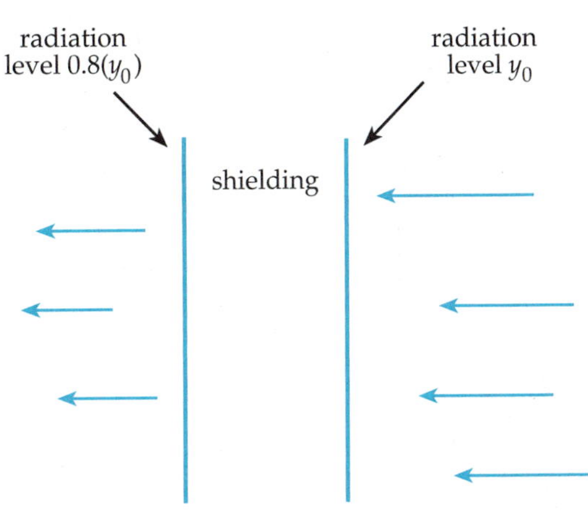

FIGURE 2.37.

18. The engineer in Exercise 17 wants a rule of thumb for estimating the required thickness for each of the materials. This will allow quick estimates for the weights and expenses of construction materials for safe construction.

   The amount of reduction for a small thickness of each material can be measured in the lab. Let r be the measured fraction reduction in intensity per millimeter of the shielding material. Then the model $y(x) = y_0 \cdot e^{-rx}$ approximates the radiation level y behind any thickness x (in millimeters) of the material. Here y_0 is the radiation level at the outside surface of the shielding. For example, if 99.8% of the radiation remains behind 0.1 mm of shielding, then $r = 0.002/0.1 = 0.02$ per mm, and $y(x) = y_0 \cdot e^{-0.02x}$.

   Find a rule of thumb for quickly estimating from measured r values the thickness of shielding material needed to reduce radiation by 50%. Check your rule of thumb by estimating the thickness needed if $r = 0.223$, corresponding to the values in Exercise 17.

Exercises 2.6

Age (yrs)	Diameter (in)
4	2.0
5	2.0
8	2.5
8	5.0
8	7.5
10	5.0
10	8.8
12	12.3
13	8.8
14	6.3
16	11.3
18	11.5
20	13.8
22	14.5
23	11.8
25	16.3
28	15.0
29	11.3
30	15.0
30	17.5
33	20.0
34	16.3
35	17.5
38	12.5
38	17.5
40	18.8
42	18.8

TABLE 2.54.
Ponderosa Pine data.

19. In Chapter 1 you developed a model for predicting the volume of usable wood from the diameter of the tree. (That model was revisited in Lesson 2.5.) One model you examined was the power model $g(x) = 0.003877x^{3.137187}$.

Suppose a timber-harvesting company has collected the data in **Table 2.54** on Ponderosa Pines.

a) Construct a model f that predicts the diameter of a tree from its age.

b) Refer back to earlier volume-and-diameter models, and use composition to write a model (equation) that predicts the volume of usable wood from the age of the tree. How could this model help the timber-harvesting company?

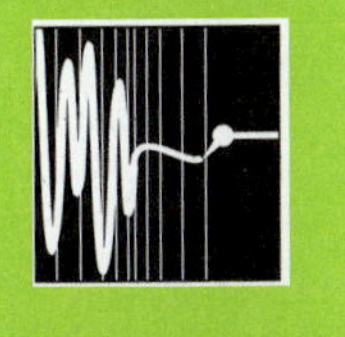

Chapter 2 Review

1. Write a summary of the important mathematical ideas in this chapter.

2. Without the use of your calculator, complete **Table 2.55** for the following functions, where x is restricted to the interval $[-2, 2]$.

Function	Value of function when $x = -2$	Value of function when $x = 2$	Description of function on the interval $[-2, 2]$	Sketch of function on $[-2, 2]$
$i(x) = 2^x$				
$j(x) = 2^{-x}$				

TABLE 2.55.
Describing exponential functions.

3. Without the use of your calculator, complete **Table 2.56** for the following functions, where x is restricted to the interval $(0, 8]$.

Function	Value of function when $x \rightarrow 0$	Value of function when $x = 8$	Description of function on the interval $(0, 8]$	Sketch of function on $(0, 8]$
$i(x) = \log_2 x$				
$j(x) = \ln x$				

TABLE 2.56.
Describing logarithmic functions.

4. Match the graphs in **Figures 2.38–2.45** to the following equations. For each match, identify the features of the graph that permitted its identification.

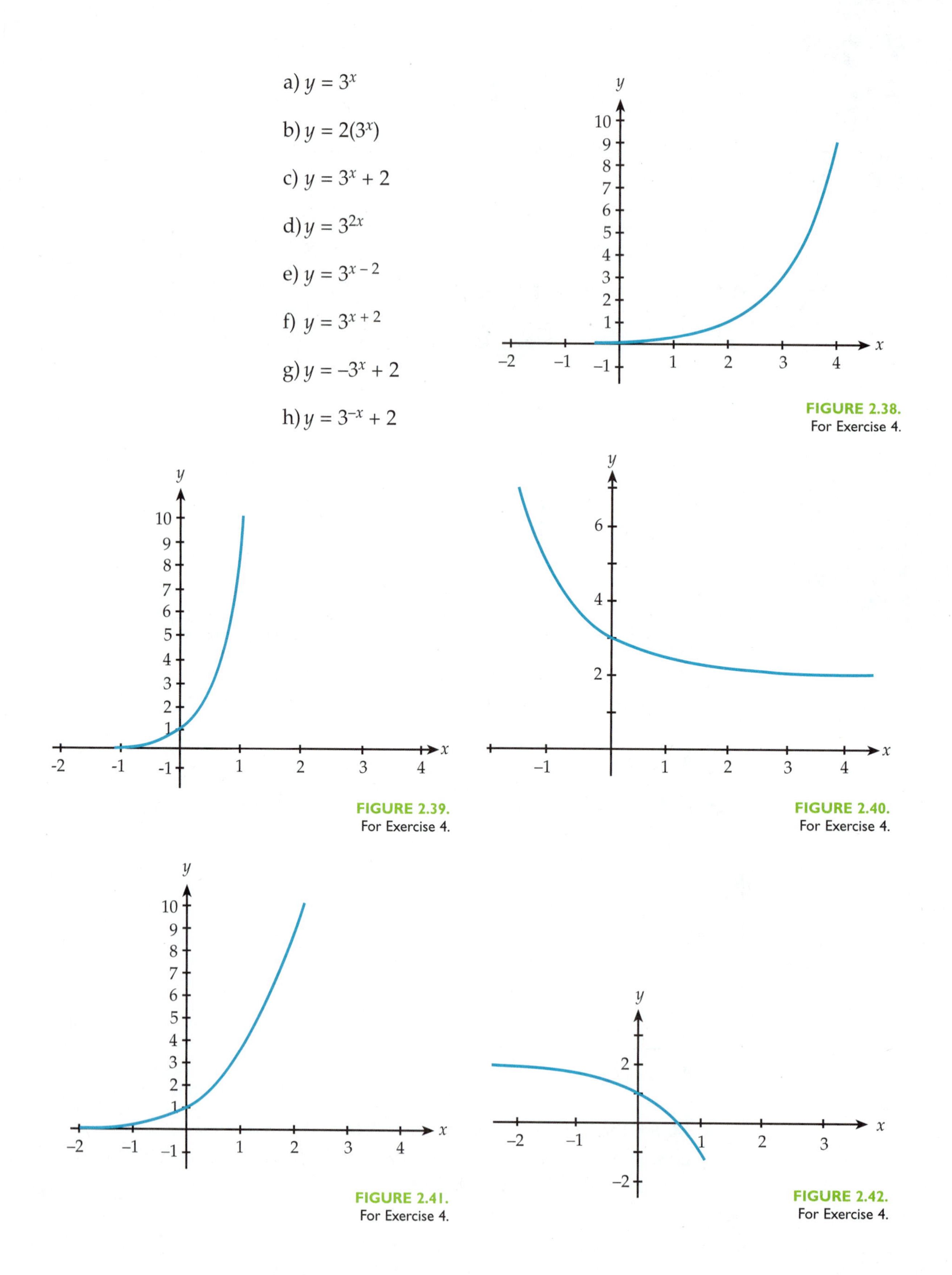

FIGURE 2.38. For Exercise 4.

FIGURE 2.39. For Exercise 4.

FIGURE 2.40. For Exercise 4.

FIGURE 2.41. For Exercise 4.

FIGURE 2.42. For Exercise 4.

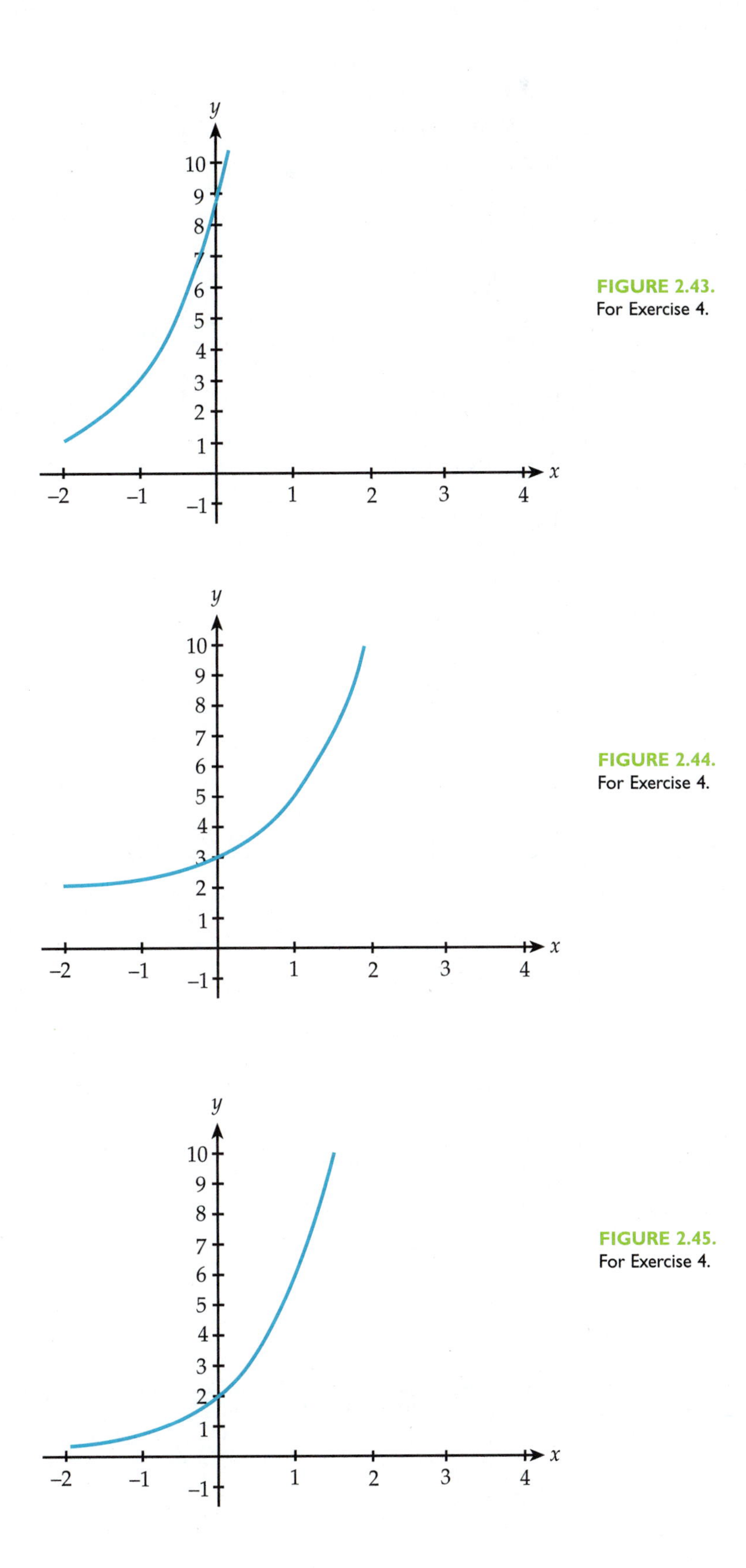

FIGURE 2.43.
For Exercise 4.

FIGURE 2.44.
For Exercise 4.

FIGURE 2.45.
For Exercise 4.

5. Match the graphs in **Figures 2.46–2.52** to the equations given below. For each match, identify the features of the graph that permitted its identification.

a) $y = 2 \cdot \log_3 x$

b) $y = \log_3 x + 2$

c) $y = \log_3 2x$

d) $y = \log_3 (x - 2)$

e) $y = \log_3 (x + 2)$

f) $y = -\log_3 x + 2$

g) $y = \log_3 (-2x)$

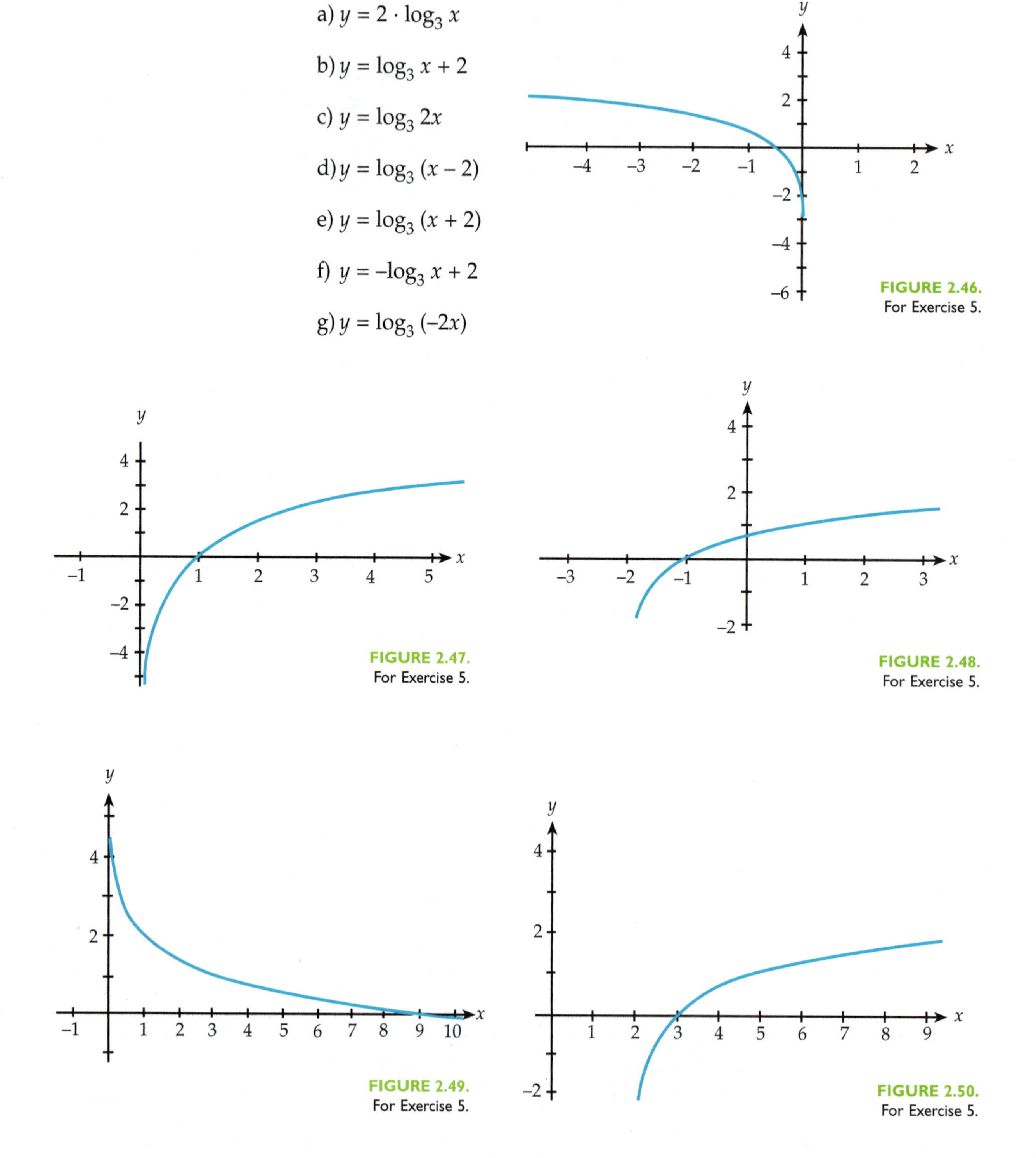

FIGURE 2.46. For Exercise 5.

FIGURE 2.47. For Exercise 5.

FIGURE 2.48. For Exercise 5.

FIGURE 2.49. For Exercise 5.

FIGURE 2.50. For Exercise 5.

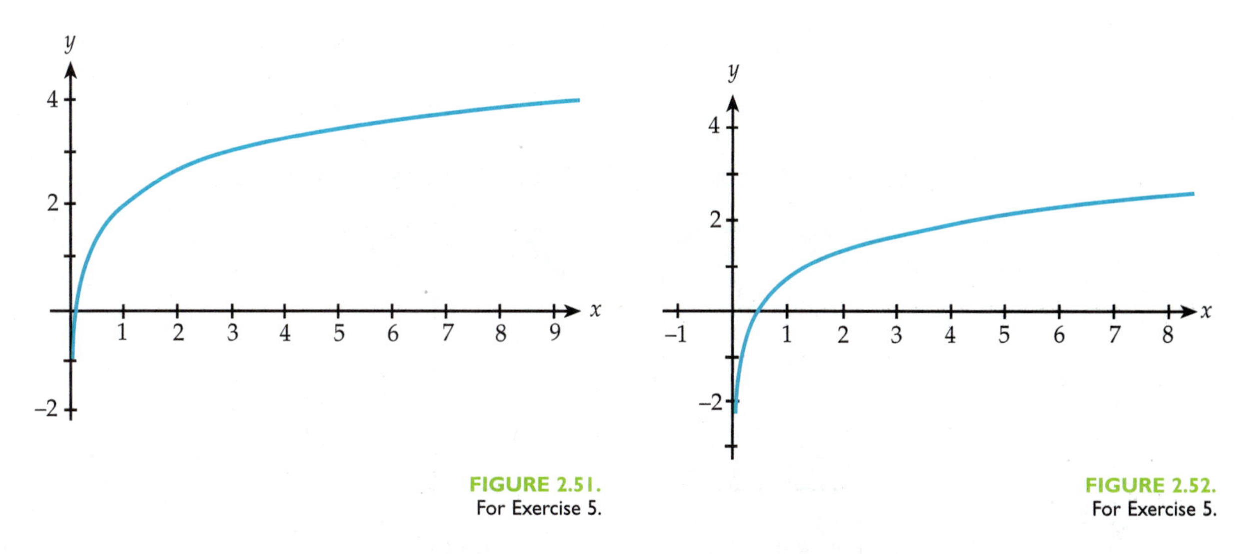

FIGURE 2.51.
For Exercise 5.

FIGURE 2.52.
For Exercise 5.

6. Solve each of the following equations and evaluate your solution as a decimal approximation. Explain your methods.

   a) $x = \log_5 33$

   b) $6 = \log_4 x$

   c) $\ln 20 = \log_x 20$

   d) $3 = \log_x 28$

   e) $12 \cdot e^x = 4.96$

   f) $128 \cdot (x)^{20} = 1300$

   g) $128 \cdot (1.12)^x = 1300$

   h) $\log_x 18 = \log_4 22$

   i) $15 \cdot 2^x = e^{3.2}$

   j) $1.32 \cdot x^{2/3} = 16.35$

7. If $f(x) = 2x^2 - 1$, $g(x) = \ln x$ (for $x > 0$) and $\mathrm{h}(x) = 4^x$, find:

   a) $(h \circ f)(x)$

   b) $(f \circ g)(x)$

   c) $f^{-1}(x)$. (Restrict the domain of f as needed.)

d) $h^{-1}(x)$. (Restrict the domain of h as needed.)

e) $(g \circ g^{-1})(x)$. What is the domain of the composition?

8. For each of **Tables 2.57–2.60**, write y as a function of x.

a)

x	y
1	0
3	1
9	2
27	3
81	4

TABLE 2.57.
For Exercise 8(a).

b)

x	y
0	1
1	3
2	9
3	27
4	81

TABLE 2.58.
For Exercise 8(b).

c)

x	y
1	4
2	16
3	64
4	256
5	1024

TABLE 2.59.
For Exercise 8(c).

d)

x	y	log x	log y	Δ log y/Δ log x
1	0.2	0	–0.699	1.751
2	0.673	0.301	–0.172	1.750
3	1.368	0.477	0.136	1.752
4	2.263	0.602	0.355	1.742
5	3.344	0.699	0.524	

TABLE 2.60.
For Exercise 8(d).

9. Consider the equation $y = \log(e^x)$. Predict and explain what its graph will look like. Then use your calculator to display the graph to verify your prediction.

10. Given the expression: $\log_2(25/8) + \log_4(5/16)$.

a) Re-write the expression in terms of base 2 only.

b) Rewrite both the original expression and your work from part (a) into base 10, and verify by evaluating both using a calculator. What is the value of the original expression?

11. Use your understanding of linear equations and re-expression to derive an equation for the natural logarithmic function $y = A\ln(x) + B$ that goes through the points (x_1, y_1) and (x_2, y_2).

12. Write a function for each of the following situations or descriptions:

a) $y_2/y_1 = 1/y_2 = y_3/1 = y_4/y_3 = 0.5$. See **Table 2.61**.

x-values	y-values
–2	y_1
–1	y_2
0	1
1	y_3
2	y_4
---	---

TABLE 2.61.
For Exercise 12(a).

b) The mystery function is logarithmic ($y = a \log(x) + b$) and passes through the points (10, 3) and (24, 5).

c) An amount of $2500 is deposited in a bank that compounds quarterly at the rate of 1.5% each quarter.

d) An amount of $8000 is deposited in a bank that compounds continuously at the rate of 5% per year.

13. The softest measurable sound has intensity $I_0 = 10^{-12}$ watt/m^2. A local rock concert is measured at 125 dB, while the noise from a motor-driven lawn mower is measured at 102 dB. How many times louder is the music than the lawn mower?

14. Johann Kepler was credited with three laws of planetary motion. One of them involves a model that relates the time it takes an object in orbit to go around a celestial body to its average orbital radius. **Table 2.62** contains actual data for the planets in our solar system.

Planet	Distance (1000000 km)	Time (days)
Mercury	58	88
Venus	108.2	225
Mars	228	687
Jupiter	778	4333
Saturn	1427	10759
Uranus	2870	30685
Neptune	4497	60190
Pluto	5899	90465

TABLE 2.62.
Data for planetary motion.

a) Based only on your intuition, which tool kit function do you think will best describe the relationship between orbital distance (x) and time for one orbit (y)?

b) Analyze the data to construct a model for predicting the time it takes to orbit the sun from the average radius of a planet's orbit.

c) Test your model, knowing that Earth has an average radius of orbit of 93,000,000 miles and orbits the sun every 365 days. (Watch your units.)

15. **Table 2.63** displays data from a cup of coffee left out to cool one morning (room temperature is 68°F).

Time (min)	Temp (°F)
0	178
6	156
12	141
18	132
24	124
30	113
40	107
51	100
140	75
165	72
177	71

TABLE 2.63.
Coffee cooling data.

a) Newton's Law of Cooling states that the difference between room temperature and coffee temperature should decay exponentially. Construct a model that predicts the temperature of the coffee from the time it has been sitting out.

b) Construct a model that uses temperature to predict how long the coffee has been standing out.

Chapter 3

POLYNOMIAL MODELS

CHAPTER INTRODUCTION

At 7:30 on the morning of November 18, 1996, Australian Prime Minister John Howard received a phone call from President Clinton. Mr. Clinton warned Mr. Howard that the Russian Mars 96 space probe, which carried radioactive plutonium, was out of control and predicted to crash to the Earth somewhere in Australia. As a result of the call, Australia began preparing for an emergency. The United States offered help if nuclear contamination should occur.

Although most objects that fall from space are not radioactive, concern exists about the potential damage caused by impacts. In 1998, the United States was tracking 8500 pieces of "space junk" that had accumulated in various orbits around Earth since space exploration began in the 1950s. An additional 70,000–150,000 pieces are too small to track. The larger each of these objects is, the higher the probability that it will survive re-entry into Earth's atmosphere.

The Mars 96 probe overshot Australia and plunged into the Pacific about 900 miles southeast of Easter Island. Predicting when an object will re-enter the atmosphere and where it will impact Earth is difficult. The atmosphere itself varies according to location, solar activity, and other factors, all of which make the atmospheric drag on a re-entering object difficult to determine. Sometimes an object skips across the atmosphere like a stone across water.

The United States and other governments carefully monitor objects in unstable orbits and post lists of those likely to re-enter the atmosphere in the near future. The best available models are used to predict when and where these objects will fall should they survive re-entry. However, as the Russian 96 probe shows, existing models could be improved.

The first step in developing a model to predict when an object in orbit will fall to Earth is to create a model that ignores air resistance and describes the fall of an object acting solely under the influence of gravity.

LESSON 3.1

Modeling Falling Objects

The simplest models for falling objects are polynomials, which are among the most common and useful of mathematical functions. This lesson introduces you to polynomials and some of their uses in mathematical modeling.

Activity 3.1

Your task in this activity is to develop a model that can be used to predict when an object falling from a given height will land. The model should assume that air resistance is negligible. Here are two approaches to consider:

1. A theory-driven model. Research the factors that affect a falling body. Develop a model that describes its fall.

2. A data-driven model. Perform one or more experiments on falling bodies. You might, for example, use calculator- or computer-based lab equipment to gather data, then fit a model to the data.

When you have finished, prepare a report summarizing your results. Be sure to discuss strengths and weaknesses of your model.

Also be sure to discuss assumptions you made in developing your model. How might the assumptions need to change to model the fall of an object re-entering Earth's atmosphere? If you developed a theory-driven model, did you gather any data to confirm your model? If you developed a data-driven model, did you conduct research to explain the relationships in your model?

GRAVITY AND POLYNOMIALS

One of the most important and historic applications of polynomials involves falling objects.

Galileo Galilei

The first serious study of the effects of gravity is credited to Galileo Galilei (1564–1642), who began his experiments in 1595 and continued them for about 10 years. Since precise measurements of objects in free fall were impossible at the time, Galileo designed a clever experiment to measure the effects of gravity by rolling a ball down a groove in an inclined ramp. Galileo's experimental work led to the theoretical advances of Isaac Newton and Albert Einstein.

Galileo estimated the acceleration that gravity imparts to a falling object. He concluded that without air resistance, all objects would fall at the same rate and that the distance a falling body has traveled is proportional to the square of the time that has lapsed since it began falling. That is, if d is the distance an object has fallen, and t is the time it has fallen, then the ratio d/t^2 is constant. (Mathematicians say that d varies directly with t^2.)

Since the height of a falling body is partially determined by the square of the time it has fallen, models that describe the height of falling bodies are usually quadratic functions of the form $h(t) = at^2 + bt + c$. The coefficient a describes the effect of gravity, the coefficient b describes how fast the object was moving initially, and the constant c describes where it began moving.

A quadratic is a type of **polynomial**, a mathematical expression of the form $a_nx^n + a_{n-1}x^{n-1} + a_{n-2}x^{n-2} + ... + a_1x^1 + a_0$. You can think of a polynomial function as one that is built by adding several power functions. Each power function in the sum is called a **term** of the polynomial. It is customary, but not necessary, to write the terms of a polynomial in descending order of the variable's powers. The highest power of the variable is the polynomial's **degree**, and its coefficient is the **leading coefficient**. The term having no variables, a_0, is known as the **constant term**.

OPERATIONS ON POLYNOMIALS

The operations of addition, subtraction, and multiplication of polynomials produce other polynomials. Division of polynomials sometimes produces another polynomial.

EXAMPLE 1

Consider the polynomials $A(x) = 2x^2 + 7x - 15$ and $B(x) = x + 5$. Determine the sum $A(x) + B(x)$, the difference $A(x) - B(x)$, and the product $A(x) \cdot B(x)$.

SOLUTION:

Addition produces another polynomial:

$(2x^2 + 7x - 15) + (x + 5) =$

$2x^2 + 7x + x - 15 + 5 =$

$2x^2 + 8x - 10.$

Subtraction also produces another polynomial:

$(2x^2 + 7x - 15) - (x + 5) =$

$2x^2 + 7x - 15 - x - 5 =$

$2x^2 + 6x - 20.$

(Note that in the case of subtraction, the sign of every term changes if the order of the polynomials is reversed.)

Likewise, multiplication produces another polynomial:

$(2x^2 + 7x - 15)(x + 5) =$

$(2x^2 + 7x - 15)x + (2x^2 + 7x - 15)5 =$

$2x^3 + 7x^2 - 15x + 10x^2 + 35x - 75 =$

$2x^3 + 17x^2 + 20x - 75.$

Division of polynomials will be addressed in more detail in Lesson 3.5. However, division does not always produce another polynomial, though in the case of Example 1 above, it does if polynomial A is divided by polynomial B:

$$\frac{2x^2 + 7x - 15}{x + 5} = \frac{(2x - 3)(x + 5)}{x + 5} = 2x - 3.$$

(Note that in this case, the division does not make sense if $x = -5$ since division by 0 is undefined.)

Exercises 3.1

1. **Table 3.1** gives hypothetical data on the fall of an object.

Time (seconds)	Distance fallen (meters)
1	4.9
2	19.5
3	44.1
4	78.2

TABLE 3.1.

a) Show that the distance the object has traveled is proportional to the square of the time it has fallen. Identify the constant of proportionality.

b) Write a quadratic function that describes the distance fallen as a function of time.

c) How would your answers to (a) and (b) change if the distances were measured in feet? (1 meter is approximately 3.30 feet.)

2. a) Sketch a graph of the quadratic $y = x^2 + x - 6$.

b) Locate the points at which the graph intercepts the axes. Discuss how these intercepts could be found without graphing.

3. The quadratic model $h(t) = -5t^2 + 80$ describes the height in meters of a falling body after t seconds. Find the intercepts of the model's graph and interpret them in this context. Explain the role of the model's constants –5 and 80.

4. **Table 3.2** contains hypothetical data for the height of a projectile fired upward from ground level at an initial velocity of 150 meters per second. Height measurements are to the nearest meter.

Time (seconds)	Height (meters)
1	145
2	280
3	406
4	522
5	628

TABLE 3.2.

a) One way to determine whether a quadratic model is appropriate is to examine differences. For equally-spaced time values, constant second differences suggest a quadratic model. Add two columns to a copy of Table 3.2. In the first of the new columns, write differences in successive values from the height column (i.e., 280 – 145 = 135). In the second of the new columns, write differences in successive values from your first difference column.

b) Use a calculator or a spreadsheet to find a quadratic regression model for these data.

c) Does the quadratic regression model fit the data well? Explain.

d) Interpret each term of the model.

e) Discuss the sensitivity of your model to changes in the data and to changes in the model's coefficients. For example, if one of the measurements is off slightly, what is the effect on the model and its predictions?

5. The function $d(v) = 0.054v^2 + 1.1v$ is sometimes used to model the relationship between an automobile's velocity (measured in miles per hour) and its stopping distance (measured in feet). Stopping distance has two components: reaction distance (the distance the car travels while the driver is reacting) and braking distance (the distance the car travels once the brakes are applied).

   a) Interpret the model's constants. (You may want to evaluate the model for a few reasonable values of a car's velocity.)

   b) It is possible to determine the driver reaction time that this model assumes. Explain.

6. Discuss the relationship between a polynomial's degree and the number of terms.

7. a) Demonstrate that the sum, difference, and product of the polynomials $2x^3 - 7x^2 - 4x$ and $2x + 1$ are polynomials.

   b) Is the quotient of these two polynomials a polynomial? Explain. If you find that the quotient is a polynomial, give an example of two polynomials for which the quotient is not another polynomial.

   c) When an operation on two polynomials produces another polynomial, how is the degree of the new polynomial related to the degrees of its "parents?"

8. a) Is a linear function a polynomial function? Explain.

   b) Is a constant function a polynomial function? Explain.

   c) What can you say about the graph of a polynomial function that does not have a constant term?

9. Drawing conclusions from quadratic models sometimes requires solving quadratic equations. Among the methods you may have used to solve quadratic equations in previous courses are:

   A graphing calculator procedure such as graphing and zooming;

   A spreadsheet procedure such as zooming on a suitable table;

Exercises 3.1

Factoring (when possible);

The quadratic formula;

Completing the square.

The first two methods produce approximate solutions; the last three give exact solutions.

Solve the quadratic equation $6x^2 + 17x = 10$ by at least two different methods. (See Appendices C and D for a review of these methods.)

10. The x-intercepts of a function's graph are also called the **zeros** of the function, or the **roots** or **solutions** of an associated quadratic equation.

    a) How many x-intercepts can the graph of a quadratic function have? Explain.

    b) Suppose a quadratic function's zeros are 2 and 7. Is it possible to find the function? Explain.

11. **Table 3.3** shows the speed of an automobile and the braking distance needed to bring the automobile traveling at that speed to a full stop.

    a) Create a scatter plot of the relationship between speed and braking distance. Use your knowledge of functions and data analysis to find an equation that reasonably fits the data. Record your equation.

    b) Make a residual plot. Describe your residuals by making note of the range (smallest and largest y-coordinates), as well as any pattern you see.

    c) If your residuals have a trend or pattern, find a regression equation that fits the residuals. Use this equation to revise your model from (a). Record the regression equation you fit to the residuals and your new revised model.

    d) Does your new equation provide a better fit to the original data? Check by examining the graph on your scatter plot. Also check the new residuals.

Speed (mph)	Breaking distance (ft)
20	20
25	28
30	40.5
35	52.5
40	72
45	92.5
50	118
55	148.5
60	182
65	220.5
70	266
75	318
80	376

TABLE 3.3. Stopping data for automobiles.

Source: U.S. Bureau of Public Roads

LESSON 3.2

The Merits of Polynomial Models

Polynomial models are quite versatile; so much so, in fact, that one must be careful about using them too often. This lesson demonstrates the versatility of polynomial models.

Activity 3.2

Table 3.4 gives data on a hypothetical object that has re-entered Earth's atmosphere. From these early observations, scientists must predict when the object will hit Earth.

Time (seconds)	Altitude (feet)
0	400,000
10	399,000
20	393,000
30	384,000

TABLE 3.4. Falling body data.

Your objective in this activity is to model the data with several polynomial functions, and to evaluate their usefulness in predicting the time at which the object will reach Earth's surface. Here are a few points to remember as you work:

- Using a data-driven model to predict beyond the range of the data is risky. However, as this case demonstrates, sometimes there is no choice. Such predicaments emphasize the importance of explaining data-driven models.

- Polynomial functions are sensitive to slight changes in the values of the polynomial's coefficients. Maintain coefficients to as many digits of accuracy as possible (but keep the precision of the original data in mind when you use the model to make predictions).

- Models should be evaluated before they are used to make predictions. Check residuals for both size and pattern. When comparing two data-driven models, residuals alone do not determine which model is better, particularly if one model can be explained but not the other.

NASA

1. Find a quadratic regression model for the data in Table 3.4. Use it to predict when the object will reach the ground. Are you comfortable with the prediction?

2. Find a cubic (3rd-degree polynomial) regression model. Compare it to the quadratic model you found in Item 1. Which do you think is the better model? Explain.

3. Add a fifth pair (40, 375000) to Table 3.4. Compare quadratic, cubic, and quartic (4th-degree polynomial) regression models.

4. Summarize what you have learned in this activity about polynomial functions as models.

FITTING A POLYNOMIAL FUNCTION TO DATA

Polynomial regression is a technique that produces a polynomial function of given degree to fit a given set of data. For example, cubic regression produces a third-degree polynomial function for which the sum of the squares of the residuals is smaller than the corresponding sum for any other cubic. Many calculators do linear, quadratic, cubic, and quartic regression.

Fitting a Polynomial Exactly

Polynomial functions have a modeling capability possessed by no other simple function: They are capable of passing through every point in a set of data exactly. (Of course, a polynomial function, like any other function, cannot pass through two distinct points with the same x-coordinate.) If lack of error were the only criterion for modeling data, modelers would need no other functions.

In general, a set of n data pairs can be modeled with no error by a polynomial function of degree no more than $n - 1$. Exact fit polynomials can be found with the aid of a calculator program that implements the following mathematical procedure.

A function $f(x)$ captures data perfectly if every data point (a, b) satisfies the function; that is, if $f(a) = b$. (Such a function is called a **spline** for the data.) By substituting each data pair into a general polynomial function of degree one smaller than the number of data points, a system of equations is obtained. Since the number of equations matches the number of variables in the system, the system can be solved. The solution of the system gives the coefficients of the polynomial.

EXAMPLE 2

Find the equation of a polynomial passing through each of the points (1, 7), (3, 14), (6, 2), and (8, 11). (Note that since this example uses four pairs, a calculator with a cubic regression feature can give the same result automatically. For simplicity, the number of pairs in this example has been kept small.)

SOLUTION:

Since a third-degree polynomial has four coefficients (a, b, c, and d in $f(x) = ax^3 + bx^2 + cx + d$), substituting each of the four pairs into the general third-degree polynomial creates the system (1) of four equations with four unknowns:

$$(1)\quad \begin{aligned} 7 &= a(1)^3 + b(1)^2 + c(1) + d \\ 14 &= a(3)^3 + b(3)^2 + c(3) + d \\ 2 &= a(6)^3 + b(6)^2 + c(6) + d \\ 11 &= a(8)^3 + b(8)^2 + c(8) + d. \end{aligned}$$

Thus, the specific cubic can be found by solving the system:

$$(2)\quad \begin{aligned} 7 &= a + b + c + d \\ 14 &= 27a + 9b + 3c + d \\ 2 &= 216a + 36b + 6c + d \\ 11 &= 512a + 64b + 8c + d. \end{aligned}$$

There are a variety of methods for solving systems of equations. The symbolic method reduces the system to one with one less variable and one less equation, then repeats the process until a simple linear equation in one variable is obtained. If the first equation in (2) is subtracted from each of the others, the system becomes:

$$(3)\quad \begin{aligned} 7 &= 26a + 8b + 2c \\ -5 &= 215a + 35b + 5c \\ 4 &= 511a + 63b + 7c. \end{aligned}$$

System (3), in turn, can be reduced to a system with two equations and two variables by selecting one of the equations and adding a multiple of it to the second, then adding another multiple of the same equation to the third equation. If the proper multiples are chosen, the result is a system of two equations and two variables.

Multiplying the first equation by $\frac{-5}{2}$ and adding the result to the second eliminates the variable c in the second equation. Similarly, multiplying the first equation by $\frac{-7}{2}$ and adding the result to the third eliminates the variable c in the third equation. The system that results is:

$$(4)\quad \begin{aligned} -22.5 &= 150a + 15b \\ -20.5 &= 420a + 35b. \end{aligned}$$

Multiplying the first equation of system (4) by $-\frac{35}{15} = -\frac{7}{3}$ and adding it to the second equation eliminates b and reduces the system to a single linear equation in one variable, which can be solved:

$$32 = 70a$$

$$a = \frac{32}{70}.$$

Substituting this value into one of the two equations in the two-variable system (4) gives the value of b. The values of a and b can then be substituted into one of the three equations in system (3) to determine c, and a final substitution of the values of a, b, and c into one of the original four equations of system (2) determines d.

Needless to say, for a system of any size, the symbolic procedure is difficult to implement since the many multiplications and additions and the frequent occurrence of non-integer values make calculation errors likely.

Most calculator programs that solve systems do so by applying one of two matrix techniques that are discussed in detail in Chapter 7 of this book. For the purposes of this chapter, systems can be solved by using a calculator program.

Two TI-83 programs are provided for this lesson on the CD-ROM that accompanies this book. The program Polyfit finds a polynomial of degree $n - 1$ that captures n data points. The program Syssolv solves a system of n equations in n variables.

Application of the Polyfit program to the example above gives $a \approx 0.4571$, $b \approx -6.0714$, $c \approx 21.8429$, $d \approx -9.2286$. (If rounded at all, these coefficients should be maintained to several decimal places beyond the precision of the original data to ensure that predictions are accurate.) Thus the polynomial function is $y = 0.4571x^3 - 6.0714x^2 + 21.8429x - 9.2286$.

Checking the function against the original data is advisable. For example, evaluating for $x = 3$ gives 13.9992, or approximately 14.

Polynomial Precautions

Since polynomials can provide error-less models, it is tempting to apply them without thought. However, unless you can explain why a particular polynomial model is appropriate, be cautious: Although a polynomial function may capture the data perfectly, it may fail to capture the trend the data represent. **Figure 3.1** shows two ways in which a polynomial function of higher-degree may fail to capture a trend that a linear or quadratic function could capture well.

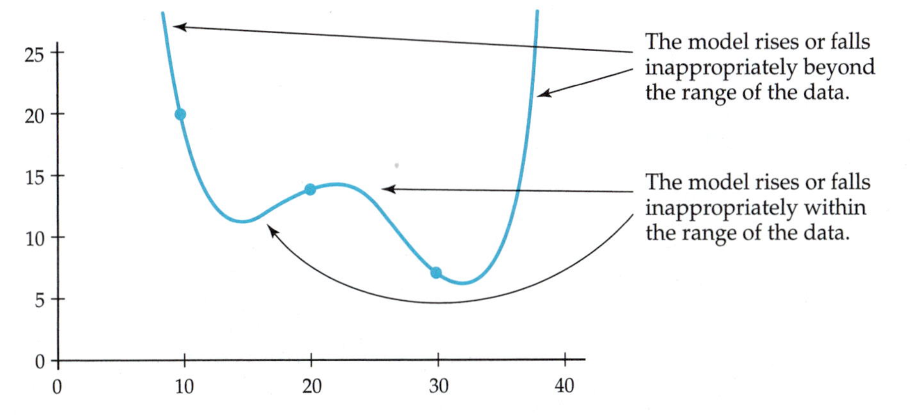

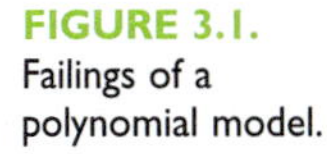

FIGURE 3.1. Failings of a polynomial model.

In situations in which a polynomial function is an appropriate model, care must be taken to avoid round-off error since predictions obtained from a polynomial model are quite sensitive to changes in the polynomial's coefficients. Therefore, it is advisable to maintain as many digits as possible in coefficients. Of course, predictions made with such a model should be rounded to the precision of the data used to generate the model.

Exercises 3.2

1. During the 1991 war in the Persian Gulf, American forces successfully used several high-tech weapons systems. Among them was a portable radar that detected enemy artillery fire so quickly that return fire was often in the air before the enemy projectile landed. Basically, the radar coordinatizes the plane in which the projectile is traveling, samples several points in the projectile's path, fits a model to the points, and uses the model to determine the source of the fire. Since the system is computerized, it all happens very quickly.

   Suppose the radar, which is at the origin of the coordinate system, detects an artillery shell headed in its direction. The coordinates of three points in the projectile's path are (200, 25.4), (252.63, 30.16), and (300, 20.63), as shown in **Figure 3.2**.

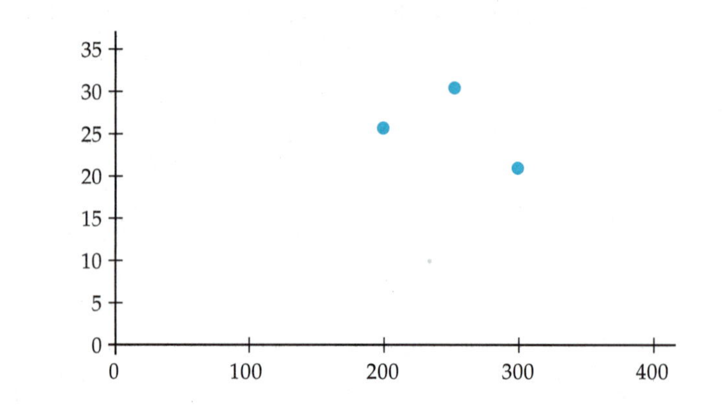

FIGURE 3.2.

   a) Explain why a quadratic function would be a good model in this context.

   b) Write a system of equations for the coefficients of a quadratic model.

   c) Using two different methods, find a quadratic model for these data.

   d) How well does your model fit these data? Explain.

   e) Use your model to determine the coordinates of the projectile's source and its point of impact.

Exercises 3.2

2. Surveyors use compasses to do their work. Unfortunately, compasses point to the magnetic pole, not the true pole, and the magnetic pole is not stationary. That means that surveys conducted many years ago must be rotated to fit modern maps. To determine the proper amount of rotation, it is necessary to know the declination, which is the difference (measured in degrees) between true north and magnetic north, at the time of the survey and at the present (see **Figure 3.3**).

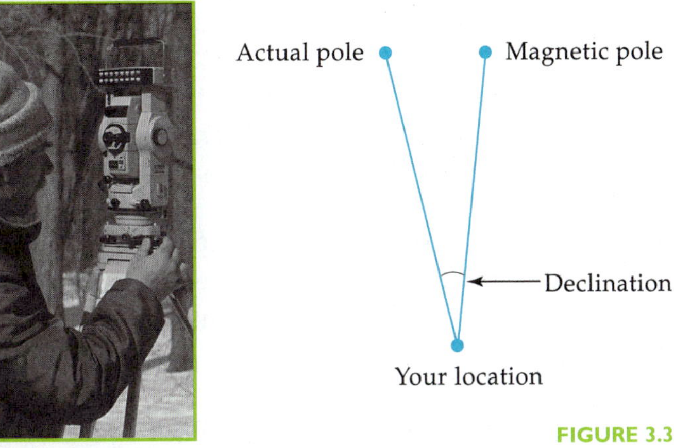

© Susan Van Etten

FIGURE 3.3.

Declination varies with time and location, but the factors that cause changes in declination are not well understood. (Information on declination is available from the National Geophysical Data Center at their web site.) **Table 3.5** gives declination for Charleston, West Virginia at 10-year intervals.

Year	Declination
1800	2° 48' East
1810	2° 56' East
1820	2° 53' East
1830	2° 42' East
1840	2° 21' East
1850	1° 54' East
1860	1° 21' East
1870	0° 44' East
1880	0° 4' East
1890	0° 36' West
1900	1° 11' West
1910	1° 39' West
1920	2° 2' West
1930	2° 30' West
1940	2° 31' West
1950	2° 25' West
1960	2° 41' West
1970	3° 25' West
1980	4° 45' West
1990	5° 52' West
2000	6° 58' West

TABLE 3.5.

a) Analyze the data and use your tool kit of functions to develop a model that you feel would be appropriate to use to estimate the declination at Charleston between 1800 and 2000. (Suggestions: Convert degrees and minutes to decimals and associate negative with either east or west, and positive with the other.)

b) Would you advise the use of your model to predict declinations before 1800 or after 2000? Explain.

3. Consider the data in **Table 3.6**. How well does the polynomial function $y = 0.0018x^4 - 0.205x^3 + 7.78x^2 - 121x + 740$ capture the trend in the data? Explain. If possible, find a better quartic (fourth-degree) polynomial model.

x	y
12	88.4
15	71.7
19	76.5
22	83.7
28	72.3

TABLE 3.6.

Exercises 3.2

x	y
7.4	171
8.1	170
8.9	166
9.8	160
10.1	158
10.4	154

TABLE 3.7.

4. Evaluate the cubic $y = -1.56x^3 + 39.22x^2 - 330.4x + 1101.3$ as a model for the data in **Table 3.7.**

5. One characteristic of polynomial functions that makes them useful modeling tools is that they can capture peaks and valleys in data.

   a) Experiment with various polynomial functions and their graphs. What can you conclude about the relationship between the degree of a polynomial and the number of peaks and valleys it can model? (For example, can a third-degree polynomial's graph have four peaks/valleys?)

   b) You can form a quadratic function that has zeros at 1 and 5 by multiplying the linear factors $(x - 1)$ and $(x - 5)$. How are the zeros related to the peak/valley of the quadratic's graph?

   c) One way to form a polynomial of degree n is to multiply n linear factors (factors of the form $ax - b$). How does the number of linear factors a polynomial can have support the answer you gave in (a)?

6. Use symbolic methods or a calculator program to solve the systems of equations in (a) and (b).

   a) $$\begin{aligned} 2x + 3y + 11 &= 0 \\ 5x - 4y - 30 &= 0 \end{aligned}$$

   b) $$\begin{aligned} x - 3y + 2z + 2 &= 0 \\ 2x + 3y - 4z + 7 &= 0 \\ x - y + 6z - 6 &= 0 \end{aligned}$$

   c) Some systems of equations have no solutions. Use a graph to explain why this system has no solution.

   $$\begin{aligned} 3x - 2y + 5 &= 0 \\ 6x - 4y - 9 &= 0 \end{aligned}$$

7. For each of the polynomial functions (a)–(c), build a table of values for $x = 1$ through 6. Investigate the rate at which y changes by including columns of first differences, second differences, and so forth until the differences become constant. What can you conclude about rates of change of polynomial functions? (If necessary, experiment with a few additional polynomial functions of your own.) How are differences related to the degree of a polynomial?

Exercises 3.2

a) $y = x^2 + 2x - 1$.

b) $y = 2x^3 - 4x^2 - 3$.

c) $y = x^4 - 2x^3 - 5x^2 + 1$.

8. If you drop all but the first term of the polynomial function $y = 2x^4 - 3x^3 + 4x^2 - 7$, you have a power function. Compare the graphs of the polynomial function and the associated power function in larger and larger windows. (Use your calculator's zoom out feature to do this quickly.) Do the same for other polynomial functions and the corresponding power functions obtained by dropping all terms except the one with the highest power. What can you conclude about the end behavior of a polynomial and its related power function?

Background for Exercises 9 and 10. The ability of polynomial functions to pass through every point in a data set has many important applications. One of them is in the field of error-correction codes—codes that correct errors caused by static or other kinds of noise. These codes are used to transmit satellite photos to Earth and to allow you to listen to a scratched compact disc. The codes do their work by representing photos and music as a collection of numbers and adding enough additional data to the message to make recovery of corrupted data possible.

The most common type of polynomial error-correcting code is known as a Reed-Solomon code. In practice, Reed-Solomon codes work with binary numbers—0s and 1s—and often encode several hundred numbers at a time, which requires a polynomial of very high degree.

The error correction in your compact disc player is probably done with a Reed-Solomon code. That's mathematics—music to your ears!

9. Consider a simple example. Suppose you want to send a three-number code composed of the values 5.2, 6.8, and 7.9, in that order. A quadratic can be passed through all three pairs (1, 5.2), (2, 6.8), and (3, 7.9), and quadratic regression gives the model $y = -0.25x^2 + 2.35x + 3.1$. To avoid errors that might occur in transmission, add several new values to the information to be coded by evaluating the polynomial for, say, 4, 5, 6, 7, 8, 9, and 10. (The actual number depends on the probability of error—the higher the probability, the more extra information you need.) Together with the new information, you now have the numbers 5.2, 6.8, 7.9, 8.5, 8.6, 8.2, 7.3, 5.9, 4, 1.6.

Exercises 3.2

All ten of these numbers are transmitted in the order in which they are listed. Suppose that two of them, the third and the sixth, have errors, and the transmission comes through as 5.2, 6.8, 9.9, 8.5, 8.6, 6.2, 7.3, 5.9, 4, 1.6. The decoding device, which knows that a quadratic model was used but not the specific quadratic, detects that these two do not fit a quadratic model. The decoding device determines a quadratic model for the remaining eight, and uses the model to correct the two that are in error.

a) Explain why the two data with errors do not fit a quadratic model.

b) Show how a quadratic model can be found for the other eight.

c) Show how to use the model to correct the errors.

10. a) Show how to use a polynomial function to encode the numbers 2.5, 4.9, 1.6, and 6.8.

    b) Encode four extra values.

11. **Investigation.** Return to the data of Activity 3.2. The data appear to have been rounded to three significant digits. Thus, for example, the value recorded as 400,000 feet may actually have been anywhere between 400,500 feet and 399,500 feet. Similar statements can be made for the other values. Rework Items 1 and 2 of Activity 3.2 using different combinations of possible actual values for the heights, and comment on the implications for the predictions you made in the activity.

LESSON 3.3

The Power of Polynomials

This lesson examines some of the properties of polynomials that are useful in modeling situations. The lesson begins with an activity in which you will create a polynomial model that can help guide the administration of medical tests.

TWO-SAMPLE POOLING

Medical testing for abuse of substances such as steroids is common today. Athletes, for example, are often screened for steroid use. Since steroid testing is expensive, sample pooling is sometimes used to reduce costs. The logic is simple: Pool several samples and make one test. If the test is negative, everyone passes and money is saved. If the test is positive, additional testing is done to determine which sample(s) caused the positive result.

The simplest case involves pooling two samples. Read the following development of a theory-driven model for this situation.

In the long run, money is saved if fewer than 2 tests are required on average. **Figure 3.4** summarizes the various results when samples of two individuals A and B are pooled, with A tested first if more than one test is required. For example, the upper branch represents a situation in which both A and B are steroid users. Therefore, the pooled sample tests positive. That's one test. A's sample is then tested individually. That's the second test. Since A's test is positive, no conclusion can be drawn about B, and a third test is performed. If A's test was negative, a third test is not necessary because the positive pooled sample can only be the fault of B.

A's actual status | B's actual status | Number of tests

A positive — B positive — 3

A positive — B negative — 3

A negative — B positive — 2

A negative — B negative — 1

FIGURE 3.4. Outcomes when two samples are pooled.

If x represents the incidence of steroid use in the population being tested, the probability that A and B will both test positive is x^2. The probability that a specific test is negative and the other positive is $x(1 - x)$. The probability that neither will test positive is $(1 - x)^2$. The average number of tests over the long run is $3x^2 + 3x(1 - x) + 2x(1 - x) + (1 - x)^2 = -x^2 + 3x + 1$.

To determine when pooling two samples saves money, the equation $-x^2 + 3x + 1 = 2$ must be solved. For some solution methods, it is helpful to transform the equation by subtracting 2 to obtain $-x^2 + 3x - 1 = 0$. The advantage of the latter equation is that solving it is identical to finding zeros of the quadratic function $f(x) = -x^2 + 3x - 1$. In fact, solving any equation is equivalent to finding zeros of a related function. As indicated, such a function can be formed by subtracting the right side of the original equation from the left side. Since solving equations is a major part of algebra, much of algebra reduces to finding zeros of functions.

The pooled-sample equation can be solved by completing the square, by applying the quadratic formula, or by applying technology. Of the two solutions, one is meaningless in this context since it exceeds 1. The other is approximately 0.38, which demonstrates that pooling two samples is cost-effective when the incidence of steroid use is under 38%.

Activity 3.3

Your task in this activity is to conduct a similar investigation into the strategy of pooling three samples. If the pooled sample tests positive, test individual samples (as opposed to pooling two of the three). Prepare a report showing how you developed your model, the solution you obtained from it, and the method you used to obtain the solution.

ONE-TERM POLYNOMIALS

The simplest of polynomials have one term and are usually referred to as **monomials**. Since one-term polynomial functions are also *power functions*, it is fairly easy to describe the end behavior of these simple polynomials. Their graphs have one of two basic shapes, depending on whether the power is even or odd (**Figures 3.5** and **3.6**, respectively); the higher the power,

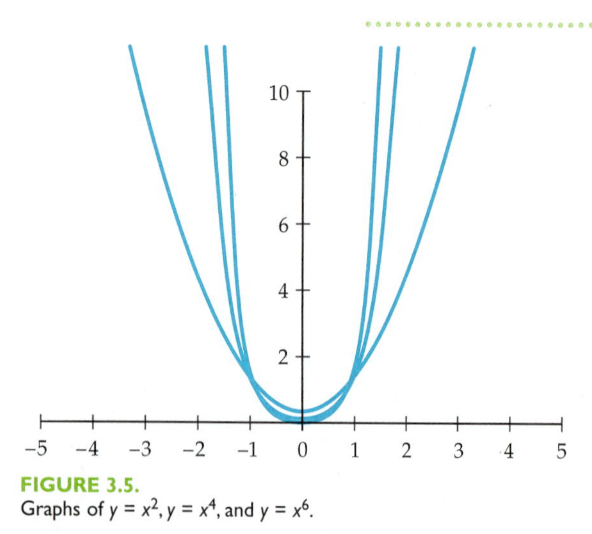

FIGURE 3.5.
Graphs of $y = x^2$, $y = x^4$, and $y = x^6$.

the more rapidly the graph rises or falls. Recall from Chapter 1 that if the coefficient of a power function is negative—$y = -x^4$, for example—the basic shape is reflected in the x-axis.

Another important characteristic of monomials is that their graphs demonstrate symmetry. Monomials of even degree produce graphs that are symmetric with respect to the y-axis; monomials of odd degree produce graphs that are symmetric with respect to the origin. For this reason, any mathematical function whose graph is symmetric with respect to the y-axis is called an **even function**; any function whose graph is symmetric with respect to the origin is called an **odd function**. For example, the absolute value function has a graph symmetric with respect to the y-axis and is therefore considered an even function even though it is not a polynomial.

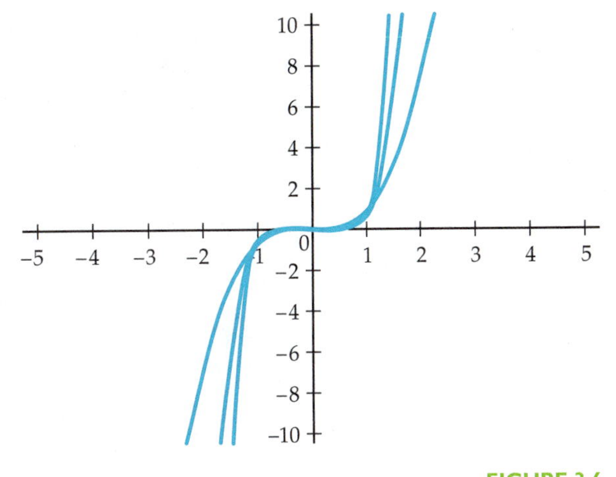

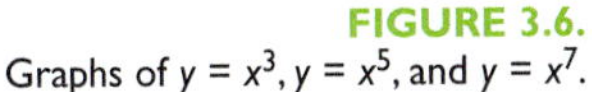
FIGURE 3.6.
Graphs of $y = x^3$, $y = x^5$, and $y = x^7$.

One-term polynomials can also be characterized by their behavior near the origin. With the exception of constant functions, their graphs pass through the origin: the only zero of these functions is 0. (When additional terms are introduced, additional zeros may appear.)

POLYNOMIALS WITH MORE THAN ONE TERM

Exercise 5 of Lesson 3.2 showed that the end behavior of a polynomial is like that of its highest-degree term. Thus, the leading coefficient and the degree of any polynomial completely determine its end behavior. However, multi-term polynomials can exhibit local behavior that differs considerably from that of one-term polynomials: graphs with several x-intercepts or graphs that rise, then fall, then rise again, for example.

You have long known that if a product of numbers is zero then one of the factors in that product must itself be zero. This property extends to polynomials as what is known as the **Factor Theorem**. Put simply, if the number c is a zero of a polynomial, then the expression $(x - c)$ must be a factor of the polynomial. This simple idea provides the basis for understanding much of the local behavior of polynomials.

Finding a Polynomial with Given x-intercepts

An important modeling characteristic of multi-term polynomial functions is the fact that it is possible to create a polynomial function that captures any collection of x-intercepts. Since a function's x-intercept is also one of its zeros, a polynomial with an x-intercept of, say, 3, must have a factor of $(x - 3)$. Therefore, to create a polynomial with a given set of x-intercepts, write a factor with each intercept as a zero and multiply the factors. Note that multiplying such factors by a constant produces a vertical stretch, which does not affect the x-intercepts.

EXAMPLE 3

a) Write an equation of a function that intercepts the x-axis at 3, 6, and 7.

b) Write an equation of a function with zeros 3, 6, and 7, and with y-intercept 10.

SOLUTION:

a) Multiply the factors $(x - 3)$, $(x - 6)$, and $(x - 7)$ to obtain the function $f(x) = (x - 3)(x - 6)(x - 7) = x^3 - 16x^2 + 81x - 126$. Since constant factors do not affect zeros, a function such as $g(x) = 2(x^3 - 16x^2 + 81x - 126) = 2x^3 - 32x^2 + 162x - 252$ has the same x-intercepts, as the graphing calculator graph of $f(x)$ and $g(x)$ in **Figure 3.7** shows.

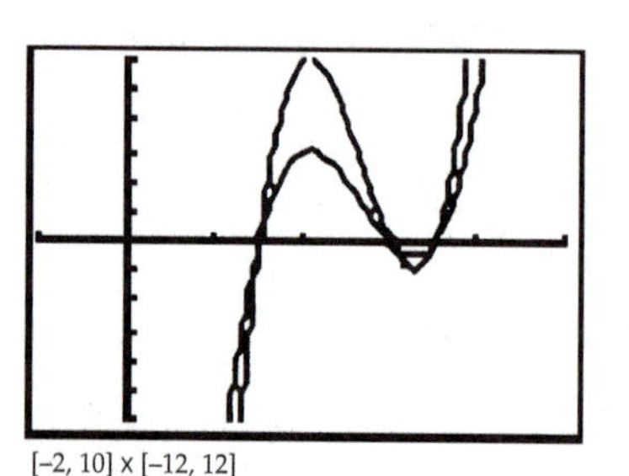

[–2, 10] x [–12, 12]

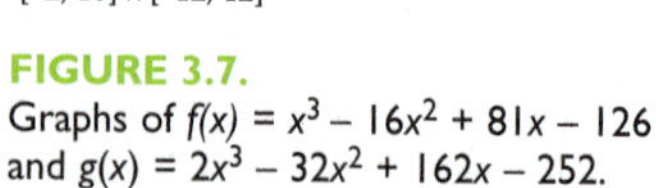

FIGURE 3.7.
Graphs of $f(x) = x^3 - 16x^2 + 81x - 126$ and $g(x) = 2x^3 - 32x^2 + 162x - 252$.

b) Since $f(x) = x^3 - 16x^2 + 81x - 126$ has y-intercept –126, a vertical shrink using a factor of $\frac{-10}{126}$ pulls the intercept to 10 and leaves the zeros unchanged: $h(x) = \frac{-10}{126}(x^3 - 16x^2 + 81x - 126)$.

Turning Points and Zeros

Since polynomial functions have smooth graphs with no breaks or jumps, the graph of the cubic function $f(x)$ in the previous example must turn twice in order to return to the x-axis after crossing it for the first time at 3 (see **Figure 3.8**). To the left of the first intercept and to the right of the last intercept, the first term (x^3) so dominates the others that the graph continues to fall or rise, never again turning toward the x-axis. Note that the number of zeros can be decreased by adding a constant to the function, an action that translates the graph vertically. However, because of the graph's shape, no vertical translation can remove all three zeros—at least one must remain.

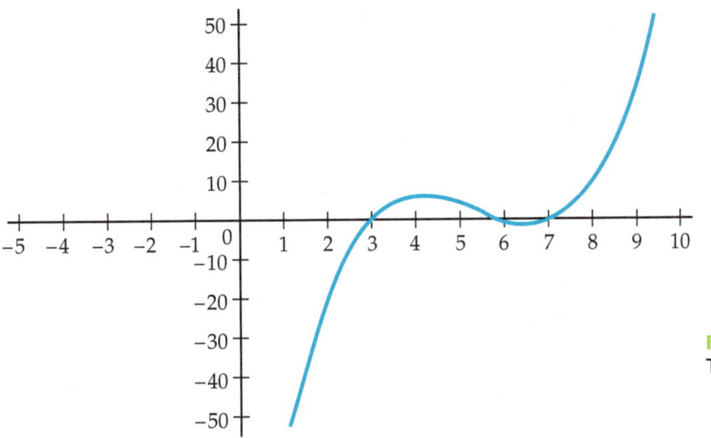

FIGURE 3.8.
The graph of $y = x^3 - 16x^2 + 81x - 126$.

In general, the graph of an nth degree polynomial function can intercept the x-axis no more than n times and can turn no more than $n - 1$ times.

> **TAKE NOTE**
> The graph of an nth-degree polynomial function can intercept the x-axis no more than n times and can turn no more than $n - 1$ times.

Finding the General Form of a Polynomial

Computer and calculator graphing utilities accept polynomial functions in any form, including the **factored form** $(x - 3)(x - 6)(x - 7)$ or the **general form** $x^3 - 16x^2 + 81x - 126$.

The general form $x^3 - 16x^2 + 81x - 126$ in the previous example is obtained by **expanding** the factored form $(x - 3)(x - 6)(x - 7)$; that is, by multiplying three linear (first-degree) factors. In general, expanding a polynomial can be difficult. However, it should be clear that the degree of the polynomial is the sum of the degrees of its factors and that the leading coefficient is the product of the leading coefficients of its factors.

Some calculators and computer software have symbol-manipulation algorithms that can expand the factored form of polynomials (see **Figure 3.9**).

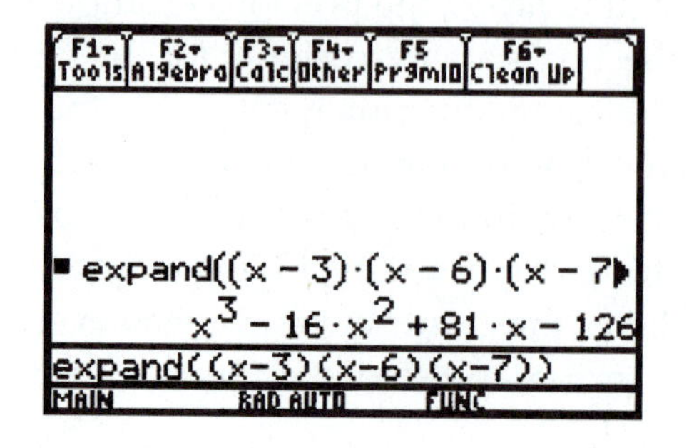

FIGURE 3.9. Multiplication of three linear factors on a calculator.

You can also apply the graphing calculator program Polyfit to the points (3, 0), (6, 0), and (7, 0) to obtain the general form.

The multiplication can be done by hand, but this approach becomes more tedious as the number of factors increase.

EXAMPLE 4

Expand the factored form $(x - 3)(x - 6)(x - 7)$.

SOLUTION:

$(x - 3)(x - 6)(x - 7) =$

$(x - 3)(x^2 - 6x - 7x + 42) =$

$(x - 3)(x^2 - 13x + 42) =$

$x(x^2 - 13x + 42) - 3(x^2 - 13x + 42) =$

$x^3 - 13x^2 + 42x - 3x^2 + 39x - 126 =$

$x^3 - 16x^2 + 81x - 126.$

Finding Zeros

Just as modeling can require the creation of a polynomial function with certain zeros, it can also require finding unknown zeros. Approximations of zeros are acceptable for modeling purposes, but it is sometimes possible to find the exact form of a polynomial's zeros. However, the level of difficulty involved in finding zeros in exact form increases rapidly as the degree of the polynomial increases. Factoring and the quadratic formula are two commonly used techniques; another method is discussed in Lesson 3.5's Exercises.

Example 5

Find all zeros of the cubic polynomial $f(x) = 2x^3 + 2x^2 - 12x$.

Solution:

Since $f(x)$ lacks a constant term, it can be factored initially as $2x(x^2 + x - 6)$ and finally as $2x(x + 3)(x - 2)$. Applying the factor theorem, the cubic's zeros may be found by setting each factor equal to 0:

$2x = 0$; $x + 3 = 0$; $x - 2 = 0$.

Therefore, the zeros are 0, –3, and 2.

Exact values of zeros of some polynomials that do not submit to factoring can be obtained by formula. For example, the general quadratic $y = ax^2 + bx + c$ has two zeros given by the quadratic formula

$$x = \frac{-b \pm \sqrt{b^2 - 4ac}}{2a}.$$

(Exact solutions for quadratics can also be obtained by completing the square.)

EXAMPLE 6

Apply the quadratic formula to $y = 3x^2 - 2x - 4 = 0$, to find both x-intercepts.

SOLUTION:

Use the substitutions $a = 3$, $b = -2$, and $c = -4$, and simplify:

$$x = \frac{-(-2) \pm \sqrt{(-2)^2 - 4(3)(-4)}}{2(3)}$$

$$x = \frac{2 \pm \sqrt{4 + 48}}{6}$$

$$x = \frac{2 \pm \sqrt{52}}{6}$$

Therefore, the zeros are $\frac{2+\sqrt{52}}{6}$ and $\frac{2-\sqrt{52}}{6}$, or about 1.54 and –0.87.

In the sixteenth century, formulas for zeros of general 3rd- and 4th-degree polynomials were derived by Italian mathematicians Nicolo Tartaglia (1500–1557) and Lodovico Ferrari (1522–1565), respectively. The formulas appeared in The *Ars Magna* written by the Italian mathematician and physician Jerome Cardan (Girolamo Cardano, 1501–1576), who was also Ferrari's teacher. The fascinating story surrounding the development of the formulas is one of the most dramatic episodes in the history of science.

The formulas that resulted from the work of Tartaglia, Ferrari, and Cardan are difficult to use and have been rendered obsolete for modeling purposes by modern technology. Two centuries after these formulas were developed, two mathematicians whose lives were cut tragically short—the Norwegian Niels Abel (1802–1829) and the Frenchman Evariste Galois (1811–1832)—proved that zeros of general polynomials of degrees higher than 4 could not be found by means of formulas involving roots and the four basic operations of arithmetic. The work of Abel and Galois revolutionized algebra.

Since zeros are equivalent to x-intercepts, approximations of zeros sufficiently precise for modeling purposes are easily obtained via technology. Most graphing calculators, for example, have a zero-finding algorithm that requires the user to identify a point to the left and another to the right of the zero and then an approximate location of the unknown zero. The calculator then locates the zero to many significant figures (see **Figure 3.10**). Such built-in zero finders are generally more reliable than using a "graph-and-trace" method, which is precise only to the width of screen pixels in the selected window.

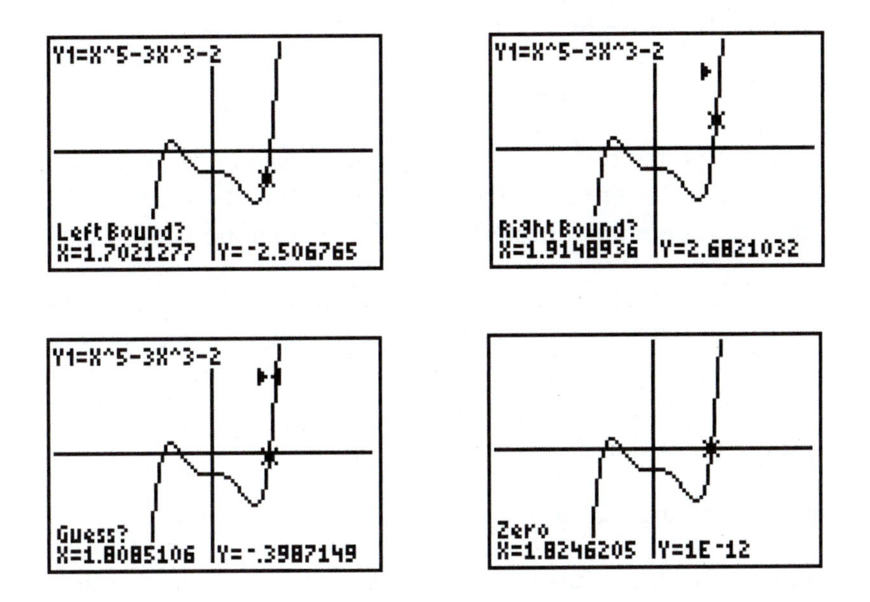

FIGURE 3.10. Using a graphing calculator to approximate a zero.

This calculator requires the user to select a point to the left of the zero (that is, the x-intercept), a point to the right of the zero, and a guess somewhere between the two. (Note that the y-coordinate in the final screen is not given as 0, but as 1×10^{-12}.)

Exercises 3.3

1. Examine the graphs of the fourth-degree polynomial functions in (a)–(d). Discuss each graph; include the number of x-intercepts, the number of turning points, and the end behavior.

   a) $(x-2)^4$

   b) $(x-2)^3(x-5)$

   c) $(x-2)^2(x-5)(x-8)$

   d) $(x-2)(x-5)(x-8)(x-11)$

   e) Can any of the functions in (a)–(d) be translated to change the number of x-intercepts? Explain how this can be done in each case, and give the numbers of intercepts that are possible.

   f) Construct a table of values for the function in (d). Use steps of 1 from 0 through 12 for x-values. Include first and second difference columns. What connections can you make between the signs of the numbers in the difference columns and the graph's direction and concavity?

2. Each of the functions in Exercise 1 has a zero at $x = 2$.

   a) For each of these functions, write the factor or factors in the equation that is responsible for the $x = 2$ zero.

   b) Each of your answers to part (a) defines a power function that has been translated horizontally. Graph each of these power functions. Then describe the behavior near $x = 2$ both for the power functions and for the functions in Exercise 1.

   c) The degree of the factor associated with a particular zero is called the **multiplicity** of the zero. For example, $x = 2$ is a zero of multiplicity 4 for the polynomial in Exercise 1(a). For the polynomial in Exercise 1(b), $x = 2$ is a zero of multiplicity 3, and $x = 5$ is a zero of multiplicity 1. Identify the multiplicity of each zero for the functions in Exercises 1(b)–(d).

3. Sometimes polynomial functions are expressed as translations of the simplest polynomial function of that degree. For example, the graph of $f(x) = 0.5(x-3)^3 + 2$, can be obtained from the graph of $y = x^3$ by shifting right three, compressing by a factor of 0.5, and

Exercises 3.3

shifting up 2. (Note that the general form of this polynomial can be found by cubing $x - 3$, distributing 0.5, and adding 2 to obtain $0.5x^3 - 4.5x^2 + 13.5x - 11.5$.) When polynomial functions are written as translations of power functions, their x-intercepts are relatively easy to find by symbolic means. Find the x-intercepts of the polynomial functions in (a)–(c).

a) $g(x) = 2(x - 3)^2 - 8$

b) $f(x) = 0.5(x - 3)^3 + 4$

c) $h(x) = 0.5(x - 3)^4 + 6$

d) What general conclusions can you draw about the x-intercepts of polynomial functions of the form $f(x) = a(x + b)^n + c$?

4. a) Find a fifth-degree polynomial function whose only x-intercepts are at –2, 3, and 7. Show a graph displaying the key features and describe those features. Construct a table of values with first and second differences and describe the connections between the table and the geometric features of the graph.

   b) Write another fifth-degree polynomial that has identical zeros.

5. The polynomial $x^3 - 16x^2 + 81x - 126$ is discussed in one of this lesson's examples. Use the factored form of the polynomial to explain why the graph of $f(x) = x^3 - 16x^2 + 81x - 126$ must continue to rise for all values of x larger than 7. (Mathematicians would say that $f(x_1) > f(x_2)$ whenever $x_1 > x_2$ and $x_2 > 7$.)

6. a) It's been said that from a distance, the graphs of all polynomial functions of a given degree look alike. Explain.

   b) The end behavior of a polynomial is completely determined by its leading coefficient and its degree. Describe how to identify the end behavior for a polynomial from just these two numbers. Justify your answer.

   c) Explain why every odd-degree polynomial must have at least one x-intercept, no matter how it is translated.

7. For each of the following polynomials, (i) identify the leading coefficient, (ii) state the degree of the polynomial, (iii) list all its x-intercepts and their multiplicities, (iv) describe its end behavior, and (v) sketch a qualitatively correct (end behavior and behavior at zeros) graph. Do not use the expanded form to find your answers.

Exercises 3.3

(a) $f(x) = (x - 2)^4$

(b) $g(x) = 3(x - 2)^2(x - 5)(x + 8)$

(c) $h(x) = (x - 2)^3(3x + 4)(5x - 2)$

(d) $j(x) = 5(x - 2)(4 - 3x)(x + 3)$

(e) $k(x) = 2(x - 2)(2x - 1)^2(x^2 + 1)$

8. You saw in this lesson that solving an equation is equivalent to finding zeros of a related function. It is also true that solving inequalities is equivalent to reading qualitative graphs. Use your graphs from Exercise 7 to solve the following inequalities. Explain how you use each graph.

a) $(x - 2)^4 > 0$

b) $3(x - 2)^2(x - 5)(x + 8) \geq 0$

c) $(x - 2)^3(3x + 4)(5x - 2) < 0$

d) $5(x - 2)(4 - 3x)(x + 3) \leq 0$

e) $2(x - 2)(2x - 1)^2(x^2 + 1) < 0$

9. Draw qualitative graphs (graphs in which the axes are not scaled) to answer the following questions.

a) When two polynomials are added, what happens to the zeros of the original functions? Explain.

b) When a polynomial is multiplied by a constant, what happens to the zeros of the original function? Explain.

c) Find a fifth-degree polynomial function whose only x-intercepts are at –2, 3, and 7, and whose y-intercept is 84.

10. Recall that the graph of an even function is symmetric with respect to the y-axis, and the graph of an odd function is symmetric with respect to the origin.

a) Explain why the graph of any polynomial function in which every term is of even degree (for example, $y = 3x^4 - 5x^2 + 7$) must be symmetric with respect to the y-axis. (Hint: What happens when a value of x is replaced with its opposite?)

Exercises 3.3

b) Explain why the graph of any polynomial function in which every term is of odd degree (for example, $y = 4x^5 - 6x^3 + 9x$) must be symmetric with respect to the origin.

c) Use function notation to characterize odd and even functions. That is, compare $f(x)$ and $f(-x)$ for an even function. Repeat for an odd function. Generalize your observations.

11. The fourth-degree polynomial $(x + 5)(x + 1)(x - 2)(x - 6)$ has four zeros.

a) Identify the four zeros of this polynomial.

b) Find the general form of the polynomial.

c) State a new polynomial function whose graph is a 100-unit downward translation of the graph of $y = (x + 5)(x + 1)(x - 2)$ $(x - 6)$. Find approximations for the x-intercepts of the new polynomial function.

d) Explain why none of the zeros of the translated function are the same as any of the zeros of the original function.

12. There is a connection between integer zeros of a polynomial and the polynomial's leading coefficient and constant term. There is another connection between rational zeros and the leading coefficient and constant term. These connections are sometimes called the **Integer Zero Theorem** and the **Rational Zero Theorem**. This exercise explores these connections.

a) Explain how to find the constant term for a polynomial in factored form without carrying out the complete expansion of its equation.

b) How are the constant term of a polynomial and the y-intercept of its graph related? Justify your answer.

c) If c is an integer and a zero of a polynomial, how is c related to the constant term? (Assume that the general form of the polynomial has only integer coefficients and is factored so that all its factors also have only integer coefficients.)

d) Suppose now that c is a rational zero of the polynomial. That is, c can be written as the ratio of two integers (with no common

Exercises 3.3

factors): $\frac{n}{m}$. Then $\left(x-\frac{n}{m}\right)$ is a factor (as is $mx - n$). How is c (that is, how are n and m) related to the constant term and leading coefficient of the polynomial? Explain. (As in part (c), assume that all coefficients are integers.)

e) Explain why $x = 6$ cannot be a solution to the equation $y = x^4 - 2x^3 - 31x^2 + 32x - 40$.

13. Reconsider the three-sample pooling situation of Activity 3.3.

a) Consider the following strategy. Rather than pooling all three samples, pool only two of them. If the result is negative, test the third individually. If the result is positive, test the first individually. If the first tests positive, pool the second with the third. Develop a model for this strategy. When does it save money?

b) Compare the strategy in part (a) with the one you analyzed in Activity 3.3. When is one strategy better than the other?

14. A company has 30 in × 50 in rectangular pieces of scrap sheet metal from which it wants to construct containers by cutting square pieces from each corner and bending the resulting flaps upward to form sides, as shown in **Figure 3.11**. (For added strength, the square pieces can also be left attached as tabs.)

© Susan Van Etten

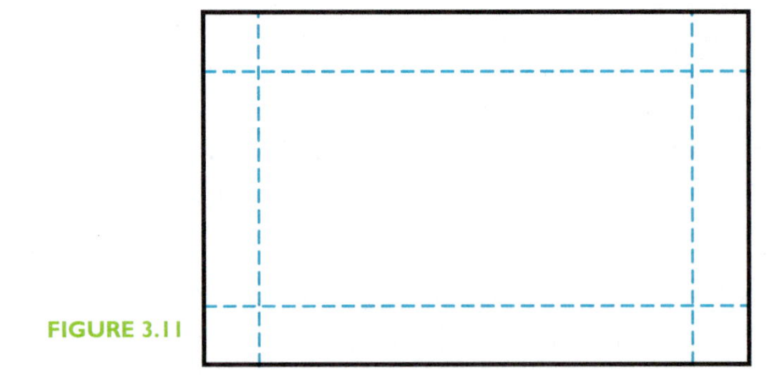

FIGURE 3.11

a) Explain why the corner pieces (tabs) need to be squares.

b) Develop a model for the volume of the container in terms of the side of the removed square.

Exercises 3.3

c) If the company wants the containers to hold 3,000 in^3 of a granular substance, what size square should it cut from each piece? Explain how you arrived at your answer.

15. **Investigation. Table 3.8** shows the data you used in Activity 3.2 of Lesson 3.2. This investigation develops another method for fitting a cubic polynomial through all four points.

a) Write four separate cubic polynomials in factored form: $f(x)$ having zeros 10, 20, and 30; $g(x)$ having zeros 0, 20, and 30; $h(x)$ having zeros 0, 10, and 30; and $k(x)$ having zeros 0, 10, and 20. Note that each of these polynomials is zero at all but one of the data values in the table.

b) Evaluate $f(0)$, $g(10)$, $h(20)$, and $k(30)$.

c) Let $p(x) = f(x) + g(x) + h(x) + k(x)$. Evaluate $p(0)$, $p(10)$, $p(20)$, and $p(30)$. Explain your result.

d) Multiplying $f(x)$ by a constant will change $f(0)$ and $p(0)$, but will not affect any of the other values you computed in parts (b) and (c) above. Explain why this is true, then find a constant multiplier M_1 for $f(x)$ to make $p(0) = 400{,}000$.

e) The polynomial $M_1f(x) + g(x) + h(x) + k(x)$ passes through (0, 400000). Find appropriate multipliers M_2, M_3, and M_4 so that $M_1f(x) + M_2g(x) + M_3h(x) + M_4k(x)$ is the cubic that passes through all four data points.

Time (seconds)	Altitude (feet)
0	400,000
10	399,000
20	393,000
30	384,000

TABLE 3.8. Falling body data.

Zeroing in on Polynomials

The history of mathematics is rich with stories about attempts to find zeros of polynomials. This lesson considers some of that history and shows how a cast of colorful characters opened the door to a new branch of mathematics.

THE CUBIC EQUATION COMPETITIONS

In this lesson's activity, you will consider some of the historical results associated with the search for ways to find zeros of polynomials of degree 3 and higher. Before beginning the activity, read the following historical background information.

Academic life in the 1500s contrasted sharply with that of today. Often knowledge was kept secret because university positions were awarded to winners of academic contests between two individuals. Thus, if you aspired to someone's position, you were likely to be offered it if you challenged and defeated that person on some intellectual issue.

Imagine, then, the predicament that faced Italian mathematician Nicolo Tartaglia when he received a challenge from Antonio Fior. Fior was reputed to possess a method for solving cubic equations of the form $x^3 + ax = b$. Fior's challenge may have resulted from his disbelief of a claim by Tartaglia that he could solve cubics of the form $x^3 + ax^2 = b$. In the time between the challenge and the appointed date, Tartaglia, whose claim was true, succeeded in finding a way to solve the type of cubic Fior could solve. The contest was a rout, with Tartaglia solving all 30 of Fior's problems and Fior solving none of Tartaglia's.

Word of Tartaglia's stunning performance reached Jerome Cardan, who at the time was engaged in lecturing and writing about science and mathematics, in part because he had been refused admission to the College of Physicians in Milan upon graduation from medical school. (Cardan, who at various times in his life was a gambler, an astrologer, a mathematician, and one of the most famous medical doctors in Europe, is also one of the most unusual characters in the history of mathematics.) Upon hearing of Tartaglia's performance, Cardan began a correspondence with Tartaglia and invited him to visit him at his home, where Cardan obtained Tartaglia's methods under an oath of secrecy.

At the time of Tartaglia's visit, Cardan had as a student a young man named Lodovico Ferrari. Just a few years before, Ferrari had knocked at Cardan's door asking for work. He soon was attracted to Cardan's mathematics and quickly became his best pupil. Working together, Cardan and Ferrari extended Tartaglia's methods to the solution of the general cubic equation $ax^3 + bx^2 + cx + d = 0$. Ferrari adapted the process used to derive the cubic result to the quartic (general fourth-degree polynomial).

In 1545 Cardan, who published over 100 books in his lifetime and left over 100 unpublished manuscripts at his death, published the cubic and quartic results in his *Ars Magna*, a treatise on algebra. Tartaglia was furious about Cardan's violation of the secrecy oath, and a bitter dispute arose between Tartaglia on one side and Cardan and Ferrari on the other. The matter appears to have culminated in a debate between Tartaglia and Ferrari in which the latter was victorious.

In the *Ars Magna*, Cardan derives the formula

$$x = \sqrt[3]{\frac{b}{2} + \sqrt{\frac{b^2}{4} + \frac{a^3}{27}}} - \sqrt[3]{-\frac{b}{2} + \sqrt{\frac{b^2}{4} + \frac{a^3}{27}}}$$

for cubics of the type $x^3 + ax = b$, then shows how the formula can be adapted to other types of cubics by means of clever substitutions.

Activity 3.4

1. As an example, Cardan applied the formula to $x^3 + 6x = 20$. Find Cardan's solution by substituting the appropriate values into his formula. Simplify any fractions, but leave the roots.

2. Use a calculator or computer to find the zeros of $x^3 + 6x = 20$. Also use the calculator to evaluate the solution you found in Item 1. What can you conclude?

3. Apply Cardan's formula to $x^3 - 6x = 4$. Again, simplify fractions when possible, but leave roots.

4. Solve $x^3 - 6x = 4$ by a method other than Cardan's. Compare the result to your answer in 3.

5. In the *Ars Magna*, Cardan gave a geometric proof of his formula in which various parts of the cubic were related to parts of a dissected cube. His choice of method was no doubt influenced by the primitive state of algebraic symbolism in the 1500s and by the prominent role that geometry played in the mathematics of the time. Perhaps because of his reliance on geometry, Cardan examined many cases that are today considered essentially the same. For example, Cardan didn't solve $x^3 - 6x = 4$; he solved $x^3 = 6x + 4$. Give some reasons why Cardan might have avoided negatives.

NUMBERS AND ALGEBRA

The history of mathematics has many examples of results that waited decades or even centuries before they found application to real-world modeling problems. Einstein, for example, applied theoretical mathematics developed in the previous century in his theory of relativity.

Few things are more basic to mathematics than numbers, and the development of new kinds of numbers was often prompted by the occurrence of strange results obtained from theoretical questions in algebra. For example, mathematicians in ancient times who solved the equation $2x = 6$ could not help but wonder about the meaning of solutions to $2x = 5$ or $2x + 11 = 5$ even though the concepts of fractions and negatives had no application at the time. When civilization advanced to the point of bank accounts and thermometers (Gabriel Fahrenheit developed the first accurate thermometer in 1714), mathematicians had negative numbers ready and waiting.

Well before the time of Christ, mathematicians encountered quadratic equations in the search for answers to problems in geometry. When they solved an equation like $x^2 - 2 = 0$, they expected the solution to be a number that could be expressed as the ratio of two whole numbers—in other words, a rational number. When they discovered that $\sqrt{2}$ could not be expressed as such a ratio, irrational numbers were born. However, nearly 2000 years passed before mathematicians were certain whether some numbers were rational or irrational (π, for example, was proved irrational in 1741).

Table 3.9 summarizes the kinds of numbers that were recognized at the time Cardan began his study of algebra.

Type	Description	Examples
Integers	Positive and negative whole numbers	–2, 0, 1, 3
Rational Numbers	Numbers expressible as the ratio of two integers	$\frac{1}{2}, \frac{-3}{4}, -3, 5, 0, 1, \sqrt{\frac{4}{9}}, \sqrt[3]{8}$
Irrational Numbers	Numbers not expressible as the ratio of two integers	$-\sqrt{3}, \sqrt[3]{-2}, \pi, \sqrt{5}$

TABLE 3.9.
Numbers recognized in the 16th century.

Although the quadratic formula had existed for centuries, square roots of negative numbers were largely ignored until Cardan encountered them. Cardan's formula for the solution of the cubic in some cases produced expressions involving square roots of negative numbers when Cardan knew the solutions to be rational numbers. This strange circumstance puzzled him, and he was unable to explain it. However, because of his work, mathematicians began taking square roots of negatives seriously.

In 1637, René Descartes (1596–1650) called numbers such as $\sqrt{-4}$ **imaginary** and used the term **real** to describe rational and irrational numbers. In 1777, Leonhard Euler (1707–1783) used the letter i to represent $\sqrt{-1}$ and expressed other imaginary numbers in terms of i. Since $i = \sqrt{-1}$, $i^2 = -1$.

Frank and Ernest

EXAMPLE 7

Express $\sqrt{-4}$ as an imaginary number in terms of i.

SOLUTION:

$\sqrt{-4} = \sqrt{4}\sqrt{-1} = 2i.$

To check the answer, calculate $(2i)^2 = 4i^2 = (4)(-1) = -4$.

COMPLEX NUMBERS AND THE FUNDAMENTAL THEOREM OF ALGEBRA

Carl Gauss

In 1797, Carl Gauss (1777–1855), a 20-year-old mathematics student in Germany, presented his doctoral dissertation. In it he proved that every polynomial could be factored into first- and second-degree factors. Gauss, whom some consider the greatest mathematician of all time, proved his theorem several times during his long and productive career, giving his final proof when he was 70. By that time, he had invented the term **complex number** to describe any "number" of the form $a + bi$, where a and b are real numbers, and extended his theorem to include complex numbers. (Note that since either a or b can be 0, real numbers and imaginary numbers are also complex numbers.)

The Fundamental Theorem of Algebra

Any polynomial $a_nx^n + a_{n-1}x^{n-1} + a_{n-2}x^{n-2} + \ldots + a_1x^1 + a_0$, where each coefficient a is a real number, can be expressed as a product of n linear factors $a_n(x - c_1)(x - c_2) \ldots (x - c_n)$, where each c is a complex number.

Note that the Fundamental Theorem does not restrict coefficients to integers or rational numbers; any real number is acceptable.

EXAMPLE 8

Express the complex number $2 + \sqrt{-9}$ in terms of i.

SOLUTION:

$2 + \sqrt{-9} = 2 + \sqrt{9}\sqrt{-1} = 2 + 3i.$

EXAMPLE 9

Factor the polynomial $x^3 - 3x^2 + 4x - 12$ into linear factors.

SOLUTION:

$x^3 - 3x^2 + 4x - 12 =$

$(x - 3)(x^2 + 4) =$

$(x - 3)(x - 2i)(x + 2i)$.

The Fundamental Theorem of Algebra helps modelers understand solutions obtained from polynomial models. However, applying the theorem can be difficult because factoring is a trial-and-error process. Some computer software and some calculators are capable of factoring polynomials. However, as the calculator screen in **Figure 3.12** demonstrates, they may not find complex factors.

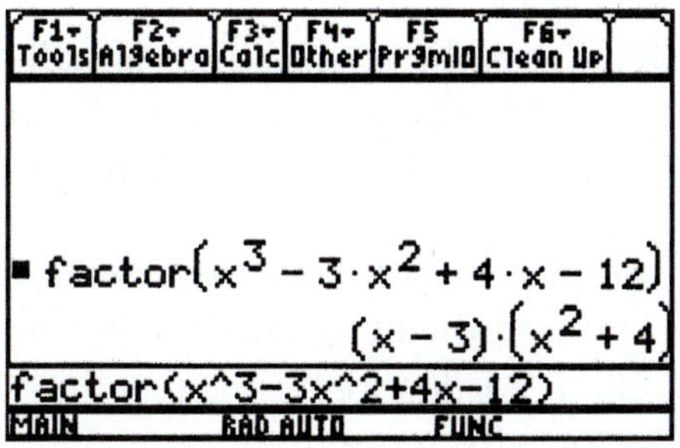

FIGURE 3.12. Factoring a polynomial on a calculator with symbol manipulation capabilities.

Since each linear factor of a polynomial produces a zero, the Fundamental Theorem of Algebra can be interpreted as saying that a polynomial of degree n has exactly n complex zeros, although some of them can be equal (for example, when two factors are identical).

Although Carl Gauss was a modeler who made important contributions to astronomy and other areas, his work with complex numbers was theoretical. Applications of complex numbers to problems in physics and engineering, for example, became common in the late 19th century.

Exercises 3.4

1. Gauss's original statement of the Fundamental Theorem of Algebra said that every polynomial (with coefficients that are real numbers) can be factored into linear and quadratic factors. For example, the factored form of a cubic might be $(x + 1)(x^2 - 2x + 2)$.

   a) Find all the zeros of $(x + 1)(x^2 - 2x + 2)$ by finding the zeros of each factor separately. Give any non-real complex zeros in the form $a + bi$. (Note that the phrase "non-real complex" is necessary since real numbers are considered complex numbers with $b = 0$.)

   b) Non-real complex numbers often occur in pairs: $a + b$i and $a - b$i. Each of these is called the **conjugate** of the other. (The basic idea is that to find the conjugate of a complex number change the sign of the imaginary term.) The Fundamental Theorem of Algebra implies that non-real complex zeros of polynomials occur in conjugate pairs. Explain why.

2. The 2 in $2 + 3i$ is often called the **real part** and the 3 is called the **imaginary part**. (You don't need to say $3i$ when you talk about the imaginary part because the i is implied by the word imaginary.)

   a) Just like real numbers, complex numbers can be added. All you do is add the real parts and add the imaginary parts. Find the sum of the complex numbers $2 + 3i$ and $4 - 5i$. Also find the difference (subtract the second from the first).

   b) Complex numbers can also be multiplied, but it's a bit more work. Two complex numbers are multiplied the same way two linear factors are multiplied: $(x + 3)(x - 2) = x^2 - 2x + 3x - 6 = x^2 + x - 6$. Multiply the complex numbers $2 + 3i$ and $4 - 5i$. (Remember that $i^2 = -1$.)

   c) What happens when you add two complex numbers that are conjugates? What happens when you multiply them? Explain.

3. In his final proof of the Fundamental Theorem of Algebra, Gauss changed from linear and quadratic factors to all linear factors, but allowed complex numbers in the linear factors. For example, in place of the factor $x^2 - 4x + 5$, Gauss would have written $(x - (2 + i))(x - (2 - i))$.

Exercises 3.4

a) Show that $2 + i$ and $2 - i$ are zeros of $x^2 - 4x + 5$.

b) Find the sum and the product of $2 + i$ and $2 - i$. Compare the sum and product to the coefficients of the equation.

c) It isn't difficult to find complex zeros for a quadratic if you're comfortable with the quadratic formula or completing the square. Finding the quadratic from the complex zeros can be messy, however, since it can require multiplying two factors with three terms: $(x - (2 + i))(x - (2 - i)) = (x - 2 - i)(x - 2 + i)$. Instead of doing this calculation, take a look at what happens when the factors $(x - c)$ and $(x - d)$ are multiplied. That is, generalize the observation you made in part (b) to show it is true of all quadratics.

d) From your results in (b), state a short cut for finding a quadratic from a conjugate pair of complex zeros. Show how to apply your short cut to a quadratic that has $2 + 3i$ as a zero.

e) Another way to obtain a quadratic from a complex zero is to write a simple equation stating that the complex number is a zero: $x = 2 + i$. The quadratic can then be obtained by moving the 2 to the left side and squaring. Show how to do this. Also show that the same equation results if the other zero is used.

4. a) Explain why the Fundamental Theorem of Algebra implies that a polynomial (with real coefficients) with 3 as a real zero and $3 - i$ as a complex zero must have a degree of at least 3.

   b) Find a polynomial that has 3 and $3 - i$ as zeros.

   c) Is your answer to (b) unique? That is, is there only one polynomial possible? Explain.

5. a) Cube each of the complex numbers $1 + i$ and $-1 + i$. That is, find $(1 + i)^3$ and $(-1 + i)^3$.

   b) Go back to your answers to Items 3 and 4 of Activity 3.4. Use your answer to Exercise 5(a) to explain what Jerome Cardan could not explain.

6. Square roots of positive numbers can be multiplied: $\sqrt{4}\sqrt{9} = \sqrt{36}$. Similarly, a square root of a positive number and the square root of a negative number can be multiplied: $\sqrt{4}\sqrt{-9} = \sqrt{-36}$.

Exercises 3.4

a) Show that $\sqrt{4}\sqrt{-9} = \sqrt{-36}$ by writing each imaginary part of the equation in terms of i.

b) You may be surprised to learn that this type of multiplication cannot be applied to square roots of two negative numbers. Show that $\sqrt{-4}\sqrt{-9} \neq \sqrt{36}$.

7. Complex numbers may have been nothing more than an interesting mathematical concept without application were it not for an important idea that occurred to three individuals around the same time. One of them was Gauss; the others were Caspar Wessel (Norwegian, 1745–1818) and Jean Robert Argand (Swiss, 1768–1822). Independently they found a way to represent complex numbers geometrically. They realized that just as every real number corresponds to a point on a line, every complex number corresponds to a point in a plane. Think of a coordinate system in which one axis represents real numbers and the other represents imaginary numbers, as shown in **Figure 3.13**.

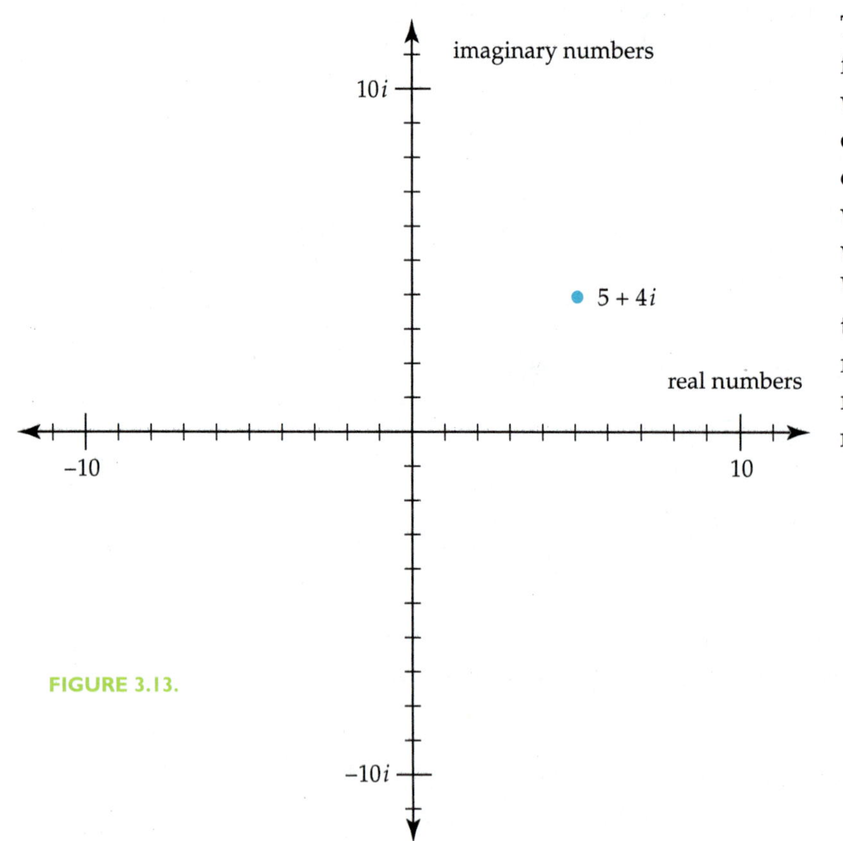

FIGURE 3.13.

The complex number $5 + 4i$, for example, is associated with the point (5, 4). (Be careful that you do not confuse this representation with the coordinate plane in which you graph functions. When you graph functions, the vertical axis represents real numbers; here it represents imaginary numbers.)

Exercises 3.4

a) Use graph paper to locate the complex numbers $7 + 2i$ and $1 + 4i$. Find their sum and locate it in the same diagram.

b) Form a quadrilateral by connecting the origin to $7 + 2i$ to the sum to $1 + 4i$ and back to the origin. What type of quadrilateral is it? Explain.

c) Your answer to (b) explains how complex numbers solve a problem in physics. If you have taken or are taking physics, explain. If not, ask someone who has studied physics to explain.

d) Describe the location of a real number (3, for example) and the location of an imaginary number ($3i$, for example) in the complex plane.

8. Describe the geometric relationship between a complex number and its conjugate.

9. a) Locate the complex number $3 + 4i$ in the complex plane. Multiply it by i and locate the answer in the same diagram. Repeat the process: multiply each answer by i until you return to $3 + 4i$.

   b) Repeat this experiment with another complex number instead of $3 + 4i$. Can you conclude that four successive multiplications by i always yield a similar result? Explain.

   c) Describe the geometric effect of multiplying a complex number by i. You may find it helpful to connect the complex numbers you located in (a) to the origin. Why might this property of complex numbers be useful?

10. The geometric representation of complex numbers as points in a plane leads quite naturally to new ways of thinking about complex numbers. The concepts of distance and direction, for example, have meaning for complex numbers, just as they do for real numbers. Fractal artwork is one application of complex numbers that uses the concept of distance. For example, the process that created the fractal in **Figure 3.14** involves calculations with complex numbers and decisions based on distance from the origin in the complex plane.

FIGURE 3.14. Fractal artwork

Exercises 3.4

a) Show how to find the distance from $4 + 4i$ to the origin of the complex plane. What geometric ideas did you use in your calculation?

b) A complex number's direction is an angle measured counter-clockwise from the positive real axis, as shown in **Figure 3.15**. (The clockwise direction is considered negative.) Find the direction of the complex number $4 + 4i$.

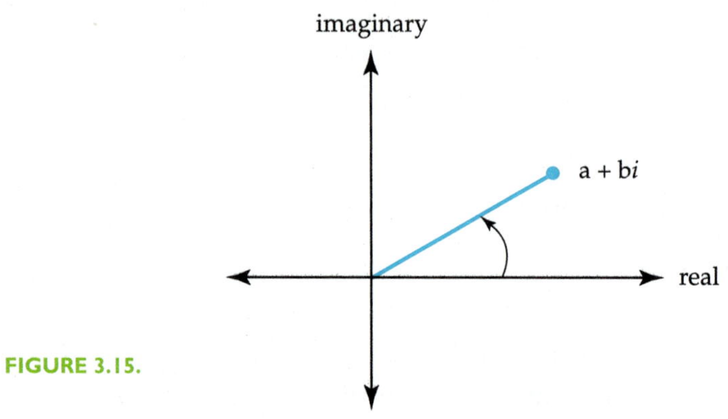

FIGURE 3.15.

Polynomial Divisions

Polynomials, you could say, have children: functions created from two polynomials. Although these functions may not be polynomials themselves, they are closely related and have properties that make them, like polynomials, useful modeling tools.

Activity 3.5

In this activity, you will examine the effect of dividing one polynomial by another. Unlike the operations of addition, subtraction, and multiplication, division of two polynomials may not produce another polynomial.

The function $f(x) = \frac{x^4 - 13x^2 + 36}{x^2 - 2x}$ represents the division of the polynomial $x^4 - 13x^2 + 36$ by the polynomial $x^2 - 2x$. Your task in this activity is to investigate and describe the behavior of this function.

1. Graph the function. Describe key features of the graph, including any characteristics that you find unusual.

2. Division by 0 is undefined. Identify all values of x for which the denominator is 0. Then examine the behavior of this function for values of x that are near the values you identified. For example, you might build a table showing values of the function for values of x near those that make the denominator 0. Discuss your observations.

3. Describe the domain and the range of this function.

4. Create a power function by dividing the term of highest degree in the numerator by the corresponding term of the denominator. Compare the graph of this simple function with the graph of $f(x) = \frac{x^4 - 13x^2 + 36}{x^2 - 2x}$. Discuss the similarities and differences.

RATIONAL FUNCTIONS AS RATIOS

Earlier in this chapter you saw that two polynomials produce another polynomial under addition, subtraction, and multiplication. They do not, however, produce another polynomial under division unless one polynomial divides the other exactly. In general, the ratio of two polynomials is a **rational function**.

Graphs of rational functions exhibit one or more of several types of asymptotes. For example, **Figure 3.16** shows the graph of the rational function $y = \frac{3x-5}{x-2}$.

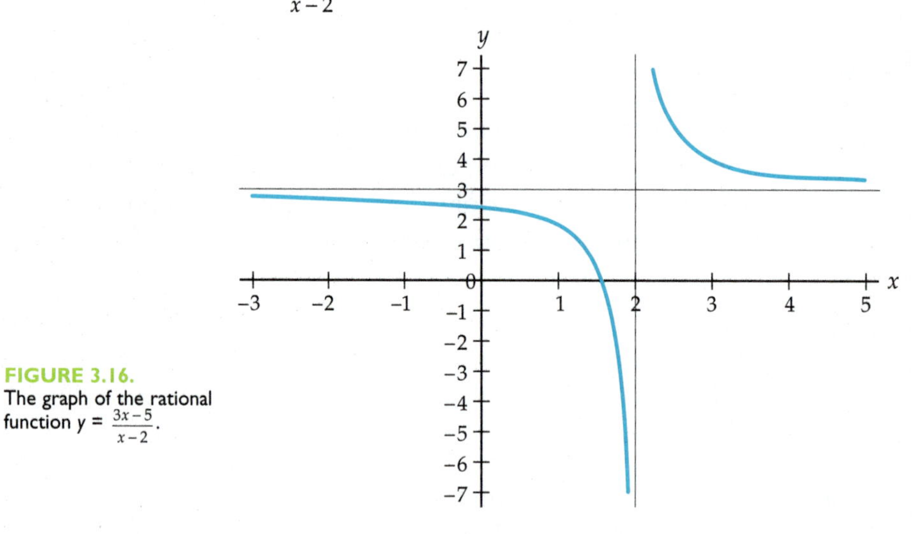

FIGURE 3.16.
The graph of the rational function $y = \frac{3x-5}{x-2}$.

Since the numerator, $3x - 5$, is 0 when $x = \frac{5}{3}$, the graph has an x–intercept at $x = \frac{5}{3}$.

x	y	x	y
0	2.5	4	3.5
1	2	3	4
1.5	1	2.5	5
1.8	–2	2.2	8
1.9	–7	2.1	13
1.99	–97	2.01	103

TABLE 3.10.
Values of the function $y = \frac{3x-5}{x-2}$ near the vertical asymptote.

The graph has a **vertical asymptote** at $x = 2$. That is, as x approaches 2, y becomes infinitely large (in a positive direction as x approaches 2 from the right; in a negative direction as x approaches 2 from the left).

The same results can be seen in **Table 3.10.**

The graph also has a **horizontal asymptote** at $y = 3$. That is, as x becomes infinitely large (in either a positive or negative direction), y approaches 3. Item 4 of Activity 3.5 provides a quick test for this end behavior: dividing the leading terms of the numerator and denominator. For $y = \frac{3x-5}{x-2}$, that ratio is $\frac{3x}{x} = 3$.

The same results can be seen in **Table 3.11**.

x	y	x	y
0	2.5	3	4
–1	2.6667	5	3.3333
–5	2.8571	10	3.125
–10	2.9167	50	3.0208
–100	2.9902	100	3.0102
–1000	2.999	1000	3.001

TABLE 3.11. Values of the function $y = \frac{3x-5}{x-2}$ showing end behavior.

Since the function has no value when $x = 2$, 2 is not in the function's domain. That is, the domain is all real numbers except 2. The function never takes on the value 3, and so the range is all real numbers except 3.

Note that neither the vertical asymptote nor the horizontal asymptote is part of the graph. That is, no point on either asymptote that satisfies the equation $y = \frac{3x-5}{x-2}$. Rather, asymptotes serve as guidelines to help you sketch accurate graphs and to describe the main features of the graph more easily.

The graph of a rational function can also exhibit a linear asymptote that is neither horizontal nor vertical, as shown in **Figure 3.17**. In this case, the asymptote is the line $y = x + 2$. In other words, as x becomes infinitely large (in a positive direction), y approaches $x + 2$. Asymptotes of this type are sometimes called **slant** or diagonal **asymptotes**.

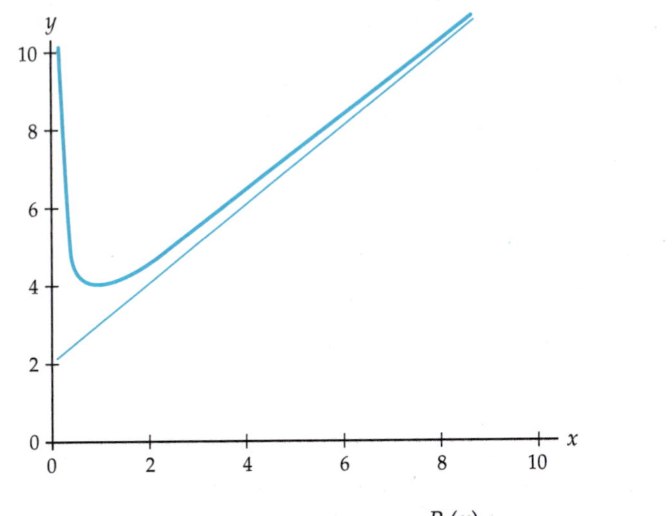

FIGURE 3.17. An asymptote that is neither horizontal nor vertical.

For rational functions written in the form $\frac{P_1(x)}{P_2(x)}$, where $P_1(x)$ and $P_2(x)$ are polynomials, the types of asymptotes and the conditions that produce them can be summarized as follows:

- Vertical asymptotes occur only at zeros of the denominator.
- Horizontal asymptotes occur when the degree of the numerator is less than or equal to the degree of the denominator. When the degree of the denominator is higher than that of the numerator, the x–axis is the asymptote.

- Linear asymptotes that are neither vertical nor horizontal occur when the degree of the numerator is 1 more than the degree of the denominator.

A vertical asymptote is an important characteristic of a rational function's local behavior at a particular point. When they occur, horizontal and diagonal asymptotes are important characteristics of a rational function's end behavior.

RATIONAL FUNCTIONS AS QUOTIENTS AND REMAINDERS

Since a rational function is the ratio of two polynomials, and since a ratio is equivalent to a division, it is often possible to re-express a rational function by performing a division.

First, consider the relationship between ratios and divisions of whole numbers. For example, the ratio $\frac{6}{3}$ is equivalent to $6 \div 3$, which is 2. Moreover, 6 (the dividend) is equal to the product of 3 (the divisor) and 2 (the quotient): $6 = 3 \times 2$.

When the result of a whole number division is not a whole number, it can be expressed as a quotient and remainder: $\frac{7}{3}$ produces a quotient of 2 and a remainder of 1; in other words, $\frac{7}{3} = 2 + \frac{1}{3}$.

Now, consider the rational function $f(x) = \frac{x^5 + 3x^3 - 4x^2 - 7}{x^3 - 2x}$. This rational function can be expressed in terms of a quotient polynomial and a remainder polynomial; in other words, there are polynomials $Q(x)$ and $R(x)$, for which $\frac{x^5 + 3x^3 - 4x^2 - 7}{x^3 - 2x} = Q(x) + \frac{R(x)}{x^3 - 2x}$.

Just as division cannot be performed on whole numbers when the numerator is smaller than the denominator, division of polynomials cannot be performed when the numerator is of smaller degree than the denominator. Thus, the degree of the remainder is always smaller than the degree of the divisor (denominator).

An advantage of the quotient and remainder form of a rational function is that the quotient, $Q(x)$, completely describes the end behavior (horizontal asymptote, slant asymptote, or polynomial growth) of the rational function, and the term containing the remainder determines the behavior of the rational function near its vertical asymptotes.

EXAMPLE 10

Consider the rational function $\frac{x^5+3x^3-4x^2-7}{x^3-2x} = x^2 + 5 + \frac{-4x^2+10x-7}{x^3-2x}$.

Use the form $x^2 + 5 + \frac{-4x^2+10x-7}{x^3-2x}$ to describe the behavior of the rational function $y = \frac{x^5+3x^3-4x^2-7}{x^3-2x}$.

SOLUTION:

The degree of the denominator of $\frac{-4x^2+10x-7}{x^3-2x}$ is larger than the degree of the numerator, so $\frac{-4x^2+10x-7}{x^3-2x}$ is near 0 for large values of x. Therefore, the end behavior of the graph of $y = \frac{x^5+3x^3-4x^2-7}{x^3-2x}$ resembles the graph of $y = x^2 + 5$.

Since $x^3 - 2x = x(x - \sqrt{2})(x + \sqrt{2})$ is 0 when $x = -\sqrt{2}$, $x = \sqrt{2}$, or $x = 0$, but $-4x^2 + 10x - 7 \neq 0$ for these same values, there are vertical asymptotes at $x = -\sqrt{2}$, $x = \sqrt{2}$, and $x = 0$.

Whether $\frac{x^5+3x^3-4x^2-7}{x^3-2x}$ is positive or negative near each vertical asymptote depends on $x^3 - 2x$ and $-4x^2 + 10x - 7$. If both are positive or both are negative for a value, say, slightly larger than 0, then $\frac{x^5+3x^3-4x^2-7}{x^3-2x}$ is positive just to the right of 0. If $x^3 - 2x$ and $-4x^2 + 10x - 7$ have opposite signs for a value slightly larger than 0, then $\frac{x^5+3x^3-4x^2-7}{x^3-2x}$ is negative just to the right of 0.

Most graphing utilities have difficulty with vertical asymptotes. In dot mode, a graphing calculator shows very little of the behavior near the vertical asymptotes. In connected mode, the calculator draws vertical lines that can be interpreted as approximations of the asymptotes, but should not be considered part of the function's graph. (See **Figure 3.18.**)

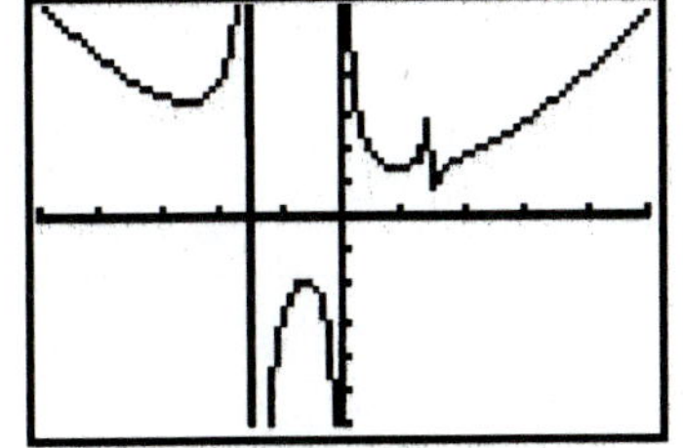

FIGURE 3.18. Graphing calculator graphs of $y = \frac{x^5+3x^3-4x^2-7}{x^3-2x}$ in dot mode (top) and connected mode (bottom).

```
F1▾ Tools | F2▾ Algebra | F3▾ Calc | F4▾ Other | F5 PrgmIO | F6▾ Clean Up
■ propFrac((x^5 + 3·x^3 − 4·x^2 ...)/(x^3 − 2·x))
    −(4·x^2 − 10·x + 7)/(x·(x^2 − 2)) + x^2 + 5
propFrac((x^5+3x^3−4x^2−7...
MAIN          RAD AUTO          FUNC
```

FIGURE 3.19.
A calculator with a proper fraction feature.

In the previous example, the quotient and remainder were given. There are several methods that can be used to find $Q(x)$ and $R(x)$. Some calculators with symbol manipulation features can express a rational function as a proper fraction, as shown in **Figure 3.19.**

The calculator display in Figure 3.19 shows that the division of $x^5 + 3x^3 - 4x^2 - 7$ by $x^3 - 2x$ produces a quotient of $x^2 + 5$ and a remainder of $-4x^2 + 10x - 7$. Thus, $\frac{x^5 + 3x^3 - 4x^2 - 7}{x^3 - 2x} = x^2 + 5 + \frac{-4x^2 + 10x - 7}{x^3 - 2x}$.

Calculators that do not have symbol manipulation features can be programmed to divide polynomials. The TI–83 program Polydiv on the CD–ROM that accompanies this text is an example of such a program.

Division of polynomials can also be done by hand using the symbol manipulation algorithm described in the following example.

EXAMPLE 11

Divide $x^5 + 3x^3 - 4x^2 - 7$ by $x^3 - 2x$.

SOLUTION:

$$x^3 + 0x^2 - 2x + 0 \overline{\big)\, x^5 + 0x^4 + 3x^3 - 4x^2 + 0x - 7}$$

To begin the process, write the terms of the divisor (denominator) and the dividend (numerator) in descending powers of the variable. Write a 0 for any missing term.

$$\begin{array}{r} x^2 \\ x^3 + 0x^2 - 2x + 0 \overline{\big)\, x^5 + 0x^4 + 3x^3 - 4x^2 + 0x - 7} \\ \underline{-(x^5 + 0x^4 - 2x^3 + 0x^2)} \qquad \\ 5x^3 - 4x^2 + 0x - 7 \end{array}$$

Divide the first term of the dividend by the first term of the divisor; this gives you the first term of the quotient. Multiply each term of the divisor by this first term, write the results under the dividend, and subtract to obtain a remainder.

Repeat this process, using the remainder as the new dividend, until the power in the remainder is less than the power of the divisor.

$$
\begin{array}{r}
x^2+5 \\
x^3+0x^2-2x+0\overline{)x^5+0x^4+3x^3-4x^2+0x-7} \\
\underline{-(x^5+0x^4-2x^3+0x^2)} \\
5x^3-4x^2+0x-7 \\
\underline{-(5x^3+0x^2-10x+0)} \\
-4x^2+10x-7
\end{array}
$$

For rational functions written in the form $Q(x) + \frac{R(x)}{D(x)}$, the types of asymptotes and the conditions that produce them can be summarized as follows:

- Vertical asymptotes occur only at zeros of the denominator, $D(x)$. The signs of $R(x)$ and $D(x)$ determine whether the graph ascends or descends near vertical asymptotes. The case in which the numerator and denominator have common zeros is the subject of Exercise 3.
- Horizontal asymptotes occur when $Q(x)$ is a constant.
- Linear asymptotes that are neither vertical nor horizontal occur when $Q(x)$ is linear.
- The end behavior of the rational function exhibits polynomial growth when $Q(x)$ is a polynomial.

Building Rational Functions

Rational functions with certain properties can be built from linear and quadratic factors, just as polynomial functions can. For example, to form a rational function with a vertical asymptote at $x = 5$, use $x - 5$ as a denominator: $y = \frac{1}{x-5}$. Since the denominator has a higher power than the numerator, the x–axis is a horizontal asymptote. Any other rational function in which $x - 5$ is a factor of the denominator will have a vertical asymptote at 5 unless, of course, $x - 5$ is also a factor of the numerator. In general, there are many rational functions that have a given collection of asymptotes.

EXAMPLE 12

Write the equation for a rational function with a vertical asymptote at $x = 5$ and a horizontal asymptote at $y = 1$. Repeat with the horizontal asymptote at $y = 2$. Modify the function to change the horizontal asymptote into a slant asymptote.

SOLUTION 1:

In quotient and remainder form, use a quotient of 1 and a constant remainder with a denominator of $x - 5$: $y = 1 + \frac{c}{x-5}$. To make the horizontal asymptote 2, change the quotient to 2: $y = 2 + \frac{c}{x-5}$. To obtain a slant asymptote, any equation of the form $y = mx + b + \frac{c}{x-5}$ will do.

SOLUTION 2:

The solution can also be given as a ratio. Place an x in the numerator and an $x - 5$ in the denominator: $y = \frac{x}{x-5}$.

For a horizontal asymptote at $x = 2$, give the x in the numerator a coefficient of 2: $y = \frac{2x}{x-5}$.

Changing the degree of the numerator to 2:
$y = \frac{2x^2}{x-5}$ produces a slant asymptote.

Exercises 3.5

1. a) Identify the x– and y–intercepts of the rational function $f(x) = \frac{2x+1}{x-3}$.

   b) Identify the asymptotes of $f(x)$. Describe the end behavior and the behavior near any vertical asymptotes.

   c) State the domain and the range of the function.

2. A rational function has a vertical asymptote at $x = 2$ and a horizontal asymptote at $y = 1$. Sketch at least two different graphs for the rational function.

3. The rules for locating asymptotes of rational functions must be modified if a single value of x is a zero of both the numerator and the denominator. For example, 2 is a zero of both numerator and denominator of $y = \frac{x-2}{(x+3)(x-2)}$. **Figure 3.20** is a graph of this function. Note the open dot at $x = 2$. This indicates that the function does not have a value when $x = 2$. However, unlike the situation when x is near –3, the function does not become infinitely large, as **Table 3.12** shows.

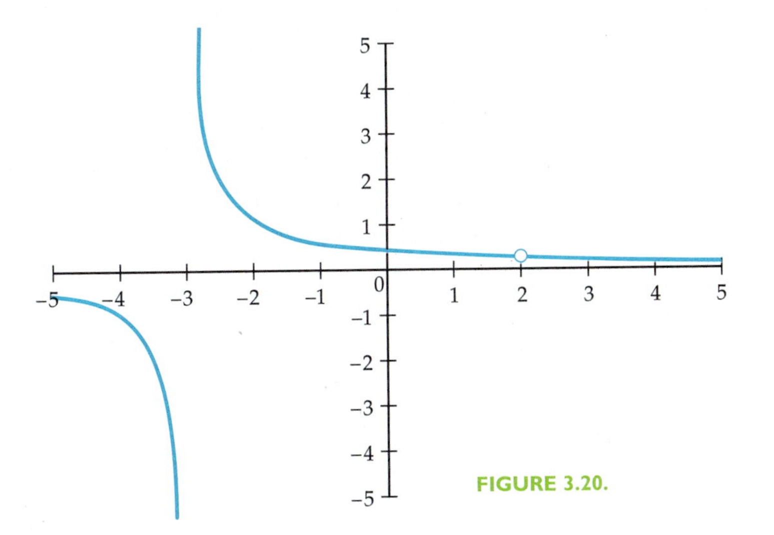

FIGURE 3.20.

x	y	x	y
–4	–1	1	0.25
–3.5	–2	1.5	0.2222
–3.2	–5	1.8	0.2083
–3.1	–10	1.9	0.2041
–3.01	–100	1.99	0.2004
–2.99	100	2.01	0.1996
–2.9	10	2.1	0.1961
–2.8	5	2.2	0.1923
–2.5	2	2.5	0.1818
–2	1	3	0.1667

TABLE 3.12.

   a) State the domain and the range of this function.

   b) Division is defined in terms of multiplication. For example $\frac{6}{3} = 2$ because 6 = 3 x 2. Compare the fractions that result when evaluating this function for $x = -3$ and $x = 2$. Restate these divisions in terms of multiplication and explain the differences.

Exercises 3.5

c) Unlike the situation when $x = -3$, there is a natural choice of a value to give the function when $x = 2$. What is that value?

d) Write a piecewise equation to define the function so that the open dot in the graph is filled.

4. a) Write an equation for a rational function having a vertical asymptote at $x = 2$, a horizontal asymptote at $y = 1$, and passing through the point (4, 5).

b) Laura claims that every equation of the form $y = k + \frac{c}{x-h}$ can be sketched quickly and accurately without any real work. Here's her method. The horizontal asymptote is $y = k$, and $x = h$ is the vertical asymptote. To locate other points on the graph, she says that the point (a, b) is on the curve if the area of the rectangle with opposite corners at (a, b) and (h, k) is exactly c units. (Rectangles upward from and to the right of (h, k) or downward from and to the left of (h, k) have positive area. Rectangles upward from and to the left of (h, k) or downward from and to the right of (h, k) have negative area.) See **Figure 3.21**. Check Laura's method on your equation in part (a).

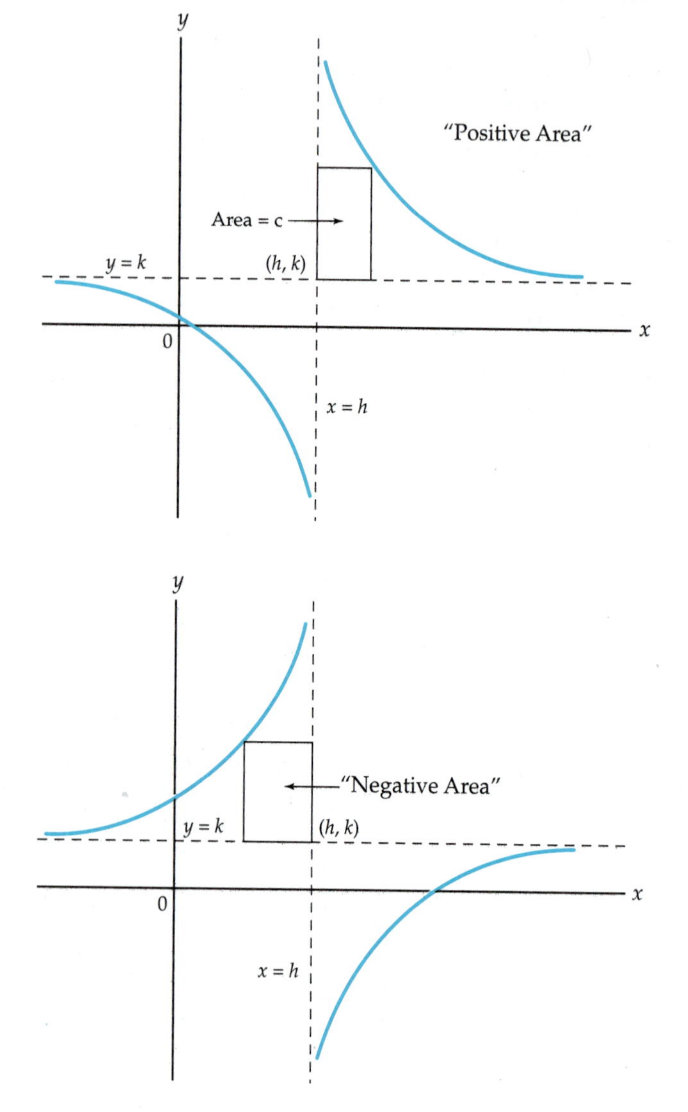

FIGURE 3.21.
Laura's method for graphing $y = k + \frac{c}{x-h}$.

Exercises 3.5

c) Explain how the function $y = k + \frac{c}{x-h}$ is the translation of a power function. That is, identify the power function to which it is related, and specify the translations that relate the two functions.

d) Use your understanding of inverse variation and power functions to explain why Laura's area rule is correct.

5. It's fairly easy to create a rational function with a desired slant asymptote. If, for example, the slant asymptote is $y = 2x + 1$, and you would also like a vertical asymptote at $x = -1$, think of $2x + 1$ as the quotient and $x + 1$ as the divisor. Choosing any constant (say, 1) as a remainder, create a function by adding : $y = 2x + 1 + \frac{1}{x+1}$.

To make this look more like a typical rational function, multiply $2x + 1$ by $x + 1$ and add the fractions:

$$\frac{(2x+1)(x+1)}{x+1} + \frac{1}{x+1} =$$
$$\frac{(2x+1)(x+1)+1}{x+1} =$$
$$\frac{2x^2+3x+1+1}{x+1} =$$
$$\frac{2x^2+3x+2}{x+1}$$

This addition can also be performed on a calculator or computer with symbol manipulation features.

Create rational functions for each of the following sets of conditions. State the domain and the range of each function. (Estimate values you cannot determine exactly.)

a) Vertical asymptote at –4; horizontal asymptote at 3.

b) Slant asymptote $y = x - 1$; vertical asymptote $x = 0$.

c) Vertical asymptotes at –2 and 2; horizontal asymptote at 0.

d) Vertical asymptote at –1; slant asymptote $y = 2 - x$.

Exercises 3.5

6. Write the equation of a function whose graph approximates the graph shown.

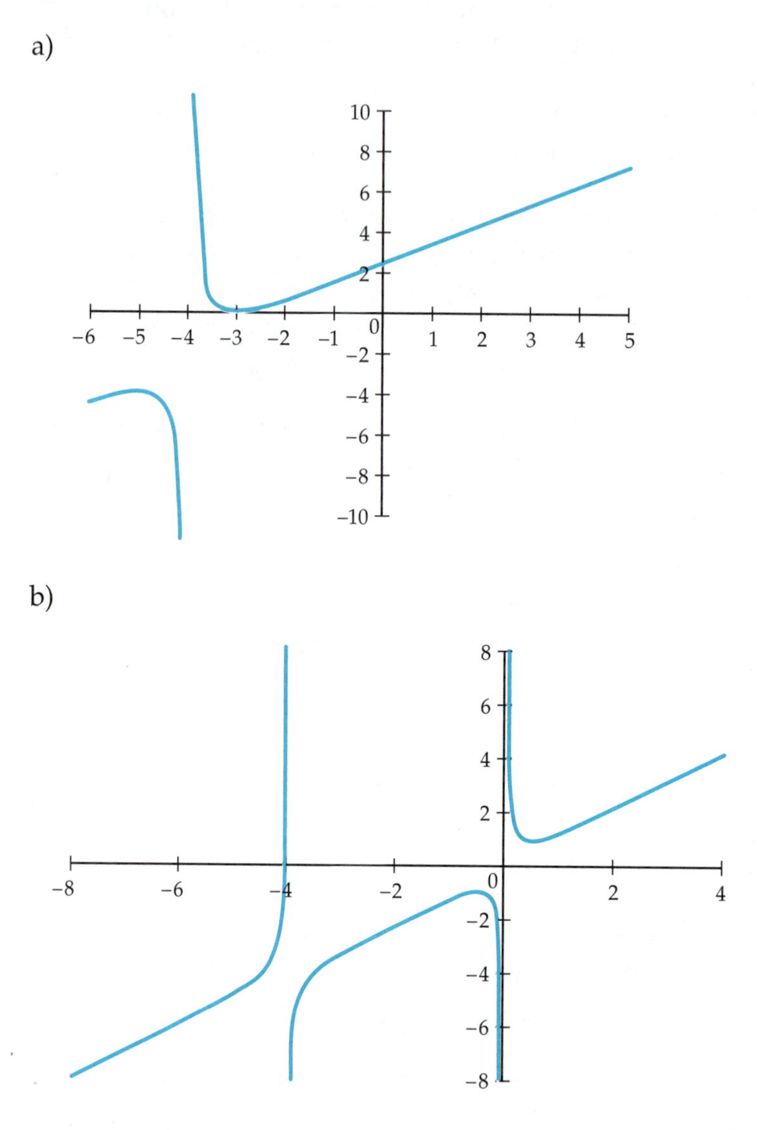

7. Some rational functions have denominators with no real zeros. Investigate each of the following rational functions. Describe any asymptotes and intercepts, and state the domain and range.

a) $y = \dfrac{x}{x^2+1}$

b) $y = \dfrac{x^2}{x^2+1}$

c) $y = \dfrac{x^3}{x^2+1}$

Exercises 3.5

d) Some people think that a function's graph can never cross an asymptote. Discuss whether this is correct.

e) Write the equation for a rational function that has no vertical asymptote and does not pass through (0, 0).

8. Use a calculator or computer symbol manipulator, a polynomial division program, or the polynomial long division algorithm to rewrite each rational function. Discuss the function's end behavior, then verify your description by using a graphing utility.

a) $y = \dfrac{3x^4 + x^3 - 4x^2 + 5}{x^3 + 2x - 3}$

b) $y = \dfrac{5x^2 + x + 3}{x^2 - 4}$

9. Exercises 12 (c) and (d) of Lesson 3.3 show that for any rational zero, $\frac{n}{m}$, of a polynomial having only integer coefficients, n must be a divisor of the constant term and m must be a divisor of the leading coefficient. This provides a way, coupled with a graphing utility, to identify possible rational zeros of any polynomial with integer coefficients. The corresponding factor must be $(mx - n)$. Polynomial division provides a way to verify a guessed factor, and the quotient of a successful division must be the source of any remaining zeros.

Use these ideas to find all (real and non–real complex) zeros of the polynomial $y = 77x^4 + 6x^3 + 300x^2 + 24x - 32$.

10. Rational functions are sometimes used to model costs of manufacturing.

a) Suppose the manufacturing process for a particular item has fixed costs of \$1500 and per unit costs of \$2.50. Write a rational function to describe the average cost per unit in terms of the number of items manufactured.

b) Identify the asymptotes and interpret them in this context.

11. Polynomial functions and rational functions are sometimes used to model a business's profits. The following are facts about a business.

- The company's sales decline from 300 items by about 0.2 items for each additional dollar it charges for its product.

- The company has fixed manufacturing costs of \$1500 and per item costs of \$92.

Exercises 3.5

- The company manufactures the same number of items that it sells.

a) Model the company's revenue, costs, and profit in terms of the price it charges. Graph the profit model and interpret any key features of the graph in this context.

b) Profit margin, which is used to measure the health of a business, is the ratio of profits to revenues. Write an equation to model the profit margin. Graph the model and interpret any key features in this context.

LESSON 3.6

Polynomial Approximations

In a sense, nearly every mathematical model is an approximation because assumptions are made that eliminate some aspects of reality. Therefore, whenever you use polynomials to create a model, you are approximating reality. But there is another way in which polynomials are used to create approximations, and these approximations originally had nothing to do with the real world. The appearance of computers and calculators changed that, however: were it not for the ideas you are about to explore, calculators would do little more than add, subtract, multiply, and divide.

Activity 3.6

Most of the polynomials with which you have worked have only a few terms. Some polynomials—for example, those used in the codes that protect your compact disc player from errors—have hundreds of terms. There are important polynomials that have an infinite number of terms. For example, consider the following polynomial:

$$(*)\ f(x) = 1 + x + \frac{x^2}{2!} + \frac{x^3}{3!} + \ldots + \frac{x^n}{n!} + \ldots .$$

(Recall that 3! is the **factorial** of 3: 3 x 2 x 1 = 6.)

When you evaluate a finite polynomial, you need only concern yourself with the value of x. Since you cannot evaluate all the terms of an infinite polynomial, you must approximate the polynomial's value by deciding the value of x and the value of n. (Note that in this case, the number of terms is 1 more than the value of n. That's because it is customary to think of the first term of some series as the "0th" term.)

1. Approximate $f(1)$ by creating a table showing the value of polynomial (*) for various values of n when $x = 1$.

2. It may seem paradoxical, but some infinite polynomials give finite results for some or even all values of x. Based on your table and, if necessary, additional calculations, do you think

the value of this polynomial for $x = 1$ is infinite or finite? If you think it is finite, what value do you think is the exact value of $f(1)$?

3. Graph a finite polynomial function using the first six terms ($n = 5$) of the infinite polynomial $1 + x + \frac{x^2}{2!} + \frac{x^3}{3!} + ... + \frac{x^n}{n!} + ...$. In the same graph, graph another function using the first seven terms. Compare the two graphs.

4. Compare the differences in values for the functions in Item 3 when $x = 1$ and when $x = 5$.

5. Compare the graphs you made in Item 3 and the values you found in Item 4 to those for the function $f(x) = e^x$. What conclusions can you draw?

POWER SERIES

The branch of mathematics known today as calculus was invented in the 17th century by Isaac Newton (1642–1727). Historians also give some credit to the German mathematician Gottfried Leibniz (1646–1716), who may have developed some of the ideas independently of Newton, although somewhat later. Calculus solved a number of important scientific problems of the day and continues to solve important problems today. In fact, it is difficult to imagine a world without calculus.

One of the important results derived by Newton, Leibniz, and others is that certain mathematical functions can be approximated by infinite polynomials called power series. A **power series** is a polynomial of the form $a_0 + a_1x + a_2x^2 + \cdots + a_nx^n + \ldots$, where each coefficient a is a real number. The expression a_nx_n is called the **general term** of the series.

Although calculators and computers amaze people with their blinding speed and impressive graphics, they are simple machines: they can perform basic arithmetic operations, display results, and little else. A scientific calculator appears to do much more than simple arithmetic because other mathematical functions can be approximated by power series or rational functions, and because calculating a power series or rational function requires only addition, subtraction, multiplication, and division. The fact that adequate precision may require thousands of such calculations is relatively unimportant because of the machine's speed.

Calculus yielded so many results so quickly that some people, including non-mathematicians such as Bishop George Berkeley (1685–1753), attacked its methods. That the methods of calculus would be questioned is understandable since calculus deals with the infinite. Although mathematicians have developed careful methods of analysis to show that power series do indeed approximate other functions, the study of those methods is best left for a calculus course. The discussion here is primarily conceptual.

A concept essential to any study of power series is convergence. A power series **converges** for a particular value of x if it gets closer and closer to some finite value as the number of terms in the series becomes infinite. (If a power series doesn't converge, it is said to **diverge**.)

EXAMPLE 13

Consider the power series $1 - x + x^2 - x^3 + x^4 + \ldots + (-1)^n x^n + \ldots$. (Notice that the signs alternate—each coefficient is either –1 or 1.) Determine whether this series converges or diverges for $x = 0.2$ and for $x = 2.0$.

SOLUTION:

Table 3.13 shows the results for these two values of x.

This power series converges for $x = 0.2$, but diverges for $x = 2$. (To prove these results conclusively requires methods that you will study in a calculus course.)

Power	$x = 0.2$	$x = 2$
0	1	1
1	0.8	–1
2	0.84	3
3	0.832	–5
4	0.8336	11
5	0.83328	–21
6	0.833344	43
7	0.8333312	–85
8	0.83333376	171
9	0.83333325	–341
10	0.83333335	683
50	0.83333333	7.506×10^{14}
100	0.83333333	8.451×10^{29}

TABLE 3.13.
Power series results.

EXAMPLE 14

Determine whether the series $1 + x + \frac{x^2}{2!} + \frac{x^3}{3!} + \ldots + \frac{x^n}{n!} + \ldots$ converges or diverges for $x = 0.2$ and for $x = 2.0$.

SOLUTION:

The series is an example of a power series that converges for every value of x: in fact, it converges to e^x. However, it converges more slowly for larger values as **Table 3.14** shows. (The error column contains the difference between the approximation and the actual value of e^x.) To nine decimal places $e^{0.2} \approx 1.221402759$, and $e^2 \approx 7.389056099$.

Power	$x = 0.2$	Error	$x = 2$	Error
0	1	0.22140276	1	6.3890561
1	1.2	0.02140276	3	4.3890561
2	1.22	0.00140276	5	2.3890561
3	1.22133333	6.9425×10^{-5}	6.33333333	1.05572277
4	1.2214	2.7582×10^{-6}	7	0.3890561
5	1.22140267	9.1494×10^{-8}	7.26666667	0.12238943
6	1.22140276	2.6046×10^{-9}	7.35555556	0.03350054
7	1.22140276	6.4932×10^{-11}	7.38095238	0.00810372
8	1.22140276	1.4397×10^{-12}	7.38730159	0.00175451
9	1.22140276	2.8866×10^{-14}	7.38871252	0.00034358
10	1.22140276	6.6613×10^{-16}	7.38899471	6.139×10^{-5}

TABLE 3.14. Power series approximation to $e^{0.2}$ and e^2.

Exercises 3.6

1. What is the minimum number of terms needed for $1 + x + \frac{x^2}{2!} + \frac{x^3}{3!} + ... + \frac{x^n}{n!} + ...$ to be within 0.001 of $e^{0.2}$? To be within 0.001 of e^2?

2. Consider the power series $1 + x + x^2 + x^3 + ...$.

   a) Does this series converge or diverge when $x = 0.5$? Justify your answer.

   b) Does this series converge or diverge when $x = 2$? Justify your answer.

   c) For this series, call the sum S: $S = 1 + x + x^2 + x^3 + ...$. Write this equation twice, multiply both sides of one of the equations by x, subtract the two equations, and solve for S. What can you conclude?

   d) Can you apply your result in (d) to the series $2 + 2x + 2x^2 + 2x^3 + ...$? In general, how does your result apply to the infinite geometric series $a + ar + ar^2 + ...$? (A geometric series is one in which a constant multiplier is used to generate a term from its predecessor.)

3. Since the pattern in a power series is usually clear, it is not necessary to give a formula for the general term. However, such formulas are commonly given. Consider the power series $1 - \frac{x^2}{2!} + \frac{x^4}{4!} - \frac{x^6}{6!} + ...$ Finding a formula for the general term is a bit tricky because the signs alternate and because the powers increase by 2. **Table 3.15** shows the essential information. (Remember that in a power series, the "first" term is considered term 0.)

Term number	Sign	Power
0	+	0
1	–	2
2	+	4
3	–	6

TABLE 3.15.

   Note that each power is double the term number. The alternating signs can be described using powers of –1. Thus, the general term is $(-1)^n \frac{x^{2n}}{(2n)!}$.

   a) Does $(-1)^{n+2} \frac{x^{2n}}{(2n)!}$ also give the general term?

   Does $(-1)^{2n} \frac{x^{2n}}{(2n)!}$?

b) Find a general term for $-1 + \frac{x^2}{2!} - \frac{x^4}{4!} + \frac{x^6}{6!} - \frac{x^8}{8!} + ...$

Exercises 3.6

c) Powers of –1 are cyclic. That is, when you raise –1 to successive powers, you get the same two numbers over and over. That's what makes –1 a handy tool for describing power series in which signs alternate. Investigate powers of i. Are they cyclic? Explain.

4. Consider the series $4 - \frac{4}{3} + \frac{4}{5} - \frac{4}{7} + \ldots.$

 a) Is this a power series? Is it a geometric series?

 b) Write an expression for the general term of this series if the first term is generated by $n = 1$. Do the same if the "first" term is generated by $n = 0$.

 c) Does the series appear to converge or diverge? Explain.

5. Since no single method always determines whether series converge, mathematicians rely on a collection of facts and tests.

 a) What can you say about an infinite series of constants (for example, the series $\frac{1}{2} + \frac{1}{2} + \frac{1}{2} + \ldots$)? Explain.

 b) Mathematicians have proved that the terms of any convergent series must approach zero. However, the fact that the terms approach zero is not a guarantee of convergence. For example, consider the series $1 + \frac{1}{2} + \frac{1}{3} + \frac{1}{4} + \ldots.$ If the terms are grouped in the following way, what can you conclude? (From one group to the next, the number of terms doubles).

$$1 + \frac{1}{2} + \left(\frac{1}{3} + \frac{1}{4}\right) + \left(\frac{1}{5} + \frac{1}{6} + \frac{1}{7} + \frac{1}{8}\right) + \ldots.$$

6. Although computers are fast, some problems are so large that time can be a deterrent. The time it takes a computer to solve a problem of a given type is a function of the size of the problem. In other words, solution time = f(size of problem). Computer scientists prefer to design algorithms for which the time function grows as slowly as possible: logarithmic time is preferable to linear time, for example. Although some people find it difficult to believe, polynomial time is preferable to exponential time. Consider **Table 3.16**, which shows the values of the polynomial x^{10} and of the exponential 2^x for values of x from 1–10.

Exercises 3.6

x	x^{10}	2^x
1	1	2
2	1024	4
3	59,049	8
4	1,048,576	16
5	9,765,625	32
6	60,466,176	64
7	282,475,249	128
8	1,073,741,824	256
9	3,486,784,401	512
10	10,000,000,000	1024

TABLE 3.16.

Since 2^x is exponential with a base larger than 1, it will eventually exceed any polynomial, including x^{10}.

a) Determine when 2^x exceeds x^{10}.

b) In general, when will B^x be greater than x^A? Explain.

Chapter 3 Review

1. Summarize the important mathematical ideas of this chapter.

2. a) If there were *no gravity* or air resistance on Earth, what model would describe the height of an object (above Earth's surface) propelled straight up from Earth's surface at a velocity of 150 feet per second? That is, model the relationship between the object's height and the time that it has been in motion.

   b) Modify the model to describe the height of the object if it is propelled upward from the top of a building 100 feet tall.

   c) Modify the model to account for the effect of gravity.

   d) Which of your three models are polynomial functions?

   e) Find the zeros of your model in (c) and interpret them in this context.

   f) Mathematical models almost always exclude some aspects of reality. Name one or more simplifying assumptions for the models you created in this exercise.

3. A scientist develops this data-driven model for the height of an object above Earth's surface: $h(t) = -4.9t^2 + 220t + 20$, where the height is in meters, and t is in seconds. Interpret each of the model's constants.

4. The data in **Table 3.17** are the results of experiments to determine the effect of fertilizer on lettuce crops.

   a) Choose a function from your tool kit to model the relationship between the amount of fertilizer and the relative yield. Which variable is the explanatory variable? Why?

   b) Graph your model and interpret key features in this context.

Pounds of fertilizer per acre	Relative Yield (%)
0	55
100	68
150	72
200	77
250	81
300	86
350	94
400	92
450	97
500	98
550	96
600	93
650	92
700	89
800	84

TABLE 3.17.

c) Explain why you think your model is appropriate in this situation.

d) What degree polynomial would be needed to fit these data exactly? Would such a polynomial be appropriate as a model?

5. Find a polynomial function that passes through each of the points whose coordinates are given in **Table 3.18**.

x	y
5	66
8	73
12	52
16	70
22	47

TABLE 3.18.

6. Use a calculator program or symbolic procedures to solve the following system. Round answers to the nearest hundredth.

$2x + 3y - z = 17$

$5x + y + 2z = 11$

$x - 4y = 7 + 6z$

7. Discuss the merits of quadratic and cubic models for the data in **Table 3.19**.

10	200
20	190
25	175
30	165

TABLE 3.19.

8. a) State the factored form of a polynomial function of lowest degree with x-intercepts at –3, 0, 2, and 6.

b) Graph the polynomial function and discuss the shape of the graph.

c) Create a table of values from $x = -4$ through $x = 7$. Include first- and second-difference columns. Discuss the relationship between the difference columns of the table and the graph.

d) Give the general form of the polynomial function.

e) Write the equation of a polynomial having the same degree and the same zeros, but passing through (1, 40).

9. Is it possible to know the symmetry of a polynomial function's graph from its symbolic form (without seeing its graph)? Explain.

10. Reconsider the sample-pooling of Activity 3.3 and Exercise 8 in Exercises 3.3. This time the strategy is to pool three samples. If the resulting test is positive, then test the first individual only. If the first individual tests positive, then pool the second and third samples. But if the first individual tests negative, then continue to test individual samples. Develop an equation to model this strategy and determine when it is cost-effective.

11. Find all complex zeros of $x^3 - 4x^2 + 5x$.

12. Find a polynomial with $2i$ and $3i$ as zeros.

13. Given the complex numbers $c_1 = 2 - i$ and $c_2 = 3 + 4i$, find:

    a) $c_1 + c_2$

    b) $c_1 \times c_2$

    c) The distance from c_1 to the origin in the complex plane.

14. The Fundamental Theorem of Algebra implies that the polynomial $(x - 4)^2(x^2 + 4)$ has four zeros. Find all four.

15. Evaluate i^2, i^3, and i^4.

16. a) Identify the intercepts and asymptotes of the rational function $f(x) = \frac{x+2}{x^2 - 4}$.

    b) What are the domain and the range of this function?

17. There are two numbers that are not in the domain of the function $y = \frac{2x}{x^2 + x}$.

    a) What are those two numbers?

    b) Is there a natural choice of a value to assign the function for either of these numbers? Explain.

18. a) Find a rational function with a vertical asymptote at –4 and a slant asymptote $y = x - 5$.

    b) What are the domain and the range of your function?

19. a) When $x = 1$, how many terms of the power series $x - \frac{x^3}{3!} + \frac{x^5}{5!} - \frac{x^7}{7!} + \ldots$ are necessary for the series to be within 0.0001 of the function it approximates?

    b) Is the same number of terms required when $x = 2$? Explain.

20. a) Does the infinite series $3 + \frac{3}{4} + \frac{3}{16} + \frac{3}{64} + \ldots$ have a finite sum? Explain.

    b) Write an expression for the general term of the series if the first term is generated by $n = 1$. Do the same if the first term is generated by $n = 0$.

Chapter 4

TRIGONOMETRIC FUNCTIONS

Corbis

CHAPTER INTRODUCTION

Many companies planning for the future must predict how particular quantities vary over time. Often these quantities are affected by the seasons. For example, owners of department stores expect to do brisk business between Thanksgiving and New Year's. Holiday sales must be sufficient to offset sales slumps that occur at other times of the year. Understanding the oscillating pattern of sales and slumps helps stores plan their economic futures.

Seasonal Affective Disorder (SAD) affects approximately 10 million Americans. For some SAD sufferers, shorter hours of daylight can trigger severe depression, while longer hours can produce a heightened sense of well being. Understanding the oscillating pattern of hours of daylight can help some SAD sufferers plan for and control their moods.

To avoid brownouts and blackouts, electric companies must predict their customers' demand for electricity on a weekly, daily, and even hourly basis and then plan how they will supply enough electricity to meet that demand.

Each of the previous examples involves planning. Sometimes this planning is formal, and sometimes it is just a memory from the years before. This chapter focuses on deepening your understanding of oscillating patterns at both a formal and informal level. At the formal level, you will learn to build models describing oscillating behavior and use them to make predictions.

LESSON 4.1

Oscillating Phenomena and Periodic Functions

A pendulum inside an old grandfather clock oscillates steadily back and forth. In some places, temperatures oscillate between hot in the summer and frigid in the winter. Waves in the ocean appear to oscillate up and down. The amount of daylight oscillates between short days in the winter and long days in the summer.

So, what does it mean to oscillate? According to the *American Heritage College Dictionary* it means:

1. To swing back and forth with a steady uninterrupted rhythm.
2. To waver, as between conflicting opinions or courses of action.
3. To vary between alternate extremes, usually within a definable period of time.

In this lesson, you will explore several phenomena that can be characterized by oscillation. Later in the chapter, you will learn how to model oscillating patterns.

Activity 4.1

Understanding oscillating patterns can help people plan for the future. Accurate predictions of future outcomes of oscillating phenomena require a mathematical description of the phenomena. As you know, mathematical descriptions come in several varieties: concise verbal descriptions, tables of data, graphs, and symbolic descriptions such as equations.

This activity focuses on developing mathematical descriptions of oscillating phenomena and using these descriptions to make predictions. You'll begin by considering an oscillating phenomenon that is familiar to everyone.

Most of the time (except during a lunar eclipse), about half the moon is illuminated. However, from the earth, you can usually see only a portion of the illuminated half. The shape of the visible portion and its apparent size change from night to night. For example, **Figure 4.1** shows how the moon looked at midnight on 5, 10, 15, and 20 January 1997.

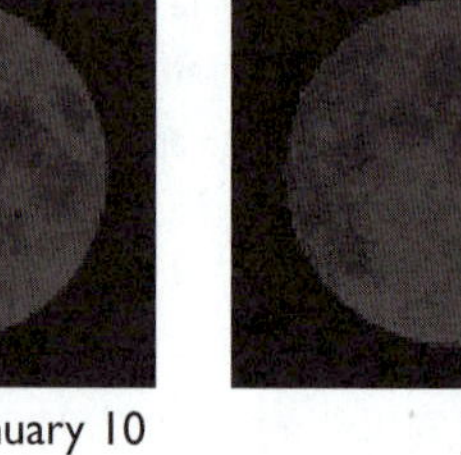
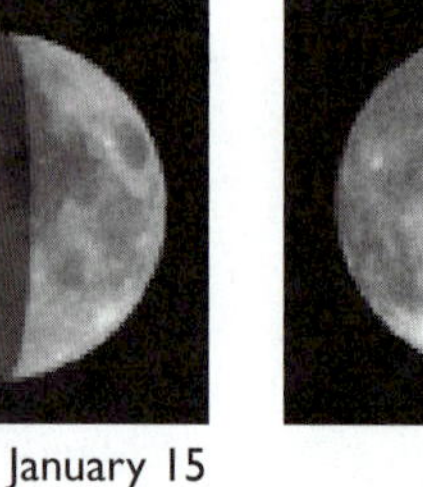

January 5 January 10 January 15 January 20

FIGURE 4.1. Illumination of the moon on four nights in January.

1. Describe how the size of the illuminated visible surface of the moon oscillates over time.

2. The visible portions of the moon's illuminated surface for the first three months in 1997 are presented in **Table 4.1**.

TABLE 4.1. Visible portion of the moon illuminated 5 January 1997–31 March 1997.

Date	1/5	1/10	1/15	1/20	1/25	1/30	2/4	2/9	2/14
Day of year	5	10	15	20	25	30	35	40	45
Visible portion illuminated	0.20	0.02	0.43	0.89	0.98	0.66	0.17	0.04	0.48

Date	2/19	2/24	3/1	3/6	3/11	3/16	3/21	3/26	3/31
Day of year	50	55	60	65	70	75	80	85	90
Visible portion illuminated	0.91	0.97	0.63	0.12	0.07	0.52	0.92	0.96	0.57

Graph the visible portion of the moon that is illuminated versus the day of the year. If you think it is appropriate in this context, draw a smooth curve that follows the pattern of your data. Interpret your graph in the context of the changing moon.

3. Use your graph to predict the visible portion of the moon's surface that is illuminated on 9 and 18 February 1997.

4. Extend your graph so that you can use it to predict the portion of the moon's surface that is illuminated on 15 April (105th day of the year) and 27 April (117th day of the year).

5. Graphs of oscillating phenomena have special characteristics that are different from graphs of mathematical functions discussed in previous chapters of this book. For example, these new graphs repeat some basic pattern over and over again. If you know what happens over this basic interval, you pretty much know the whole thing. Describe the basic interval over which your graph repeats.

6. Another characteristic of graphs of oscillating phenomena is that they often have a limited range. Describe the range of your graph.

7. As a second example of an oscillating phenomenon, consider the revolution of bicycle pedals. Suppose each pedal is 5 inches from the ground at its lowest point and 18 inches from the ground at its highest. You start pedaling with the right pedal at its lowest point and pedal at a constant rate of 1 revolution per second.

   a) Prepare a mathematical analysis of the height of the right pedal over 4 seconds. Include a table and a graph and information similar to your answers to 5 and 6 above.

   b) The speed at which your pedals turn can also be given in, say, inches per minute or miles per hour. Explain how to use the circumference of the circular path of the pedals to convert 1 revolution per minute to inches per minute.

PERIODIC FUNCTIONS

A mathematical function is called **periodic** if it repeats itself over intervals of equal length. Such functions are useful because the world is replete with oscillating phenomena. Every human being encounters them daily.

Date	12/31	1/30	3/1	3/31	4/30	5/30	6/29	7/29
Day number	0	30	60	90	120	150	180	210
Length (hours)	9.1	9.9	11.2	12.7	14.0	15.0	15.3	14.6

Date	8/28	9/27	10/27	11/26	12/26	1/25	2/24	3/26
Day number	240	270	300	330	360	390	420	450
Length (hours)	13.3	11.9	10.6	9.5	9.1	9.7	11.0	12.4

TABLE 4.2.
Data on length of day.

Consider, for example, the length of a day (the number of hours from sunrise to sunset) as a function of the number of the day from a fixed date. **Table 4.2** shows the length of day every 30 days for Boston, Massachusetts, from 31 December 1997 to 26 March 1999. Day numbers are measured from 31 December 1997.

Figure 4.2 shows a graph of the length of the day versus the day number.

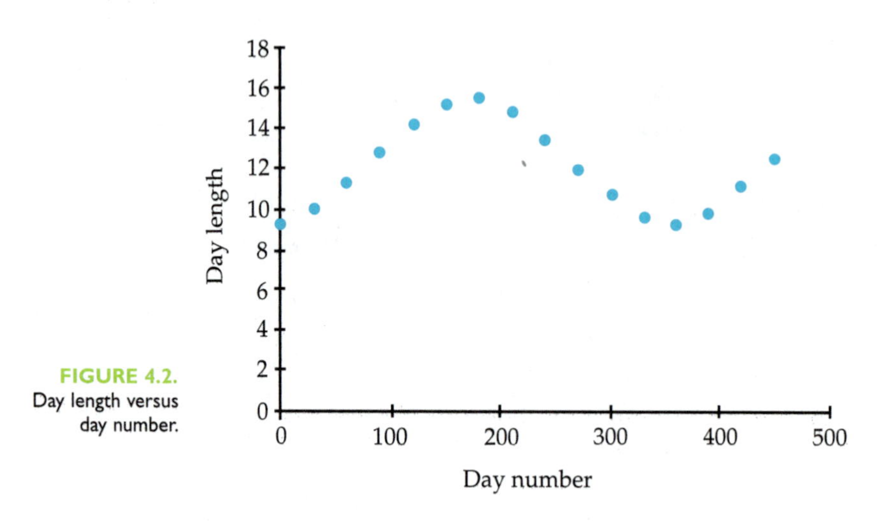

FIGURE 4.2. Day length versus day number.

Figure 4.3 shows the same graph drawn with a smooth curve. The curve represents a graph of a periodic function. (Symbolic representations of such functions are considered in Lesson 4.2.)

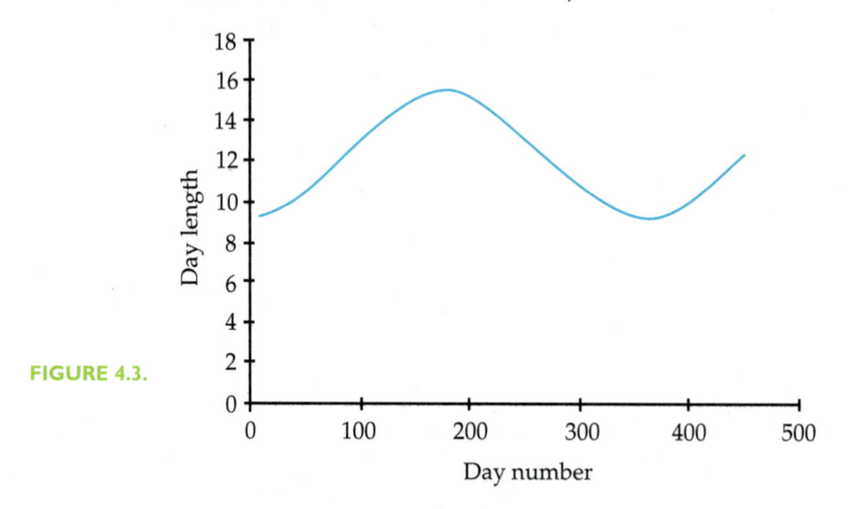

FIGURE 4.3.

In this context, however, a curve might be misleading since it can be interpreted to mean that fractional day numbers make sense. They do not; only integer values are meaningful day numbers. (Negative integers can be interpreted as days before 31 December 1997.) In this situation, a line seems acceptable since individual points for each day would be so close together as to appear to form a continuous curve.

The graph can be used to make predictions, although they will not be very accurate. In this case, the number of hours of daylight can be predicted from the number of the day (counted from 31 December 1997), or days having a given amount of daylight can be estimated. For example, the graph in **Figure 4.4** and the fact that the graph repeats the same pattern show that there are many days for which the day length is approximately 12 hours, but it is difficult to give the day numbers exactly.

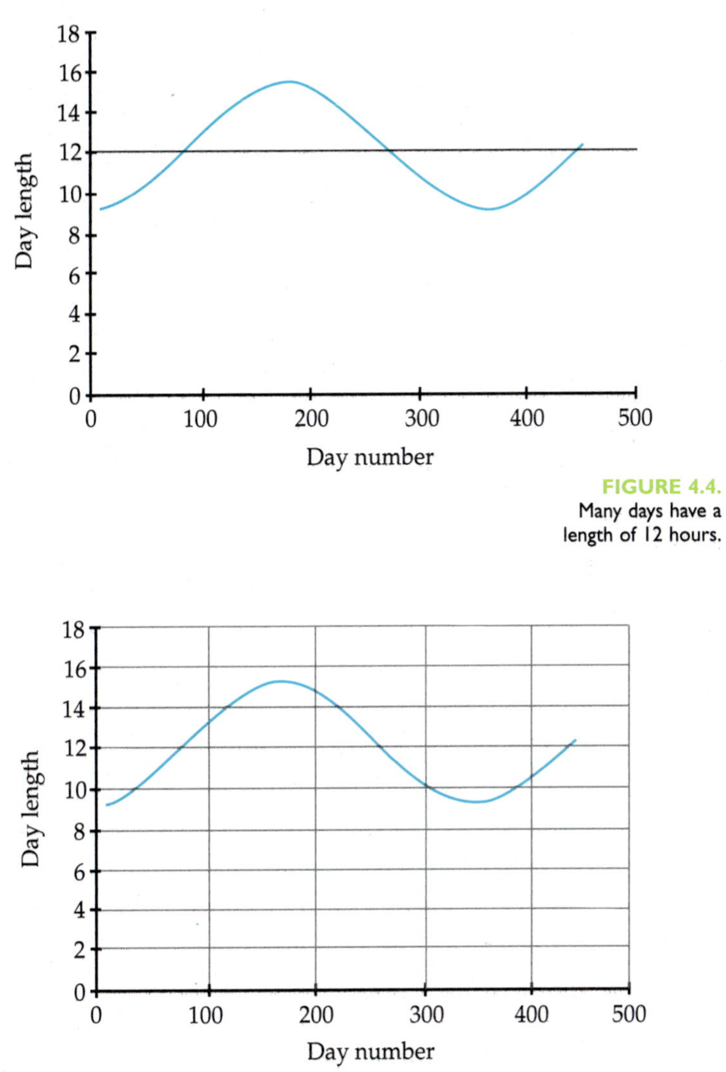

FIGURE 4.4.
Many days have a length of 12 hours.

FIGURE 4.5.
One period extends from 0 to approximately 360.

Accurate predictions are better made with a symbolic representation such as an equation. You will develop symbolic models for periodic phenomena later in this chapter.

PERIOD AND AMPLITUDE

Graphs of periodic functions have a repeating shape. The shortest horizontal length of the basic repeating shape is called the fundamental period, or just the **period**. **Figure 4.5** shows the period of the graph of day length versus day number.

Of course, if you are familiar with the context, you may be able to identify the period without a graph or a table of data. That is the case here. You know that day length oscillates annually, so the period is 365 days.

Another characteristic of periodic functions is that they often have a limited range. The data in this example range between 9.1 and 15.3 hours. The **amplitude** of a periodic graph is half the vertical height of the basic repeating shape. It can be calculated numerically by finding half the difference between the largest and smallest range values:

$$\frac{15.3 - 9.1}{2} = 3.1.$$

A simple graphical interpretation of the amplitude is the distance between the graph's horizontal centerline, called the **axis of oscillation,** and either the highest or lowest point of the graph (**Figure 4.6**). Since the axis of oscillation is centered between the highest and lowest points of the curve, its height can be calculated by averaging the largest and smallest values of the oscillating quantity. (Its height can also be calculated by adding the amplitude to the height of the graph's low point or by subtracting the amplitude from the height of the graph's high point.)

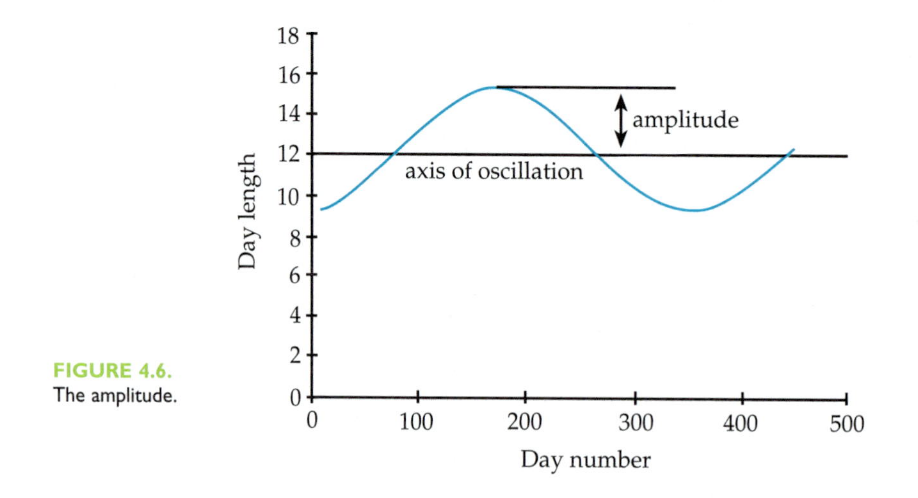

FIGURE 4.6. The amplitude.

Many oscillating phenomena can be approximated with periodic functions. In such cases, predictions are possible, but within some reasonable amount of error.

EXAMPLE 1

Table 4.3 shows a homeowner's heating costs over a two-year period. Graph these data and discuss the period, amplitude, and axis of oscillation of the graph. (Use the number of months as the independent variable.) Predict the heating cost for April of the third year and discuss the accuracy of your prediction.

TABLE 4.3. Monthly heating costs over a two-year period.

Month	Jan	Feb	Mar	Apr	May	June	July	Aug	Sept	Oct	Nov	Dec
Cost ($)	85	87	72	44	25	20	18	17	19	35	79	87

Month	Jan	Feb	Mar	Apr	May	June	July	Aug	Sept	Oct	Nov	Dec
Cost ($)	75	84	58	50	23	15	16	17	20	36	65	85

SOLUTION:

The graph is shown in **Figure 4.7**. In this case, connecting the points with a curve seems inappropriate since fractional month numbers do not make sense.

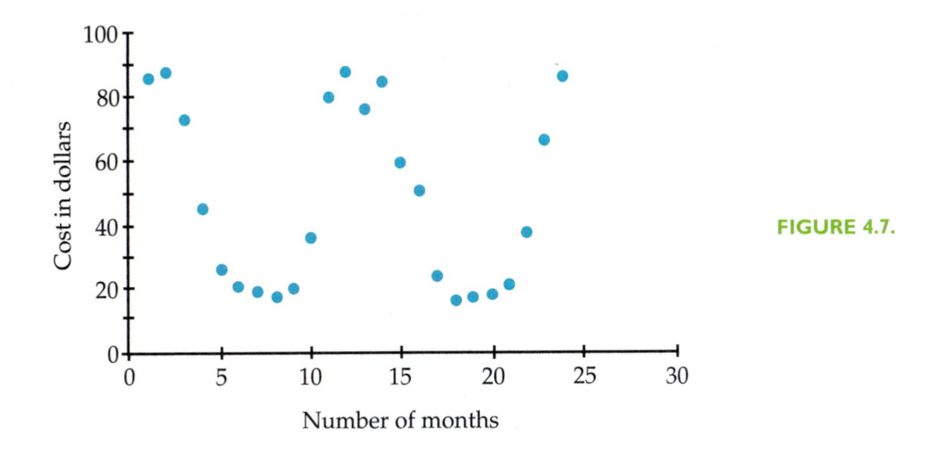

FIGURE 4.7.

The graph is approximately periodic. It repeats over a period of 12 months. The lowest value is at 17 and the highest at 87. Therefore, the amplitude is approximately $\frac{87-17}{2} = 35$. The axis of oscillation is a horizontal line at approximately $y = 17 + 35 = 52$.

Based on these data, a prediction of about $47 seems reasonable for April of the third year, but an error of about $5 also seems reasonable. However, major changes in the weather and/or the price of fuel could cause even greater error. Note, for example, the fluctuation between March of the first year and March of the second.

Exercises 4.1

1. **Figure 4.8** shows graphs of three periodic functions. Estimate the period, amplitude, and axis of oscillation of each.

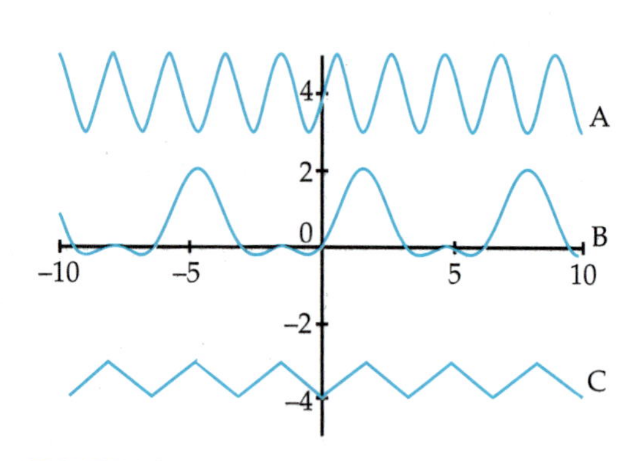

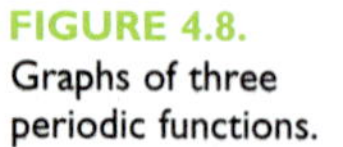
FIGURE 4.8.
Graphs of three periodic functions.

2. Sketch graphs that represent each of the following situations. Tell whether each graph is periodic. If a graph is periodic, discuss its period and amplitude.

   a) Your height above the ground as you ride a roller coaster.

   b) The height of the first car on a roller coaster if it is plotted all day. Assume that the roller coaster is a popular ride and runs continuously all day.

   c) Explain the difference between situation (a) and situation (b).

   d) Your distance from the ticket booth as you ride a horse on a carousel.

3. Carousel horses move up and down as the carousel rotates. Suppose the back hooves of a horse are 5 inches above the floor of the carousel at their lowest point and 2.5 feet above the floor at their highest point. It takes 5 seconds for the horse to move from its lowest level to its highest.

   a) Prepare a table and graph showing the height of the back hooves during a 30-second portion of a carousel ride. Assume the horse starts at its lowest point.

   b) Discuss the period, amplitude, and axis of oscillation of your graph.

   c) When you ride a carousel, in addition to moving up and down, you get to spin around. The Dentzel carousel at Glen Echo Park in Maryland is 48 feet in diameter and turns counterclockwise, making about 5 revolutions per minute. Suppose you choose to ride the intricately carved lead horse, located in the outer row approximately 2 feet from the edge. Each time the carousel makes a complete turn, how far (in feet) have you ridden around? (Do not include any up-and-down distance.)

d) How fast, in feet per minute, are you riding? How fast would this be in mph?

Exercises 4.1

e) Your friend is riding on an inside horse that is approximately 6 feet from the edge. In terms of miles per hour, how much faster are you moving than your friend?

4. **Table 4.4** gives the total electric consumption by bakeries in 1988–1989. Power consumption is recorded as the percentage of January 1987 usage. (This method of recording enables the bakery to compare its power consumption with a benchmark level.)

Month	Jan	Feb	Mar	Apr	May	June	July	Aug	Sept	Oct	Nov	Dec
1988	94.6	92.6	93.7	92.6	96.2	106.3	109.2	114.7	114.1	105.1	101.0	99.5
1989	99.5	93.6	95.5	98.3	99.1	110.8	114.7	110.8	115.0	109.7	98.9	100.6

TABLE 4.4. Electric power consumption by bakeries, 1988–1989.

a) Graph these data and discuss whether the graph is approximately periodic.

b) If you think the graph is periodic, estimate the period and amplitude.

c) Describe how to use your graph to predict the power consumption for March 1990. How much error do you think is in your prediction? Explain.

5. Data on housing starts are of interest to city planners, to real-estate agents, and most of all to building contractors and construction workers. Housing-start data (in thousands) for one year beginning March 1983 are presented in **Table 4.5**.

Month	Mar 83	Apr	May	June	July	Aug
Starts (x1000)	124.3	122.1	161.5	160.1	148.0	159.8

Month	Sept	Oct	Nov	Dec	Jan 84	Feb
Starts (x1000)	139.6	147.8	122.1	103.2	102.7	120.2

TABLE 4.5. Housing starts March 1983–March 1984.

a) Graph these data. Then assume that the pattern is periodic and extend your graph to represent the pattern of housing starts for a two-year period.

Exercises 4.1

b) Based on your graph, in what months do you expect the fewest housing starts? In what months do you expect the largest number of housing starts? Do your answers to these two questions make sense from what you know about home building? Explain.

c) Can you offer an explanation for why the number of housing starts in July might be less than for June or August?

d) Do you think that it is reasonable to assume that housing starts are approximately periodic? Why or why not? What information would you need to check the reasonableness of this assumption?

6. Assume that the Ferris wheel pictured in **Figure 4.9** rotates at a constant speed of 30 seconds for one complete rotation. The ride lasts approximately 5 minutes and 3 seconds from the time it picks up the last rider until it begins stopping to let riders off.

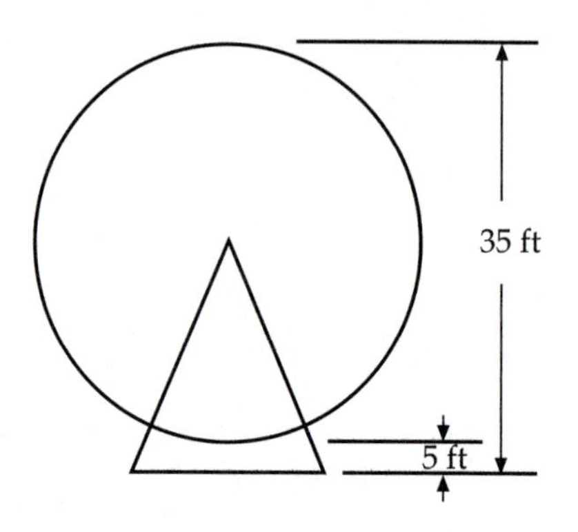

FIGURE 4.9.
Ferris wheel rotating at a constant speed.

a) How fast are the riders spinning in revolutions per minute (rpm)? How fast are they moving in feet per second?

b) Your friend is the last person to board the Ferris wheel. Plot her height above the ground every 7.5 seconds for the first 2 minutes of the ride. Then draw a smooth curve (no corners) connecting the points on your graph.

c) At approximately what times during the first half-minute of the ride is your friend 30 feet above the ground?

d) Use the periodic nature of your graph to describe all the times when your friend is 30 feet above the ground during the entire ride. How are these times related to the times in (c)?

e) In part (b) you were asked to draw your curve with no corners. Explain why sharp corners are unreasonable.

7. How would your graph in 6(b) change under each of the following conditions? In each, sketch a graph of height versus time for the altered situation on a copy of your answer graph in 6(b).

a) The Ferris wheel is slower and makes one revolution every minute instead of every half minute.

b) The Ferris wheel is faster and makes one revolution every 15 seconds instead of every half minute.

c) The Ferris wheel is turning at the original speed, but is 10 feet off the ground instead of 5 feet.

d) The Ferris wheel is turning at the original speed, is 5 feet off the ground, but has a diameter of 50 feet instead of 30 feet.

e) The Ferris wheel is making 2 revolutions per minute, is 2 feet off the ground, and 40 feet in diameter.

8. Holland-based Nauta-Bussink sells amusement park rides that include giant 33 meter, 44 meter, and 55 meter Ferris wheels. These dimensions refer to the height of the top of the wheel. Assume that the wheels rotate at constant velocities (ignoring start-up and slow down times).

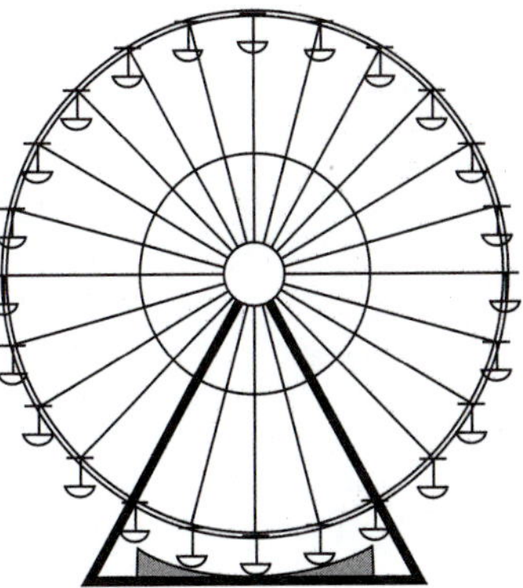

a) The 33 meter Ferris wheel has a wheel diameter of 29 meters and turns at a rate of 2.6 rpm. How fast do its riders move, in meters per second, around the circle?

b) The 44 meter Ferris wheel has a wheel diameter of 40.7 meters and turns at a rate of 1.5 rpm. In terms of meters per second, would a rider be moving faster or slower on the 44 meter Ferris wheel than on the 33 meter Ferris wheel? How much faster or slower?

c) The 55 meter Ferris wheel has a wheel diameter of 52.0 meters and turns at a rate of 1.5 rpm. How fast do its riders move (in meters per second) around the circle?

d) If you plotted a rider's height versus time for each of these Ferris wheels, what would be the period and amplitude of each of your graphs?

e) Suppose that one of the lights on the outside of the 55 meter Ferris wheel has burned out. Imagine starting your stopwatch the instant the light is at the bottom of the wheel. How high off the ground is the light at the start of the ride? Approximately how high off the ground will this light be 10 seconds later? 20 seconds later? 30 seconds later? Explain.

Exercises 4.1

f) If the ride in (e) continues for another 4 minutes after you start your stopwatch, at what times will the burned-out light be at the top of the wheel? How are your answers to this question related?

9. Eclipsing binary stars are two stars in close orbit around each other. The stars are often so close that they appear to be one point of light in the night sky. If the plane of their orbit is oriented to contain the earth, the brightness of the star system will vary over time. A light curve is a graph of the brightness of the star system over time. **Figure 4.10** is a plot of hypothetical brightness data measured daily. (The larger the Luminosity number, the brighter the star system.)

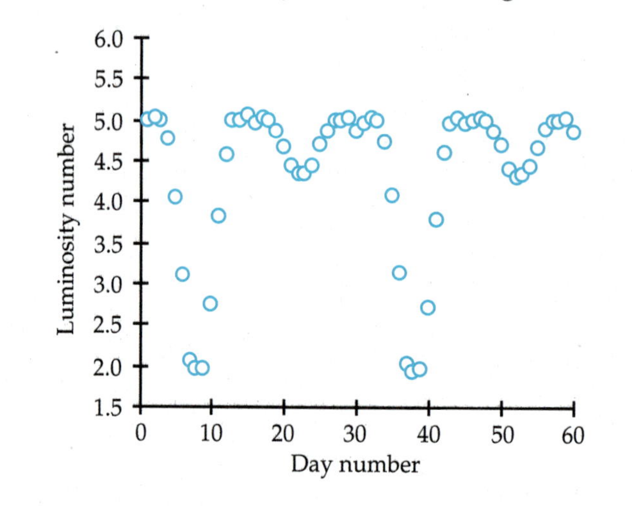

FIGURE 4.10.
Luminosity versus day.

a) Would you consider these data to be (approximately) periodic? If so, what are the period and amplitude?

b) Use the pattern of the graph to predict the Luminosity number of the star system on days 80 and 100. How can your answer to (a) be used to help you answer this item?

c) What do you think might explain the shape of this light curve (the graph of Luminosity versus time)?

10. You may have noticed that portions of periodic graphs resemble parabolas. This exercise considers parabolic approximations of portions of periodic graphs. The graph in **Figure 4.11** represents the height above the floor of a carousel horse's back hooves versus time during a carousel ride.

Exercises 4.1

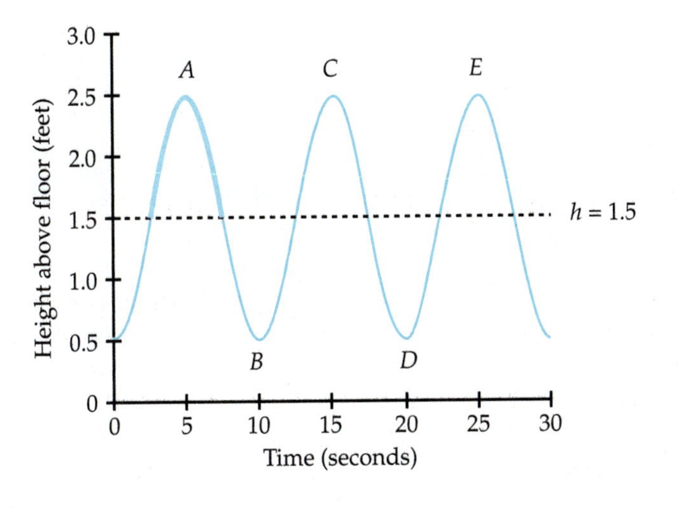

FIGURE 4.11.
The height of the carousel horse's hooves versus time.

a) Write an equation in the form $y = a(t - h)^2 + k$ that describes loop A (the approximately parabolic section labeled A). For what values of t does this equation apply? How did you determine your equation?

b) What modifications would you make to your equation in (a) in order to describe loops C and E? Over what intervals would each of these equations apply?

c) Write an equation in the form $y = a(t - h)^2 + k$ that describes loop B. For what values of t does this equation apply?

d) Adapt your equation describing loop B to describe loop D. What constants in the equation $y = a(t - h)^2 + k$ did you have to change? Over what interval does your adapted equation apply?

e) Explain how to use your equations to estimate times at which the height is 2 feet.

LESSON 4.2

The Sine Function

In the previous lesson, you investigated the height above the ground of someone on a Ferris wheel over time. In drawing your graphs, you plotted only a few points per period and then joined them using a smooth curve. Ideally you could avoid the data step and use some kind of equation (function) to describe such motion. A quadratic can be used to represent a portion of a periodic graph, but quadratics have at least two difficulties. First, you have not verified that a quadratic is accurate. And second, quadratics are cumbersome to use since each portion of the periodic graph requires a separate equation.

The key to dealing with these two problems is looking more closely at the context—more data! Of course, you can't fit a full-size Ferris wheel in your classroom, so you need something smaller to model a Ferris wheel rider's height above the ground versus time.

In this lesson, you will use a circular object as the physical model. After studying this model, you will return to the Ferris wheel and apply your new knowledge to model someone's height above ground over time.

Courtesy Carnegie Library of Pittsburgh

The original Ferris wheel was built for the 1893 Chicago World's Fair by George Ferris, a Pittsburgh bridge builder.

Activity 4.2

It is difficult to time a revolving object accurately. Since Ferris wheels move at a relatively constant speed, the time a rider travels and the distance the rider travels are directly related. Therefore, this activity explores the relationship between a rider's height and the distance traveled.

For this activity you need a circular object such as a large coffee can, a piece of string, a ruler, and a tape measure. You need enough string to wrap around the circular object several times.

The can represents the Ferris wheel. To determine the distance a person riding the wheel has traveled, you will wrap the string to a point representing a person's location and then measure the string. Use the ruler to measure the height of the point.

To begin, secure the tape measure and the string to an arbitrary starting point with a piece of tape. Use one or more pieces of tape to secure the tape measure to your table or the floor. Use a marker to mark 10 or more points around the circumference of your circular object.

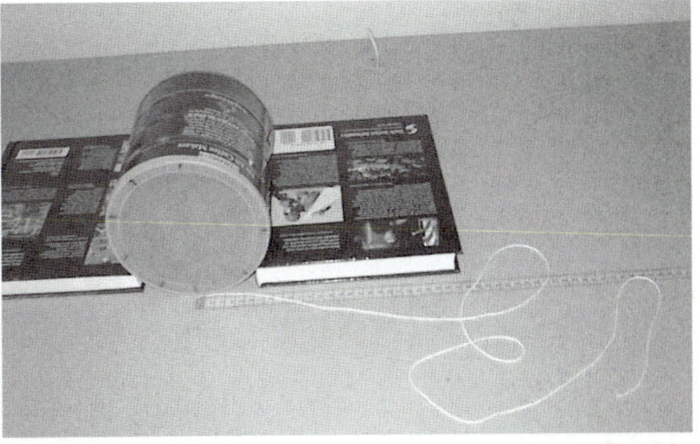

FIGURE 4.12.
The equipment assembled.

Choose one of the marked points to represent the starting position of the person riding the Ferris wheel and position the can so this point is directly over the tape measure's zero mark. Use two books to keep the circular object from moving. When you are finished setting up your experiment, it should resemble **Figure 4.12.**

FIGURE 4.13.
Measuring the height of a point.

Figures 4.13–4.15 show how data are collected. Select a point, measure its height (Figure 4.13), wrap the string to the point (Figure 4.14), then measure the length of the string (Figure 4.15).

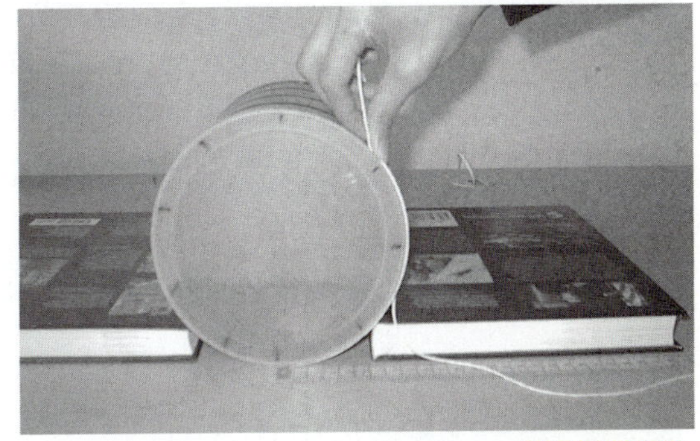

FIGURE 4.14.
Wrapping the string to the point.

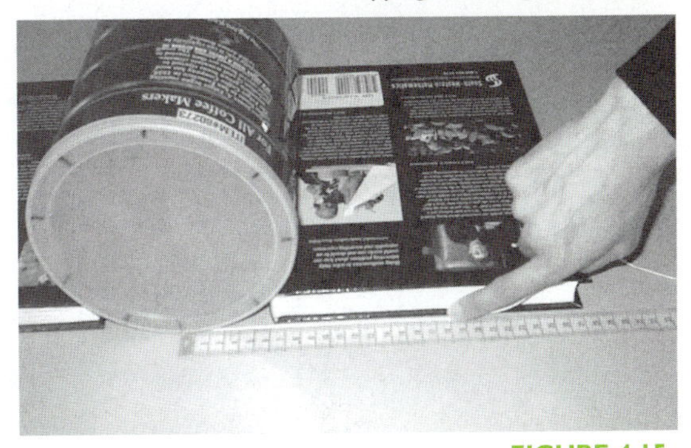

FIGURE 4.15.
Measuring the length of the string.

1. Collect data representing two complete revolutions of the Ferris wheel. That means you will have two pairs for each point. For the second pair, wrap the string once around the circular object, then up to the point. Collect your data in a table.

2. Plot the height of the point versus the length of the string. Sketch a smooth graph that you think represents the relationship between the two variables.

3. Explain why these data should be periodic. Find the period and the amplitude and discuss their meaning in this setting.

4. Your graph can be extended to the left of the origin. Explain the meaning of negative values of the independent variable in this setting.

5. Mathematicians often use a function called the sine function to model periodic phenomena. (It is customary to shorten sine

to sin.) Your calculator is capable of evaluating sine for different units (modes) of the independent variable. For this item, put your calculator in the radian mode. Graph the function $y = \sin(x)$. By experimenting with different windows and by using the trace feature of your calculator, identify the period and amplitude. Show your graph for at least two complete periods.

6. Compare your graph of the sine function in Item 5 to the curve you sketched in Item 2. Discuss transformations (vertical and horizontal shifts and stretches) that would turn the sine graph into your sketch in 2.

7. Experiment with variations of the sine function on your graphing calculator. Try to find a function that demonstrates one or more of the transformations you described in Item 6. Record the function and describe the transformations of $y = \sin(x)$ that it demonstrates.

RADIAN MEASURE

When a carousel turns, the rate at which a person rotates is the same regardless of the distance from the center. Yet people who are at different distances from the center cover different distances in equal amounts of time. Mathematical modelers often simplify discussions of circular motion by assigning a value of one unit to the radius of a circle. Standardizing the radius in this way leads to a measurement system called radians. Radians are a dimensionless measure that can be applied to arcs of a circle or to angles at the circle's center. To begin as simply as possible, you will consider radian measure as it applies to arcs of the simplest of circles. You will extend radian measure to arcs of other circles and to angles in this lesson's exercises.

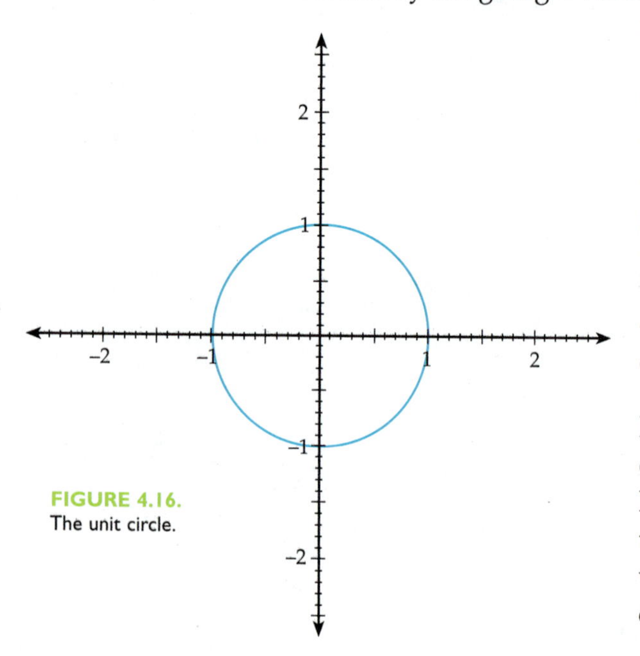

FIGURE 4.16. The unit circle.

If your class is typical, the size of circular objects used in Activity 4.2 varied. Mathematicians use a circle with radius one unit to define periodic functions. A circle with radius one unit centered at the origin is called the **unit circle (Figure 4.16)**.

In Activity 4.2, you wrapped string from a point at the bottom of your circular object. Mathematicians wrap from a point on the right of the unit circle and measure arcs in radians. As applied to the unit circle, **radian measure** is the directed length of an arc that begins at (1, 0). Positive radians are measured in a counterclockwise direction; negative radians are measured in a clockwise direction.

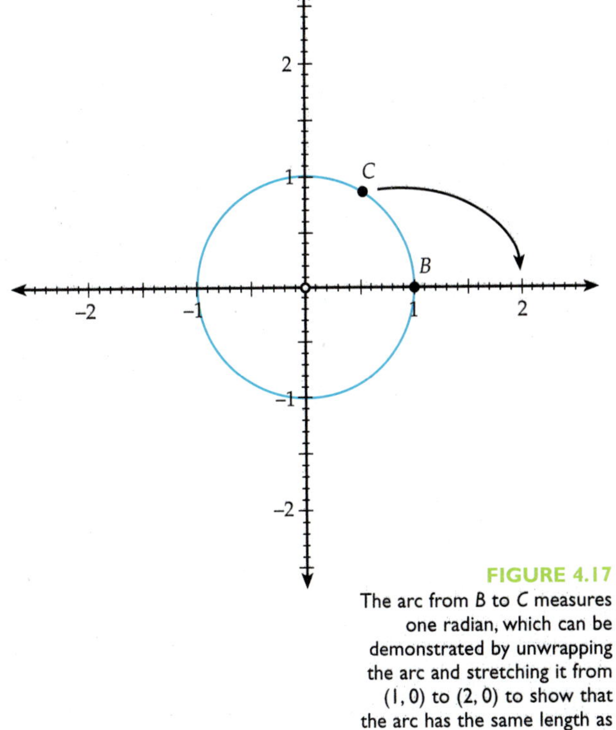

FIGURE 4.17
The arc from *B* to *C* measures one radian, which can be demonstrated by unwrapping the arc and stretching it from (1, 0) to (2, 0) to show that the arc has the same length as the circle's radius.

The size of the unit chosen for a coordinate system is arbitrary. It might be an inch, a centimeter, a foot, or a mile. Any circle can be a unit circle if the coordinate system is chosen so that one unit is equal to the circle's radius. Thus, 1 radian measures an arc beginning at (1, 0) whose length is equal to the circle's radius, which equals one unit (**Figure 4.17**) of the coordinate system.

Since the circumference of a circle is 2π times the length of the radius, the exact radian measure associated with some points on a unit circle can be found.

EXAMPLE 2

Find several radian measures associated with the point (0, 1) on a unit circle.

SOLUTION:

The circumference of the unit circle is $2\pi(1) = 2\pi$ units. Wrapping from (1, 0) to (0, 1) in a counterclockwise direction is one quarter of the unit circle, so a radian measure associated with (0, 1) is $\frac{2\pi}{4} = \frac{\pi}{2}$, or approximately 1.57 radians.

If the wrapping is done in a clockwise direction, it covers three-quarters of the circle before reaching (0, 1), so the radian measure is $\frac{-3}{4}(2\pi) = \frac{-3\pi}{2}$, or about –4.71 radians.

Since wrapping can progress completely around the circle one or more times, other radian measures include $\frac{\pi}{2}+2\pi=\frac{5\pi}{2}$, $\frac{\pi}{2}+4\pi=\frac{9\pi}{2}$, $\frac{-3\pi}{2}-2\pi=\frac{-7\pi}{2}$, and $\frac{-3\pi}{2}-4\pi=\frac{-11\pi}{2}$. There are infinitely many radian measures associated with (0, 1).

EXAMPLE 3

Sketch a unit circle showing an arc measuring 3 radians.

SOLUTION:

Since the circumference of the unit circle is $2\pi \approx 6.28$ units, halfway around the circle is $\pi \approx 3.14$ units. Therefore, an arc in a counterclockwise direction from (1, 0) to a point on the circle slightly above (–1, 0) has a measure of approximately three radians. In **Figure 4.18**, that is the counterclockwise arc from B to C.

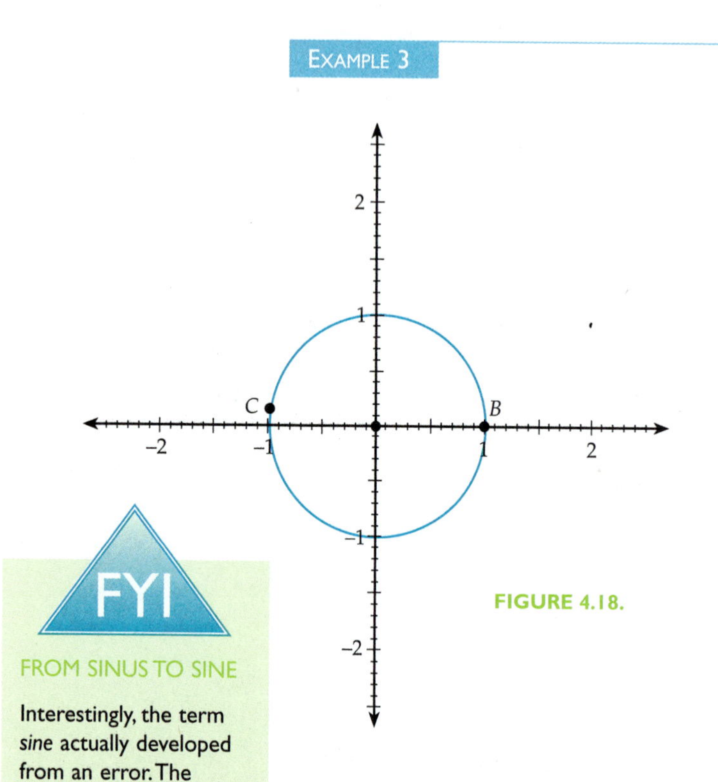

FIGURE 4.18.

FYI

FROM SINUS TO SINE

Interestingly, the term *sine* actually developed from an error. The Hindu term *jya*, meaning half-chord became *jyb* in Arabic. A translator, translating from Arabic to latin, incorrectly read *jyb* as *jayb* (Arabic for pocket). Thus, the latin *sinus* was used by mistake and we now use this in the form *sine*.

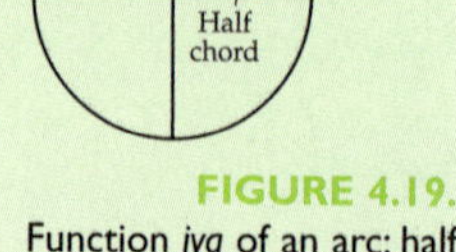

FIGURE 4.19.
Function *jya* of an arc: half chord in a circle of radius *r*.

The Sine Function

The **sine function** pairs radian measure with the vertical displacement from the horizontal axis of the point at which the arc of the unit circle terminates. Since vertical displacement is negative for points below the horizontal axis, the sine is negative when wrapping terminates in Quadrant III or IV.

EXAMPLE 4

Use the unit circle to find $\sin\left(\frac{\pi}{2}\right)$. Use a calculator to verify the answer.

SOLUTION:

On the unit circle, a directed arc clockwise from (1, 0) with length $\left(\frac{\pi}{2}\right)$ terminates at (0, 1). So, the vertical displacement is 1, and $\sin\left(\frac{\pi}{2}\right) = 1$.

Figure 4.20 shows how to verify the answer with a calculator. First put the calculator in the radian mode (left), then evaluate the expression on the calculator's home screen (right).

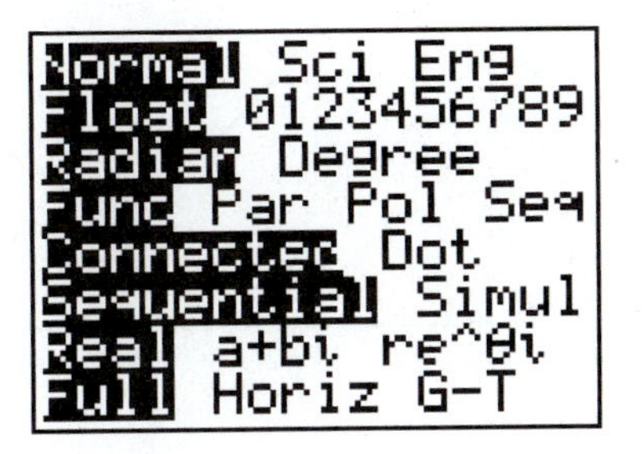

sin(π/2)
1

FIGURE 4.20.

The sine function performs the same action you used to collect data in Activity 4.2, but it uses a circle of radius 1 centered at the origin. The radian measure is similar to the length of the string, and the value of the sine is similar to the height of the point. Your data did not generate a true sine graph because your string lengths and heights were not taken from a unit circle and because your circle was not centered at the origin. In this lesson's exercises, you will extend radian measure to non-unit circles and to angles.

To use the sine function to model your data in Activity 4.2, you must first consider its graph, then learn to transform it to fit your data. **Figure 4.21** shows two cycles of the graph of $y = \sin(x)$. The period is 2π, the circumference of the unit circle. The amplitude is 1, the radius of the unit circle.

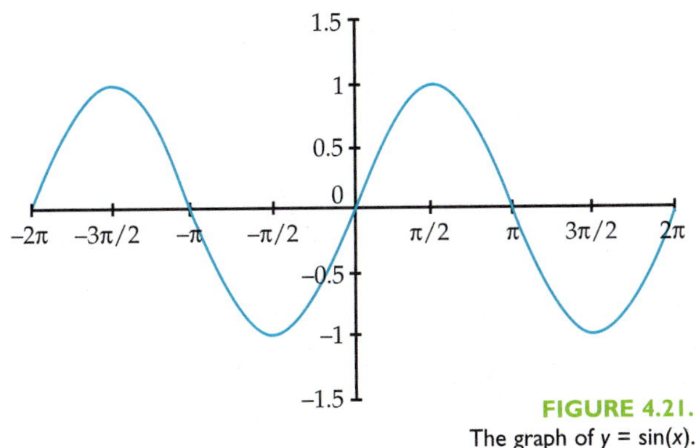

FIGURE 4.21.
The graph of $y = \sin(x)$.

TRANSFORMING THE SINE FUNCTION

The graph of the sine function can be transformed to model many periodic data. The essential transformations are ones with which you are already familiar: vertical shifts, vertical stretches, horizontal shifts, and horizontal stretches.

A **sinusoidal function** has the form $y = A\sin(B(x - C)) + D$. Each control number is responsible for one of the basic transformations, as shown in **Figures 4.22–4.25**.

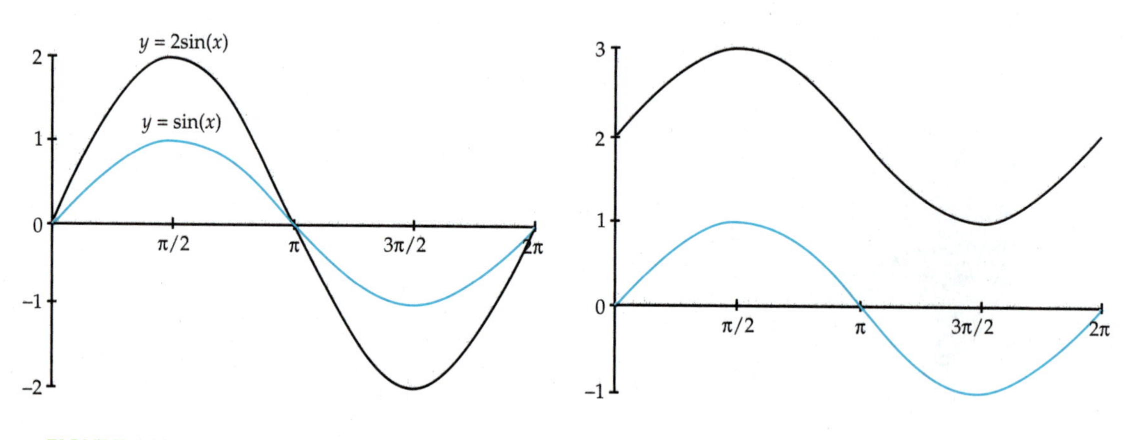

FIGURE 4.22.
The graphs of $y = \sin(x)$ and $y = 2\sin(x)$. The effect of the 2 is to stretch the graph vertically. The amplitude has been doubled.

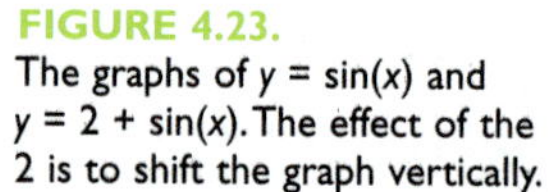

FIGURE 4.23.
The graphs of $y = \sin(x)$ and $y = 2 + \sin(x)$. The effect of the 2 is to shift the graph vertically.

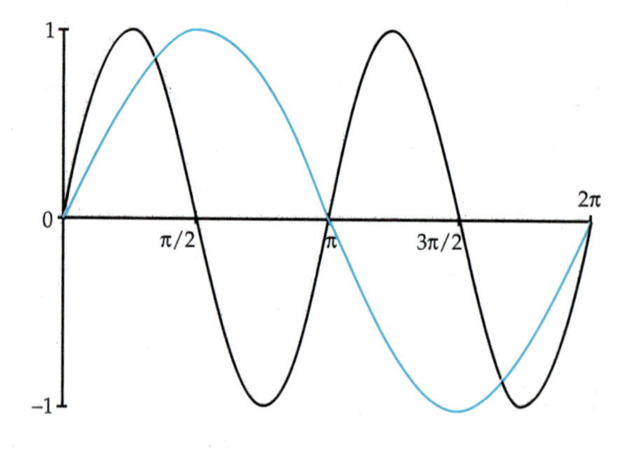

FIGURE 4.24.
The graphs of $y = \sin(x)$ and $y = \sin(2x)$. The effect of the 2 is to compress the graph horizontally. The period has been halved.

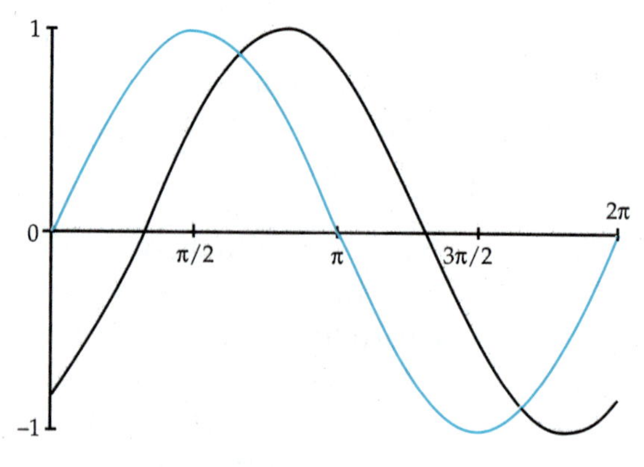

FIGURE 4.25.
The graphs of $y = \sin(x)$ and $y = \sin(x - 1)$. The effect of the 1 is to shift the graph horizontally.

Care must be taken when combining transformations. For example, a horizontal shift of 1 unit followed by compression by a factor of 2 is not the same as compression by a factor of 2 followed by a horizontal shift of 1 unit, as **Figure 4.26** shows.

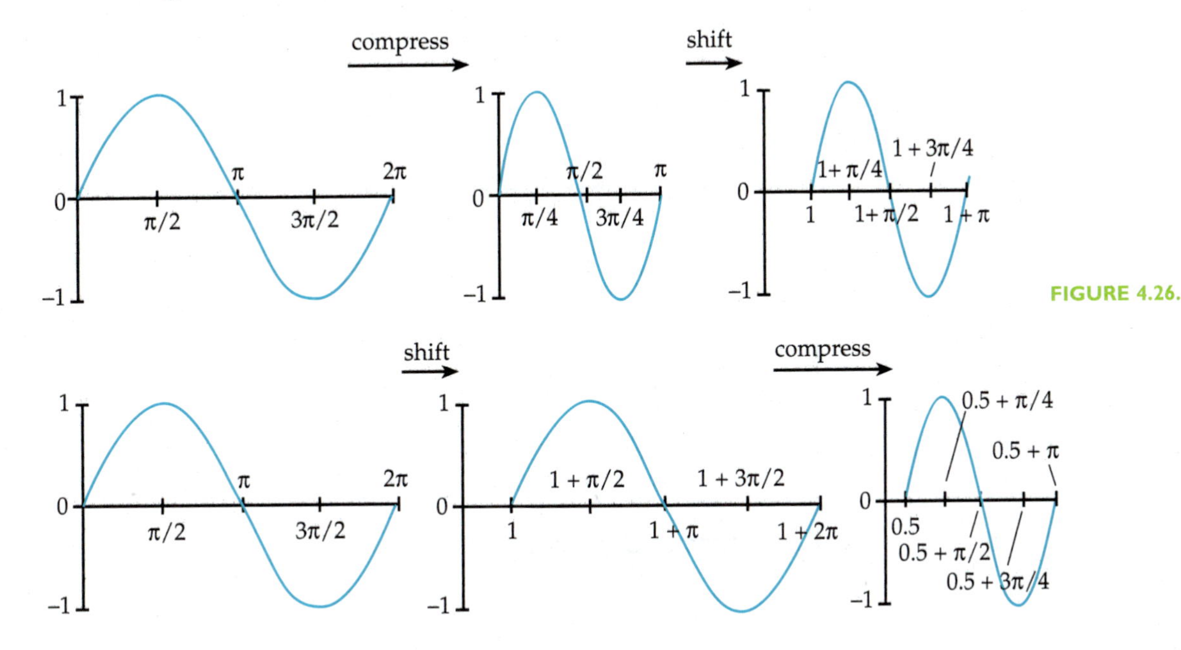

FIGURE 4.26.

In $y = \sin(B(x - C))$, C is the amount the graph of $y = \sin(Bx)$ must be shifted to produce the graph of $y = \sin(B(x - C))$ and is called the **phase shift** of $y = \sin(B(x - C))$.

Example 5

Discuss the transformations of the graph of $y = \sin(x)$ produced by $y = 2 + \sin(0.5(x - 1))$. Verify your answers by using a calculator or computer to graph $y = 2 + \sin(0.5(x - 1))$.

Solution:

The control number 0.5 stretches the graph of $y = \sin(x)$ horizontally by a factor of 2, which is the reciprocal of 0.5. The control number 2 shifts the graph of $y = \sin(0.5x)$ to the right 1 unit. The control number 2 shifts the graph of $y = \sin(0.5(x - 1))$ upward 2 units. All three transformations are verified by the graphs in **Figure 4.27**, one of which was done on a computer, the other on a graphing calculator.

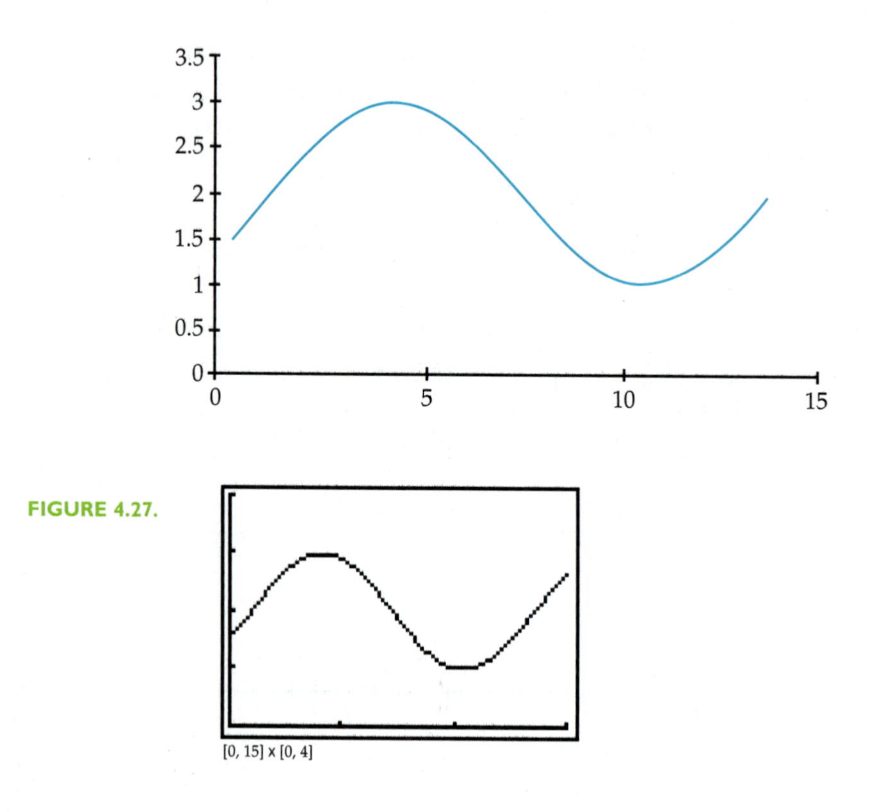

FIGURE 4.27.

Exercises 4.2

1. In (a)–(c), two radian measures are given. For each pair, sketch a unit circle and draw arcs that have each radian measure. Indicate which arc corresponds to each measure.

   a) $\frac{3}{4}\pi$ and $-\pi$

   b) 3π and $\frac{\pi}{2}$

   c) –2.3 and 6.0

2. Radian measure can be extended to circles with radii that do not measure 1 unit. The key idea is that a specific number of radians measure the same amount of rotation on any circle. For example, **Figure 4.28** shows that an arc of 3 radians covers nearly half a rotation, regardless of the circle's size; and an arc of $\frac{-3\pi}{4} \approx -2.4$ radians extends $\frac{3}{8}$ of the way around the circle in the opposite direction, regardless of the circle's size.

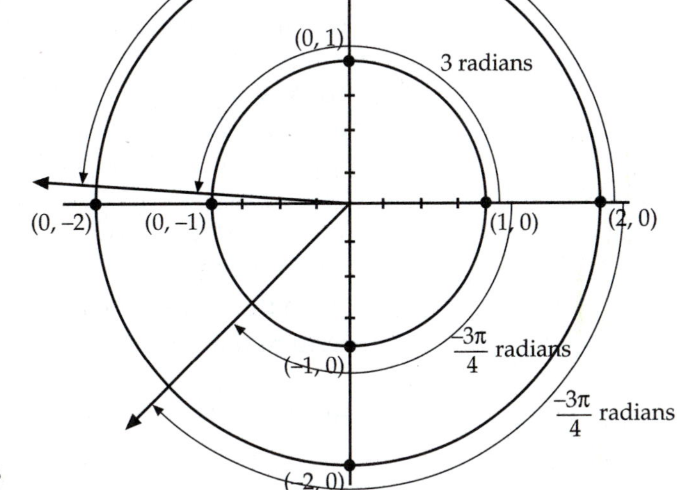

FIGURE 4.28. Arcs measuring 3 radians and $\frac{-3\pi}{4}$ radians on circles with different radii.

   a) When a circle's radius is not 1 unit, an arc's length can be converted to radian measure by making a simple calculation using the length of the arc and the length of the radius. Explain.

   b) Return to your data from Activity 4.2. Convert all your string lengths to radian measures. Explain how you made your conversions.

   c) What effect does the conversion have on the period of your data?

   d) The calculation you made to find radian measures on non-unit circles can be used to show that radian measure is dimensionless; that is, that radians are not measured in the same units as the circle's radius. Explain.

Exercises 4.2

3. In question 2 you showed that radians measure the same amount of rotation regardless of a circle's size. Therefore, radian measure, like degrees, is a form of angle measure.

   a) Explain why π radians and 180° measure the same amount of rotation.

   b) Use your answer to (a) to find a radian measure that is equivalent to 60°.

   c) Give a decimal approximation for the radian measure you found in (b). An often-asked question by people studying radian measure is, "What is the degree measure that corresponds to 1 radian?" Give an approximate answer to this question.

4. How would you convert your height data in Activity 4.2 to match values produced by the graph of the sine function (that is, an axis of oscillation at 0 and amplitude 1)? Explain.

5. Suppose you gathered data by measuring the distance traveled by a carousel horse and the distance from the horse to the ticket booth. Explain the transformations that could standardize these periodic data to radian input, axis of oscillation 0, and amplitude 1.

6. **Figure 4.29** shows several points on a circle of radius 2.

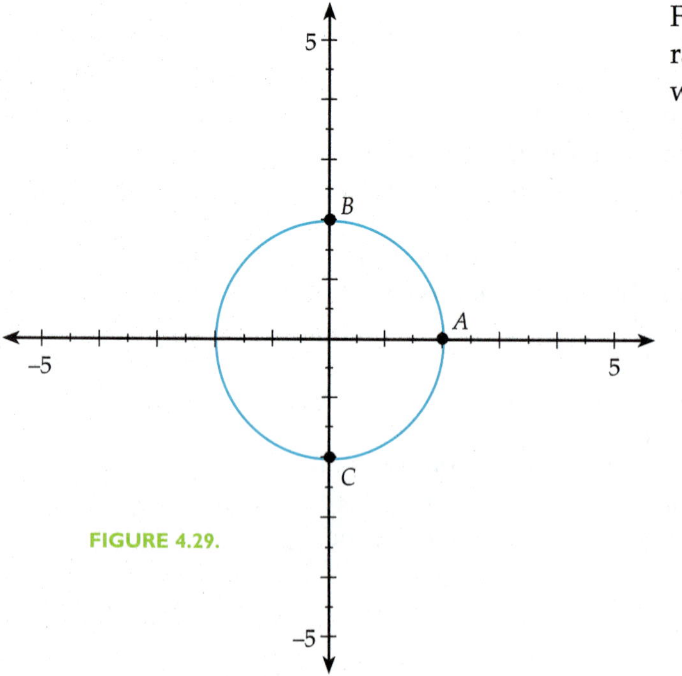

FIGURE 4.29.

Find the length of each arc and the radian measure. Assume that the arc wraps less than once around the circle.

   a) The arc from A to B in a counterclockwise direction.

   b) The arc from A to B in a clockwise direction.

   c) The arc from A to C in a counterclockwise direction.

   d) The arc from A to C in a clockwise direction.

Exercises 4.2

7. In each of the following pairs, describe the effect of the control numbers. Contrast the graphs with the graph of $y = \sin(x)$ and describe any changes in the amplitude or period. Verify your answers by using a graphing calculator or other graphing utility.

   a) $y = \sin(x - 2)$ and $y = \sin(x + 2)$

   b) $y = \sin(3x)$ and $y = \sin\left(\frac{1}{3}x\right)$

   c) $y = 3\sin(x)$ and $y = \frac{1}{3}\sin(x)$

   d) $y = 3 + \sin(x)$ and $y = -3 + \sin(x)$.

8. How would you modify $y = \sin(x)$ so the period of the modified function is:

   a) 4π?

   b) 3π?

   c) 2?

   d) 10?

9. Describe how each of the following modifications to the sine function affects the graph. (In other words, does the modification change the amplitude, period, vertical, or horizontal position of the sine wave?) Then sketch the graph of the modified function and $y = \sin(x)$ on the same set of axes.

   a) $y = 3\sin(x) + 2$

   b) $y = 0.5\sin\left(x - \frac{\pi}{2}\right)$

   c) $y = \sin\left(2\left(x + \frac{\pi}{2}\right)\right)$

10. a) Discuss the symmetry of the graph of $y = \sin(x)$.

    b) Use the symmetry of the graph of $y = \sin(x)$ to express $\sin(-x)$ in terms of $\sin(x)$. What type of function is the sine function?

    c) Compare the graph of $y = -\sin(x)$ to the graph of $y = \sin(x)$. What is the amplitude of the graph of $y = -\sin(x)$?

Exercises 4.2

11. Even if you use a calculator or a computer to produce graphs of sinusoidal functions, you will find it helpful to have a rough idea of what the graph will look like. You can, for example, save yourself some time by getting a good window setting on the first try. In this exercise, you will make observations about $y = 3 + 2\sin\left(2\left(x - \frac{\pi}{3}\right)\right)$ that will help you get a good window setting.

    a) Explain how you can identify the axis of oscillation from the equation.

    b) Explain how you can use the axis of oscillation and one of the control numbers to determine the highest and lowest points on the graph.

    c) Explain how to use two of the control numbers to determine the extent of one cycle of the graph.

    d) Here is another way to determine the extent of one cycle of the transformed graph. One cycle of $y = \sin(x)$ starts at $x = 0$ and ends at $x = 2\pi$. Therefore, one cycle of $y = 3 + 2\sin\left(2\left(x - \frac{\pi}{3}\right)\right)$ starts at $\left(2\left(x - \frac{\pi}{3}\right)\right) = 0$ and ends at $\left(2\left(x - \frac{\pi}{3}\right)\right) = 2\pi$. Solve these two linear equations to determine the values of x that produce the start and end of one cycle.

    e) Set a window on your graphing calculator or other graphing utility that will give a graph of at least one cycle of $y = 3 + 2\sin\left(2\left(x - \frac{\pi}{3}\right)\right)$. Show the graph.

12. **Table 4.6** is the moon phase data you used in Activity 4.1. You may want to revisit your results from that activity.

TABLE 4.6. Visible portion of the moon illuminated 5 January 1997–31 March 1977.

Date	1/5	1/10	1/15	1/20	1/25	1/30	2/4	2/9	2/14
Day of year	5	10	15	20	25	30	35	40	45
Visible portion illuminated	0.20	0.02	0.43	0.89	0.98	0.66	0.17	0.04	0.48

Date	2/19	2/24	3/1	3/6	3/11	3/16	3/21	3/26	3/31
Day of year	50	55	60	65	70	75	80	85	90
Visible portion illuminated	0.91	0.97	0.63	0.12	0.07	0.52	0.92	0.96	0.57

a) What vertical transformations of $y = \sin(x)$ would be necessary to model these data?

b) Use a plot of the moon data to identify a day of the year that is near the beginning of one cycle of a sinusoidal model. Explain how to transform $y = \sin(x)$ so one of its cycles begins at the same value.

c) Explain how to alter the period of $y = \sin(x)$ to match the period of the data.

d) Put it all together. Write the equation of a sinusoidal model that approximates the moon data. Compare a graph of the sinusoidal model and a plot of the data. Do you think your model is a good one?

e) Use the model you developed in (d) to predict the portion of the moon visible on 2 February 1997.

13. Exercise 6 of Lesson 4.1 described a Ferris wheel that completed one revolution in 30 seconds and that was 5 feet off the ground at its lowest point and 35 feet off the ground at its highest.

   a) Develop a sinusoidal model for a rider's height above ground as a function of the time the rider has been riding. Explain how you obtained your model.

   b) Explain how to use your model to determine the rider's height above ground 10 seconds into the ride.

   c) Prepare a calculator or computer graph of your model that represents a ride lasting 2 minutes. Explain how to use the graph to determine approximate times at which the rider is 20 feet above ground.

14. Explain how to transform $y = \sin(x)$ to model the data you collected in Activity 4.2.

15. Several groups of students are collecting data by spinning a bicycle wheel with a diameter of 26 inches. They spin the wheel so it turns at a rate of 1 revolution every 4 seconds. They videotape the wheel as it spins and stop the tape periodically to measure the height of a

Exercises 4.2

point that began at the bottom of the wheel. The models developed by three of the groups appear different:

$y = 13 + 13\sin\left(\frac{\pi}{2}(x-1)\right)$, $y = 13 - 13\sin\left(\frac{\pi}{2}(x+1)\right)$, and

$y = 13 - 13\sin\left(\frac{\pi}{2}(x-3)\right)$.

Examine these models and comment on the thinking of the students who developed them.

16. Sections of sinusoidal graphs resemble parabolas. As you know, one characteristic of quadratic data is that constant intervals produce constant second differences. Prepare a table of values of $\sin(x)$ from 0 to π in steps of 0.01. Add columns for first and second differences. What can you conclude?

LESSON 4.3

The Cosine Function

In Lesson 4.1, you examined periodic data, and in Lesson 4.2 you developed sinusoidal models of the form $y = A\sin(B(x - C)) + D$ that described some of these periodic data. The sine function, the basis for sinusoidal models, is defined as the vertical displacement (height) as you wrap around a unit circle. Many of the real-world situations described in Lesson 4.1 involved circular motion, so it is not surprising that a function defined in terms of circular motion can model these situations. The orbital motions of the earth around the sun, the moon around the earth, and the spinning of the earth around its axis are examples of phenomena associated with periodic data about daily temperature or visible portion of illuminated moon surface over time that can be modeled using sinusoidal functions. Sinusoidal functions are also used to model phenomena, such as sound frequencies, that have no apparent connection to circular motion or that may not be obviously periodic.

In this lesson, you investigate the rates at which sinusoidal functions change. In the process, you develop a new periodic function.

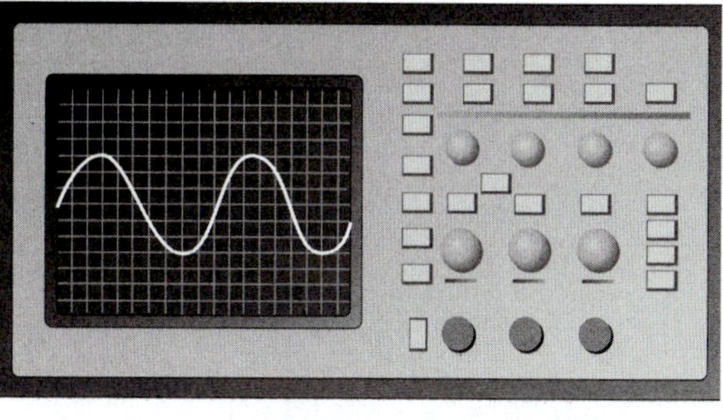

An oscilloscope is an instrument that can be used to visualize sound frequencies.

Activity 4.3

In this activity you investigate the rate at which the sine function changes. The basic idea is that the average change in one variable with respect to a second variable is the amount of change in one variable divided by the amount of change in the other. For example, if your car travels 5 miles in 0.1 hours, then the average change in distance with respect to time is $\frac{5 \text{ miles}}{0.1 \text{ hours}} = 50$ miles per hour.

Before doing the number crunching, give some thought to a situation that a sinusoidal function models. Consider a Ferris wheel ride and discuss Items 1 and 2 with others in your class.

1. As you ride the Ferris wheel, does your height above ground with respect to time change at a constant rate? If not, when is your height changing most slowly? Most rapidly?

2. Interpret your answer to 1 in terms of a graph of height versus time.

3. Use a spreadsheet or the table feature of a calculator to build a table of values for the sine function in steps of 0.1 from 0 to approximately 2π. Add a third column that gives the average rate of change over each interval.

4. Plot the sine data and the rate of change data on the same grid.

5. What do you notice about the plot of the average rates of change? Can you think of a function that would model them?

6. Another periodic function used by mathematicians is the cosine function. Use a calculator or other graphing utility to graph both the sine and cosine functions. Record your observations.

THE COSINE FUNCTION

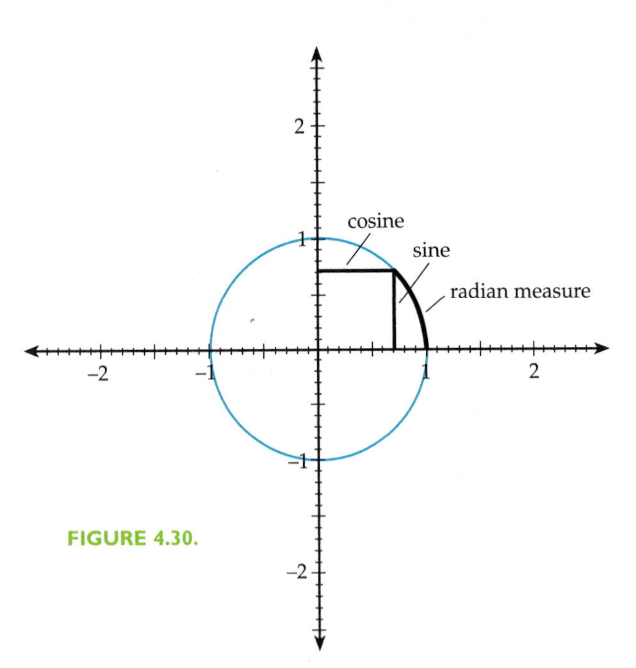

FIGURE 4.30.

The **cosine function** pairs radian measure with the horizontal displacement from the horizontal axis of the point at which the arc of the unit circle terminates.

Figure 4.30 shows the relationship that exists among radian measure, sine, and cosine.

The cosine function also measures the rate at which the sine function changes. You will explore the reason for this surprising result when you study calculus.

IDENTITIES

The graphs of the sine and cosine functions are shown in **Figure 4.31**.

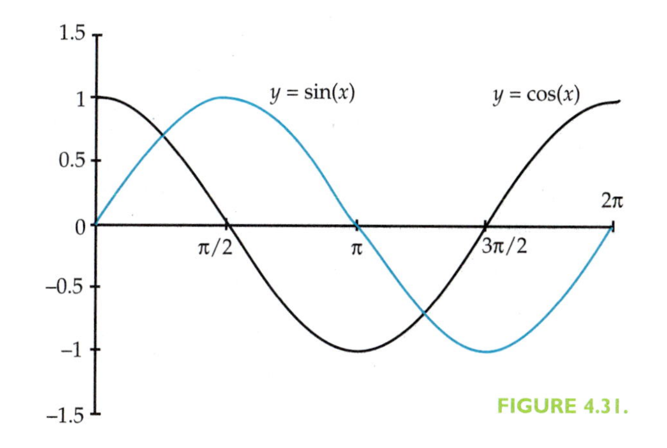

FIGURE 4.31.

The graphs have the same period, 2π, and the same amplitude, 1. The graphs have the same shape. The only difference is their position; shifting one $\frac{\pi}{2}$ to either the right or the left gives you the other.

Expressed symbolically, $\sin\left(x+\frac{\pi}{2}\right) = \cos(x)$. That is, shifting the sine graph $\frac{\pi}{2}$ units to the left gives you the cosine graph. The equation $\sin\left(x+\frac{\pi}{2}\right) = \cos(x)$ is an **identity**—an equation that is true for all values of the variable. To put it more specifically, it is an identity that expresses $\sin\left(x+\frac{\pi}{2}\right)$ in terms of $\cos(x)$.

EXAMPLE 6

Use graphs to find an identity that expresses $\cos\left(x+\frac{\pi}{2}\right)$ in terms of $\sin(x)$.

SOLUTION:

$\cos\left(x+\frac{\pi}{2}\right)$ shifts the cosine graph $\frac{\pi}{2}$ units to the right, as shown in the calculator graph in **Figure 4.32**.

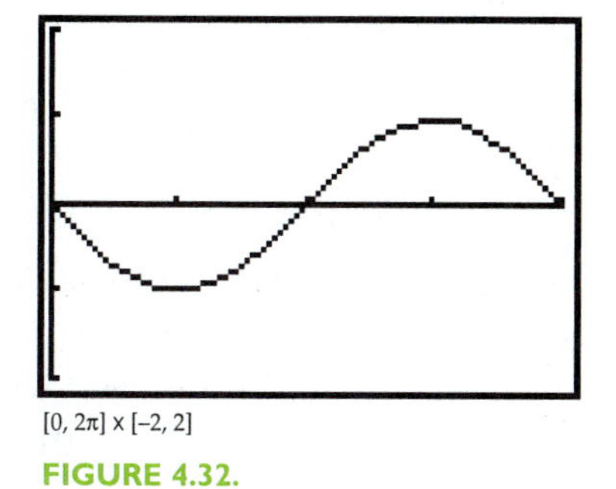

[0, 2π] x [–2, 2]

FIGURE 4.32.

This graph is the sine graph reflected in the x-axis, so $\cos\left(x+\frac{\pi}{2}\right) = -\sin(x)$.

EXAMPLE 7

Write a function of the form $y = A\sin(B(x - C)) + D$ and a function of the form $y = A\cos(B(x - C)) + D$ for the graph shown in **Figure 4.33**. What identity is implied by the functions you wrote?

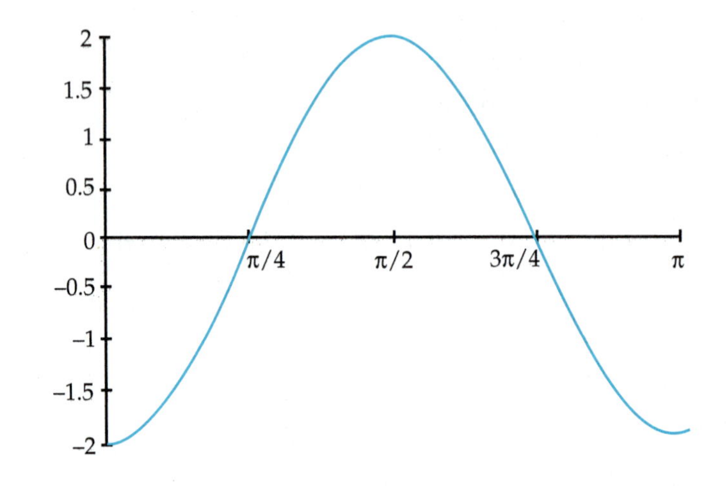

FIGURE 4.33.

SOLUTION:

The graph has amplitude 2 and period π. It resembles a cosine graph reflected in the x-axis. Therefore, one answer is $y = -2\cos(2x)$. The graph can also be thought of as a sine graph shifted $\frac{\pi}{4}$ units to the right. Therefore, another answer is $y = 2\sin\left(2\left(x - \frac{\pi}{4}\right)\right)$.

The corresponding identity is $-2\cos(2x) = 2\sin\left(2\left(x - \frac{\pi}{4}\right)\right)$, or $-\cos(2x) = \sin\left(2\left(x - \frac{\pi}{4}\right)\right)$.

TRIGONOMETRIC EQUATIONS

TAKE NOTE

If you are confused about the difference between an equation and an identity, that's understandable since an identity is really a special type of equation—one that is true for all values of the variable(s). The goal when solving an equation is to find the *relatively few values* of the variable for which the equation is true. The goal when developing an identity is to write a convincing argument that the equation is true for *all values* of the variable.

Sinusoidal models are often used to determine when a particular event occurs. For example, a model for the height of a Ferris wheel rider over time might be used to determine when the rider is 20 feet above the ground. Because of periodicity, such questions can have many answers.

Consider the trigonometric equation $\sin(x) = 0.5$. Four of the infinitely many solutions to this equation are shown in the calculator graph in **Figure 4.34**.

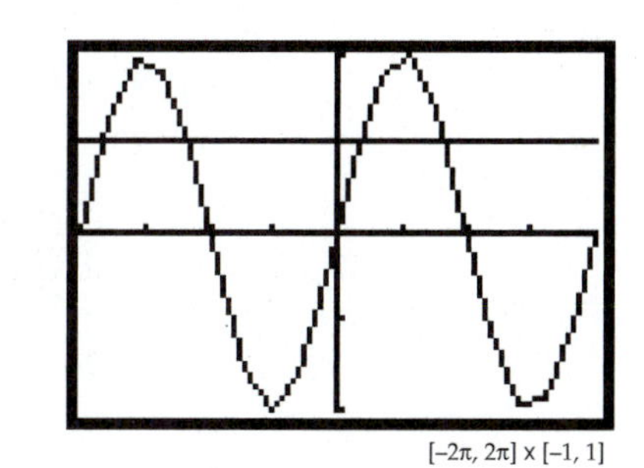

$[-2\pi, 2\pi] \times [-1, 1]$

FIGURE 4.34.

Zooming/tracing or the calculator's intersection finder can be used to approximate the first two positive solutions, $x \approx 0.52$ and $x \approx 2.62$.

Since the period of sine is 2π, other solutions can be found by adding 2π to these two solutions repeatedly or subtracting 2π from them repeatedly. For example $0.52 + 2\pi \approx 6.80$, $0.52 + 4\pi \approx 13.09$, and $0.52 - 2\pi \approx -5.76$ are all solutions.

Mathematicians have invented a shorthand to express the infinite number of solutions. They write $x \approx 0.52 \pm 2n\pi$, where n is any whole number and $x \approx 2.62 \pm 2n\pi$ to express the other. In everyday English, these say simply that you may add or subtract any integer multiple of 2π to the given solutions.

EXAMPLE 8

Find all solutions of $2\sin\left(2\left(x - \frac{\pi}{4}\right)\right) = -1$.

SOLUTION:

First note that the period of $y = 2\sin\left(2\left(x - \frac{\pi}{4}\right)\right)$ is π. Then use a calculator to determine approximate solutions (**Figure 4.35**).

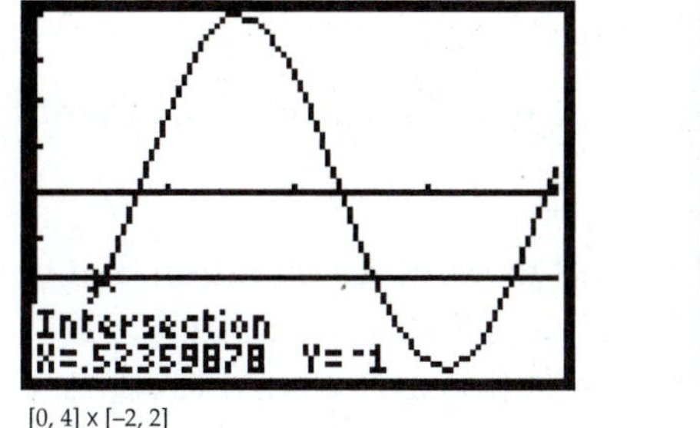

$[0, 4] \times [-2, 2]$

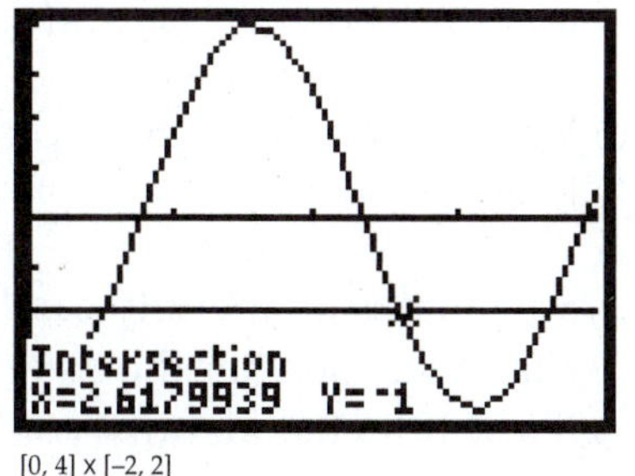

$[0, 4] \times [-2, 2]$

FIGURE 4.35.

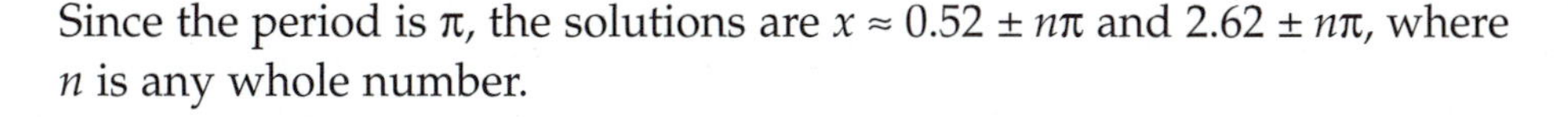

Since the period is π, the solutions are $x \approx 0.52 \pm n\pi$ and $2.62 \pm n\pi$, where n is any whole number.

Exercises 4.3

1. Use graphs to decide whether each of the following is an identity. Explain your reasoning.

   a) $\cos\left(x-\frac{\pi}{2}\right) = \sin(x)$.

   b) $\cos\left(x+\frac{\pi}{2}\right) = \sin(x)$.

   c) $\cos(x - 2\pi) = \cos(x)$.

   d) $\sin\left(x-\frac{\pi}{2}\right) = \cos(x)$.

   e) $\cos(x + \pi) = -\cos(x)$.

   f) $\sin(x + \pi) = -\sin(x)$.

2. Suppose a friend of yours who is just beginning to study trigonometric functions thinks $\cos(2x) = 2\cos(x)$. Write a brief explanation of the error.

3. Calculations with trigonometric functions sometimes involve squaring. To save parentheses, mathematicians sometimes write $(\sin(x))^2$ as $\sin^2(x)$.

   a) Explain why $(\sin(x))^2$ and $\sin^2(x)$ are different from $\sin(x^2)$.

   b) On a copy of Figure 4.30 on page 276, connect the origin to the upper endpoint of the segment representing the vertical displacement. Use a familiar geometric property to write an identity about $\sin^2(x)$ and $\cos^2(x)$.

   c) What conclusions about $\cos(x)$ can you draw from your identity if $\sin(x) = 0.6$?

   d) Sometimes an identity can be used to solve an equation. For example, consider the equation $\sin^2(x) + \sin(x) + \cos^2(x) = 1$. Show how the identity you wrote in (b) can be used to replace part of this equation with a simple numeric value. Then solve the equation.

4. Write two equations for the mystery graph in **Figure 4.36**, one using sine and the other using cosine. Check your answers with a graphing calculator or other graphing utility.

Exercises 4.3

a)

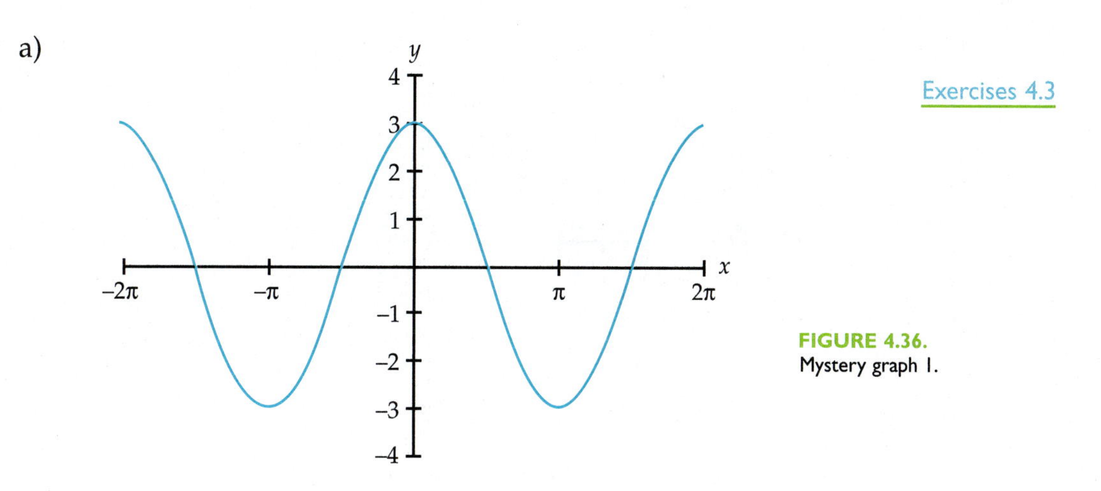

FIGURE 4.36.
Mystery graph 1.

b) Repeat for the mystery graph in **Figure 4.37**.

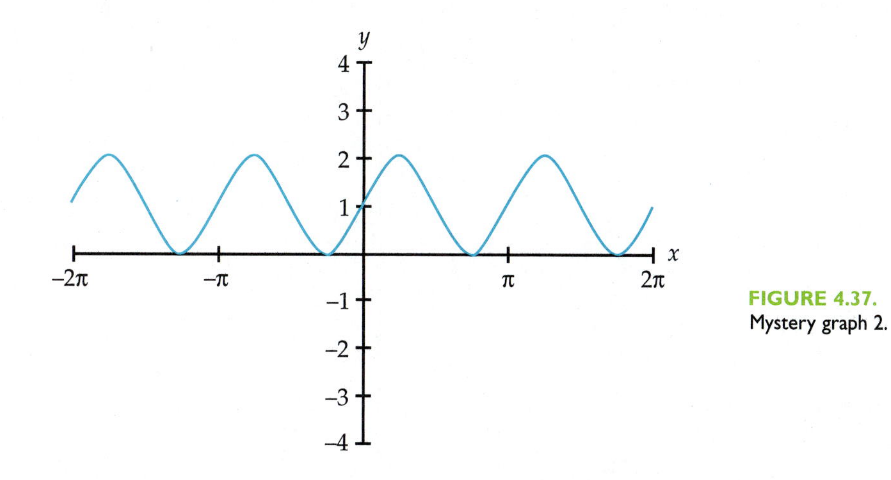

FIGURE 4.37.
Mystery graph 2.

5. For each of the graphs in **Figures 4.38–4.44**, determine a sine or cosine equation that has (approximately) the given graph. Also state the period, amplitude, and axis of oscillation.

a)

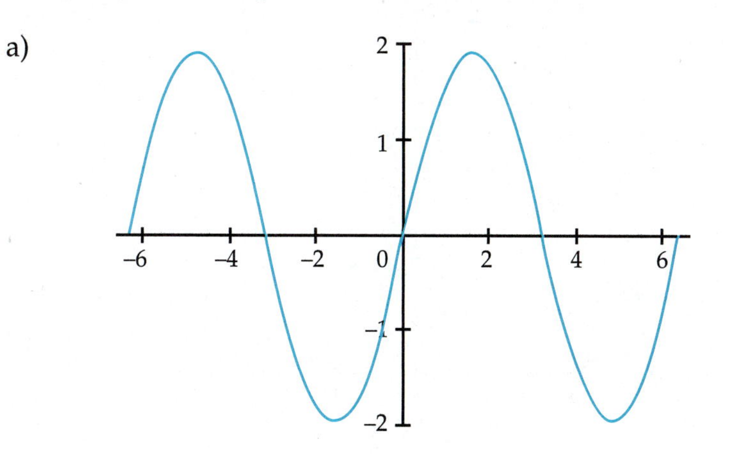

FIGURE 4.38.
Graph of a periodic function.

Exercises 4.3

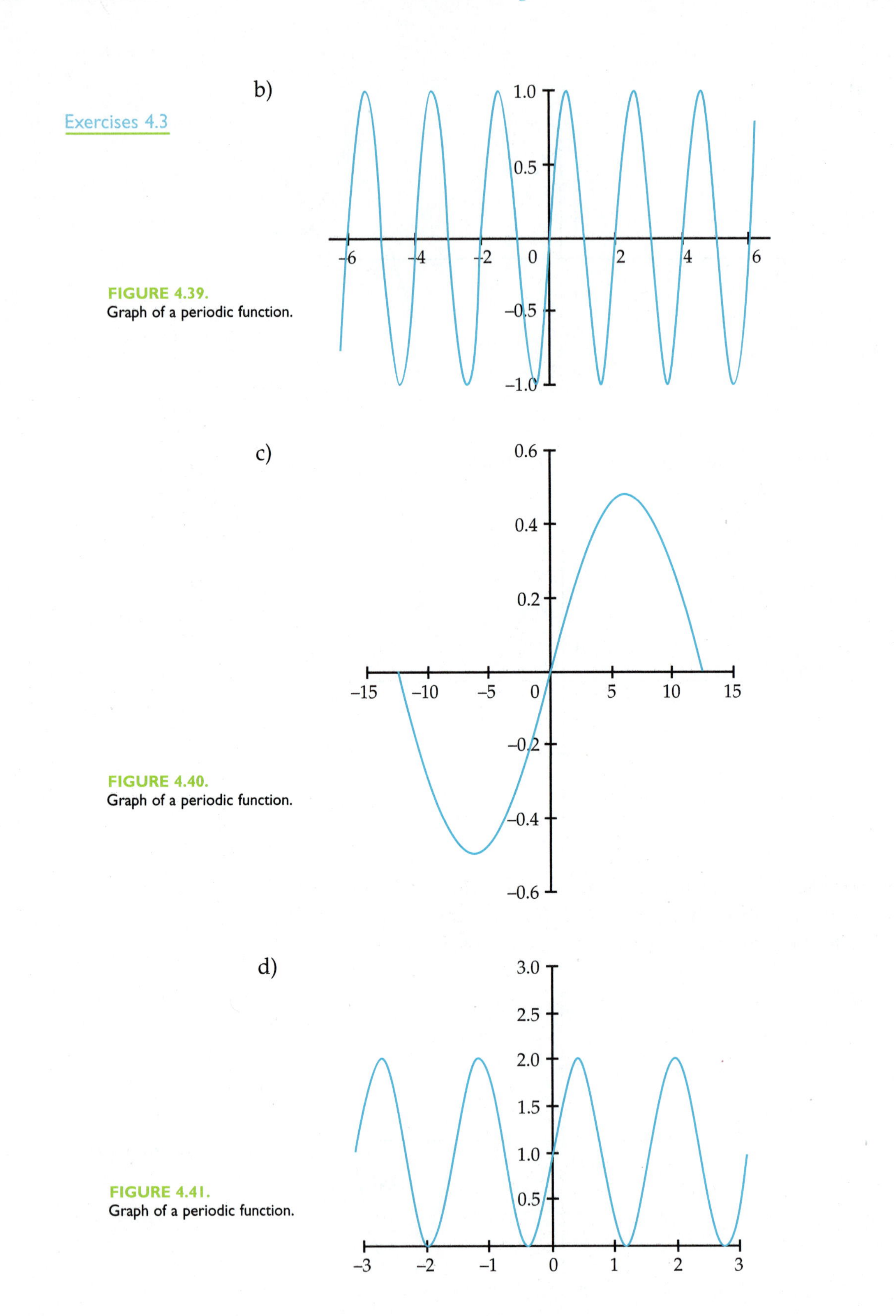

FIGURE 4.39.
Graph of a periodic function.

FIGURE 4.40.
Graph of a periodic function.

FIGURE 4.41.
Graph of a periodic function.

Exercises 4.3

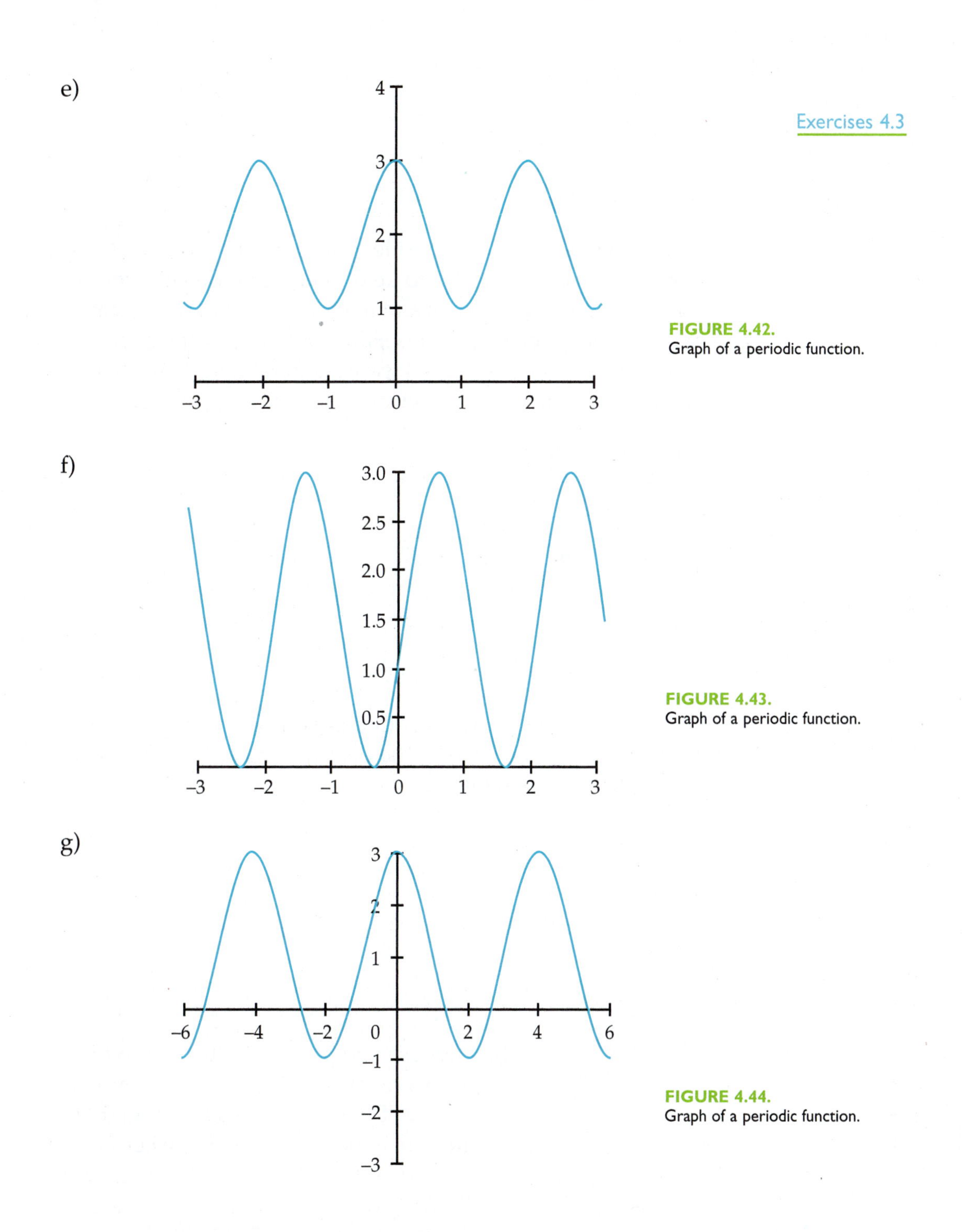

FIGURE 4.42.
Graph of a periodic function.

FIGURE 4.43.
Graph of a periodic function.

FIGURE 4.44.
Graph of a periodic function.

6. Describe in words how each equation transforms the graphs of $y = \cos(x)$. Then write a function of the form $y = A\sin(B(x - C)) + D$ that has the same graph.

Exercises 4.3

a) $y = 2\cos(2x) + 2$

b) $y = \cos(2(x - 3))$

c) $y = 0.25\cos(4(x - \pi)) - 1$

7. In this exercise, you will consider the effect of transformations on rate functions. Your goal is to determine a rate function for each of the following. There are several methods you can use: Compare the graph to the graph of $y = \sin(x)$; think about how the calculation of rates you used in Activity 4.3 would differ; build a rate table similar to the one you made in Activity 4.3. Justify your answer.

   a) $y = \sin(x) + 5$

   b) $y = 3\sin(x)$

   c) $y = \sin(x - 2)$

   d) $y = \sin(2x)$

8. Conduct an investigation into the rate at which $y = \cos(x)$ changes.

9. Use a graphing calculator or other graphing utility to solve each equation. Round answers to the nearest hundredth. Write symbolic expressions for all solutions.

   a) $3\cos(2x) + 1 = 2$

   b) $\sin(x - 1) = 1$

   c) $\cos(3x) - 3 = 0$

   d) $\sin\left(\frac{x}{2}\right) = 0.7$

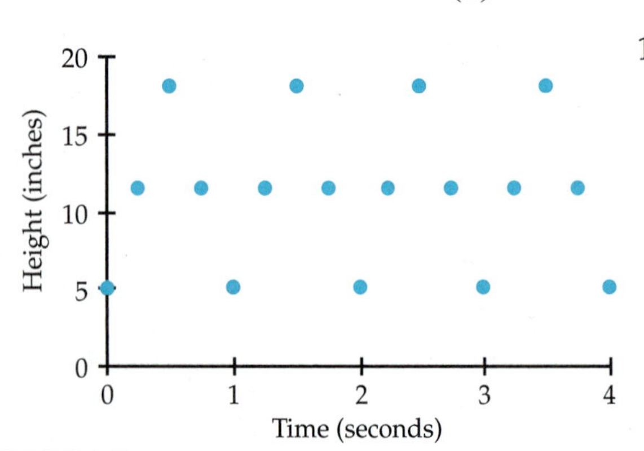

FIGURE 4.45.
Height of bicycle pedal over time.

10. Suppose you pedal your bicycle so that the pedals revolve at a constant rate once a second. The pedal is 5 inches from the ground at its lowest and 18 inches from the ground at its highest. **Figure 4.45** is a plot showing the height of the right pedal from the ground every 0.25 seconds for 4 seconds if you begin pedaling with your right pedal at its lowest point.

a) Use sine or cosine to model the height of the right pedal from the ground over time. (Use a graphing calculator or other graphing utility to check your model.)

Exercises 4.3

b) Use your equation to find and express all times at which the right pedal is 10 inches above the ground if your ride continues indefinitely.

11. The Potomac River flows through Washington, D. C. Tide levels for the Potomac are approximately sinusoidal and range from a height of 5 feet to a low of 1.5 feet when measured against a post at Harbor Place (a small plaza with a dock at which boats can land).

Suppose on 1 July, high tide is at 6:30 A.M. and low tide is at 12:45 P.M. Write an equation describing the height of the water for every hour after midnight on 1 July. Explain how you arrived at your model.

b) Estimate the height of the water at noon on 4 July.

c) At what times on 4 July will the tide levels be rising most quickly? Falling most quickly?

d) If the owners of Harbor Place need to hire someone to help people docking when the water is below 3 feet, for about how long every day should they hire help? (Assume Harbor Place is open 24 hours a day.)

12. Over the course of a year, the length of the day (the number of hours from sunrise to sunset) changes every day. **Table 4.7** shows the length of day every 30 days for Boston, Massachusetts, from 31 December 1997 to 26 March 1999.

Date	12/31	1/30	3/1	3/31	4/30	5/30	6/29	7/29
Day number	0	30	60	90	120	150	180	210
Length (hours)	9.1	9.9	11.2	12.7	14.0	15.0	15.3	14.6

Date	8/28	9/27	10/27	11/26	12/26	1/25	2/24	3/26
Day number	240	270	300	330	360	390	420	450
Length (hours)	13.3	11.9	10.6	9.5	9.1	9.7	11.0	12.4

TABLE 4.7.
Data on length of day.

Exercises 4.3

a) Many graphing calculators have a sinusoidal regression feature that fits sinusoidal models to data. If your calculator has this feature, use it to fit a model to these data. If not, use the techniques of this lesson to develop a model.

b) Rosita, who lives in Boston, suffers from SAD (seasonal affective disorder). During the winter she gets depressed, but by the first day of spring, 21 March, she feels wonderful. She has been advised to use light therapy during the winter. Assume that one hour of light therapy replaces one hour of natural daylight. Write a model that Rosita could use during the fall and winter of 1998–1999 (21 September 1998–21 March 1999) to determine how much light therapy she needs on a particular day in order to have an equivalent amount of daylight until 21 March. Draw a graph of this model over the interval corresponding to the days when she would use this model.

c) Use your model to determine how long Rosita should apply light therapy on 15 January 1999.

d) Like Rosita, Sasha lives in Boston and also suffers from SAD. However, her depression is triggered at times when the length of daylight is changing most rapidly. At what times of the year will she be most likely to suffer from depression? How did you obtain this information from your graph?

13. **Table 4.8** shows data on the homeowner's heating costs that appeared in Example 1.

TABLE 4.8. Fuel costs over a 2-year period.

Month	Jan	Feb	Mar	Apr	May	June	July	Aug	Sept	Oct	Nov	Dec
Cost ($)	85	87	72	44	25	20	18	17	19	35	79	87

Month	Jan	Feb	Mar	Apr	May	June	July	Aug	Sept	Oct	Nov	Dec
Cost ($)	75	84	58	50	23	15	16	17	20	36	65	85

a) Use a calculator to find a sinusoidal regression model for these data. (Let x = the month number. January of the first year is month 1, and so forth.)

b) How useful do you think this model would be for predicting next year's fuel costs?

Exercises 4.3

14. Pat and Terry are turning a long jump rope for Tracy, who is waiting to jump in. Pat and Terry turn the rope one time per second. The maximum height is 7 feet, while at its lowest, it just touches the ground. Assume that the rope was on the ground moving up and away from Tracy at starting time.

    a) Write an equation expressing the height of the rope as a function of time since Pat and Terry started turning it. Explain how you arrived at your equation.

    b) Tracy is 5 feet 3 inches tall. When will the rope be higher than Tracy's head? (Write an expression that would describe all the times that this occurs.)

15. Sinusoidal models are sometimes given in a form slightly different from the one used in this book. For example, some graphing calculators give models such as $y = \sin(2x - 1)$. Explain how to determine horizontal shift from this form.

LESSON 4.4

The Tangent and Other Functions

The sine and cosine functions are the primary tools for modeling periodic phenomena. The graphs of these two functions are so similar that the choice of one over the other for modeling purposes is a matter of personal preference. However, there are a total of six trigonometric functions, and the other four have graphs quite different from those of sine and cosine.

This lesson will help you develop an understanding of the tangent, cotangent, secant, and cosecant functions.

Activity 4.4

In this activity you explore the tangent function.

The tangent function pairs the radian measure of an arc of the unit circle with the ratio of the vertical to horizontal displacement of the point at which the arc terminates. Since the slope of a line is also a ratio of vertical displacement to horizontal displacement, you can think of the tangent as the slope of a line through the origin and the arc's termination point. See **Figure 4.46**.

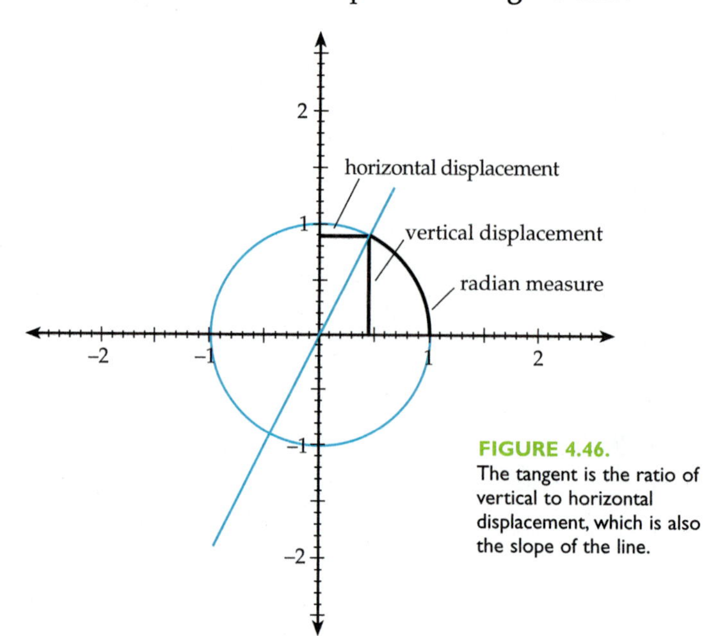

FIGURE 4.46. The tangent is the ratio of vertical to horizontal displacement, which is also the slope of the line.

When the vertical displacement is large relative to the horizontal displacement, the tangent is large. When the vertical displacement is small relative to the horizontal displacement, the tangent is small.

1. Discuss the way tangent changes as the radian measure increases from 0 to $\frac{\pi}{2}$. (You may find it helpful to think of the tangent as the slope of the line through the origin in Figure 4.46 as the line rotates counterclockwise from horizontal to vertical.)

2. Discuss the way tangent changes as radian measure increases from $\frac{\pi}{2}$ to π; from π to $\frac{3\pi}{2}$; and from $\frac{3\pi}{2}$ to 2π. (Don't forget that since vertical and horizontal displacements are measured with respect to the axes, either can be negative.)

3. Use a graphing calculator or other graphing utility to graph $y = \tan(x)$. Use your graph to check your answers to 1 and 2. Discuss the meaning of amplitude and period for this graph.

4. Graphs of sine and cosine do not have asymptotes. The graph of tangent does, and the asymptotes may appear on your calculator if it is in connected mode. Describe the locations of these asymptotes.

5. Is any symmetry apparent in the graph of tangent?

THE TANGENT FUNCTION

The **tangent function** pairs radian measure with the ratio of the vertical to horizontal displacement of the point at which the arc of the unit circle terminates.

Since tangent is a ratio, it has vertical asymptotes whenever its denominator, the horizontal displacement, is 0 (and its numerator is non-zero). By similar reasoning, the tangent is 0 whenever its numerator, the vertical displacement, is 0 (and its denominator is non-zero). These observations are visible in the graph of $y = \tan(x)$ in **Figure 4.47**.

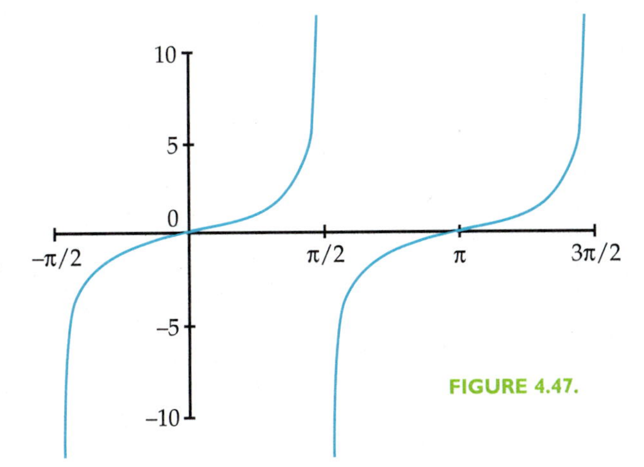

FIGURE 4.47.

Since the definition of tangent is based on the unit circle, it is reasonable to expect tangent to be periodic with period 2π. However, the graph shows that although tangent is periodic, the period is not 2π. The graph repeats twice in an interval of 2π. Since period is the smallest interval over which the graph repeats, the period is π.

Since tangent has neither a largest or smallest value, it has no amplitude.

The tangent graph is symmetric about any of its x-intercepts. It is also symmetric about the points on the x-axis where asymptotes occur.

If you draw an analogy between the unit circle and a clock that runs in reverse, you can think of the graph of tangent as telling the story of the slope of a clock's hour hand as it rotates. For example, at 3 o'clock (radian measure = 0) the hand is level (slope = 0). As the hand moves (counterclockwise) toward 12 (radian measure = $\frac{\pi}{2}$), the slope gradually increases until the hand becomes vertical (infinite slope). As the hand moves past 12, it at first has a steep downward (negative) slope, and it becomes level (slope = 0) when it reaches the 9 o'clock position (radian measure = π).

THE COTANGENT FUNCTION

The **cotangent function** pairs radian measure with the ratio of the horizontal to vertical displacement of the point at which the arc of the unit circle terminates.

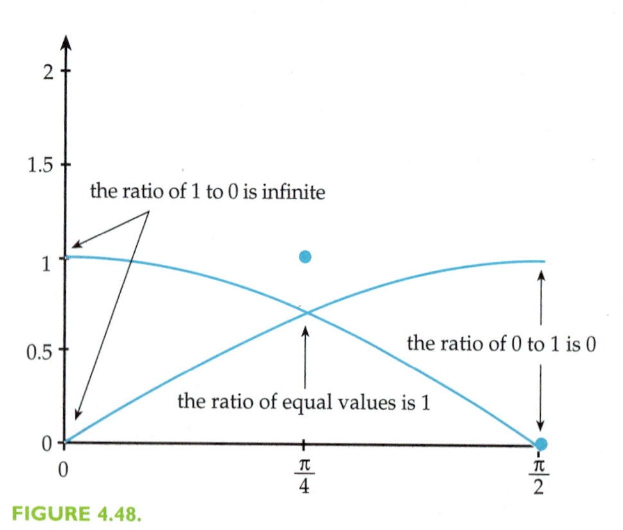

FIGURE 4.48.

Since cosine is the horizontal displacement and sine is the vertical displacement, observations about key points in the graph of cotangent can be made from the graphs of cosine and sine. **Figure 4.48** shows how the graphs of cosine and sine can be used to conclude that between 0 and $\frac{\pi}{2}$, the graph of cotangent decreases from an infinitely large value to 0.

Figure 4.49 shows the graph of the cotangent function between 0 and 2π.

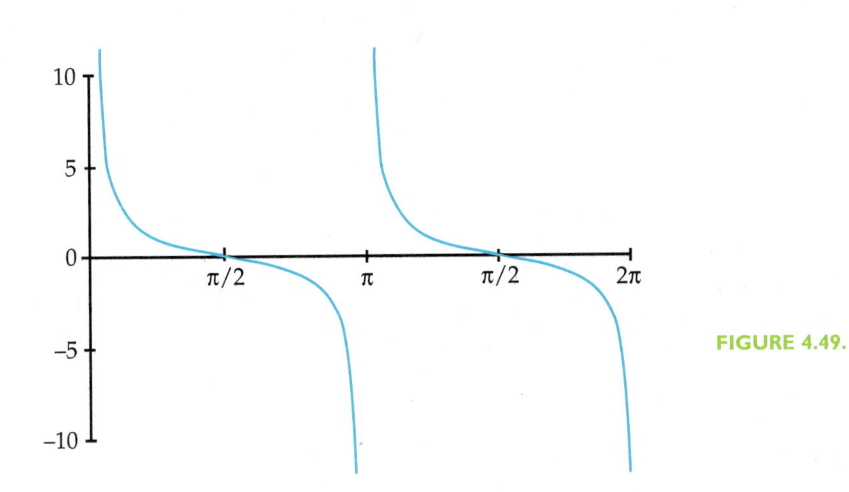

FIGURE 4.49.

Cotangent is periodic with period π. The graph has no amplitude.

Both cot and ctn are used as abbreviations of cotangent.

You may find transformations of the graphs of tangent and cotangent easiest to follow if you track a point at which the graph is 0: $x = 0$ for tangent; $x = \frac{\pi}{2}$ for cotangent.

EXAMPLE 9

Discuss transformations of the graph of $y = \tan(x)$ that produce the graph of $y = 2\tan(2(x - 0.5)) + 1$. Use a calculator or other graphing utility to verify your answers.

SOLUTION:

The graph of $y = \tan(x)$ is compressed horizontally by a factor of 2 and stretched vertically by a factor of 2. It is translated 0.5 units to the right and 1 unit upward.

Because of the horizontal and vertical shifts, the x-intercept (0, 0) is moved to (0.5, 1). Because of the horizontal compression, the period is changed from π to $\frac{\pi}{2}$.

Since asymptotes occur at half the period on either side of an x-intercept, asymptotes are at $x = 0.5 - \frac{\pi}{4} \approx -0.29$ and at $x = 0.5 + \frac{\pi}{4} \approx 1.29$. A window that uses these values should yield approximately one section of the graph, as the graphing calculator screen in **Figure 4.50** shows.

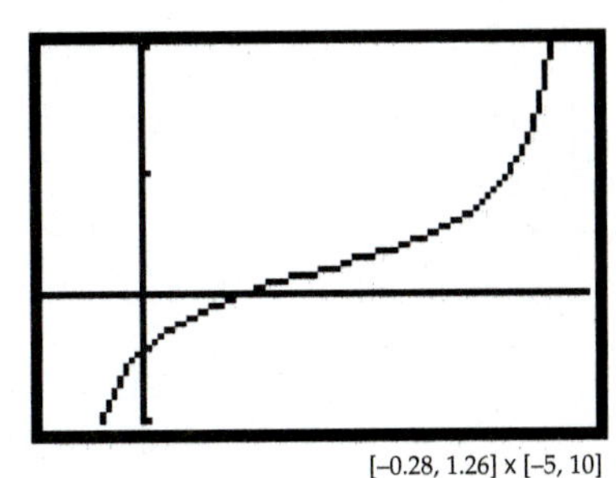

[–0.28, 1.26] x [–5, 10]

FIGURE 4.50.

The techniques used in Example 9 are also helpful in solving equations involving tangent.

EXAMPLE 10

Estimate all solutions of $2\tan(2(x-0.5))+1=2$ to the nearest hundredth. Check the first two positive solutions by direct calculation on a calculator.

SOLUTION:

The graphing calculator screen in **Figure 4.51** shows one intersection of $y=2\tan(2(x-0.5))+1$ and $y=2$.

Since the period of $y=2\tan(2(x-0.5))+1$ is $\frac{\pi}{2}$, additional solutions can be found by adding or subtracting multiples of $\frac{\pi}{2}$ to this solution. The solutions are approximately $0.73 \pm n\frac{\pi}{2}$, where n is any whole number.

The calculator screen in **Figure 4.52** shows that the first two positive solutions are approximately correct.

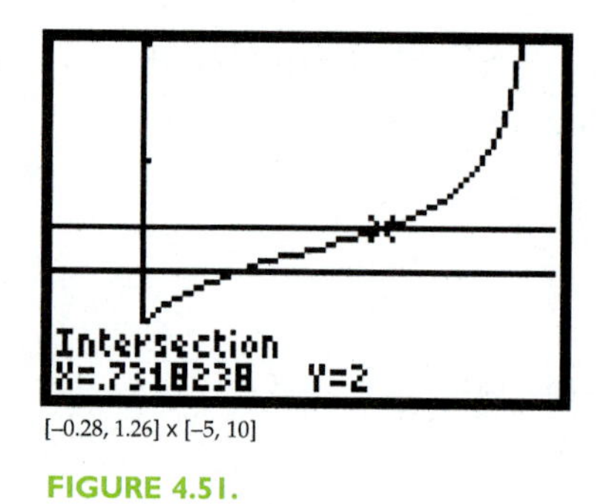

[–0.28, 1.26] x [–5, 10]

FIGURE 4.51.

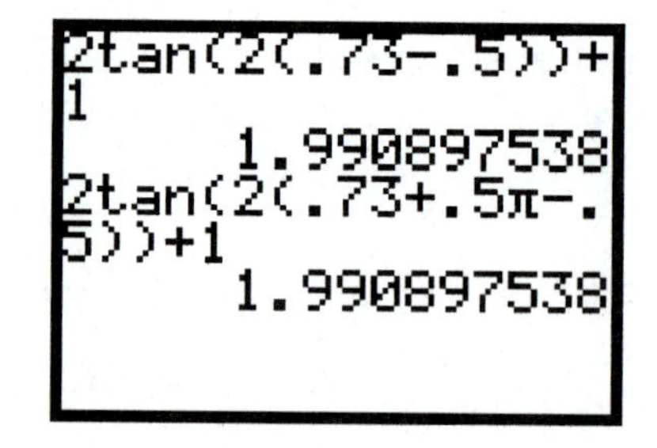

FIGURE 4.52.

Exercises 4.4

1. The remaining two trigonometric functions are the secant (sec) and the cosecant (csc). On the unit circle, secant pairs radian measure with the reciprocal of the horizontal displacement. Cosecant pairs radian measure with the reciprocal of the vertical displacement.

   a) Secant is related to cosine in the same way cotangent is related to tangent. The same is true for the relationship between cosecant and sine. Review the definitions of tangent and cotangent from this lesson and explain their relationship.

   b) Use graphs of sine and cosine to discuss the location of asymptotes for cosecant and secant.

   c) You can make a rough sketch of the graph of secant or cosecant from a graph of cosine or sine by thinking about the relationship between reciprocals. The reciprocal of 1 is 1, and the reciprocal of a number smaller than 1 is a number larger than 1. Make a rough sketch of the graph of $y = \sin(x)$ between 0 and 2π. On the same graph, sketch the graph of $y = \csc(x)$.

   d) Make a rough sketch of the graphs of $y = \cos(x)$ and $y = \sec(x)$ between 0 and 2π.

2. Determine whether tangent, cotangent, secant, and cosecant are odd or even functions or neither. Explain your reasoning.

3. Since tangent is the ratio of vertical to horizontal displacement on the unit circle and since sine is the vertical displacement and cosine is the horizontal displacement, tangent is the ratio sine to cosine. This fact can be expressed as an identity: $\tan(x) = \frac{\sin(x)}{\cos(x)}$. Write identities that express cotangent, secant, and cosecant in terms of sine and/or cosine.

4. Vertical and horizontal displacements can be negative, which explains why all trigonometric functions take on negative values. By thinking about the definitions of the functions, it is possible to tell which functions are positive from information about the radian measure. For example, if a radian measure is between $\frac{\pi}{2}$ and π, the arc wraps into Quadrant II, where vertical displacement is positive and horizontal displacement is negative. Since sine and cosecant are based on vertical displacement only, they are positive when radian measure is between $\frac{\pi}{2}$ and π.

Exercises 4.4

a) Do any of the other four trigonometric functions have positive values between $\frac{\pi}{2}$ and π? Explain.

b) Determine which of the trigonometric functions are positive for radian measures between 0 and $\frac{\pi}{2}$.

c) Determine which of the trigonometric functions are positive for radian measures between π and $\frac{3\pi}{2}$.

d) Determine which of the trigonometric functions are positive for radian measures between $\frac{3\pi}{2}$ and 2π.

e) Determine which of the trigonometric functions are positive for radian measures between $-\frac{\pi}{2}$ and 0.

5. In Exercise 3 of Lesson 4.3, you saw that the identity $\sin^2(x) + \cos^2(x) = 1$ could be used to calculate values of sine from cosine or vice versa. When this identity is combined with your answers to Exercises 3 and 4 above, you can find all five of the other trigonometric functions if you know only sine or cosine!

a) Suppose $\sin(x) = 0.6$ and x has radian measure between $\frac{\pi}{2}$ and π. Show how to use $\sin^2(x) + \cos^2(x) = 1$ to calculate two possible values for $\cos(x)$. Then explain how to use the fact that x has radian measure between $\frac{\pi}{2}$ and π to eliminate one of the two answers.

b) Use the identities you wrote in Exercise 3 to find the values of the other four functions.

c) Find all trigonometric functions of x if $\sec(x) = -3$ and x has radian measure between π and $\frac{3\pi}{2}$.

d) Sometimes people grab formulas and run without much thought. It's always a good idea to think about the given information before starting. Explain why there is no need to use any formulas if you are asked to find all the other trigonometric functions of x if $\sin(x) = \frac{2}{3}$ and x has radian measure between $\frac{3\pi}{2}$ and 2π.

e) Find all six trigonometric functions of x if $\sec(x) = 0.7$ and the radian measure of x is between 0 and $\frac{\pi}{2}$.

Exercises 4. 4

6. The identity $\sin^2(x) + \cos^2(x) = 1$ can be used to obtain two similar identities.

   a) Show how to obtain $1 + \cot^2(x) = \csc^2(x)$ from $\sin^2(x) + \cos^2(x) = 1$ by dividing by $\sin^2(x)$.

   b) Find another identity by dividing $\sin^2(x) + \cos^2(x) = 1$ by $\cos^2(x)$ and simplifying.

7. For each of the following functions, discuss intervals in $[0, 2\pi]$ for which it is increasing or decreasing. Also discuss the types of concavity.

   a) The tangent function.

   b) The cotangent function.

   c) The secant function.

   d) The cosecant function.

8. For some radian measures, it is possible to determine the trigonometric functions directly from the unit circle because the horizontal and vertical displacements are obvious or easily calculated.

   a) Explain how to find all the trigonometric functions of π directly from the unit circle.

   b) If you notice something special about the vertical and horizontal displacements of $\frac{\pi}{4}$ and use $\sin^2(x) + \cos^2(x) = 1$, you can find exact values of the trigonometric functions of $\frac{\pi}{4}$. Do so.

   c) Explain how you can find the trigonometric functions of $\frac{3\pi}{4}$ from your answers to part (b) by changing some signs.

9. The smallest positive solution of $2\tan(0.5x) = 3$ is approximately 1.97. Explain how to find all solutions.

10. Express approximations of all solutions to each equation.

   a) $\tan(2(x + 1)) = 1$

   b) $1 + \tan(\pi x) = -2$

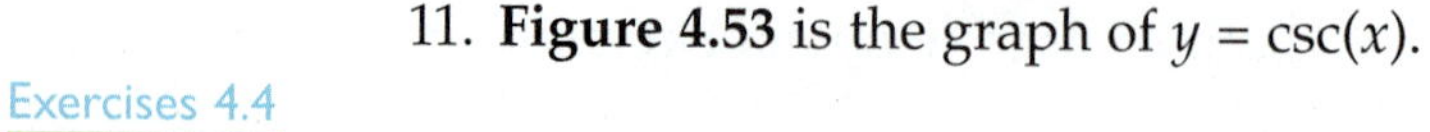

Exercises 4.4

11. **Figure 4.53** is the graph of $y = \csc(x)$.

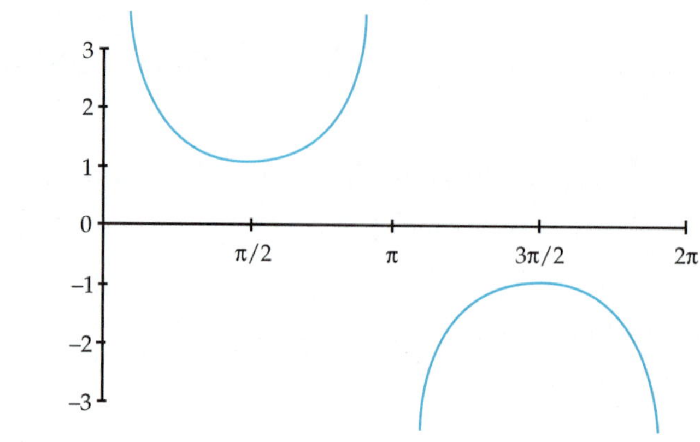

FIGURE 4.53.
The graph of $y = \csc(x)$.

a) Discuss how this graph is transformed to obtain the graph of $y = 2\csc(x - 1) + 1$.

b) Use a graphing calculator or other graphing utility to graph $y = 2\csc(x - 1) + 1$ in a window that shows the transformed sections of the graph in Figure 4.53. Since most calculators do not have a csc function, you may need to replace csc with $\frac{1}{\sin}$.

12. There are many trigonometric identities used in various branches and applications of mathematics. In this exercise, you will develop an identity for $\cos(2x)$.

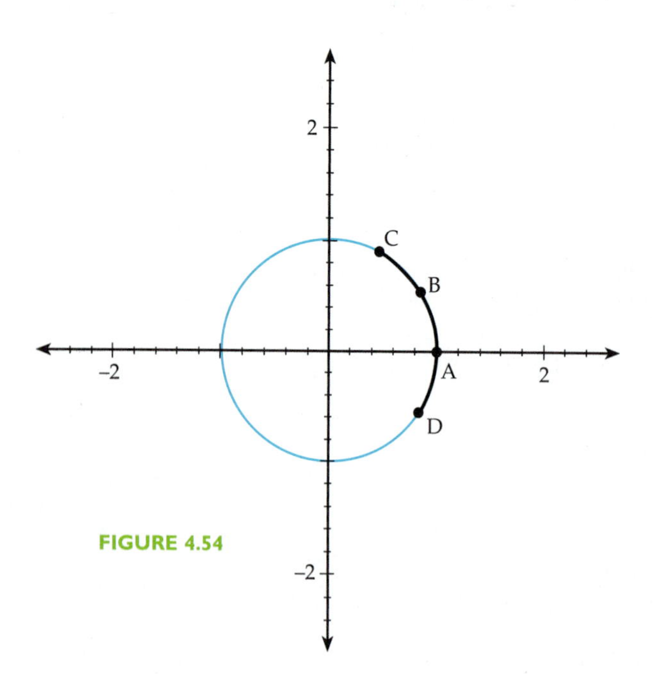

FIGURE 4.54

a) In **Figure 4.54**, the arcs between A and B, between B and C, and between A and D all have the same length. If x is the radian measure of the arc from A to B, what is the radian measure of the arc from A to C? From A to D?

b) Since the arc from A to B has radian measure x, the coordinates of B are $(\cos(x), \sin(x))$. Write the coordinates of points C and D as cosines and sines.

c) On a copy of Figure 4.54, draw segments connecting A to C and B to D. How do the lengths of these segments compare?

d) The linear distance between two points can be calculated with the formula $d = \sqrt{(x_2 - x_1)^2 + (y_2 - y_1)^2}$. Write this formula applied to your results in (b) and (c).

e) Now you need to do some symbol manipulation to isolate cos(2*x*) in the equation you wrote in (d). Start by squaring both sides to eliminate the square root signs. But before doing any additional squaring, apply your knowledge of odd and even functions to cos(–*x*) and sin(–*x*).

CHAPTER 4

TRIGONOMETRIC FUNCTIONS

Chapter 4 Review

1. **Table 4.9** shows the average monthly temperatures in degrees Fahrenheit for Boston, Massachusetts from January 1995–December 1997.

Year	Jan	Feb	Mar	Apr	May	June
1995	34.6	28.4	38.8	46.0	57.2	68.6
1996	30.1	30.9	36.5	47.9	57.4	68.1
1997	29.2	36.0	36.7	46.3	56.1	68.2

Year	July	Aug	Sept	Oct	Nov	Dec
1995	75.9	72.8	63.1	58.3	41.9	31.7
1996	71.8	70.9	64.1	53.2	40.3	39.2
1997	73.6	71.1	64.2	52.8	41.6	35.2

TABLE 4.9. Average monthly temperature for Boston, Massachusetts.

a) Make a scatter plot of these data. How did you define the explanatory variable?

b) Your plot in (a) should appear (approximately) periodic. What is the period? Approximately what is the amplitude?

c) Use your answers to (b) to determine a model of the form $y = A\sin(B(t - C)) + D$ that you think describes these data well. Explain how you determined your model.

d) Overlay a graph of your model on a scatter plot of your data. Does your model appear to do a good job in describing these data?

e) If your calculator has a sinusoidal regression feature, use it to fit a model to these data. Compare your model from (c) to the one fit by sinusoidal regression. How are the equations for these formulas alike and how are they different? Are their graphs similar?

2. Consider the function $y = 3\sin\left(2\left(x - \frac{\pi}{6}\right)\right) + 6$.

a) What are the period and amplitude of this function? What is its phase shift? Its axis of oscillation?

b) Make a rough sketch of the graph of this function. After you have completed your sketch, use your calculator to check it.

c) Write a function involving the cosine that produces the same graph as in (b).

3. What would you have to do to the graph of $y = \cos(x)$ to transform it into a graph that coincides with the graph of $y = 0.5\cos(0.5(x + 2\pi)) - 2$?

4. Describe the location of the asymptotes of $y = 3\tan(2(x + 0.5)) - 2$.

5. Approximate all solutions of the equation $3\sin(\pi(x - 1)) = 1$.

6. Water wheels have been used as a power source since ancient times. In early America, they often powered mills, and some of these wheels are still in operation.

Corbis

a) Develop a model for the vertical displacement measured from ground level over time of a point on the circumference of a water wheel. The wheel is 12 feet in diameter and turns once every 6 seconds. Assume that the wheel is mounted so that it extends four feet below ground level and the point is at the bottom of the wheel when timing starts.

b) Use your model to determine the height of the point after 2 seconds.

c) Use your model to express all times after timing began for which the height of the point will be 1 foot.

7. Find all of the trigonometric functions of x if its cosine is -0.4 and its radian measure is between 0.5π and π.

8. A carousel has a diameter of 40 feet and turns at the rate of one revolution every 8 seconds. How fast does the rider of a horse 2 feet from the carousel's outer edge turn?

9. Sales of snowblowers are seasonal. Fewer people buy snowblowers in the spring than in the fall. Suppose that snowblower dealers in one community began recording sales in November 1995. The data are presented in **Table 4.10**.

TABLE 4.10.

Nov	Dec	Jan	Feb	Mar	Apr	May	June	July	Aug	Sep	Oct
936	739	490	248	79	7	80	256	511	760	940	1008

Nov	Dec	Jan	Feb	Mar	Apr	May	June	July	Aug	Sep	Oct
960	757	513	275	90	38	88	285	533	772	957	1045

Nov	Dec	Jan	Feb	Mar	Apr	May	June	July	Aug	Sep	Oct
962	790	536	295	100	45	106	297	528	792	995	1051

a) Make a scatter plot of these data. Do they appear to be periodic?

b) Fit a sinusoidal regression model to these data. If necessary, adjust the model so that the period is exactly 12 months. Use a residual plot to decide whether your model adequately describes the pattern of the snowblower data.

c) When residual plots have a pattern, mathematical modelers sometimes fit a model to the residuals. If your residuals in (b) have a pattern, fit an appropriate model to them.

d) Build and evaluate a new model by combining your sinusoidal regression model from (b) with your residual model in (c).

e) A business is thinking of adding, in the year 2002, snowblowers to the merchandise it already sells. However, it first wants to project the number of snowblower sales in the peak selling season, August–December, in this community. Use your model to estimate the number of snowblowers that will be sold from August to December in the year 2002.

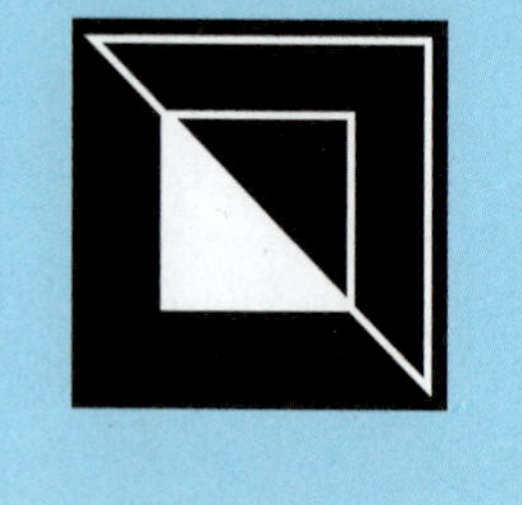

Chapter 5

TRIANGLE TRIGONOMETRY

Corbis

CHAPTER INTRODUCTION

To some people, trigonometry appears to live in two very different worlds. One of them is the function world, where trigonometry is an essential tool for modeling periodic phenomena. The other is the geometric world, where trigonometry serves a seemingly unrelated role in solving certain kinds of triangle problems.

Triangle trigonometry is an everyday tool in many vocations. Engineers and surveyors use triangle trigonometry to design roads, bridges, and underpasses. Architects and carpenters use triangle trigonometry to design and build roofs. Many scientists use triangle trigonometry to calculate angles and lengths in diverse settings.

In this chapter, you will use trigonometry to develop triangle models and examine connections between the function side of trigonometry and the geometric side.

LESSON 5.1

Right Triangles

For nearly 40 years, a photograph taken by an American satellite orbiting the moon has been a subject of controversy. The photo (see **Figure 5.1**) appears to show several objects casting extraordinarily long shadows. The objects are called the Blair Cuspids, after a scientist who studied the photo. Blair felt that the arrangement of the objects and their heights suggested they were not a natural phenomenon.

FIGURE 5.1. The Blair Cuspids.

Many people offered explanations. One suggested the cuspids were towers erected by aliens, and another worried that they were missiles aimed at the earth.

But the origins of the cuspids were not the only subject of controversy. Scientists did not agree on the geometric model with which to estimate their heights. Initially, a simple right triangle model was used. You will examine the right triangle model in this lesson, and return to the cuspids with an alternate model later in this chapter.

Activity 5.1

One of the most common uses of triangle trigonometry is in determining inaccessible heights. How the height is found depends on the measurements that can be taken. You will explore two methods in this activity.

The simplest method involves measuring the shadow of the object in question, then comparing it to the length of the shadow of an object of known height. For example, in **Figure 5.2** a person 5 feet 10 inches tall (standing upright) casts a shadow 15 feet long and an object of unknown height (standing upright) casts a shadow 60 feet long.

1. Explain why the two triangles in Figure 5.2 must be similar.

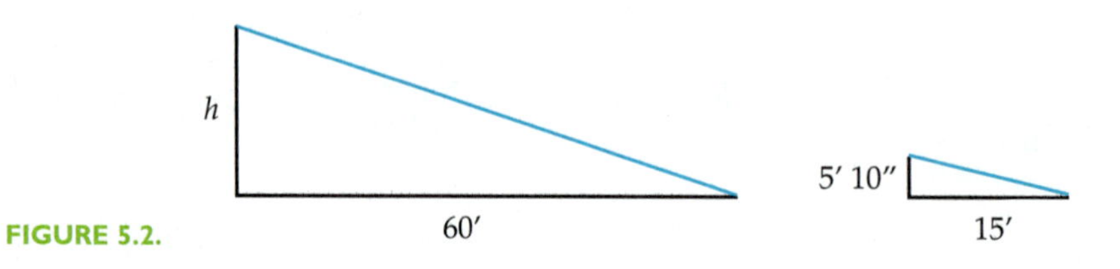

FIGURE 5.2.

2. Apply your knowledge of proportions in similar triangles to estimate the object's height.

3. Explain why this method cannot be used to determine the height of a Blair Cuspid.

To see how to adapt the previous method, consider **Figure 5.3**. In triangle ABC, angle A measures 30°, and side AC measures 100 meters. Side BC is an unknown height. A coordinate system is introduced so A is at the origin and side AC lies along the positive x-axis. The circle centered at A is a unit circle (not drawn to scale). (The introduction of the coordinate system creates several angles at A. In this discussion, angle A means the angle A of triangle ABC.)

FIGURE 5.3.

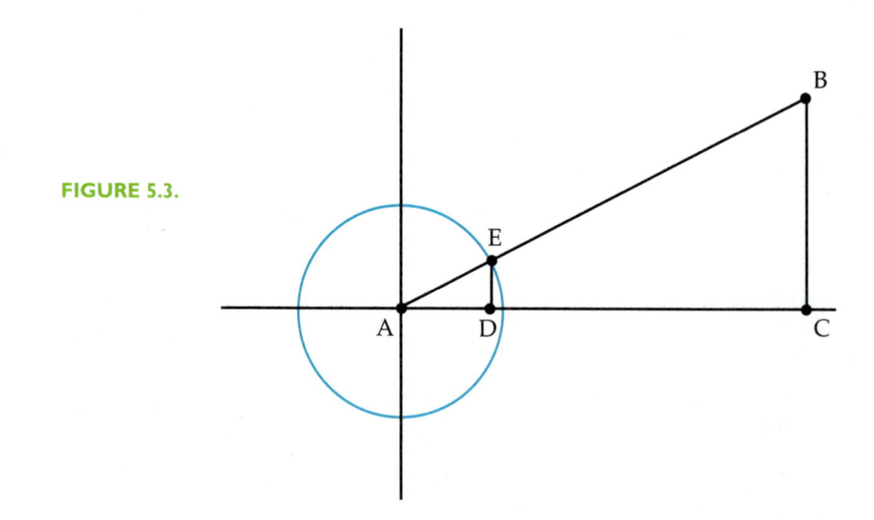

4. What is the radian measure that corresponds to angle A's degree measure? (Hint: what fraction of a full revolution is angle A, and what is the length of the arc of the unit circle enclosed by the sides of angle A?)

5. The small right triangle and the large right triangle are similar. What ratio in the small triangle equals the ratio $\frac{BC}{AC}$ in the large triangle?

6. The ratio you gave as your answer to 5 corresponds to the definition of one of the trigonometric functions for the radian measure you gave in your answer to 4. Which one? Use a calculator to find a decimal approximation.

7. Show how to use the approximation you gave in your answer to 6 and the proportion implied by your answer to 5 to find the length of side BC.

8. What information would be needed to apply this method to the cuspids? What assumptions would have to be made?

RADIANS AND DEGREES

Triangle trigonometry is usually applied to triangles with angles measured in degrees. Angles can be measured in degrees or radians, and there is a direct relationship between the two systems. To demonstrate the connection, draw an angle A so point A is at the origin of a rectangular coordinate system and one side of the angle lies along the positive x-axis (**Figure 5.4**).

Since the circle is a unit circle, angle A's radian measure is the same as the length of arc BC.

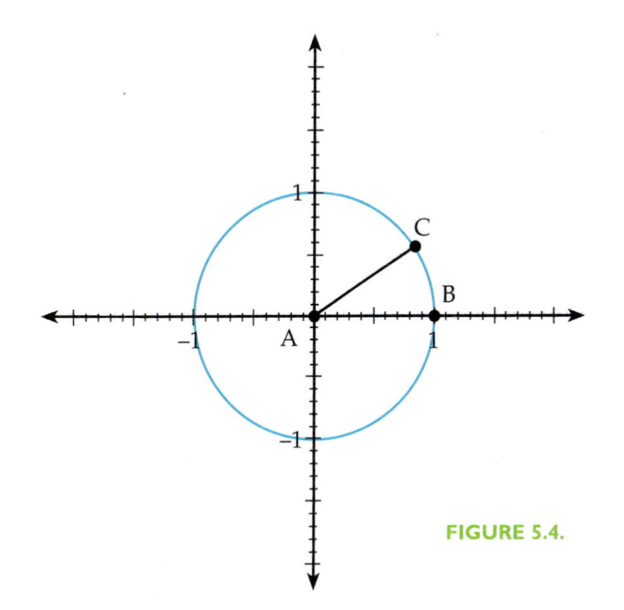

FIGURE 5.4.

EXAMPLE 1

In Figure 5.4, if angle A measures 60°, what is the corresponding radian measure?

SOLUTION:

An angle of 60° is $\frac{1}{6}$ of a complete revolution (360°), so the corresponding radian measure is $\frac{1}{6}$ of 2π, the unit circle's circumference.

$\frac{1}{6}2\pi = \frac{\pi}{3} \approx 1.05$ radians.

EXAMPLE 2

Find the degree measure that corresponds to $-\frac{\pi}{2}$ radians.

SOLUTION:

$\frac{\pi}{2}$ is a quarter of the unit circle's circumference, so $-\frac{\pi}{2}$ radians $= \frac{1}{4}(-360°) = -90°$.

As Examples 1 and 2 show, converting between radian and degree measures can require only a basic knowledge of revolutions and circumference. Since radian measure and degree measure vary directly, any proportion based on two equivalent measures will do the job if a formula is needed.

An example is: $\frac{\text{degree measure}}{360°} = \frac{\text{radian measure}}{2\pi}$.

But a simpler proportion is $\frac{\text{degree measure}}{180°} = \frac{\text{degree measure}}{180°}$.

EXAMPLE 3

Find the degree measure that corresponds to 2.5 radians.

SOLUTION:

$\frac{\text{degree measure}}{180°} = \frac{2.5}{\pi}$, so the degree measure is $\frac{2.5}{\pi}180 = \frac{450}{\pi} \approx 143.2$.

RIGHT TRIANGLE TRIGONOMETRY

The trigonometric functions can be applied to degree measures by first converting these measures to radians. However, to avoid the nuisance of making frequent conversions, it is helpful to redefine the trigonometric functions so that they apply directly to degree measures of angles of right triangles. Of course, the new definitions must be consistent with the old.

Figure 5.5 shows a right triangle positioned in a rectangular coordinate system so that the vertex of one acute angle is at the origin, and one of the angle's sides lies along the positive x-axis. The circle is the unit circle.

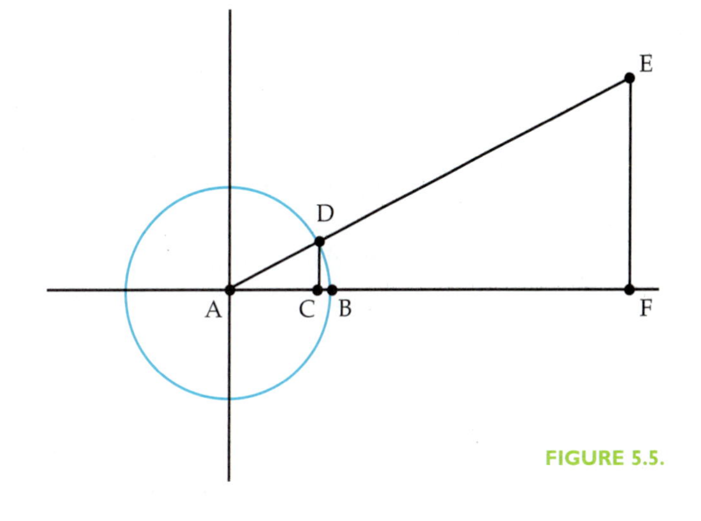

FIGURE 5.5.

The radian measure of angle A is the same as the length of the arc between B and D. The arc's horizontal displacement is AC, and its vertical displacement is DC. Thus, sine(angle A) = CD, cosine(angle A) = AC, and tangent (angle A) = $\frac{CD}{AC}$.

However, triangles ACD and AFE are similar, and it is convenient to describe angle A's trigonometric ratios in terms of triangle AFE's sides. By similar triangles,

$$\text{Sine(angle A)} = CD = \frac{CD}{1} = \frac{CD}{AD} = \frac{FE}{AE}$$

$$\text{Cosine(angle A)} = AC = \frac{AC}{1} = \frac{AC}{AD} = \frac{AF}{AE}$$

$$\text{Tangent(angle A)} = \frac{CD}{AC} = \frac{FE}{AF}.$$

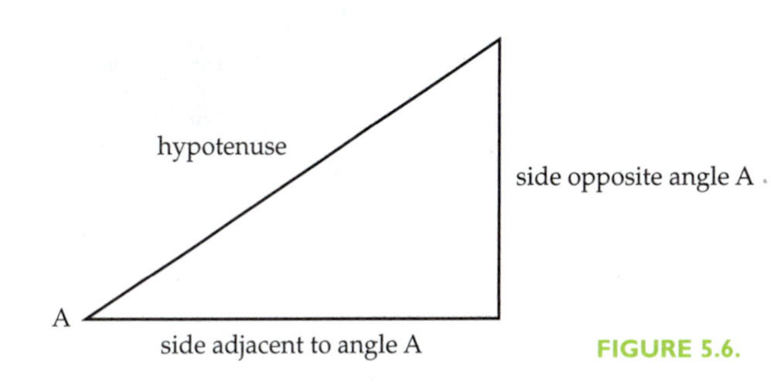

FIGURE 5.6.

Since not all triangles are labeled with the letters used in Figure 5.5, it is convenient to describe the trigonometric ratios in terms of sides relative to the chosen angle, as shown in **Figure 5.6**.

Thus, **sin(A)** = $\frac{\text{opposite side}}{\text{hypotenuse}}$, **cos(A)** = $\frac{\text{adjacent side}}{\text{hypotenuse}}$, and **tan(A)** = $\frac{\text{opposite side}}{\text{adjacent side}}$.

These commonly used definitions of the trig ratios in right triangles assume that the term *side* does not refer to the hypotenuse.

In this form, the trigonometric ratios can be used to solve many right triangle problems. Since calculators have a degree mode, using triangle trigonometry to find lengths does not require converting between radians and degrees. All that's necessary is to select a trigonometric ratio for which the angle and one of the two lengths are known, and to perform a simple calculation to find the other length.

EXAMPLE 4

For the right triangle in **Figure 5.7**, identify a trigonometric function of the measured angle that will enable you to find the length marked x. Use a calculator to determine a value for x.

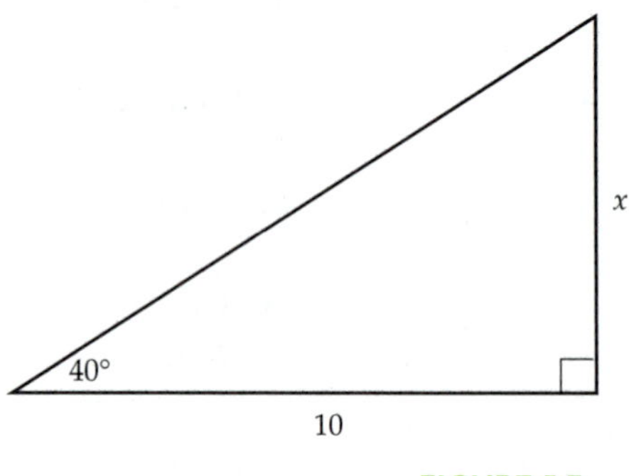

FIGURE 5.7.

SOLUTION:

The unknown length is opposite the 40° angle and the side measuring 10 is adjacent to this angle. Therefore, use the tangent ratio: tangent $(40°) = \frac{x}{10}$, or $x = 10\tan(40°)$.

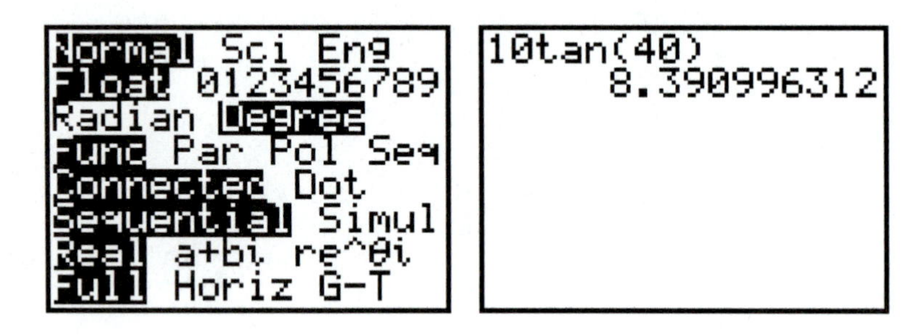

FIGURE 5.8.

Be sure the calculator is in the degree mode, then type and execute the instruction. **Figure 5.8** demonstrates for one model of calculator.

Therefore, $x \approx 8.4$.

EXAMPLE 5

The **angle of elevation** of an object is the angle between the horizontal and a line of sight to the object. Find the height of an object that casts a shadow 53 meters long when the angle of elevation of the sun is 42°.

SOLUTION:

First make a sketch of the problem (**Figure 5.9**). It need not be to scale.

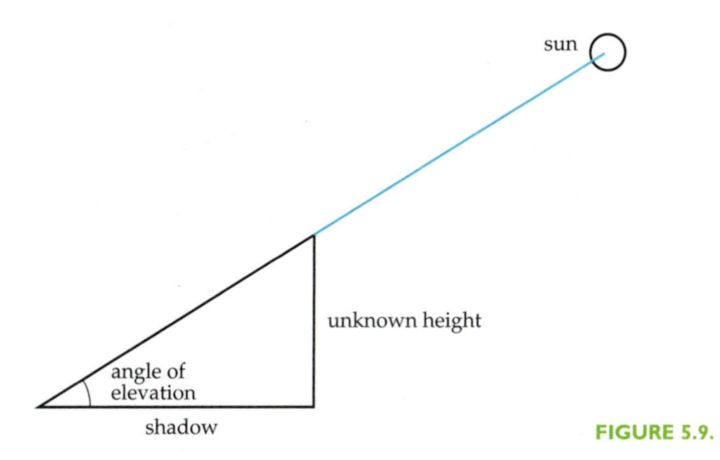

FIGURE 5.9.

The unknown height is the side opposite the known angle, and the length of the shadow is the side adjacent to the known angle. Therefore, use the tangent ratio.

$\text{Tan}(42°) = \frac{h}{53}$, or $h = 53\tan(42°) \approx 49$ meters.

EXAMPLE 6

The side opposite a 37° angle of a right triangle is 5.7 cm long. Find the length of the hypotenuse.

SOLUTION A:

Make a sketch of the situation (**Figure 5.10**).

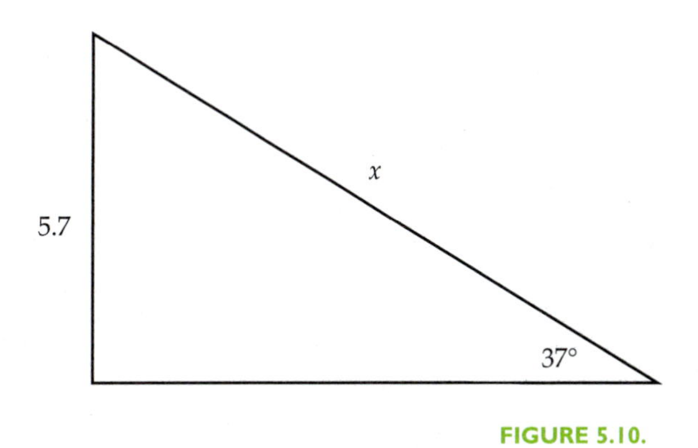

FIGURE 5.10.

The known side is opposite the known angle, and the side to be found is the hypotenuse. Therefore, use the sine ratio.

$\text{Sin}(37°) = \frac{5.7}{x}$, so $x = \frac{5.7}{\sin(37°)} \approx 9.5$ cm.

SOLUTION B:

Since cosecant is the reciprocal of sine, $\csc(37°) = \frac{x}{5.7}$, or $\frac{1}{\sin(37°)} = \frac{x}{5.7}$.
Thus, $x = \frac{5.7}{\sin(37°)} \approx 9.5$ cm.

SOLUTION C:

The other acute angle of the right triangle measures 90° – 37° = 53°. For this angle, the known side is the adjacent side and the hypotenuse is the side to be found. Therefore, use cosine.

$\text{Cos}(53°) = \frac{5.7}{x}$, so $x = \frac{5.7}{\cos(53°)} \approx 9.5$ cm.

Exercises 5.1

1. a) As accurately as possible, find the degree measure that corresponds to one radian.

   b) Discuss the radian and degree measures described by the motion of the minute hand of a clock from 8:05 to 8:10.

2. The length of the longest shadow cast by a Blair Cuspid was estimated at 110 meters, and the sun's angle of elevation at the time was estimated at 10.9°.

   a) Use a right triangle model to estimate the height of this cuspid.

   b) During what part of the lunar day do you think the photograph in Figure 5.1 was taken? Explain.

3. Right triangle models are also used to estimate the depth of craters in satellite photos of the moon. **Figure 5.11** represents a shadow cast between opposite rims of a crater. If the shadow's length is approximately 76 meters and the angle of elevation of the sun at the time the photo was taken is 34°, estimate the depth of the crater.

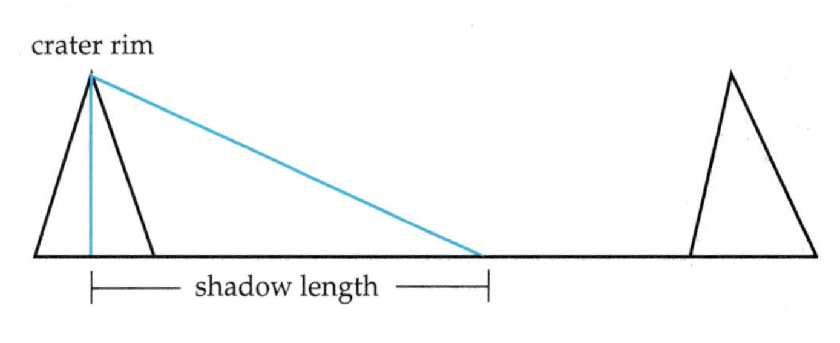

FIGURE 5.11.

4. People who fly kites or model rockets often want to estimate the height of the kite or rocket.

   a) One way to estimate the height is to measure or estimate the distance from the flier to a point on the ground directly below the kite or rocket. It's also necessary to estimate the angle of elevation of the kite or rocket from the flier. (See **Figure 5.12.**)

If the distance from the flier to the point on the ground directly below the kite is 87 feet, the angle of elevation of the kite is 51°, and the flier's eye is 5 feet from the ground, estimate the height of the kite.

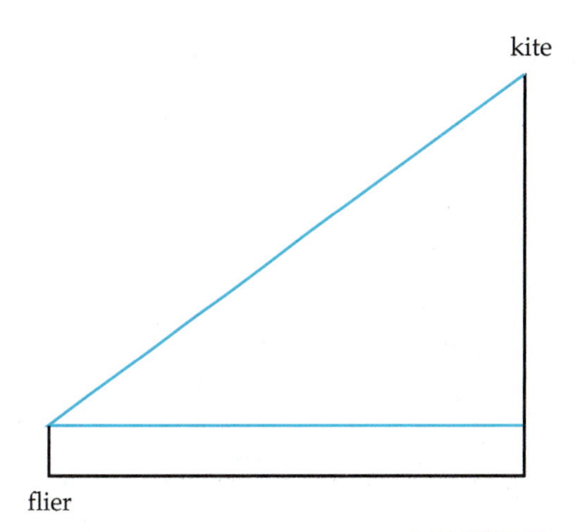

FIGURE 5.12.

Exercises 5.1

b) An alternate method can be used with kites. Instead of measuring the distance from the flier to a point on the ground, measure the length of the string. Suppose the angle of elevation is 56° and the string measures 134 feet. Estimate the height of the kite.

c) Serious fliers of kites and rockets use a manufactured device called a clinometer to measure angles of elevation. A simple clinometer can be made from a protractor, a fishing weight, and a piece of string. The weight is suspended from the protractor's center. The observer sights along the edge of the protractor and reads an angle measure at the point where the string intersects the protractor's scale. In **Figure 5.13**, how is the 70° reading related to the angle of elevation?

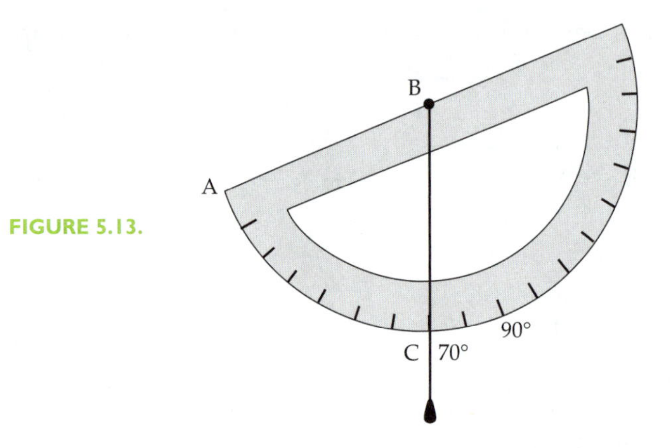

FIGURE 5.13.

d) Review the methods in (a) and (b). What modeling assumptions might affect the accuracy of either result?

5. Machinists use sine bars and sine plates to set precise angles. **Figure 5.14** demonstrates how a sine bar is used. The bar is mounted on rollers that have a fixed distance between centers. Pieces of metal (called gage blocks) of varying thickness are stacked to a height that produces the desired angle.

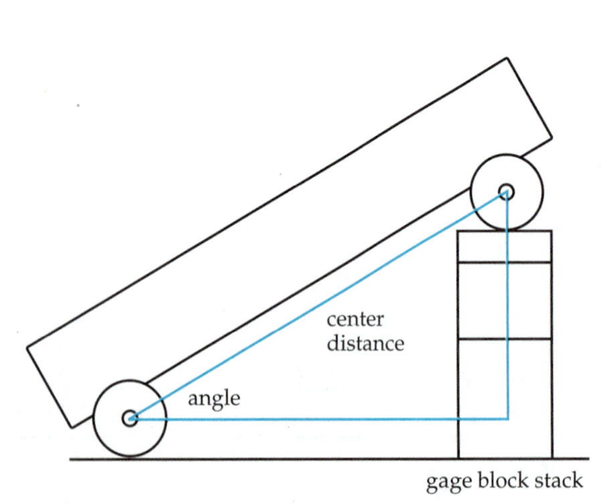

FIGURE 5.14.

a) If a machinist wants an angle of 8° 15′ and is using a 10-inch (distance between roller centers) sine bar, what should be the height of the gage block stack? (Round to the nearest thousandth of an inch.)

b) Explain why the height of the gage block stack is equal to the length of the side opposite the desired angle in the right triangle in Figure 5.14.

6. Henry Dreyfuss (1904–1972) was an American industrial designer who became famous for his extensive ergonomic studies. Among his interests were optimal viewing angles for the human eye. The Dreyfuss data suggest that the maximum comfortable vertical viewing angle for the human eye is about 27°, and the maximum horizontal viewing angle is about 60° (30° on either side of straight ahead).

Henry Dreyfus

a) If a two-story building is about 30 feet tall, how far away from the building should a person whose eyes are 5 feet 6 inches above ground level stand to comfortably view the entire building? Make a sketch to explain your method.

b) Assuming the house described in (a) is set back from the street the distance you calculated, what do the Dreyfuss data suggest for the width of the house? Explain.

c) Based on the Dreyfuss data and your calculations, what recommendations would you make for the design of a residential neighborhood of one- and two-story houses?

7. Radar is used to forecast weather. It can be especially valuable during storms, but it is not perfect. To avoid obstacles on the earth's surface, weather radar is pointed slightly upward from horizontal. This fact of radar life presents a problem for meteorologists since some weather phenomena such as tornadoes can be too low to be detected.

a) For one commonly used weather-radar system, the angle above the horizontal is one-half a degree. Use a right triangle model to estimate the height of the radar beam above the earth's surface 100 miles from the radar site.

b) Depending on atmospheric conditions, the beam described in (a) is about 9000–10,000 feet above the surface of the earth at 100 miles from the radar site. Compare this to the value you calculated in (a). What modeling assumptions might account for any discrepancy?

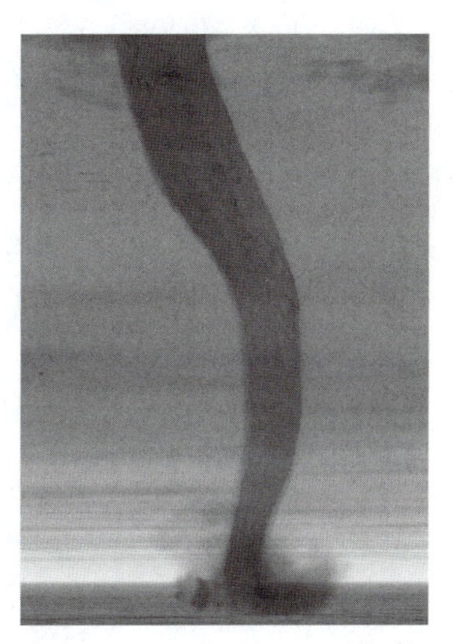
Corbis

Exercises 5.1

8. **Figure 5.15** shows a Radarsat (a Canadian remote sensing satellite) image of the Confederation Bridge, which connects Borden Point, Prince Edward Island, with Jourimain Island, New Brunswick. The inset, which is an enlargement of a section of the image on the right, gives the incorrect impression of three bridge decks. This phenomenon is caused by the reflection of the radar beam off bridge components and the water.

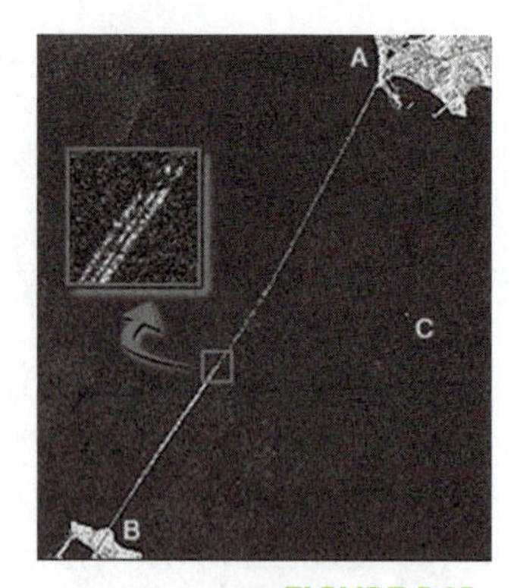

FIGURE 5.15. Radarsat image of the Confederation Bridge.

It's possible to calculate the amount by which the outside bridge decks appear to be offset from the middle deck. **Figure 5.16** shows the satellite's angle of incidence, which is the angle between vertical and a line of sight to the satellite.

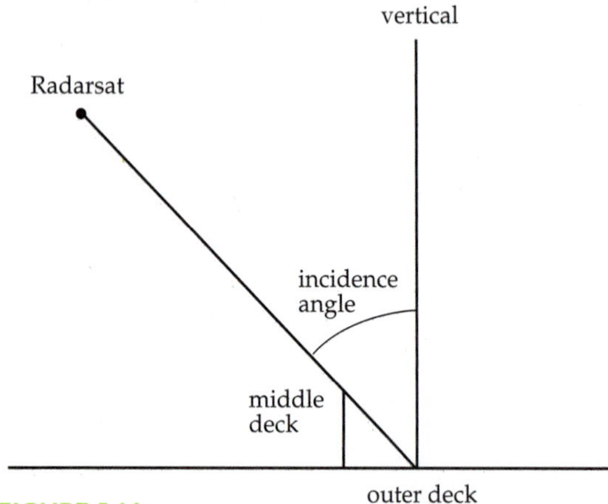

FIGURE 5.16.

At the time the bridge image was taken, Radarsat's angle of incidence was 43°. The Confederation Bridge averages about 40 meters above the water. Estimate the amount of offset.

9. Bicyclists care about the size and shape of a bike's stem (the part that holds the handlebars) because it affects a rider's position. The geometry of a bicycle stem is slightly complex. The stem angle is the angle between the portion of the stem that holds the handlebars and a line perpendicular to the stem post (S in **Figure 5.17**). The head angle (H in the right photo) is the angle between the stem post and the horizontal.

   a) If the head angle is 72°, and the stem angle is 20°, find the angles marked D and A in Figure 5.17.

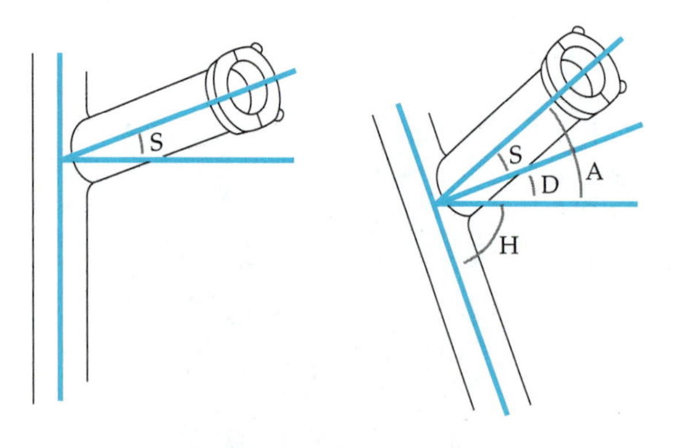

FIGURE 5.17. Geometry of a bicycle stem. The handlebars are clamped through the circular opening at the right of the stem.

Exercises 5.1

b) Suppose the stem in Figure 5.17 is a 100 mm stem. That is, the distance between the handlebars and the stem post is 100 mm. The handlebar rise is the distance between the point at which the handlebars are attached to the stem and the horizontal. Find the handlebar rise if the head angle and stem angle are those given in (a).

c) Find the handlebar forward distance; that is, the distance from the top of the stem post to a point directly below the handlebars.

d) A bicyclist wants to increase the forward distance and is contemplating replacing a 100 mm stem with a 110 mm stem. Comment.

10. To use a right triangle model to find the height of an object, you must be able to measure one other side of the right triangle. However, this isn't always possible, as **Figure 5.18** shows. In this case, surveyors are trying to determine the height of a hill. They have measured the angles of elevation of the top of the hill from two points 100 meters apart.

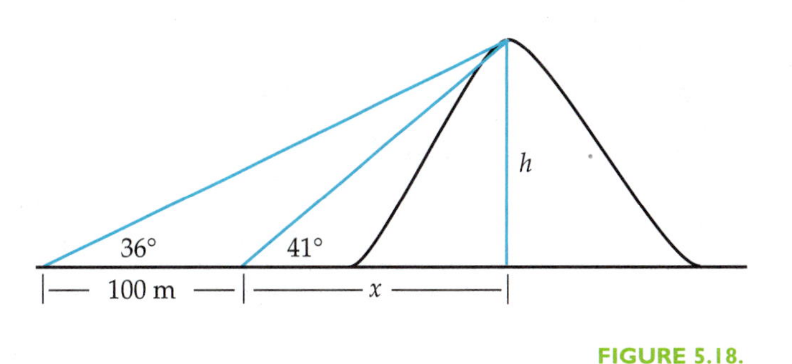

FIGURE 5.18.

a) Use a right triangle containing the 41° angle and a right triangle containing the 36° angle to write the tangent ratio for each of these angles.

b) Your answer to (a) is a system of two equations in two unknowns. Solve the system to determine the height of the hill.

11. **Figure 5.19** is a graph of $y = \sin(x)$ between 0 and 2π.

a) Replace the radian labels on the x-axis labels with the corresponding degree labels.

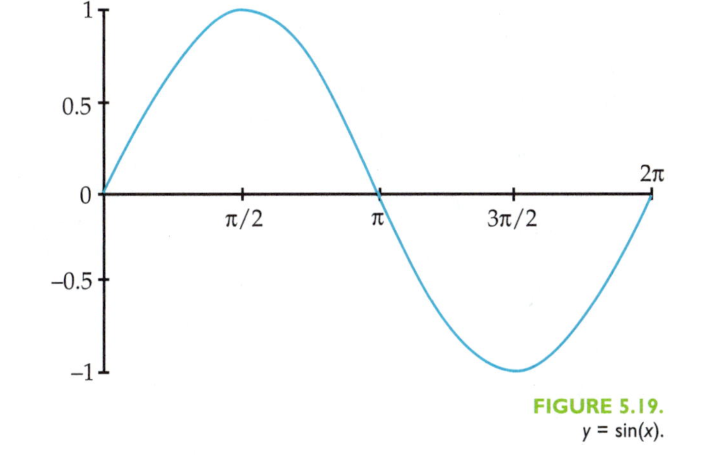

FIGURE 5.19.
$y = \sin(x)$.

Exercises 5.1

b) If a graphing calculator is in the degree mode, it will graph $y = \sin(x)$ as if x is an angle measured in degrees. Use a graphing calculator in the degree mode to graph one cycle of $y = \sin(x)$. What is the period?

c) Mathematicians sometimes indicate radians by using x for the independent variable and degrees by using θ (the Greek letter theta). What is the period of $y = 2 + 5\sin(3\theta)$?

12. a) What transformations of the graph of $y = \cos(\theta)$ produce the graph of $y = \cos(90° - \theta)$?

b) What identity is implied by the graph of $y = \cos(90° - \theta)$? Explain.

c) There is a geometric term that applies to the pair of acute angles θ and $(90° - \theta)$. What is that term?

13. Sometimes people expect an equation to be an identity, but it is not. Since an identity must be true for all values of the variable(s), you can disprove an identity by finding a **counterexample;** that is, one or more values of the variable(s) for which the identity fails.

a) Find a counterexample that shows $\cos(\alpha + \beta) = \cos(\alpha) + \cos(\beta)$ is not an identity. α (alpha) and β (beta) represent any two angles, so their values can be in either radians or degrees. You can use your calculator to approximate the cosines.

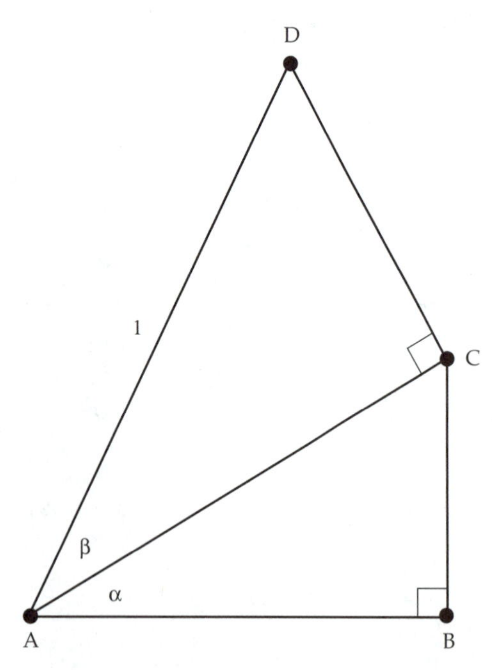

The remainder of this exercise develops a proof of an identity for $\cos(\alpha + \beta)$ in terms of sines and cosines of α and β. Since the proof uses right triangles, its conclusion is limited to acute angles, but the resulting identity can be proved for all real values of α and β.

In **Figure 5.20**, triangle ABC is a right triangle containing α, and triangle ACD is a right triangle containing β. They are drawn so that the hypotenuse of the first triangle is a leg of the second. AD has length 1.

FIGURE 5.20.

Exercises 5.1

b) Express AC and DC in terms of trigonometric ratios of β.

c) Express AB in terms of trigonometric ratios of α and β. (Caution: You cannot assume AC = 1.)

d) In **Figure 5.21**, a perpendicular has been drawn from D to AB. Explain why AE = cos(α + β).

e) Since you found an expression for AB in (b), you will have an identity for cos(α + β) if you can find an expression for EB. Consider a segment that is equal to EB. In **Figure 5.22**, a perpendicular has been drawn from F to BC. Explain why triangles FGC and DCF are similar to triangle AEF.

f) What ratio in triangle DCF equals the ratio $\frac{FG}{FC}$ in triangle FGC? Solve the resulting proportion for FG and explain why the solution is equal to sin(α)sin(β).

g) What can you conclude about cos(α + β)?

h) Verify this identity for the values you used as a counterexample in part (a).

14. The identity you proved in Exercise 13 can be used to prove other identities. For example, an identity for cos(α – β) can be proved by noting that cos(α – β) = cos(α + (–β)), then applying the identity from Exercise 13 and knowledge about odd and even functions. Complete the proof.

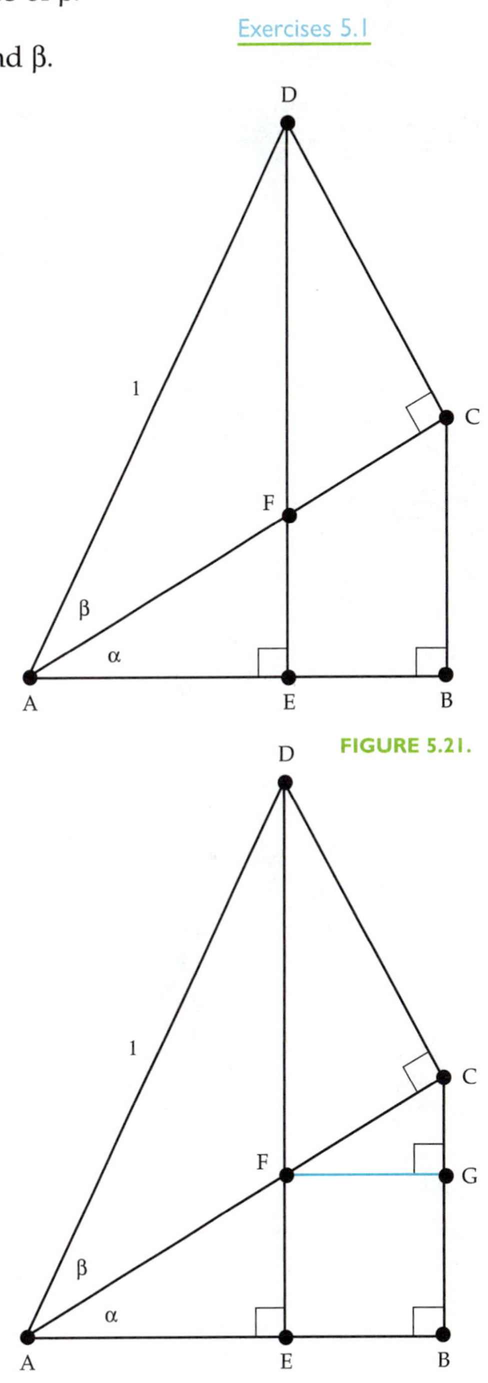

FIGURE 5.21.

FIGURE 5.22.

LESSON 5.2

Inverses

In the previous lesson you used right triangle models to find unknown lengths. In many modeling situations, two or three sides of a right triangle are known, but neither of the acute angles. For example, camera lenses come in many different sizes (**Figure 5.23**). Each lens has its own viewing angle (or a range of viewing angles in the case of zooms). Some are narrow and capture a small portion of a scene; others are wide and can capture a panoramic view of what the eye sees. A simple right triangle model can be used to calculate the angle of view of a lens from two known lengths: a dimension of the film and a dimension of the lens.

FIGURE 5.23.
Single lens reflex camera lenses of varying sizes.

In this lesson, you use right triangle models to find unknown angles. In the next lesson, you will return to the problem of the Blair Cuspids and see how right triangle trigonometry can be adapted to non-right triangle situations.

Activity 5.2

In this activity you explore the basic problem: finding an acute angle of a right triangle when two sides are known.

Consider the triangle in **Figure 5.24**.

1. If two sides of a right triangle are known, then the third can be found without the use of trigonometric functions. Explain.

2. Finding an angle in a right triangle involves nothing more than solving a simple trigonometric equation (which you did several times in Chapter 4). Write a trigonometric equation of the form tan(A) = ___ and show how it can be solved using a graph. Also, describe all solutions of this equation.

3. Your answer to 2 gives an infinity of solutions. Which ones make sense in a right triangle situation? Explain.

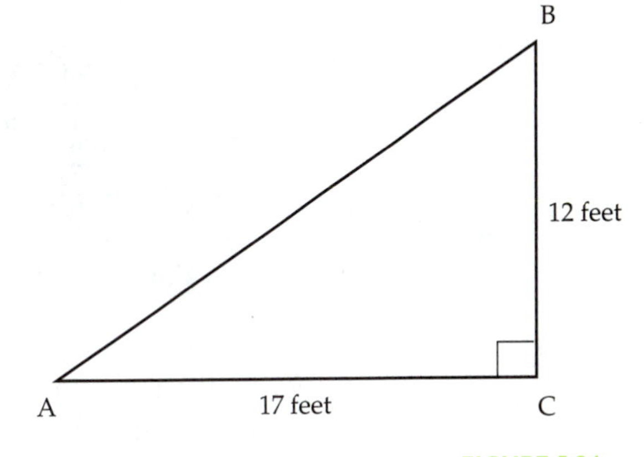

FIGURE 5.24.
A right triangle.

4. Your calculator is capable of calculating angle A directly. On your calculator's home screen, evaluate the expression $\tan^{-1}\left(\frac{12}{17}\right)$ in both the degree and radian modes. Are the results equivalent?

5. A calculator cannot produce infinitely many answers. Thus, it gives you only one solution to a trigonometric equation such as the one you wrote in 2. Explore $\sin^{-1}$, $\cos^{-1}$, and $\tan^{-1}$. Include both positive and negative values of sine, cosine, and tangent. Describe the range of degree measures that each of these calculator functions produces.

6. Describe the domain of $\sin^{-1}$, $\cos^{-1}$, and $\tan^{-1}$.

FINDING ANGLES

Finding an unknown angle in a right triangle corresponds to finding one solution of a trigonometric equation. For example, consider the right triangle in **Figure 5.25**.

To find angle A, write the trigonometric equation $\tan(A) = \frac{5}{10} = \frac{1}{2}$. This equation has an infinite number of solutions, but only one of them lies between 0° and 90°. A graph can be used to estimate the solution, but a calculator can produce this estimate directly, as shown in **Figure 5.26**.

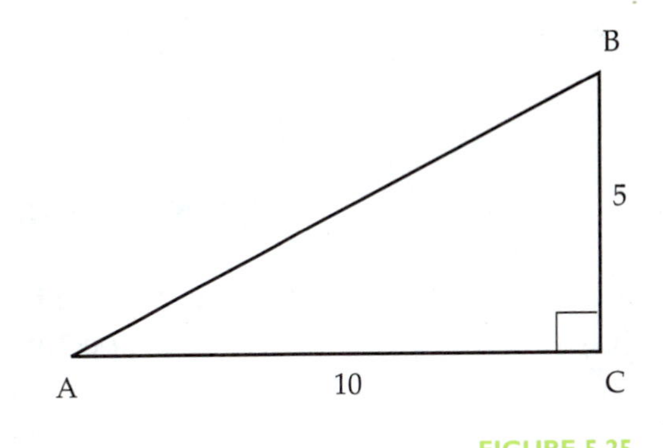

FIGURE 5.25.

```
Normal Sci Eng
Float 0123456789
Radian Degree
Func Par Pol Seq
Connected Dot
Sequential Simul
Real a+bi re^θi
Full Horiz G-T
```

```
tan⁻¹(1/2)
        26.56505118
```

FIGURE 5.26. Setting degree mode and finding an angle's measure.

Since right triangles cannot have angles measuring more than 90° (or less than 0°), this calculator method always produces the correct "angle in right triangle situations."

In right triangle situations, you can read the expression $\tan^{-1}\left(\frac{1}{2}\right)$ as the "angle whose tangent is $\frac{1}{2}$."

EXAMPLE 7

Use the value produced by $\sin^{-1}(0.5)$ to find all solutions of the equation $\sin(A) = 0.5$.

SOLUTION:

In the degree mode, calculator evaluation gives $\sin^{-1}(0.5) = 30$.

Since the period of sine is 360°, all values given by the expression $30° \pm n360°$, where n is a whole number, are solutions. However, there are more.

The symmetry of the graph of the sine function implies that the second positive solution is $180° - 30° = 150°$ (**Figure 5.27**). Therefore, all values given by the expression $150° \pm n360°$, where n is a whole number, are also solutions.

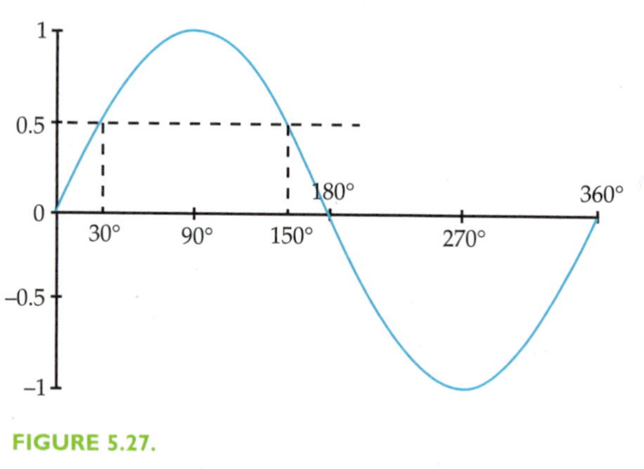

FIGURE 5.27.

INVERSE TRIGONOMETRIC FUNCTIONS

Sin^{-1}, $\cos^{-1}$, and $\tan^{-1}$ are functions that are inverses of sine, cosine, and tangent, respectively. For example, the sine function pairs 30° with 0.5, so the $\sin^{-1}$ function pairs 0.5 with 30°. However, the sine function pairs 150° with 0.5, but $\sin^{-1}$ does not pair 0.5 with 150°. If $\sin^{-1}$ paired 0.5 with both 30° and 150°, it would not be a function.

Since the inverses of trigonometric functions produce only one value out of an infinity of possibilities, it is important to know the range of values produced. Here is a summary.

Values of $\sin^{-1}$ fall between –90° and 90°, inclusive. This range can be expressed with inequalities: $-90° \le \sin^{-1} \le 90°$, or with interval notation as [–90°, 90°]. In radian measure, the corresponding range is $-\frac{\pi}{2} \le \sin^{-1} \le \frac{\pi}{2}$ or $\left[\frac{-\pi}{2}, \frac{\pi}{2}\right]$.

Values of $\tan^{-1}$ fall between –90° and 90°. The corresponding inequality is $-90° < \tan^{-1} < 90°$, and the interval is (–90°, 90°). In radian measure, the range is $-\frac{\pi}{2} < \tan^{-1} < \frac{\pi}{2}$ or $\left(\frac{-\pi}{2}, \frac{\pi}{2}\right)$.

Values of $\cos^{-1}$ fall between 0° and 180°, inclusive. The corresponding inequality is $0° \le \cos^{-1} \le 180°$, and the interval is [0, 180°]. In radian measure, the range is $0 \le \cos^{-1} \le \pi$ or $[0, \pi]$.

The values chosen for $\sin^{-1}$, $\cos^{-1}$, and $\tan^{-1}$ can be visualized in the graphs of the sine, cosine, and tangent functions (**Figure 5.28**) or on unit circle diagrams (**Figure 5.29**).

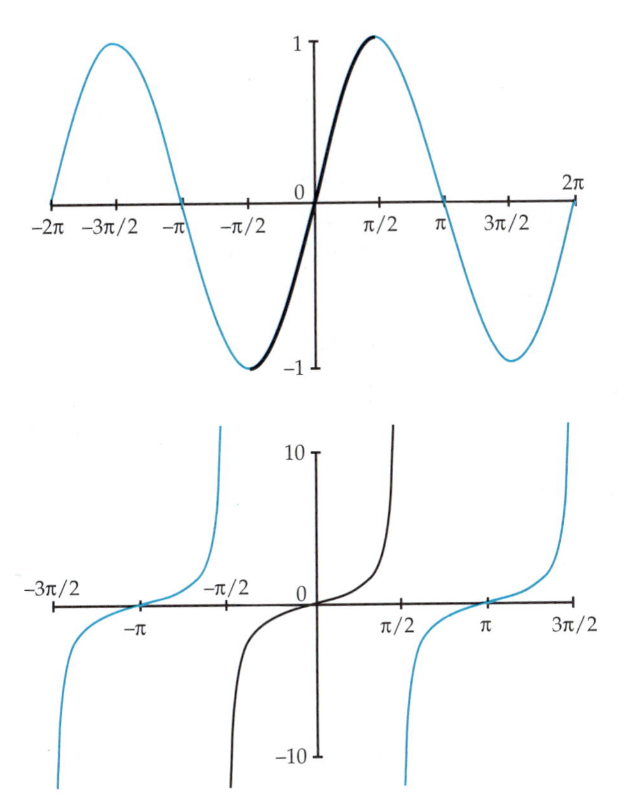

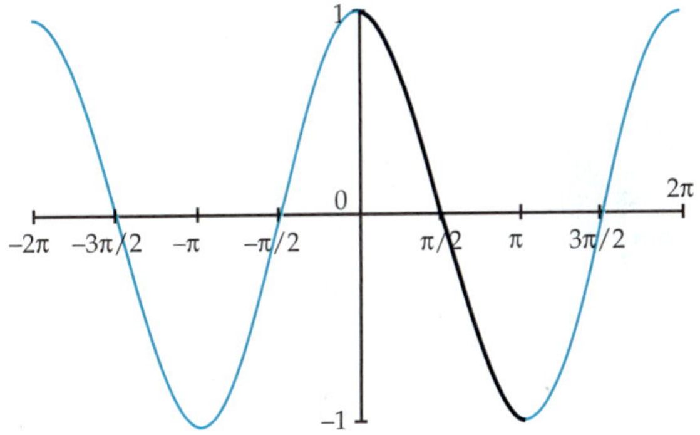

FIGURE 5.28.
The bold line shows the values chosen for the inverses of the sine, cosine, and tangent functions.

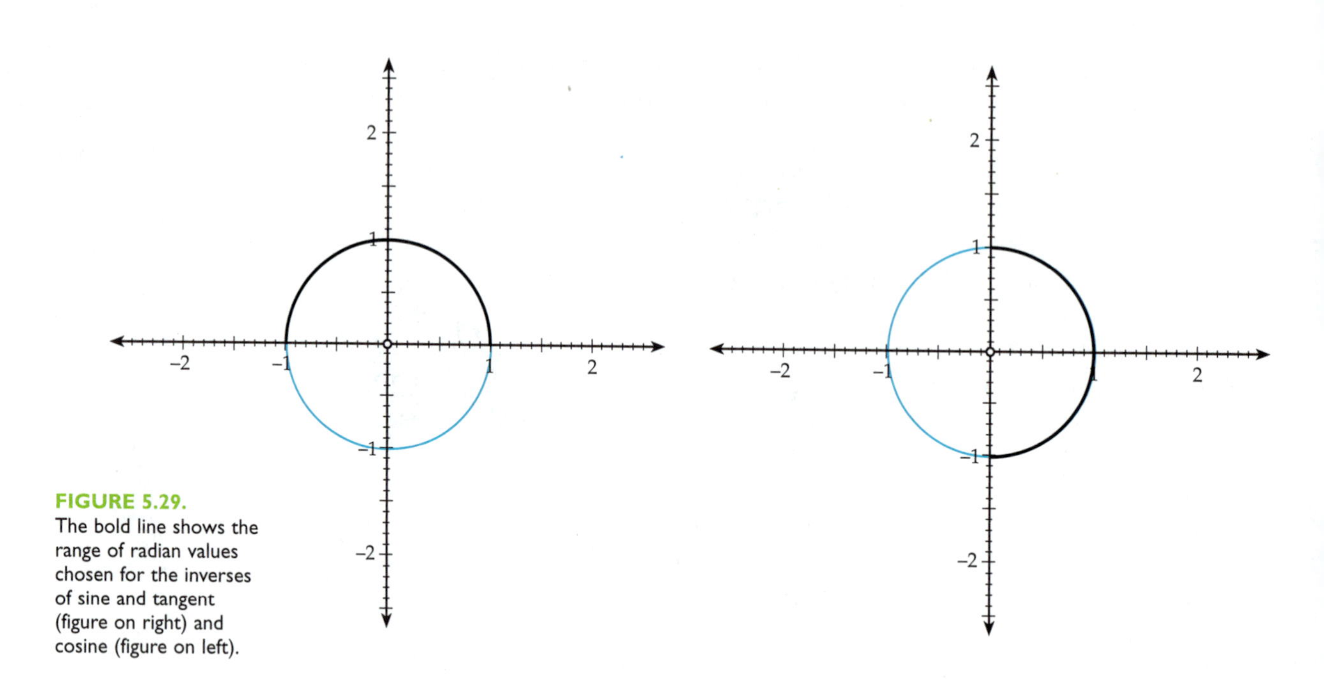

FIGURE 5.29.
The bold line shows the range of radian values chosen for the inverses of sine and tangent (figure on right) and cosine (figure on left).

It's fairly easy to remember these three ranges. Both $\sin^{-1}$ and $\tan^{-1}$ produce positive acute angles when sine and tangent are positive and the negatives of these angles when sine and tangent are negative. Cosine is an exception because, for example, cos(–30°) is not negative.

TAKE NOTE Be careful that you do not confuse $^{-1}$ as it is used for inverses with an exponent. That is, $\sin^{-1}(x)$ does *not* mean $(\sin(x))^{-1} = \dfrac{1}{\sin(x)}$.

Some books use arcsin, arccos, and arctan interchangeably with $\sin^{-1}$, $\cos^{-1}$, and $\tan^{-1}$, respectively. In a few books, arcsin, arccos, and arctan are reserved for radians only. These authors think of them as meaning "the arc whose sine is," for example.

Many computer spreadsheets use the symbols asin, acos, and atan for $\sin^{-1}$, $\cos^{-1}$, and $\tan^{-1}$, respectively.

Example 8

Use the calculator value of $\cos^{-1}(-0.4)$ to express all angles whose cosine is –0.4.

Solution:

In the degree mode, a calculator gives $\cos^{-1}(-0.4) \approx 113.6°$. The graph in **Figure 5.30** shows the relationship of this value to others.

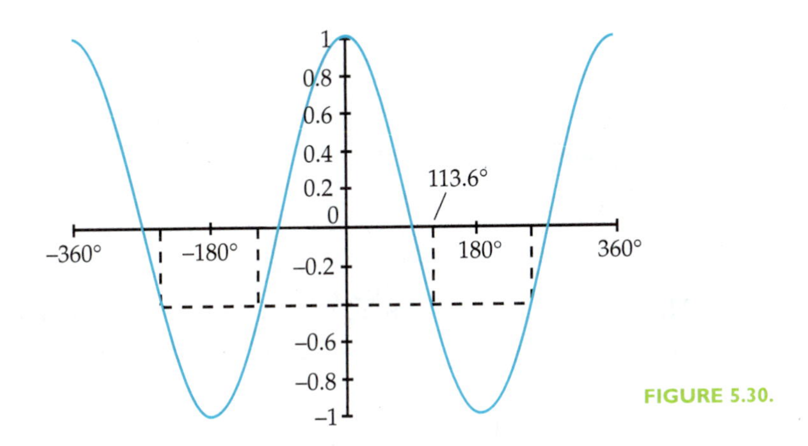

FIGURE 5.30.

Since the cosine curve is symmetric about the y-axis, the simplest way to express solutions on the ascending portions of the curve is with –113.6°.

Therefore, angles with a cosine of –0.4 are given by the expressions 113.6° ± n360° or –113.6° ± n360°, where n is a whole number.

GRAPHS OF INVERSE TRIGONOMETRIC FUNCTIONS

You can produce a graph of $y = \sin^{-1}(x)$, $y = \cos^{-1}(x)$, or $y = \tan^{-1}(x)$ on a graphing calculator or other graphing utility. To set a proper window, you need to keep in mind the domains and ranges of these functions. It's also useful to remember that the roles of the variables in the function $y = \sin(x)$ are reversed in the function $y = \sin^{-1}(x)$. In the first, x is an angle measured in degrees (or an arc measured in radians) and y is the sine; in the second, x is the sine and y is the angle (or arc).

For example, the graph of $y = \sin^{-1}(x)$ is shown in **Figure 5.31**. Note that the domain is all real numbers in the interval [–1, 1] and the range is all real numbers in the interval [–90°, 90°].

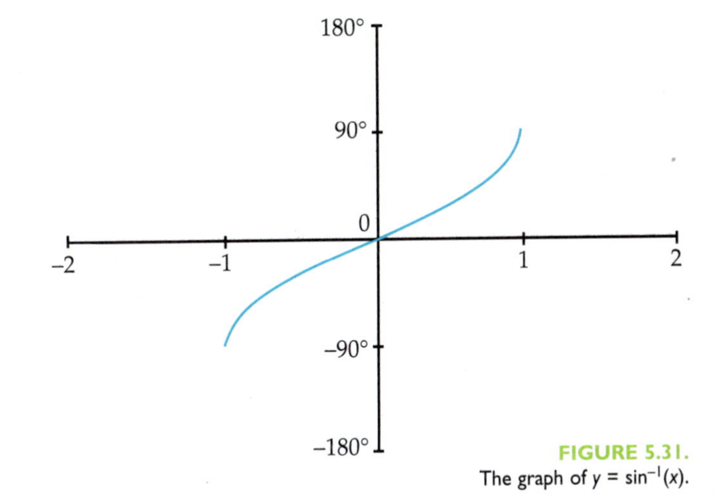

FIGURE 5.31.
The graph of $y = \sin^{-1}(x)$.

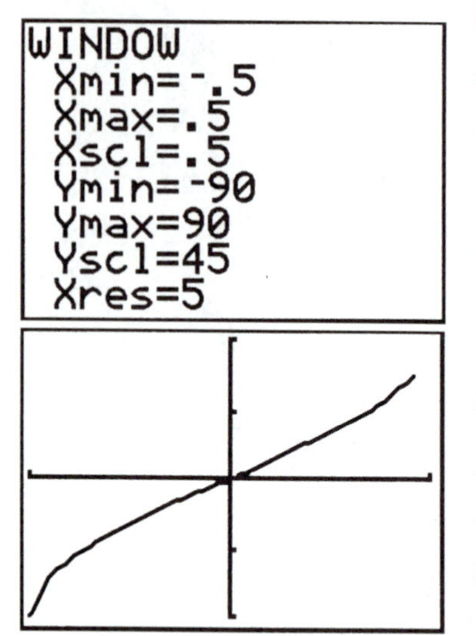

FIGURE 5.32.
Graphing $y = \cos^{-1}(x)$ on a calculator.

EXAMPLE 9

Set a suitable range and produce a calculator graph of the function $y = \cos^{-1}(x)$.

SOLUTION:

The values of $\cos^{-1}$ range between 0° and 180°, which are suitable values for the y-axis minimum and y-axis maximum, respectively.

Since cosine ranges between –1 and 1, these are suitable values for the x-axis minimum and x-axis maximum, respectively.

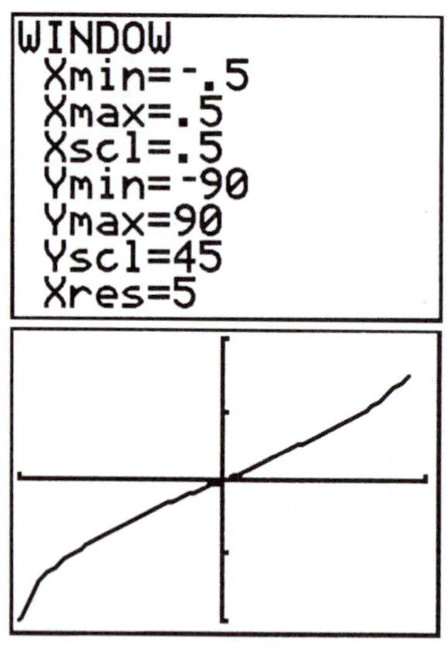

FIGURE 5.33.
Graphing $y = \sin^{-1}(2x)$ on a calculator.

EXAMPLE 10

Determine a suitable window and produce a calculator graph of $y = \sin^{-1}(2x)$.

SOLUTION:

$y = \sin^{-1}(x)$ produces values between –90° and 90°. $y = \sin^{-1}(2x)$ should do the same since no changes have been made to the outside of the function.

The domain of $y = \sin^{-1}(x)$ is [–1, 1], so the domain of $y = \sin^{-1}(2x)$ should be half that interval, or [–0.5, 0.5]. (Note that the domain can also be found by solving the equations $2x = -1$ and $2x = 1$.)

Exercises 5.2

1. Find the measure of angle B in each right triangle.

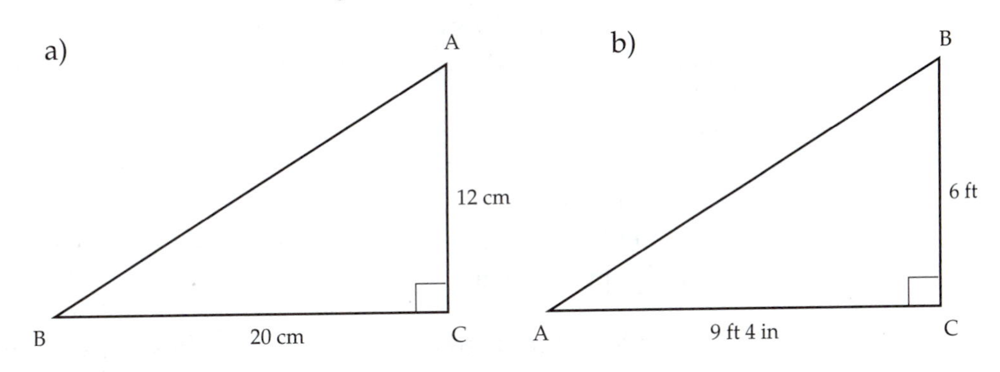

2. Skiers worry about avalanches. The likelihood of an avalanche is high if a slope is too steep. The critical angle varies with the condition of the snow, but slopes angled at 32° or more from the horizontal are considered dangerous. Some skiers carry a slope meter (**Figure 5.34**) to estimate angles.

   a) A skier who does not have a slope meter estimates the distance to the top of a slope as 160 feet and its height (above the horizontal) as 75 feet. By these estimates, is the slope dangerous? Explain.

   b) If the estimate of the height is off by 5 feet, and the estimate of the distance to the top is off by 10 feet, could these errors be serious? Explain.

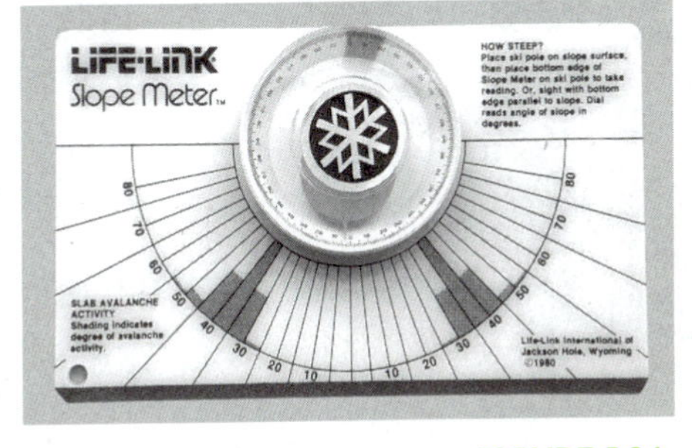

FIGURE 5.34. A slope meter.

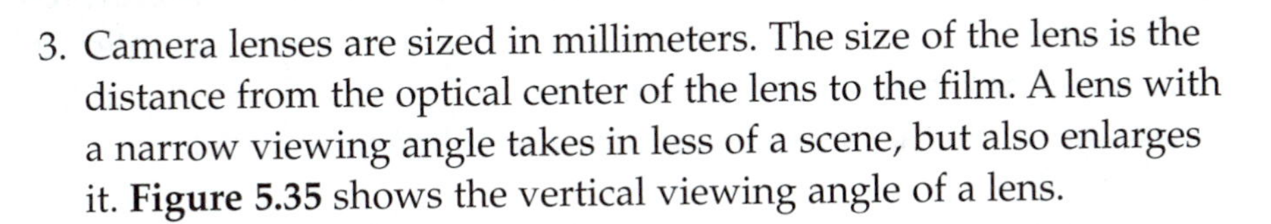

3. Camera lenses are sized in millimeters. The size of the lens is the distance from the optical center of the lens to the film. A lens with a narrow viewing angle takes in less of a scene, but also enlarges it. **Figure 5.35** shows the vertical viewing angle of a lens.

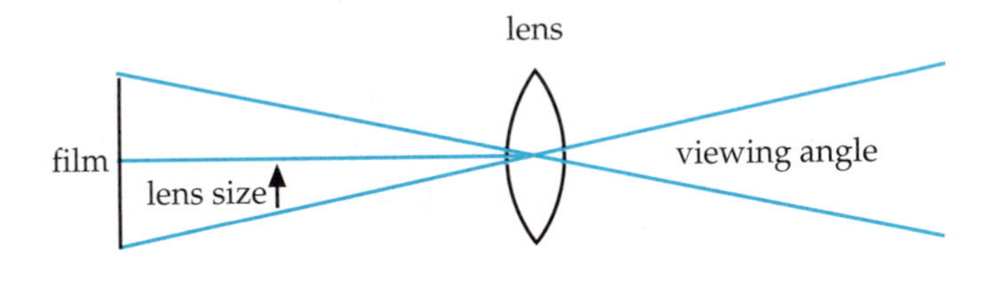

FIGURE 5.35. The vertical viewing angle of a camera lens.

Exercises 5.2

a) The most popular photography film is 35 mm. The height of 35 mm film's picture area is 24 mm. What is the vertical viewing angle of a 50 mm lens in a 35 mm camera?

b) Determine the horizontal viewing angle of a 50 mm lens used in a 35 mm camera. Note: the width of 35 mm film's picture area is actually 36 mm.

c) Another popular film is used in advanced photo system (APS) cameras. The picture frame on APS film is approximately 17 mm high and 30 mm wide. Find the vertical and horizontal viewing angles of a 50 mm lens in an APS camera.

d) Does a 50 mm lens have a greater magnification effect in a 35 mm camera or in an APS camera? Explain.

e) The smaller a negative, the more likely it is to produce enlargements that are not sharp. Compare the sizes of 35 mm and APS negatives.

4. A person is standing 15 feet from a sign. The sign is 6 feet high and its bottom edge is 8 feet from the ground. The person's eyes are 5 feet from the ground. Make a sketch of this situation and show how to use right triangles to determine the person's vertical viewing angle.

5. Sonar is an acronym for Sound Navigation and Ranging. The sound beams produced by sonar can locate fish, submarines, and wreckage. Sonar units used to investigate wreckage are trailed behind ships. The angle at which sonar strikes an object is critical in determining its shape. Since the speed of sound waves in water is known, the distance of a sonar unit from the ocean bottom and the distance from the object being investigated are also known.

   If a sonar emitter is 105 feet from the ocean bottom and 352 feet from an object, at what angle does the sonar strike the object? Make a sketch and show how you determined the angle.

6. The Uniform Federal Accessibility Standards state that wheelchair ramps in new construction should not exceed a slope of 1:12. What is the maximum ramp angle permitted by these standards?

Exercises 5.2

7. The human eye focuses at a point approximately 31 inches away when it is at rest. Thus, the ideal distance from the eye to a computer monitor is 31 inches. Long periods of work at shorter distances can cause eyestrain. If a person sits 31 inches from a monitor screen that is 10 inches wide and 8 inches high, what are the horizontal and vertical viewing angles?

8. One factor that guitarists use to judge the playability of a guitar is the neck angle. The strings of a guitar follow a straight-line path from the bridge to the nut at the end of the guitar's neck. Therefore, the neck angle can be interpreted as the angle made by the strings and the plane of the guitar body's top surface (**Figure 5.36**).

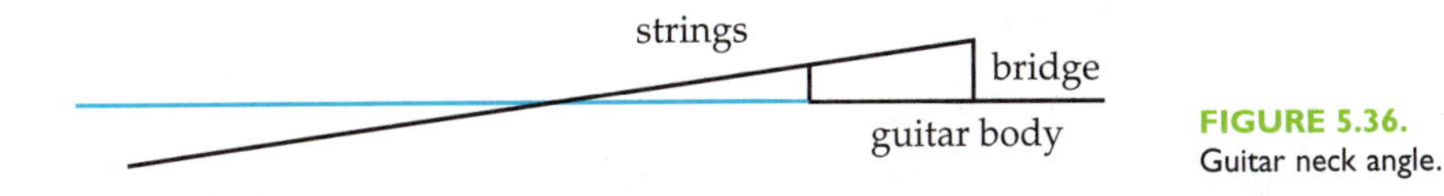

FIGURE 5.36. Guitar neck angle.

Suppose a guitar's bridge sits $\frac{5}{8}$ inch above the guitar body. The strings are $\frac{1}{4}$ inch above the guitar body at the point where the neck is attached, and the distance from this point to the bridge is 6.25 inches. What is the neck angle?

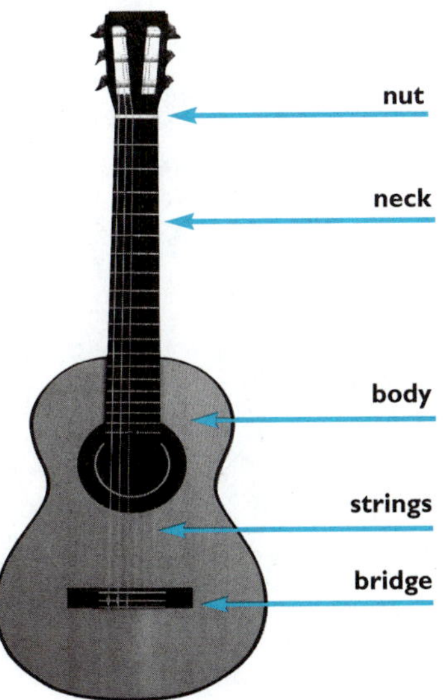

9. To the nearest tenth of a degree, a calculator approximation of $\sin^{-1}(0.65)$ is 40.5.

   a) Use this value to approximate all angles whose sine is 0.65.

   b) Use $\sin^{-1}(0.65)$ to give the radian measures of all arcs whose sine is 0.65.

   c) Use radian measure to explain the difference between $\sin^{-1}(0.65)$ and $(\sin(0.65))^{-1}$.

10. What is the value of $\cos^{-1}(2)$ in degrees? In radians? Explain.

Exercises 5.2

11. The material that precedes this set of exercises shows graphs of the $\sin^{-1}$ and $\cos^{-1}$ functions. Your task in this exercise is to sketch a graph of the $\tan^{-1}$ function.

    Here are a few steps that may help if you are unsure of how to proceed: Begin by refreshing your memory of the graph of the tangent function, then review the range of the $\tan^{-1}$ function. Finally, reverse the roles of the variables in the tangent function and sketch the graph of $y = \tan^{-1}(x)$. (Recall from Chapter 2 that reversing the roles of the variables reflects a graph across the line $y = x$.)

12. Give a graph for each of the following functions and identify the domain and range.

    a) $y = 1 + \tan^{-1}(x)$. (Note: since 1 does not carry a degree symbol, the range of this function is in radians.)

    b) $y = \cos^{-1}(3x - 1)$.

13. For each of the following functions, describe the intervals over which they are increasing, the intervals over which they are decreasing, and discuss the graph's concavity.

    a) $y = \sin^{-1}(x)$

    b) $y = \cos^{-1}(x)$

    c) $y = \tan^{-1}(x)$

LESSON 5.3

Oblique Triangles

For a right triangle model to accurately determine the height of an object, the ground on which it is standing must be horizontal. Here again is the satellite photo of the Blair Cuspids that you first saw in Lesson 5.1.

Notice that the ground on the shadow side of the cuspids is fairly dark. Many scientists who examined the photo took the darkness as an indicator that the ground sloped downward in the direction of the cuspid shadows. Therefore, they concluded that the right triangle model overestimated the heights of the cuspids.

In this lesson you will adapt right triangle trigonometry to non-right triangles, which will make trigonometry a more valuable modeling tool.

FIGURE 5.1. The Blair Cuspids.

Activity 5.3

If an object casts a shadow on ground that slopes downward, the shadow is longer than it would be if the ground is level. **Figure 5.38** demonstrates this. (Key points have been labeled to facilitate discussion.)

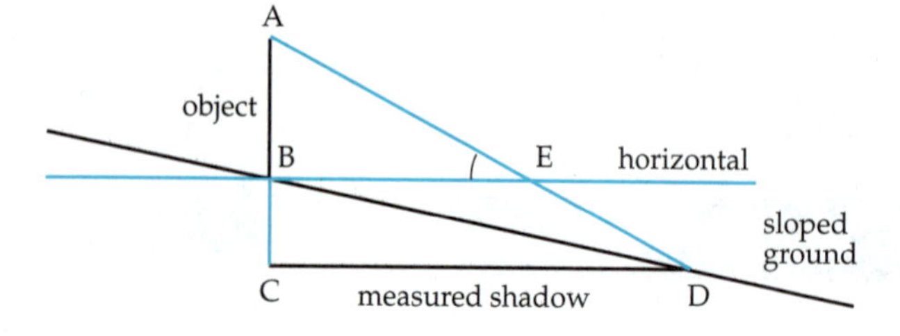

FIGURE 5.38.
An upright object on sloped ground.

The sun's angle of elevation is the marked angle at E in Figure 5.38. If the ground were level, the object's shadow would run from B to E. Since the shadow is observed from overhead, its apparent length is represented by the segment from C to D.

1. If the shadow length from C to D and the angle of elevation of the sun are known, what other segments and angles could be found using right triangle trigonometry or other geometric facts (such as the Pythagorean Theorem)?

2. If a right triangle model is used because the ground is assumed level, what length in Figure 5.38 is calculated for the height of the object?

3. At the time of the Blair Cuspids photograph, the sun's angle of elevation was about 10.9°. Suppose the ground sloped downward at an angle of 5° from horizontal. What other parts (angles or lengths) can be found in Figure 5.38?

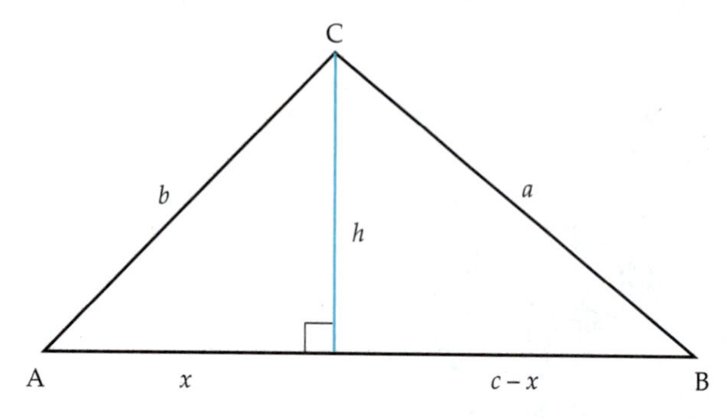

FIGURE 5.39.
A triangle and one of its altitudes.

Figure 5.38 is fairly complicated. To develop trigonometric relationships that can be used in this and other non-right triangle situations, consider a single triangle and one of its altitudes (**Figure 5.39**). To facilitate discussion, the lengths of all segments are represented by appropriate variables.

When applying trigonometry to non-right triangles, one way to proceed is to use the right triangles created by an altitude. This will introduce one or more variables that were not sides (or angles) of the original triangle. Thus, to reach conclusions about the original triangle, algebraic manipulations are used to eliminate the introduced variables.

4. The altitude in Figure 5.39 forms two right triangles. Use one of them to write the sine of angle A and the other to write the sine of angle B.

5. The equations you wrote in 4 can be combined to form a single equation by solving each of them for h. Show how to do this. What can you conclude about the original triangle ABC?

6. Use the right triangles in Figure 5.39 to write the Pythagorean formula twice—once for each triangle.

7. The equations you wrote in 6 contain two variables, x and h, that are not sides of the original triangle. Explore ways to eliminate them. You can use algebraic techniques and/or trigonometric ratios. Remember, the goal is to eliminate x and h.

OBLIQUE TRIANGLES

Many modeling situations involve non-right triangles. A triangle that does not have a right angle is called an **oblique triangle (Figure 5.40)**.

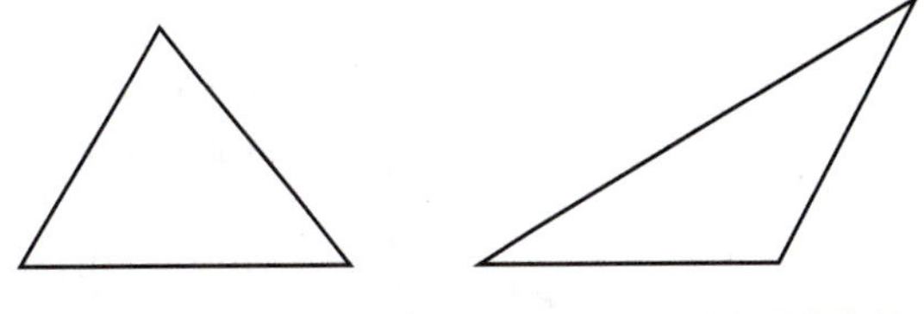

FIGURE 5.40.
Examples of oblique triangles.

One way to begin the solution of an oblique triangle problem is by dividing the triangle into two right triangles. However, the additional segments and angles that are introduced complicate the situation unnecessarily.

It is possible to derive trigonometric formulas that apply directly to oblique triangles. However, since these formulas can produce misleading results, it is a good idea to keep two geometric theorems in mind when working with oblique triangles.

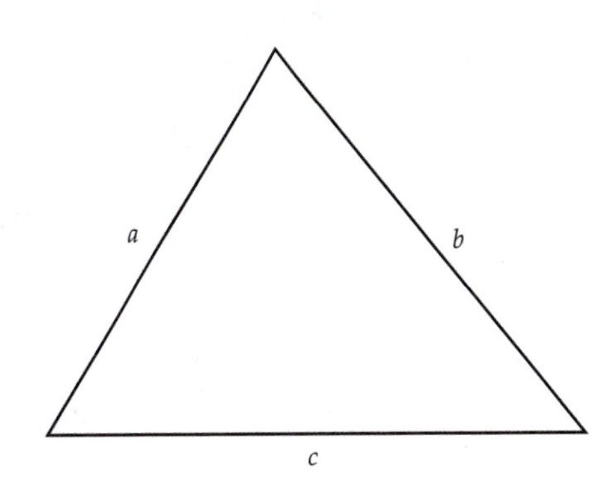

FIGURE 5.41.
The Triangle Inequality Theorem: $a + b > c$, $a + c > b$, and $b + c > a$.

1. In any triangle, the sum of the lengths of any two sides must be greater than the length of the remaining side. This result is known as the **Triangle Inequality Theorem (Figure 5.41)**.

2. In any triangle, the lengths of the sides are ordered according to the measures of the angles opposite them. In other words, the largest side is opposite the largest angle, and so forth (**Figure 5.42**).

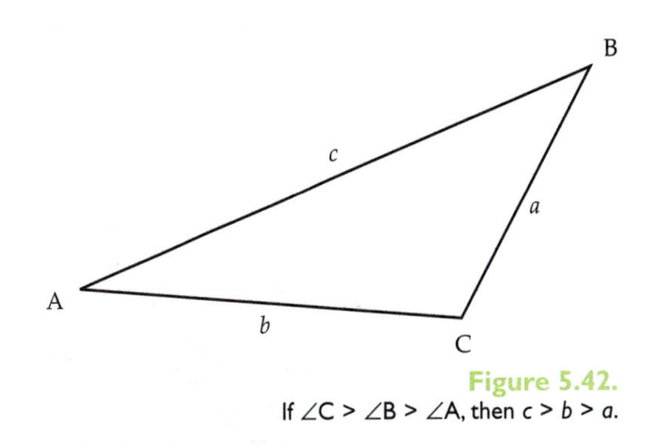

Figure 5.42.
If $\angle C > \angle B > \angle A$, then $c > b > a$.

There are two formulas that are useful when working with oblique triangle models. One is called The Law of Cosines; the other is called The Law of Sines.

THE LAW OF COSINES

The Law of Cosines resembles the Pythagorean Theorem. It says that in any triangle ABC, if a, b, and c are the sides opposite angles A, B, and C, respectively, then:

$a^2 = b^2 + c^2 - 2bc\cos(A)$; $b^2 = a^2 + c^2 - 2ac\cos(B)$; and $c^2 = a^2 + b^2 - 2ab\cos(C)$.

Of course, memorizing this collection of sides and angles is silly. There is a pattern that is fairly easy to remember: Select any side of the triangle and write the Pythagorean formula with this side as the hypotenuse. Then subtract twice the product of the other two sides and the cosine of the angle opposite the side with which you started.

The cosine formula has four variables: three sides and one angle. Whenever three of the four are known, the remaining side or angle can be found.

EXAMPLE 11

Find side a in triangle ABC (**Figure 5.43**).

SOLUTION:

$a^2 = 12.6^2 + 15.4^2 - 2(12.6)(15.4)\cos(42°)$

Use a calculator to obtain

$a = \sqrt{12.6^2 + 15.4^2 - 2(12.6)(15.4)\cos(42°)} \approx 10.4$ cm.

(See **Figure 5.44.**)

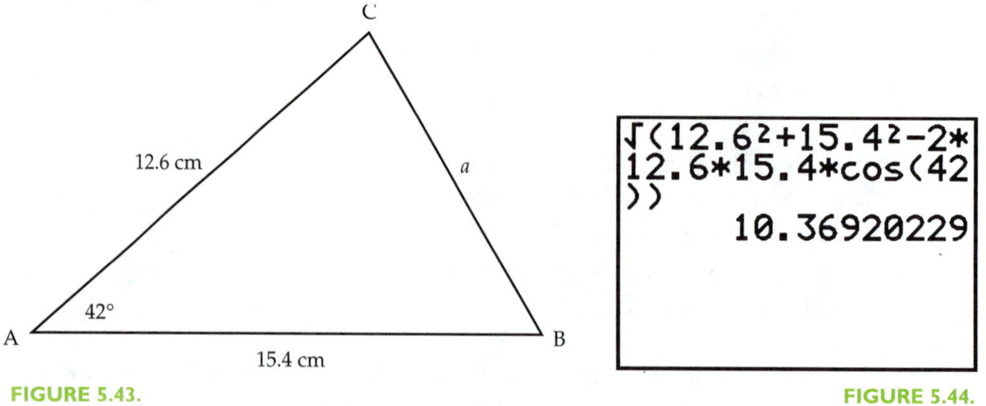

FIGURE 5.43. FIGURE 5.44.

EXAMPLE 12

Find the measure of angle B in **Figure 5.45.**

SOLUTION:

First note that 48.9 + 57.3 > 72.3. If the length of the longest side were greater than or equal to the sum of the lengths of the other two sides, the figure could not be a triangle, and it would make no sense to pursue a solution.

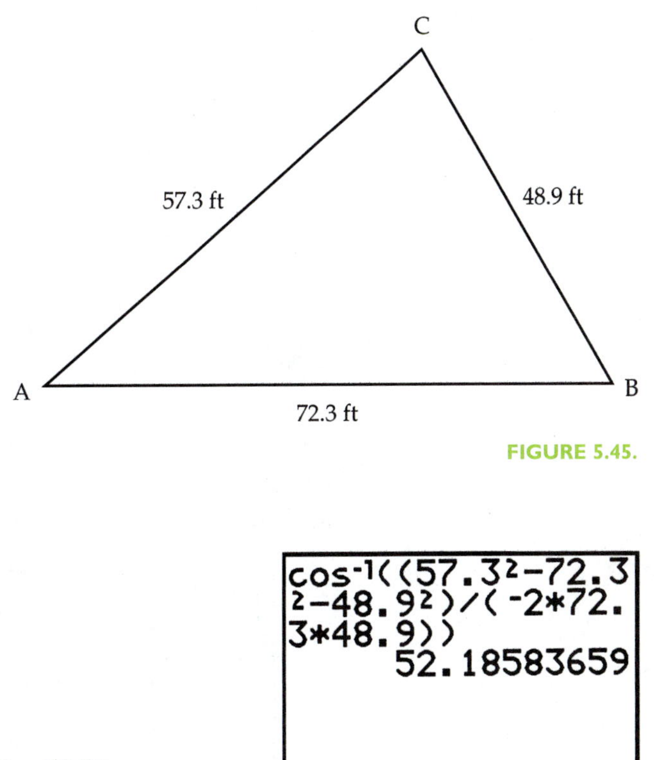

FIGURE 5.45.

To find angle B, write The Law of Cosines beginning with the side opposite angle B. (If it is written in any other way, there will be two unknowns.)

$57.3^2 = 72.3^2 + 48.9^2 - 2(72.3)(48.9)\cos(B)$.

Solve for cos(B) $\cos(B) = \dfrac{57.3^2 - 72.3^2 - 48.9^2}{-2(72.3)(48.9)}$.

Use a calculator in degree mode to obtain

$$B = \cos^{-1}\left(\frac{57.3^2 - 72.3^2 - 48.9^2}{-2(72.3)(48.9)}\right) \approx \cos^{-1}(0.613102) \approx 52.2°.$$

(See **Figure 5.46.**)

```
cos-1((57.3²-72.3
²-48.9²)/(-2*72.
3*48.9))
         52.18583659
```

FIGURE 5.46.

Find side b in **Figure 5.47.**

SOLUTION:

First, note that it is unproductive to write The Law of Cosines beginning with side b since Angle B is unknown: $b^2 = 136^2 + 189^2 - 2(136)(189)\cos(B)$.

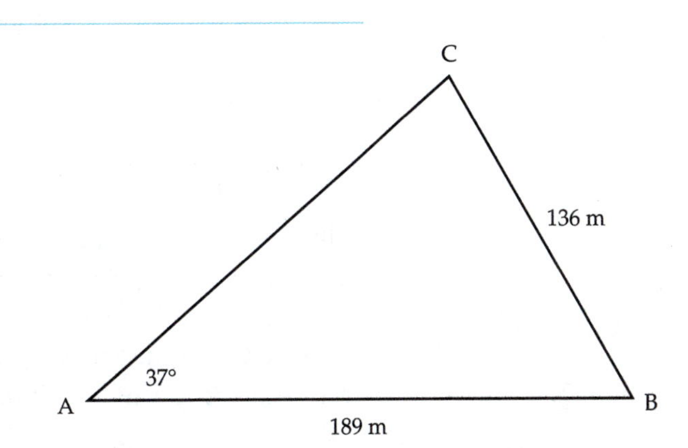

FIGURE 5.47.

To obtain an equation with a single unknown, write The Law of Cosines beginning with the known side that is opposite the known angle:

$136^2 = b^2 + 189^2 - 2(189)b\cos(37°)$.

Use the value $2(189)\cos(37°) \approx 301.884223$ and rewrite this as a quadratic equation in variable b:

$0 = b^2 - 301.884223b + 189^2 - 136^2 = b^2 - 301.884223b + 17225$.

Use a graphing calculator completing the square, or the quadratic formula to approximate the solutions of this equation.

For example, the quadratic formula says

$$b = \frac{301.884223 \pm \sqrt{301.884223^2 - 4(1)(17225)}}{2(1)}.$$

The solutions are approximately 225 m and 76 m, which can be obtained by evaluating the quadratic formula on a calculator (**Figure 5.48**). Thus, there are two possible values for side b.

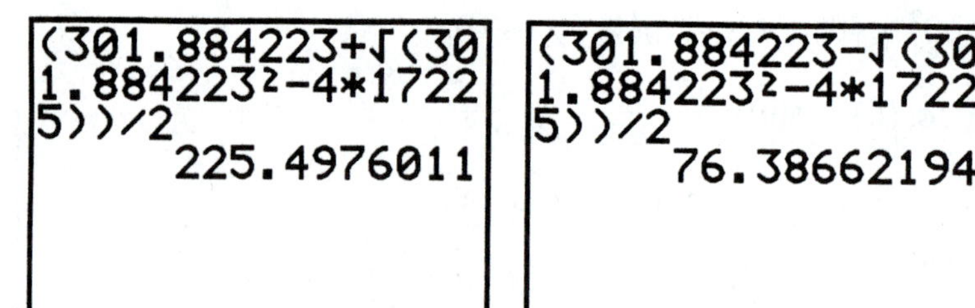

FIGURE 5.48.

THE LAW OF SINES

A simple proportion exists among the sides and angles of an oblique triangle. In general, in triangle ABC with sides a, b, and c opposite angles A, B, and C, respectively:

$$\frac{a}{\sin(A)} = \frac{b}{\sin(B)} = \frac{c}{\sin(C)}.$$

Since reciprocals of equal fractions are equal, The Law of Sines can also be stated as $\frac{\sin(A)}{a} = \frac{\sin(B)}{b} = \frac{\sin(C)}{c}$.

Some people find it convenient to use the first form when finding a side and the second when finding an angle.

When the Law of Sines is applied to a triangle, any two of the three ratios are used to form a proportion. Such a proportion has four variables: two sides and two angles. Whenever three of the four are known, the remaining side or angle can be found.

Example 14

In Example 11, use the solution and the given information to find angle B.

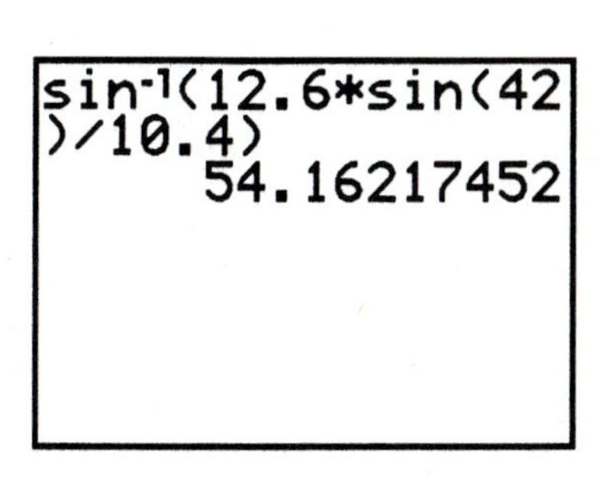

FIGURE 5.49.

Solution:

$$\frac{\sin(42°)}{10.4} = \frac{\sin(B)}{12.6}.$$

Thus, $\sin(B) = 12.6\frac{\sin(42°)}{10.4}$, or $B = \sin^{-1}\left(12.6\frac{\sin(42°)}{10.4}\right) \approx 54°$, which can be obtained by calculator evaluation (**Figure 5.49**).

Example 15

Find side b in **Figure 5.50.**

C

38°

66°

A

B

23.7 cm

FIGURE 5.50.

Solution:

Note that it is not possible to use the Law of Cosines in this situation since only one side is known.

The measure of angle C is $180° - (38° + 66°) = 76°$.

$$\frac{b}{\sin(66°)} = \frac{23.7}{\sin(76°)}, \text{ or } b = \sin(66°)\,\frac{23.7}{\sin(76°)} \approx 22.3 \text{ cm}.$$

TAKE NOTE The Law of Sines is deceptively simple to use. However, it will occasionally cause problems if it is used to find angles of a triangle because the sine ratio is incapable of distinguishing between acute and obtuse angles. To be safe, use The Law of Sines only to find angles you are certain are not obtuse.

Exercises 5.3

1. Find all remaining parts in each triangle (a)–(c). Use The Law of Cosines and/or The Law of Sines as you deem necessary.

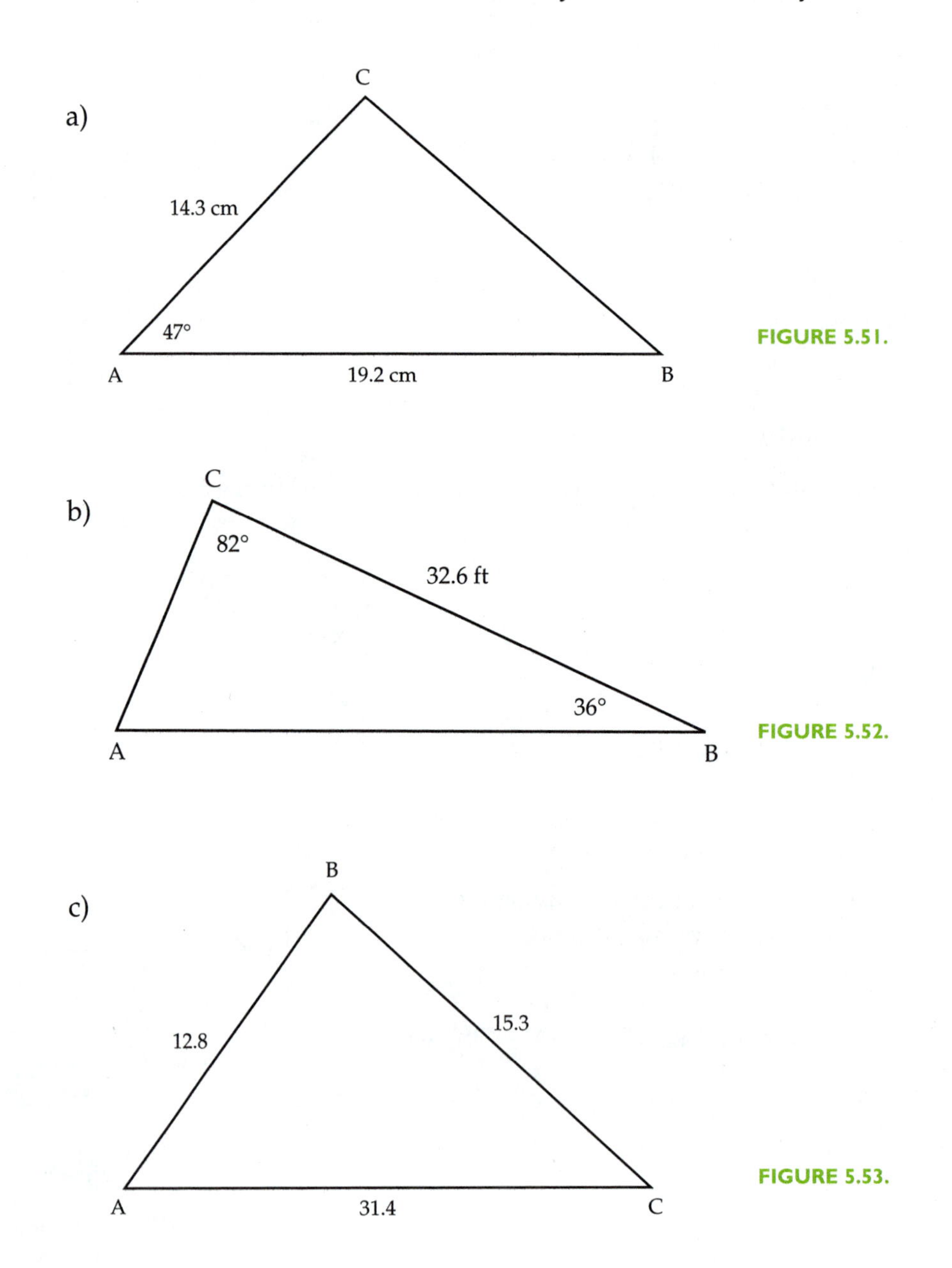

FIGURE 5.51.

FIGURE 5.52.

FIGURE 5.53.

Exercises 5.3

2. In part (c) of Exercise 1, change the length of AC to 26.0.

   a) Use The Law of Cosines to find angle C.

   b) Some people prefer The Law of Sines to The Law of Cosines because the former requires fewer calculations. Use your answer in part (a) and The Law of Sines to find angle B.

   c) The simplest way to find the remaining angle is by subtracting the two you have already found from 180°. Do so and explain why your angles do not make sense.

   d) This paradox occurs because the sine of an obtuse angle is positive. Since AC is the largest side of the triangle, B must be the largest angle. Since B is the largest angle, it might be obtuse. Find another way to determine the angles of this triangle without using the sine ratio to find angle B. (Note: The Law of Sines can be used to find angle B, but it requires some careful thinking about the triangle. This matter is considered in Exercise 11.)

3. **Figure 5.54** is the Blair Cuspid diagram that appeared in Activity 5.3. You may want to review your conclusions from that Activity before doing this exercise.

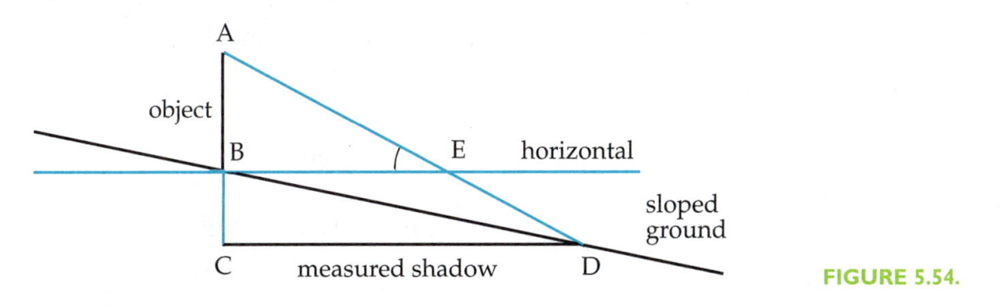

FIGURE 5.54.

   At the time of the Blair Cuspid photograph, the sun's angle of elevation was about 10.9°. The longest cuspid shadow was approximately 110 m. Some scientists who analyzed the photo estimated that the ground sloped downward 6.9° in the direction of this shadow.

   a) Explain why an estimate of the measure of angle CDB in Figure 5.54 is 4°.

   b) Use a right triangle model to estimate length BD in Figure 5.54.

Exercises 5.3

c) Explain why the measure of angle A is 79.1°.

d) In triangle ABD, which side is the longest? Which side is the shortest? Explain.

e) Apply The Law of Sines to triangle ABD to estimate the height of the cuspid.

4. Tunnel construction requires mathematical modeling to determine the length of the tunnel and its direction.

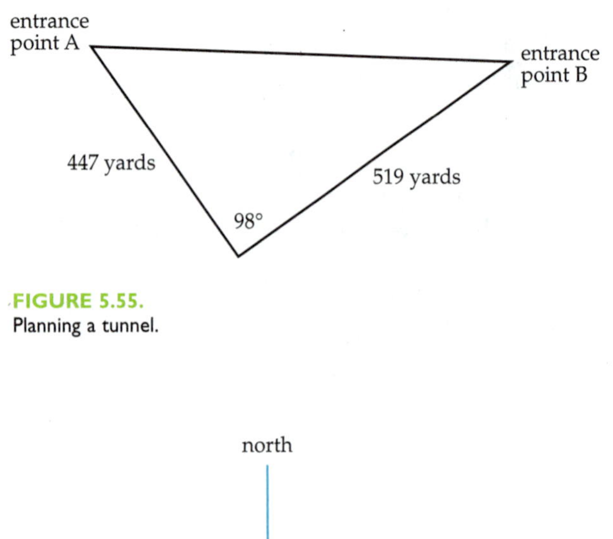

FIGURE 5.55.
Planning a tunnel.

Consider a hypothetical situation in which engineers are planning the route of a tunnel. They have selected a point of observation that is 447 yards from one tunnel entrance and 519 yards from the other entrance. They have measured the angle between the lines of sight to the entrance points as 98°. (See **Figure 5.55.**)

a) Show how to use The Law of Cosines to find the length of the tunnel.

b) To guide the machine that bores the tunnel, the engineers must calculate the angle at one of the entrance points in Figure 5.55. Find the angle at each entrance point.

5. Oblique triangle models are used to navigate planes and ships. Navigational headings are measured from north. Suppose a ship leaves port and travels 125 nautical miles on a heading of 48°, then turns and travels 87 nautical miles on a heading of 112° (**Figure 5.56**).

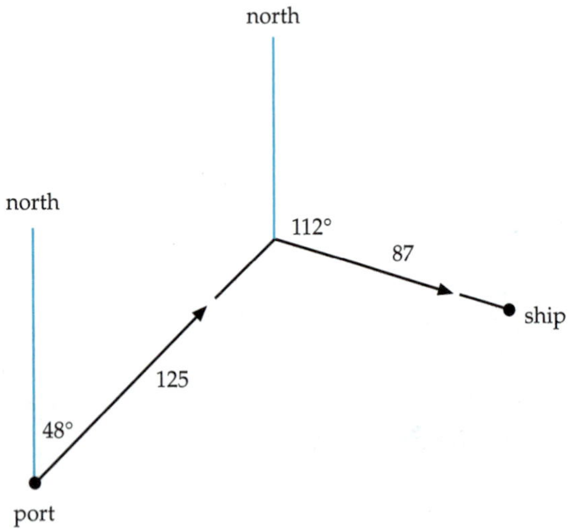

FIGURE 5.56.

a) What is the measure of the angle between the ship's first course and its second?

b) How far is the ship from port?

Exercises 5.3

c) If the ship is disabled, on what heading from the port should a rescue helicopter fly?

6. It has been common practice since ancient times to use lengths that satisfy the Pythagorean formula to establish right angles. For example, a rope divided into 12 equal sections and stretched to form a triangle with sides of 3, 4, and 5 will have a right angle since $3^2 + 4^2 = 5^2$. However, measurement error affects the reliability of such procedures. In this exercise you explore the effect of error in length measurement on angle size.

a) If the hypotenuse of a 3-4-5 right triangle is 5% long and the other two sides are each 5% short, by what percentage does the largest angle differ from 90°?

b) If the hypotenuse of a 3-4-5 right triangle is 5% short and the other two sides are each 5% long, by what percentage does the largest angle differ from 90°?

c) Will the relative error in the size of the largest angle be as large if a 5% error in all three sides is in the same direction? That is, if all sides are 5% long (or 5% short)? Explain.

7. Surveyors sometimes need to determine the area of an irregularly shaped piece of land. A common method is to divide the land into triangles and apply an appropriate formula for the area of each triangle. However, the traditional geometric formula for the area of a triangle, $\frac{1}{2}bh$, can be inconvenient because altitudes must be measured.

a) Use right triangle trigonometry in one of the right triangles formed by the altitude in **Figure 5.57** to eliminate h from the traditional formula for the area of triangle ABC.

b) Use your results in (a) to describe a general formula for finding the area of a triangle without knowing the length of any altitude.

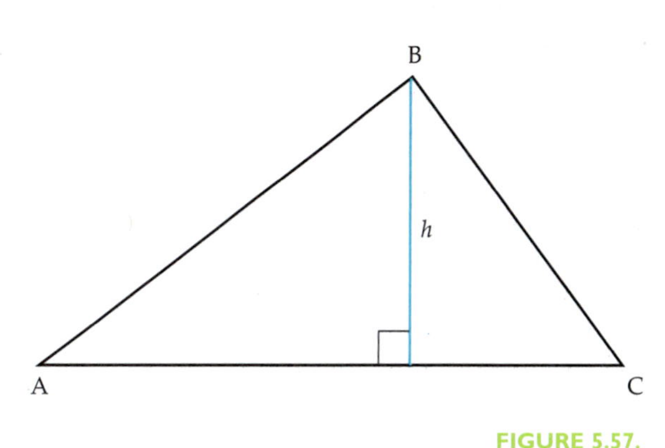

FIGURE 5.57.

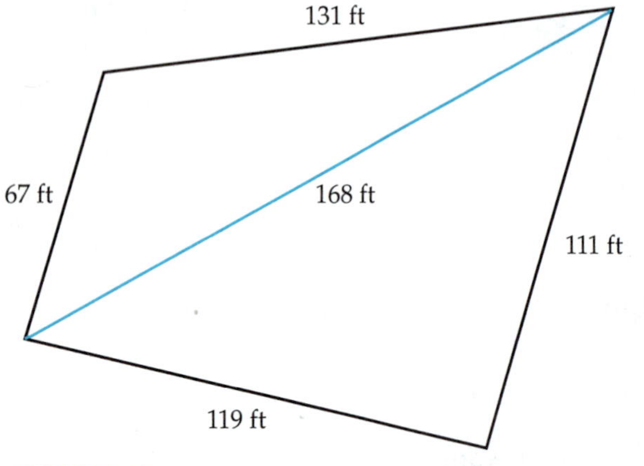

FIGURE 5.58.

c) A surveyor makes the measurements shown in **Figure 5.58**. Find any additional parts you need in order to use the formula you described in (b) to find the area of the piece of land.

8. Home plate, first base, second base, and third base of a baseball diamond form a square 90 feet on a side. The pitcher's mound is 60 feet from home plate.

   a) How far is the pitcher's mound from first base?

   b) How far is the pitcher's mound from second base?

9. Ornithologists have spotted an eagle nest in a distant treetop. They do not want to disturb the eagles, so they cannot take measurements near the tree. To pinpoint the location of the nest, they make several measurements from two points, 46.7 meters apart. From one point, lines of sight to the nest and the base of the tree form an angle of 23.4°; from the other point, the lines of sight form an angle of 19.2°. **(See Figure 5.59.)**

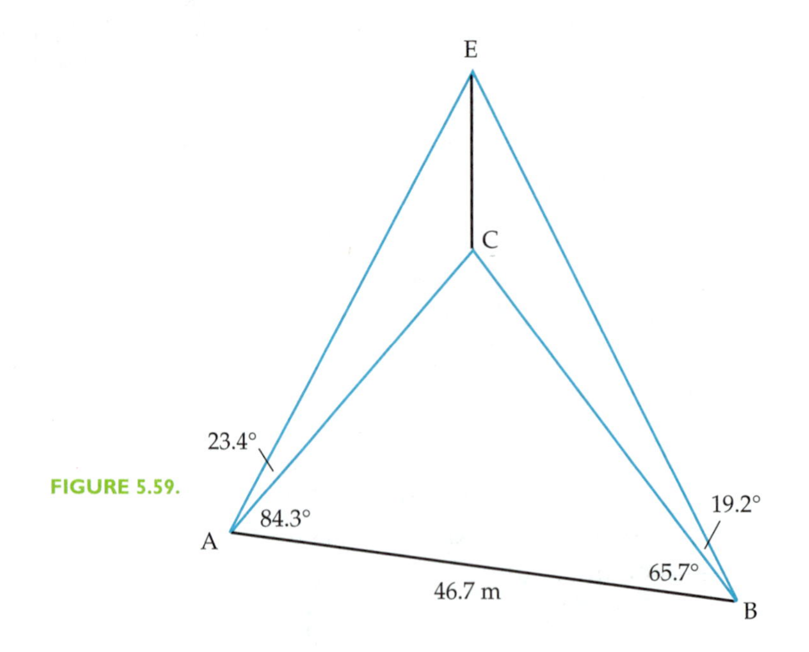

FIGURE 5.59.

a) Use The Law of Sines to estimate the measures of segments AC and BC.

Exercises 5.3

b) It is tempting to assume that triangles ACE and BCE are right triangles, but they probably are not. Explain.

c) Determine the elevation of the nest above observation point A.

d) Determine the elevation of the nest above observation point B.

10. You may recall this problem from Lesson 5.1. Surveyors make two observations of the top of a hill from points 100 meters apart (**Figure 5.60**). If only right triangle trigonometry is used, a system of two equations in x and h is required. Use oblique triangle trigonometry to find h without using a system of equations.

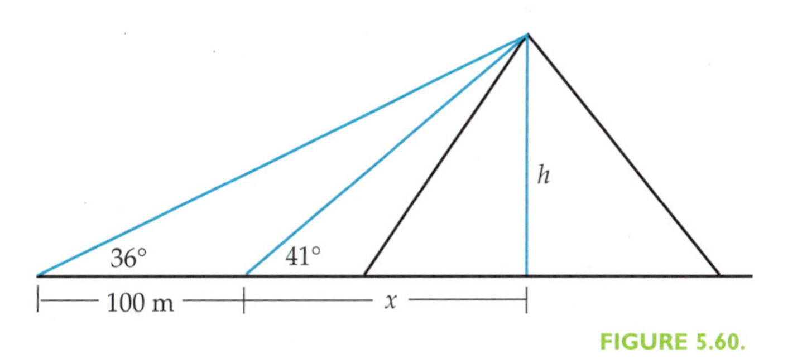

FIGURE 5.60.

11. There is one type of situation in oblique triangle trigonometry that is especially sticky. To understand it, consider triangle ABC, in which a and b are 10 cm and 15 cm, respectively. Angle A measures 26°.

a) Use a ruler and protractor to draw this triangle accurately. Explain why there are two locations possible for vertex B and make a sketch of each of the two possible triangles.

b) Use a ruler and protractor to measure the remaining parts in each of your sketches. To keep track of the triangles, label the larger triangle 1 and the smaller triangle 2.

c) Use The Law of Sines to find angle B. How could you use the result to find the other possible angle B?

d) Explain how to use The Law of Cosines to create a quadratic equation to find c.

e) Draw an altitude from C to side AB. Use right triangle trigonometry to find the length of this altitude.

f) Compare the length of the altitude to the value of a. Discuss changes in the given value of a that affect the number of triangles that are possible in this situation.

Exercises 5.3

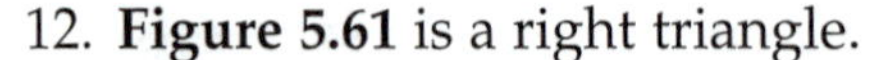

12. **Figure 5.61** is a right triangle.

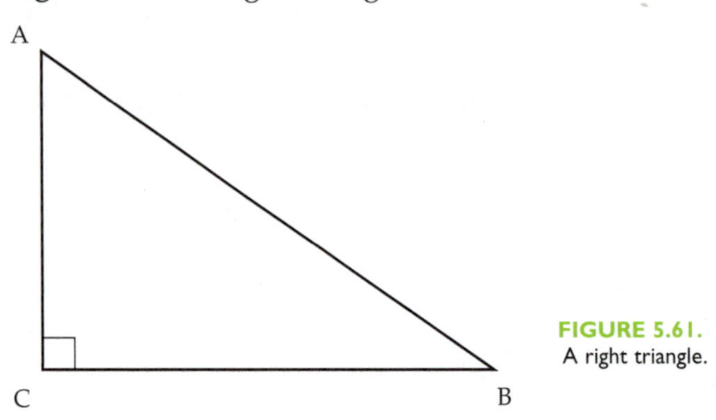

FIGURE 5.61.
A right triangle.

a) Use The Law of Sines to write a proportion involving angle C and one of the acute angles. What do you notice?

b) Write The Law of Cosines as it applies to angle C. What do you notice?

13. Recall that in a 30°–60° right triangle, the shortest side is half the hypotenuse, and the remaining side is $\sqrt{3}$ times the shortest side. Consider a 30°–60° right triangle in which the shortest side measures 5 cm.

a) Verify The Law of Sines for this triangle.

b) Some people think that since sines are proportional to the sides opposite them, cosines should be, too. Use this triangle to give a counterexample.

c) Are tangents proportional to the sides opposite them?

14. Your work in Activity 5.3 led to a proof of The Law of Cosines and The Law of Sines. However, these proofs were for a triangle with three acute angles. If one of the angles is obtuse, two of the three altitudes fall outside the triangle. Use **Figure 5.62** to develop a proof of one of the three forms of The Law of Cosines for triangle ABC.

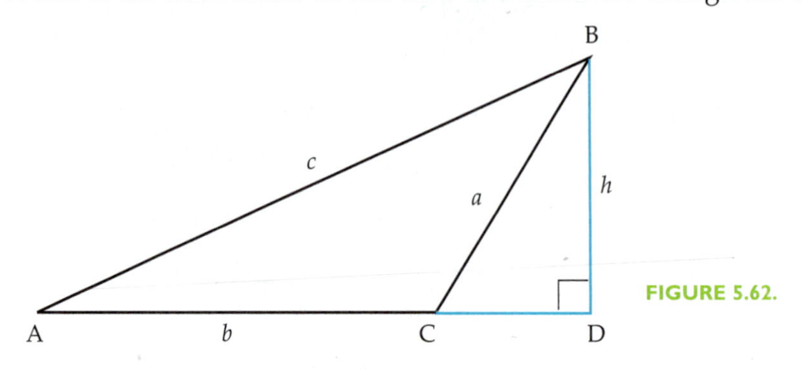

FIGURE 5.62.

TRIANGLE TRIGONOMETRY

Chapter 5 Review

1. Find the degree equivalent of 1.3 radians.

2. One company that manufactures HDTV sets recommends a viewing distance of 10.5 feet–13 feet from an HDTV screen that is 31 inches high and 56 inches wide.

   a) Find the vertical and horizontal viewing angles for a person sitting directly in front of the set 12 feet away.

   b) One conventional TV model is 21 inches wide and 28 inches high. Find the vertical and horizontal viewing angles for a person sitting directly in front of the set 12 feet away.

   c) At what distance from the conventional set in (b) should a person sit to have approximately the vertical viewing angle you found for the HDTV set in (a)?

3. Sketch a graph of $y = 1 + \sin^{-1}(2x)$. Identify the domain and range.

4. Fishing rod holders are designed to hold the rod at a certain angle. Some holders adjust to several angles or a range of angles. The rod holder in **Figure 5.63** bolts to the edge of a boat and holds rods at a fixed 22° angle to the vertical. If a rod 8 feet long is placed in this holder, how far does it extend over the side of the boat?

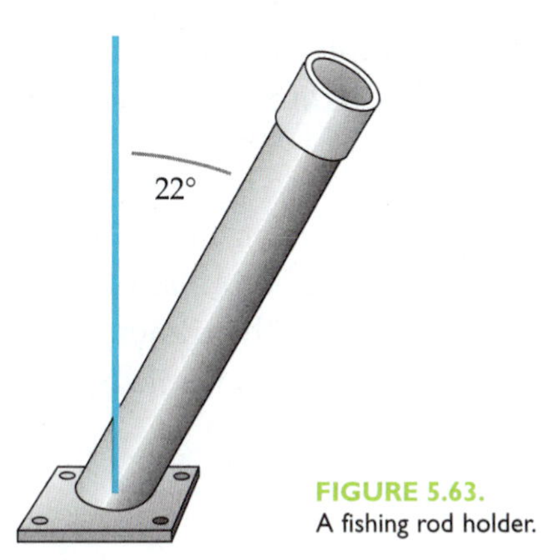

FIGURE 5.63.
A fishing rod holder.

5. A metal worker machines a quarter-inch thick piece of steel that is triangular in shape with sides of 15 inches, 21 inches, and 29 inches. What is the weight of the piece? (Steel weighs 490 lbs per cubic foot.)

6. A plane flying at an altitude of 7250 feet spots two points on the ground. The angle of depression from the plane to point A is 51°, and the angle of depression from the plane to point B is 42° (**Figure 5.64**). How far apart are points A and B?

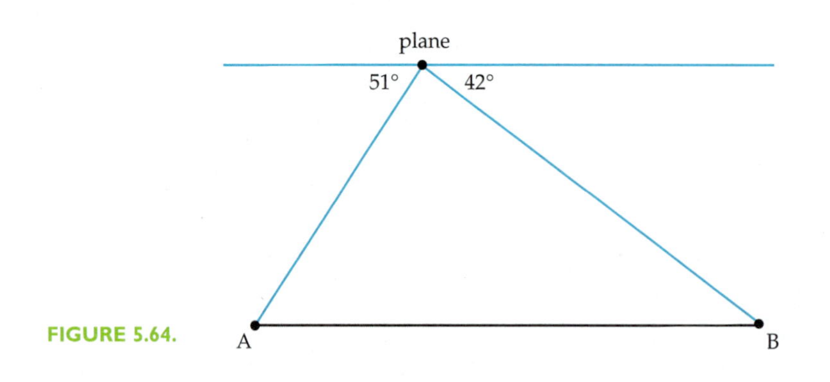

FIGURE 5.64.

7. A forest fire is spotted from observation towers that are 2.3 miles apart. From one tower, the fire's compass heading is 67°, and from the other tower the fire's compass heading is 312°. How far is the fire from each tower?

8. Find the angle, to the nearest degree, that the line $y = 2x - 1$ forms with the x-axis. Explain your method.

9. The steepness of a road is usually measured in percent grade. For example, a road with a 10% grade rises 0.10 of a foot for every foot of horizontal distance. What is the angle of inclination of such a road?

10. This news article from the Sentinel in Carlisle, PA, describes a planning commission difficulty in determining the length of a hotel. Use the dimensions given in the article to find the length of the hotel's frontage.

OFFICIALS DISAGREE ON HOW TO MEASURE INN

How does one calculate the length of a hotel? This seemingly simple question kept the Middlesex Planning Commission busy for more than an hour Monday. The township's ordinances say a hotel can be no longer than 150 feet. But developers of a proposed Hampton Inn that would have more than 90 rooms said the ordinances do not explain how that 150 feet should be measured. They plan to ask the board of supervisors for a ruling on how building length should be calculated. The proposed building would consist of two legs—one measuring 150 feet and the other 85 feet. The two legs meet at a 60-degree angle in a central lobby to form a concave shape similar to a boomerang. An attorney for the landowner argued the long leg only should be used to calculate length. Trust attorney Charles Zwally told the commission Monday most building length calculations are based on frontages.

A Hampton Inn representative showed Middlesex officials this drawing of the proposed hotel.

Chapter 6

COORDINATE SYSTEMS AND VECTORS

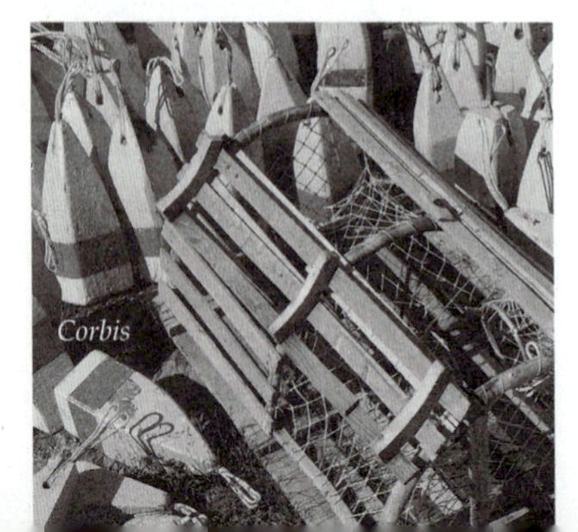
Corbis

CHAPTER INTRODUCTION

Storms are the stuff of legends. Unfortunately, those who achieve heroism against nature's fury often pay the ultimate price. Their efforts are chronicled in many ways—from a song about 15-year-old Hazel Miner, who died protecting her younger brother from a North Dakota spring blizzard, to a best-selling book about the Andrea Gail.

Efforts to predict, prepare for, avoid, and survive storms depend on radar, a technology that originated in the 1930s when British scientists discovered that radio waves bounce off objects and a portion of the wave returns to the source. Knowing the time interval between when the signal is sent and subsequently returned allows one to determine how far away the object is. In 1968, U. S. Air Force scientists found a way to use radar to detect precipitation. Today, radar is routinely used by meteorologists to track storms, and by ships at sea to avoid collisions with land, icebergs, and other ships, especially during foggy weather.

Radar operates by sending special radio signals outward from a central location such as a weather observatory or a ship. The signal's return is translated into information about the distance and direction to objects such as moisture-bearing clouds or ships.

Since distance and direction are radar's fundamental units, its coordinate system is unlike the rectangular system. Rectangular coordinates locate a point by specifying distances in two directions—one vertical, the other horizontal—from a reference point. In contrast, the **polar coordinate system** locates a point by specifying a single distance and a direction from the reference point.

The first person to contemplate polar coordinates was probably Isaac Newton (1642–1727), but it was Jakob Bernoulli (1654–1705), who proposed the system in 1691. However, the writings of Leonhard Euler (1707–1783) around 1750 made polar coordinates popular. First used to treat certain calculus problems, the system is far more important today than its originators could have imagined.

LESSON 6.1

Polar Coordinates

This lesson focuses on locating points in the plane using polar coordinates, some applications of the system, and the formulas needed to convert between the Cartesian coordinate system and this new one.

Activity 6.1

Since polar coordinates are based on distance and direction, polar graph paper shows distances from a reference point, called the **pole**, as concentric circles and directions from a reference line, called the **polar axis**, as segments radiating outward from the pole. **Figure 6.1** shows six units of distance, and directions in increments of 30°. (To avoid clutter, polar graph paper that uses finer increments of direction often omits some or all direction lines near the pole.)

It is customary to associate a point with an ordered pair in which the distance is written first and the direction second. For example, the pair (4, 30°) designates the point A in Figure 6.1.

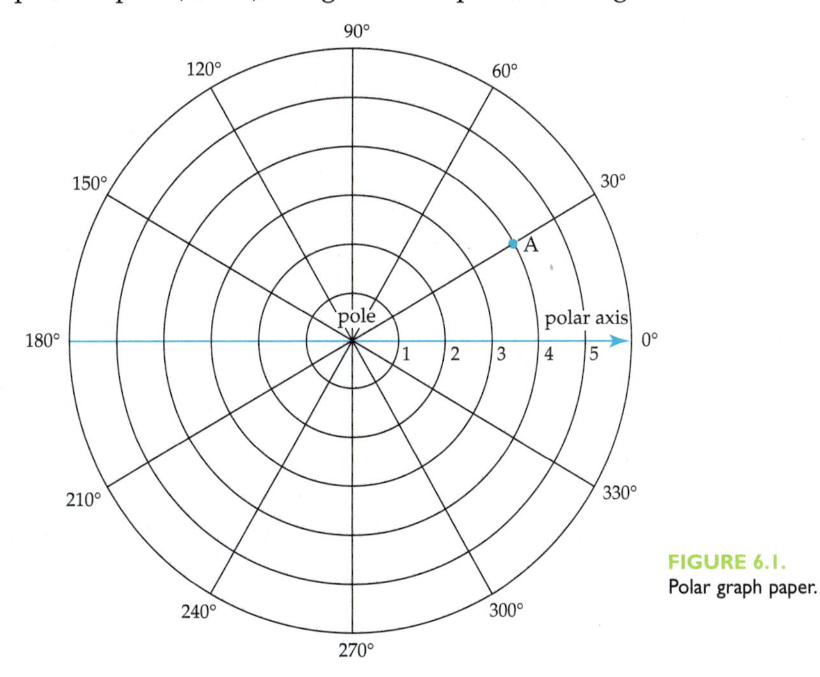

FIGURE 6.1. Polar graph paper.

Unlike rectangular coordinates, the coordinates of a point in the polar system are not unique because angles can have measures greater than 360° or less than 0° and distances can be negative. Point A in Figure 6.1, for example, is also the location specified by (4, 390°), (4, –330°) and (–4, 210°).

To acquaint yourself with this new system, you will play polar tic-tac-toe in this activity. The game is played by two people or by two teams of people on a copy of the polar graph paper in Figure 6.1. Here are the rules:

1. Flip a coin to decide who is first. (If more than one game is played, players alternate being first.)
2. The first person calls any pair that appears as the intersection of a circle and a ray in Figure 6.1, with the exception of the pole.
3. The second person marks the corresponding point with an X. If the players agree on the location of the point, the caller and the marker switch roles, with the new marker using an O, instead of an X.
4. If the players disagree about the location of the point, the disagreement is resolved by the players (or, if necessary, by consultation with an outside authority). If the person who called the point is wrong, the point is erased and the second person becomes the caller. If the person who marked the point is wrong, the mark is erased and the correct point is marked. Then the caller gets another turn calling a point.
5. The caller wins when the marker is forced to complete four adjacent marks along one of the following three (see **Figure 6.2**):

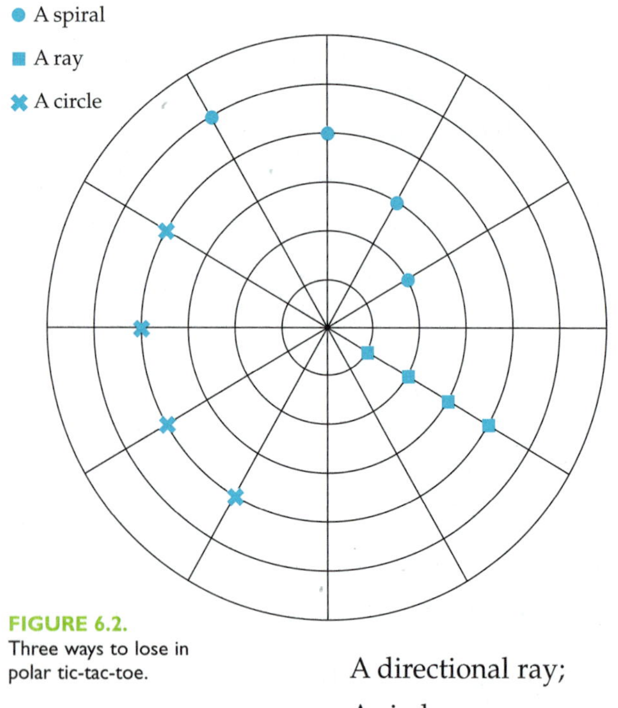

FIGURE 6.2. Three ways to lose in polar tic-tac-toe.

A directional ray;

A circle;

A spiral (clockwise or counterclockwise).

Play the game several times against one or more opponents. As you play the game, letter each point that is marked and record the pair associated with it.

POLAR COORDINATES

Polar coordinates are useful when it is convenient to specify a location in the plane with a distance from a reference point (the pole) and a direction from a reference direction (the polar axis) defined by a ray from the pole. For example, the radar system used by airports to control plane traffic measures distance from the airport and direction clockwise from a ray pointing north from the airport.

Several characteristics of polar coordinates can cause confusion at first. These include:

Differences in the way direction is measured;

Directions that exceed 360° or are negative;

The occurrence of negative directions.

Mathematicians measure direction as an angle measured counterclockwise from a horizontal ray pointing to the right. However, by virtue of the fact that compasses point north, navigational directions are often measured clockwise from north (see **Figure 6.3**) and are called **azimuth**. In this chapter, angles will be assumed to use the mathematicians' definition, unless specific reference to standard compass direction or azimuth is made.

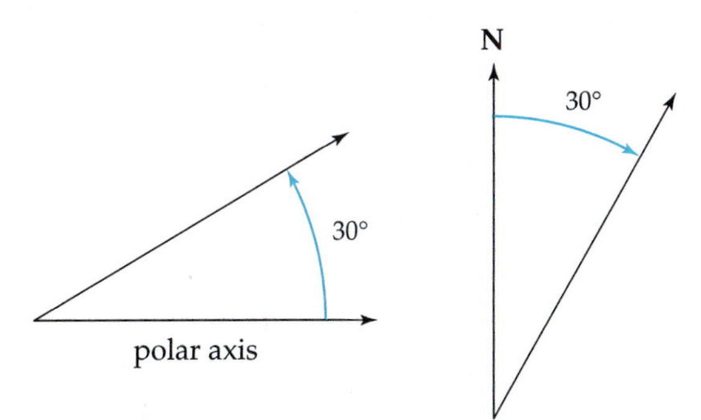

FIGURE 6.3. Mathematical direction of 30° (left) and navigational azimuth of 30° (right).

In mathematics, it is customary to use the variable r for distance and the variable θ (the Greek letter theta) for direction. Thus, (r, θ) is a general pair in polar coordinates, just as (x, y) is a general pair in rectangular coordinates.

Unlike rectangular coordinates, the pair that locates a point in polar coordinates is not unique because angle measures are not limited to a range of 0° to 360°. When an angle measure is greater than 360° or less than 0°, it specifies the same direction as a unique angle between 0° and 360°.

Consider, for example, point A in **Figure 6.4**. The simplest pair that locates this point is (4, 30°). However, an angle measure of 390° (30° + 360°) gives the exact same direction, as do 750° (30° + 360° + 360°) and –330° (30° – 360°). Since each additional full rotation leaves direction unchanged, adding or subtracting 360° to or from a direction's measure produces the same direction. Mathematicians sometimes write 30° ± n360°, where n = 0, 1, 2, ... to indicate that any integer multiple of 360° can be added to or subtracted from 30° without affecting the specified direction.

The point (–4, 30°) is the same distance from the pole as (4, 30°), but in the opposite direction (see Figure 6.4). Negative directions are measured clockwise; that is, in the opposite direction to the standard orientation. Therefore, the coordinates (4, –330°), (–4, 210°), (–4, –150°) and (–4, 570°) each locate point A.

When plotting a point having a negative distance coordinate, you may find it helpful to use a straight edge to be sure you are on the correct ray.

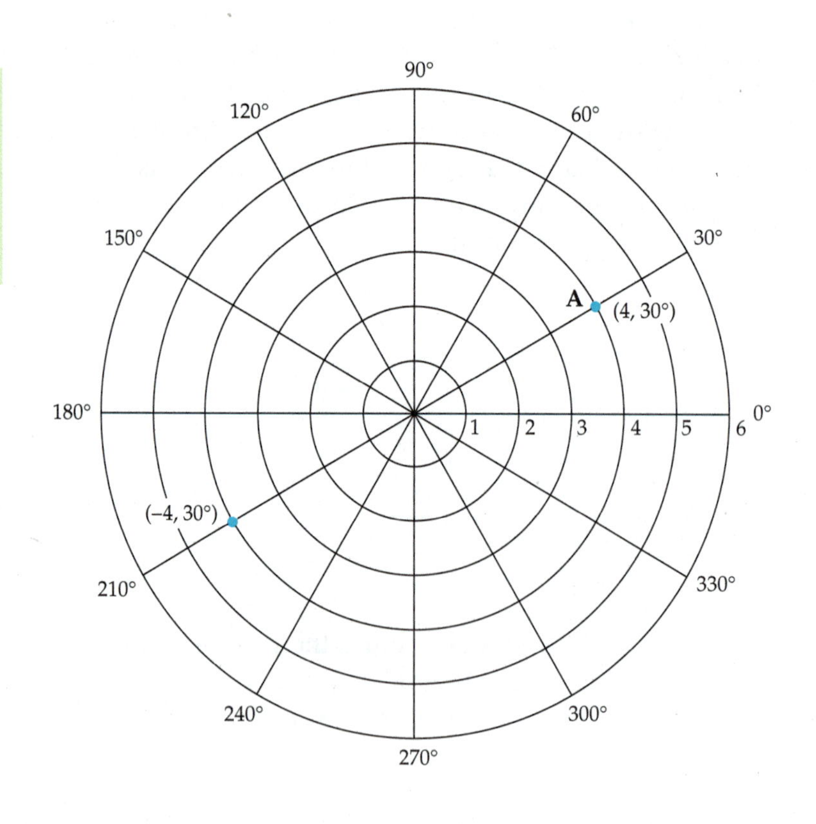

FIGURE 6.4.
Locating a point with a negative distance coordinate.

CONVERTING BETWEEN POLAR COORDINATES AND RECTANGULAR COORDINATES

Figure 6.5 shows rectangular and polar coordinate systems superimposed. The pole and the origin coincide, as do the polar axis and the positive x-axis. Point A's polar coordinates (r, θ) and rectangular coordinates (x, y) describe a right triangle.

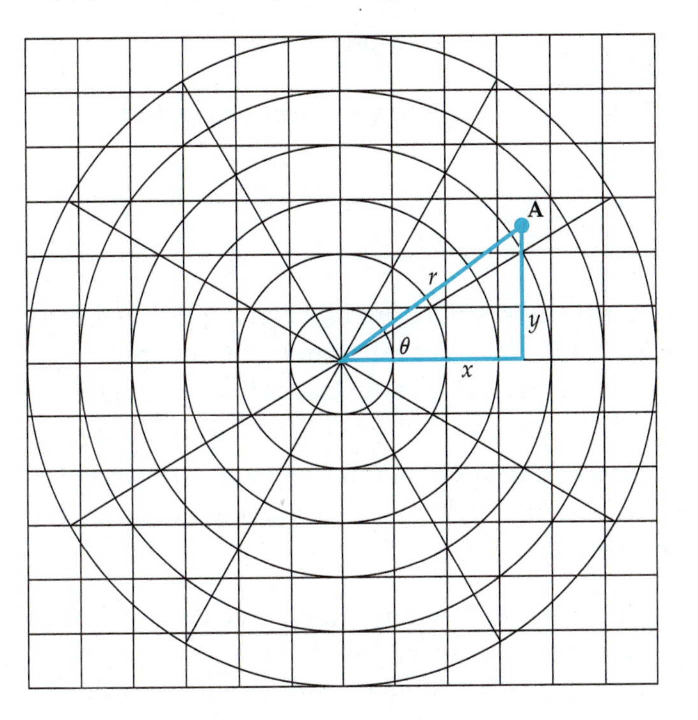

FIGURE 6.5.
Polar and rectangular grids superimposed.

Since point A is in the first quadrant, conversion between coordinate systems is accomplished with basic right triangle relationships: the Pythagorean formula and trigonometric ratios.

Converting polar to Cartesian:

$$x = r\cos(\theta) \text{ and } y = r\sin(\theta).$$

Converting Cartesian to polar:

$$r = \sqrt{x^2 + y^2} \text{ and } \tan(\theta) = \frac{y}{x}.$$

EXAMPLE 1

a) Convert the polar coordinates (4, 30°) to rectangular coordinates.

b) Convert the rectangular coordinates (2, 3) to polar coordinates.

SOLUTION:

a) $x = 4\cos 30° \approx 3.46$, and $y = 4\sin 30° = 2$.

b) $r = \sqrt{2^2 + 3^2} = \sqrt{13} \approx 3.61$, and the direction θ is found as $\tan^{-1}\left(\frac{3}{2}\right) \approx 56°$.

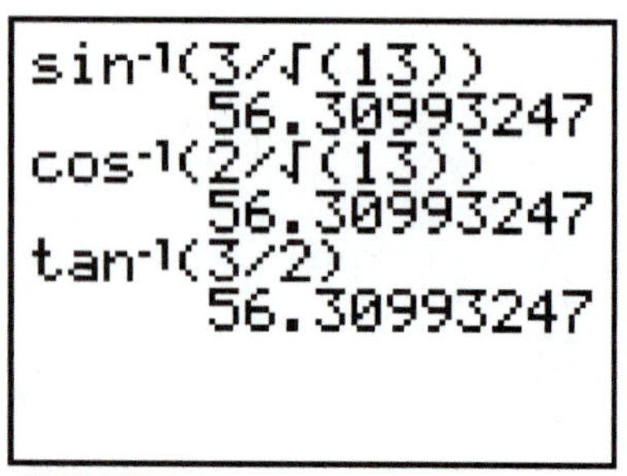

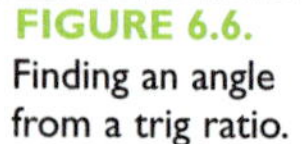
FIGURE 6.6.
Finding an angle from a trig ratio.

Note that from a diagram like Figure 6.5, the direction angle in Example 1(b) can be determined from any one of the three basic trigonometric ratios sine, cosine, or tangent: $\tan^{-1}\left(\frac{3}{2}\right)$, $\sin^{-1}\left(\frac{3}{\sqrt{13}}\right)$, or $\cos^{-1}\left(\frac{2}{\sqrt{13}}\right)$. **Figure 6.6** shows the calculator screen for all three trigonometric calculations on a typical calculator. (The calculator is in degree mode for each of these.)

Using $\tan^{-1}$ to find the direction angle for a given point produces only angles with measures in the interval [–90°, 90°]. Thus, slight complications arise for points in quadrants II or III. This difficulty is addressed algebraically in the exercises that follow. However, many calculators have built-in conversions commands that automate the process. **Figures 6.7** and **6.8** show these features on one model of calculator. Consult your calculator manual for details.

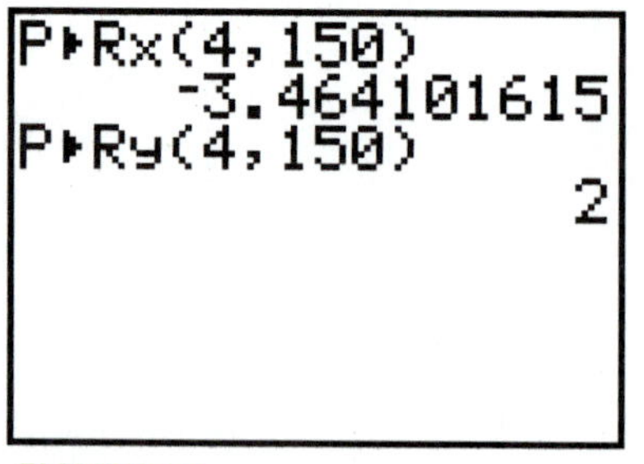

FIGURE 6.7.
Converting the polar coordinates (4, 150°) to rectangular coordinates.

R▸Pr(-3,-4)
5
R▸Pθ(-3,-4)
-126.8698976

FIGURE 6.8.
Converting the rectangular coordinates (–3, –4) to polar coordinates.

Exercises 6.1

1. Determine whether each coordinate pair represents point A, B, C, or D in **Figure 6.9** below.

   a) (5, 120°)

   b) (5, –120°)

   c) (–5, 120°)

   d) (–5, –120°)

   e) (5, 60°)

   f) (–5, –60°)

   g) (5, 240°)

   h) (–5, –240°)

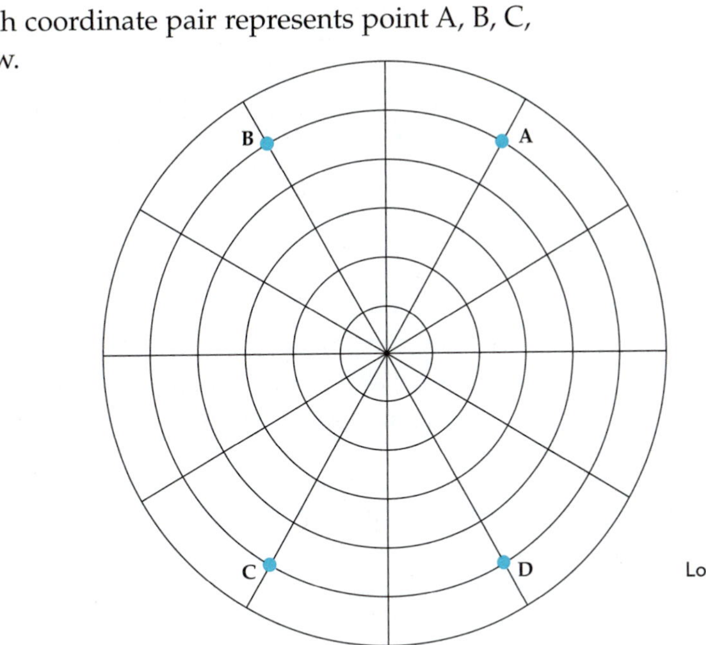

FIGURE 6.9. Locations of points for Exercise 1.

2. What do polar coordinate pairs that represent the pole have in common?

3. In general, adding or subtracting 360° to the angle in a pair does not change a point's location. That is, the point that represents (r, θ) also represents $(r, \theta \pm n360°)$, where $n = 0, 1, 2, \ldots$. Discuss the effect of adding multiples of 180° to the angle. When does adding a multiple of 180° leave location unchanged?

4. In Activity 4.1, there were three ways to lose at polar tic-tac-toe—directional ray, circle, and spiral. Describe each losing configuration as conditions on the coordinates of the losing points.

5. You can think of rectangular and polar coordinate systems as being (horizontal distance, vertical distance) and (distance, direction) pairs of information, respectively.

   a) Explain why you need two pieces of information in each case.

   b) Discuss the merits of a coordinate system using two poles, with (direction, direction) information only.

Exercises 6.1

6. Recall that angles are sometimes measured in radians rather than degrees, and that 180° is equivalent to π radians. (Note: To avoid confusion, the degree symbol must be used to indicate directions in degrees; otherwise, polar directions are in radians.)

   a) Prepare a polar coordinate grid like the one in Figure 6.1, but label the rays with radian measures.

   b) On your graph, locate and label each of $(3, -\pi/6)$, $(-4, 2\pi/3)$, and $(-5, -3\pi/4)$.

7. Given each of the following pairs as either polar or rectangular, convert the pair to the other system. When your answer is a polar pair, give at least one additional polar pair that corresponds to the same point. (Angle measures may be rounded to the nearest degree.)

   a) rectangular: $(-3, 4)$

   b) rectangular: $(0, -4)$

   c) polar: $(2, 45°)$

   d) polar: $(1, 2\pi/3)$

8. Sketch a rectangular coordinate graph of the function $y = \sin \theta$, where θ is measured in degrees, for several cycles.

   a) Use the graph to explain why a calculator cannot determine a unique angle whose sine is 0.5.

   b) With the aid of your calculator and the graph you drew, describe all polar directions that have a sine of 0.5.

   c) Use your calculator and graph to describe all directions that have a sine of –0.5.

   d) Use your calculator and graph to describe all directions that have a sine of 0.8660.

   e) Use your calculator and graph to describe all directions that have a sine of –0.7071.

f) Each polar direction angle is related to another angle called a **reference angle**. The reference angle is defined as the acute angle between the direction ray and the horizontal. Reference angles always measure between 0° and 90° (no negative values). **Figure 6.10** shows an example. Explain how to determine the reference angle if you know a direction angle.

Exercises 6.1

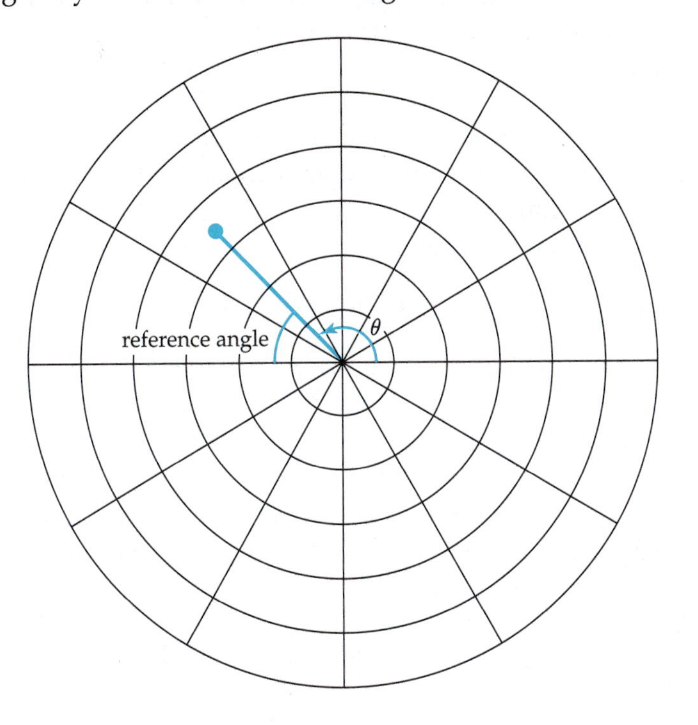

FIGURE 6.10. Reference angle and direction angle θ.

g) Choose an arbitrary angle and calculate its reference angle. Then take the sine of your first angle and of the reference angle. How do these two sines compare? Repeat the process for several other angles having the same reference angle. (Check all four quadrants.) Repeat the calculations again with the cosine and tangent values. Generalize your findings. How do the trigonometric ratios for an angle compare to the trigonometric ratios for its reference angle?

h) How does knowing the reference angle help you answer parts (b)–(e)?

9. For each of the following rectangular coordinates, find the reference angle for the direction needed to describe the same location using polar coordinates. Explain how you found your answer.

Exercises 6.1

a) (–4, 6.928)

b) (–6, –8)

c) (4.33, –2.5)

10. Review the ranges of $\sin^{-1}$, $\cos^{-1}$, and $\tan^{-1}$ from Chapter 5. In other words, describe the degree measures that a calculator produces for each function.

11. This lesson introduced formulas for converting between polar and Cartesian coordinates. Complete that work by developing formulas to find a direction angle (your choice of inverse trig function) when the coordinate (x, y) is in:

a) Quadrant II

b) Quadrant III

c) Quadrant IV

© Susan Van Etten

12. Recall that the way direction is measured in mathematics differs from azimuth as measured in navigation (refer back to Figure 6.3).

a) Investigate and describe the relationship between mathematical direction and azimuth. Develop a formula that will convert one measure to the other.

b) Make a sketch to show the results of your formula when applied to an azimuth of 120° and show that the result is correct.

c) Find horizontal and vertical components of the location of a ship whose azimuth is 120° and is a distance of 100 km from the radar station (pole). Explain how you obtained your answer, and interpret the results as north-south, east-west of the station.

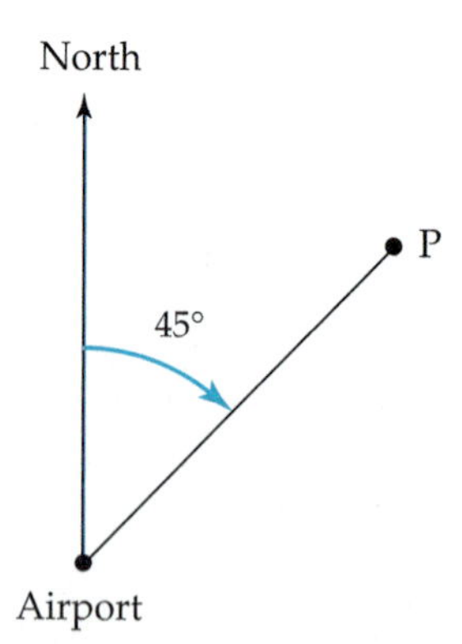

FIGURE 6.11. Diagram for use with Exercise 13.

13. Navigators use azimuth to identify their **heading** (their direction of motion), treating their plane or ship as the pole. In **Figure 6.11**, a plane is at location P, at a bearing of 45° from the airport. (**Bearing** specifies the direction from an observer to the object being observed.) What is the navigator's heading azimuth if the plane is flying toward the airport? What if it is flying away from the airport?

Exercises 6.1

14. A hodograph is a polar plot of wind speed and direction (in which the wind is traveling) at various altitudes. (Direction is measured clockwise from north.) Measurements are taken at uniform increments of altitude, the pairs (velocity, direction) plotted, and the points connected in order from lowest altitude to highest. The plot that results is used to forecast the type of thunderstorms that may develop and the potential for tornadoes and other severe weather.

a) **Table 6.1** displays hypothetical wind data. Use a sheet of polar graph paper to construct a hodograph.

Altitude (meters)	Velocity (mph)	Direction
1000	13	282°
2000	21	265°
3000	34	243°
4000	32	261°
5000	26	244°
6000	36	198°
7000	38	185°
8000	41	170°
9000	51	153°
10,000	58	138°

TABLE 6.1.
Wind data for hodograph.

b) Certain patterns in hodographs—spirals and loops, for example—indicate a potential for severe weather, especially when accompanied by certain other conditions. Describe any patterns in your hodograph.

15. A sun chart is a mathematical model of the sun's movement over a period of anywhere from one day to several months. Sun charts are used for many purposes. Architects, for example, use them to design energy-efficient buildings. **Table 6.2** shows the sun's movement on October 7, 1997 in Omaha, Nebraska. (For an astronomer, the sun's azimuth is also measured clockwise from North along the horizon, and the **altitude** is measured from the horizon upward, with 90° being straight up. They use the two angles to specify a direction

Time	Sun Altitude	Sun Azimuth
6:00	–5.9°	92.2°
7:00	5.5°	102.2°
8:00	16.1°	112.7°
9:00	25.9°	124.7°
10:00	34.3°	139.1°
11:00	40.4°	156.3°
12:00	43.1°	176.0°
13:00	41.8°	196.2°
14:00	37.0°	214.5°
15:00	29.4°	229.9°
16:00	20.0°	242.7°
17:00	9.6°	253.7°
18:00	–1.6°	263.8°

TABLE 6.2.
Sun positions at various times of the day.

in the sky from an observer on the ground.) You can obtain similar data for your area from the U. S. Naval Observatory Astronomical Applications Data Services web site.

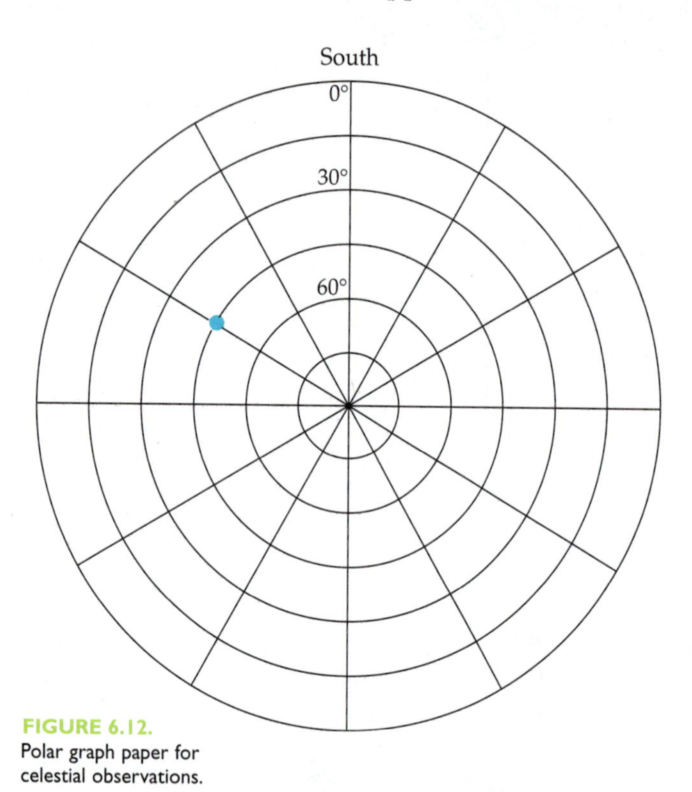

FIGURE 6.12.
Polar graph paper for celestial observations.

a) Several kinds of sun charts are used. One type plots the sun's altitude against the time of day on rectangular coordinate paper. Prepare such a chart for these data or for data you obtained for your own area.

b) Another type of sun chart plots the sun's altitude vs. its direction on the special polar graph paper shown in **Figure 6.12**. Notice that on this paper, each distance circle represents 15° of altitude in *decreasing* increments from the pole. Thus, at the sample point shown, the sun's direction is 120°, and it is 45° above the horizon.

Prepare such a chart for these data or for data you obtained for your own area.

c) Interpret each graph in terms of the sun's motion.

16. Mathematical models that use polar coordinates sometimes require graphing a function that describes the relationship between distance and direction. Typically, direction is the independent (explanatory) variable and distance is the dependent (response) variable. An example of such an application is in the field of speech synthesis, in which scientists develop models of the human tongue's movement—the tip of the human tongue follows a curved path, but the base of the tongue is fixed (the pole). These scientists find it easier to describe the movement of the tongue's tip in terms of distance and direction than in terms of two directions.

In many models, the descriptions of the relationship between distance and direction are equations involving trigonometric functions. An example is $r = 4 \sin 2\theta$. Prepare a polar graph of this function. If you do the graph by hand, choose a suitable increment of θ—say 10°—and build a table. You can also use the polar graphing mode of a graphing calculator.

17. **Investigation**. Explore the families of polar functions listed below. In each case, graph several examples by selecting different values for the parameters (control numbers) used in the equation. Describe the graphical features for that family. Then apply the coordinate conversion formulas developed in this lesson to write a corresponding Cartesian equation.

   a) $\theta = k$

   b) $r = k$

   c) $r = k\sin\theta$

   d) $r = k\cos\theta$

   e) $r = k/\sin\theta$

   f) $r = k/\cos\theta$

   g) $r = a\sin\theta + b$

   h) $r = a\cos\theta + b$

   i) $r = a\sin b\theta$

   j) $r = a\cos b\theta$

LESSON 6.2

Polar Form of Complex Numbers

The development of polar coordinates led mathematicians to a new understanding of complex numbers. In this lesson, you will look at the operation of multiplication on complex numbers and interpret the results with the aid of polar coordinates.

Activity 6.2

Recall that a complex number can be represented in rectangular coordinates by associating the vertical axis with imaginary numbers and the horizontal axis with real numbers. The complex number $a + bi$ is plotted as the rectangular pair (a, b). The complex number can also be plotted in polar coordinates by converting the rectangular coordinates to polar and plotting on polar graph paper.

1. Multiply $3 + 4i$ and $12 + 5i$, and convert the rectangular pair associated with the product to polar coordinates. Then convert the rectangular pairs associated with the individual complex numbers $3 + 4i$ and $12 + 5i$ to polar coordinates. Describe any relationship you notice between the polar pairs for the original numbers and that for their product.

2. Evaluate $(3 + 4i)^2$ and $(3 + 4i)^3$. Then find the polar pair associated with each power and describe its relationship to the polar pair associated with $3 + 4i$.

3. Mathematicians think it is easier to multiply complex numbers or raise them to powers if they use polar representations. Discuss why you think this is the case. (Repeat Items 1 and 2, above, for other complex numbers to check your ideas.) Summarize general principles you have observed.

4. Recall that roots are equivalent to powers. For example, $\sqrt{9} = 9^{\frac{1}{2}} = 3$. Apply your observations about powers of complex numbers to calculate $\sqrt[3]{2+2i}$ as a "fractional" power. Show that your answer is correct by cubing it.

5. Mathematicians have discovered that every complex number has three cube roots. Can you find the other two cube roots of $2 + 2i$? (Hint: Remember that 45° is not the only possible direction angle for $2 + 2i$.) Verify each root by cubing it to get $2 + 2i$.

THE POLAR FORM OF COMPLEX NUMBERS

All complex numbers, including real and imaginary numbers, can be represented graphically as points in the complex plane. Since any point in the plane has both rectangular and polar coordinates, a complex number has both rectangular and polar representations.

Figure 6.13 shows the complex number $a + bi$ graphed in the complex plane. The point that represents $a + bi$ has rectangular coordinates (a, b) and polar coordinates (r, θ).

Rectangular and polar representations of the complex number $a + bi$ have the standard rectangular/polar relationship. That is, $r = \sqrt{a^2 + b^2}$, $a = r \cos \theta$, and $b = r \sin \theta$. (Remember, the calculation of r uses b, not bi.) Therefore, $a + bi = r \cos \theta + ir \sin \theta = r(\cos \theta + i \sin \theta)$. Mathematicians usually shorten $\underline{c}$os $\theta + \underline{i}$ $\underline{s}$in θ to cis θ. So, the polar form of a complex number is often written r cis θ. Remember that $a + bi$ and r cis θ are equivalent ways to write the same *number*, whereas (a, b) and (r, θ) are equivalent ways to represent that number as a *point* in the complex plane.

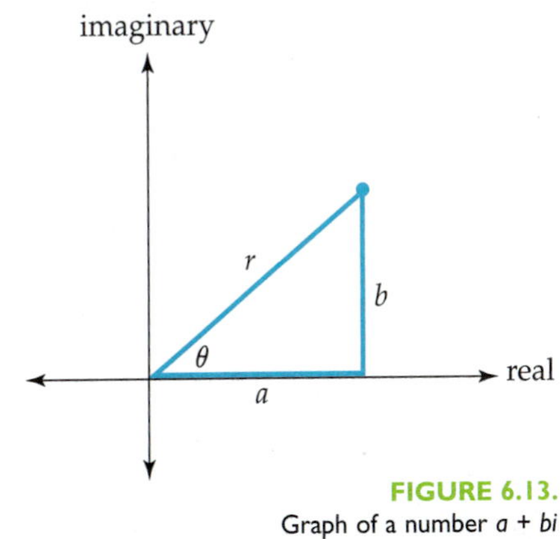

FIGURE 6.13.
Graph of a number $a + bi$ in the complex plane.

DE MOIVRE'S THEOREM

An advance in the understanding of complex numbers came in the 1700s when mathematicians made an important discovery about powers of complex numbers. The result is named after the mathematician Abraham De Moivre (1667–1754), although it was Leonhard Euler (1707–1783) who first proved the result.

Abraham De Moivre

De Moivre was born in France but moved to England when he was in his teens. He made important contributions in probability and statistics. He is also known for a fable that claims he predicted the date of his own death: on noticing that he was sleeping 15 minutes longer each night, he calculated the day he would sleep 24 hours.

Polar representations of complex numbers permit tremendous simplifications in some kinds of arithmetic. For example, you may have noticed in Item 1 of Activity 6.2 that the distance coordinate for the product was the product of the distance coordinates of the two factors and that the direction angle was the sum of the original direction angles of the factors. It turns out that this result generalizes to any product (and, thus, quotient). In general,

$[r_1 \text{ cis } (\theta_1)][r_2 \text{ cis } (\theta_2)] = r_1 r_2 \text{ cis } (\theta_1 + \theta_2)$, and

$\dfrac{r_1 cis(\theta_1)}{r_2 cis(\theta_2)} = \dfrac{r_1}{r_2} \text{ cis } (\theta_1 - \theta_2)$.

De Moivre's Theorem extends this "product rule" to powers and is the kind of result that pleases mathematicians; it is elegant in its simplicity. Raising the complex number $a + bi$ to a power is a tedious task in rectangular form, but it becomes almost trivial in polar form. One need only raise a single real number to the power and multiply another real number by the exponent. Formally De Moivre's theorem states:
$[r \text{ cis } (\theta)]^n = r^n \text{ cis } (n\theta)$.

EXAMPLE 2

Evaluate $(3 + 2i)^2$.

SOLUTION:

Change the base number into polar form: $3 + 2i \approx \sqrt{13}$ cis 33.7°. Here, $n = 2$, $r = \sqrt{13}$, and $\theta \approx 33.7°$. Therefore, compute $(\sqrt{13})^2$ and double 33.7° to get 67.4°. Thus, $(3 + 2i)^2 \approx 13$ cis 67.4° $\approx 5 + 12i$.

De Moivre's Theorem can also be used to find roots of complex numbers since each root is equivalent to a rational exponent. In addition, because every direction may be specified by multiple direction angles, every complex number has two square roots (and three cube roots, and four fourth roots, and so forth). In general, the nth roots of the complex number r cis θ (using degree measure) are $\sqrt[n]{r}\text{cis}\left(\frac{\theta}{n}\right)$, $\sqrt[n]{r}\text{cis}\left(\frac{\theta+360^\circ}{n}\right)$, $\sqrt[n]{r}\text{cis}\left(\frac{\theta+2\cdot 360^\circ}{n}\right)$, ..., $\sqrt[n]{r}\text{cis}\left(\frac{\theta+(n-1)\cdot 360^\circ}{n}\right)$.

EXAMPLE 3

Evaluate $\sqrt{3+2i}$.

SOLUTION:

Using polar form from Example 2, find the square root of r: $\sqrt{\sqrt{13}} \approx 1.90$. Then divide the angle by 2: $33.7^\circ/2 \approx 16.8^\circ$. Thus, $\sqrt{3+2i} \approx 1.90$ cis $16.8^\circ \approx 1.82 + 0.55i$.

To find the other square root of $3 + 2i$, rewrite the polar form $3 + 2i \approx \sqrt{13}$ cis 33.7° using an equivalent direction angle: $3 + 2i \approx \sqrt{13}$ cis 393.7°. Then repeat the process used to find the principal root: $\sqrt{3+2i} \approx 1.90$ cis $(393.7^\circ/2) = 1.90$ cis $196.8^\circ \approx -1.82 - 0.55i$.

Exercises 6.2

1. Since real and imaginary numbers are also complex numbers, they can be written in polar form. Write each of the following in polar form; use the cis abbreviation. (If necessary, locate the number on the complex plane to determine its distance from the pole and its direction from the polar axis.)

   a) 5
   b) -5
   c) $5i$
   d) $-5i$
   e) $2 + 5i$
   f) $-4 + i$
   g) $4 - 2i$

2. Carry out the indicated multiplications and divisions in polar form.

   a) $(3 + 2i)(2 - 3i)$
   b) $(3 + 2i)(-4 - i)$
   c) $(2 - 3i)(-4 - i)$
   d) $\frac{3+2i}{2-3i}$
   e) $\frac{3+2i}{-4-i}$
   f) $\frac{2-3i}{-4-i}$

3. Use De Moivre's Theorem to evaluate the indicated powers.

   a) $(-1 + 3i)^3$
   b) $(5 - 2i)^5$
   c) $(-2 - i)^4$

4. The equation $x^n = b$ has n solutions, including $\sqrt[n]{b}$. Find all indicated roots:

   a) cube roots of $2 + 3i$
   b) fourth roots of $2 + 2\sqrt{3}i$
   c) cube roots of $2i$

Exercises 6.2

5. De Moivre's Theorem sheds light on the Fundamental Theorem of Algebra. For example, according to the Fundamental Theorem of Algebra, the polynomial equation $x^4 + 1 = 0$ has four complex solutions, although some solutions might repeat.

   a) Rewrite the equation to show that the solutions are roots of –1.

   b) Find all solutions. (Remember that –1 is a complex number with imaginary part 0.) State the solutions in both polar and rectangular forms.

   c) Graph either the rectangular form of your solutions on rectangular graph paper or the polar form on polar graph paper. Connect them to form a polygon. What type of polygon is it? Explain.

   d) Determine the distances between pairs of solutions. Explain.

6. Find the cube roots of 1 and represent them graphically. What type of polygon do the roots form? What is the length of each side of the polygon?

7. Based on the results of Items 5 and 6, predict all square roots of i. Also predict all fifth roots of i.

8. Describe the pattern you expect for the graph of all of the nth roots of a given complex number in the complex plane.

9. The number $1 + i$ is a fourth root of some complex number.

   a) What does it mean for $1 + i$ to be a fourth root of some complex number?

   b) Find the other three roots of that complex number without first finding the number for which they are roots. Explain your method.

   c) Verify that all four are roots of the same complex number.

Exercises 6.2

10. Let $u + vi = \text{cis}\ \theta = (\cos\theta + i\sin\theta)$, where $u = \cos\theta$ and $v = \sin\theta$.

   a) By De Moivre's Theorem, $(\cos\theta + i\sin\theta)^3 = \cos 3\theta + i\sin 3\theta$. Expand the left side of this expression without using De Moivre's Theorem; that is, raise the binomial to the 3rd power. Use your results to write formulas (identities) for $\cos 3\theta$ and $\sin 3\theta$.

   b) Repeat part (a) for $(\cos\theta + i\sin\theta)^2$ to deduce identities for $\cos 2\theta$ and $\sin 2\theta$.

11. a) The direction and distance from the pole to real and pure imaginary numbers are obvious because such numbers lie directly on an axis in the complex plane. For some complex numbers not on an axis, distance or direction is also easy. Write $1 + i$ in polar form and explain your method.

   b) Find the polar form of $1 - i$ quickly.

   c) Use polar forms to multiply the numbers $1 + i$ and $1 - i$ quickly.

12. "Division" of complex numbers in rectangular form really is not division in the usual sense of the word. Rather, it consists of rationalizing the denominator so that the imaginary part of the denominator is 0. That is, the denominator of the "quotient" is a real number. To carry out this process, first write the division as a "fraction." Then multiply both the numerator and denominator by the conjugate of the denominator. (Recall that the conjugate of $a + bi$ is $a - bi$.)

   a) Simplify $\frac{1+i}{1-i}$ as described.

   b) Divide the polar forms of $1 + i$ and $1 - i$ using De Moivre's theorem, and compare your answer to your results from part (a).

   c) Use polar forms to calculate $5i \div (1 + \sqrt{3}i)$.

   d) What must be true of the quotient of two complex numbers having the same direction angle? Explain.

Exercises 6.2

13. Find the sum of the complex numbers 2 cis 45° and 2 cis 135°. Explain how you obtained your answer.

14. Describe the sum of two complex numbers that are the same distance from the pole, but in opposite directions; that is, two complex numbers of the form r cis θ and r cis $(\theta + 180°)$. Explain.

15. In general, when would you use rectangular form to perform arithmetic operations on complex numbers, and when would you use polar form? Explain.

16. A ship leaves a port (pole, or origin) and travels 200 miles on a heading (azimuth) of 45°. It then changes its direction to a heading of 180° because of engine problems. After traveling 100 miles, its engine fails.

    a) Make a sketch showing the approximate location of the ship and the path it traveled.

    b) In your sketch, indicate the path a rescue ship should follow from the port of origin.

    c) Describe the component paths taken by the ship, this time with directions using the mathematical system of measurement.

    d) Graph the complex numbers 200 cis 45° and 100 cis 270° in the complex plane. Find the sum of these two complex numbers and graph it in the same graph. Write the sum in both rectangular and polar forms.

    e) How is part (d) related to part (b)?

    f) Check your conclusion in (e) by applying geometric (non-coordinate) methods.

United States Coast Guard

LESSON 6.3

The Geometry of Vectors

Lessons 6.1 and 6.2 focused on coordinate systems for describing locations in the real and complex planes. This lesson introduces vectors, a convenient tool for describing directed quantities such as force or change in position, and explores related geometric properties.

Activity 6.3

Orienteering is a sport that has grown quite popular in the last two decades. Based on the skills of reading a map and using a compass, competitors navigate around a previously determined course, attempting to arrive at certain checkpoints in order. As a sport, the goal is to be accurate and faster than your competitors. As a mathematical activity, the focus is on accuracy; speed will come with practice!

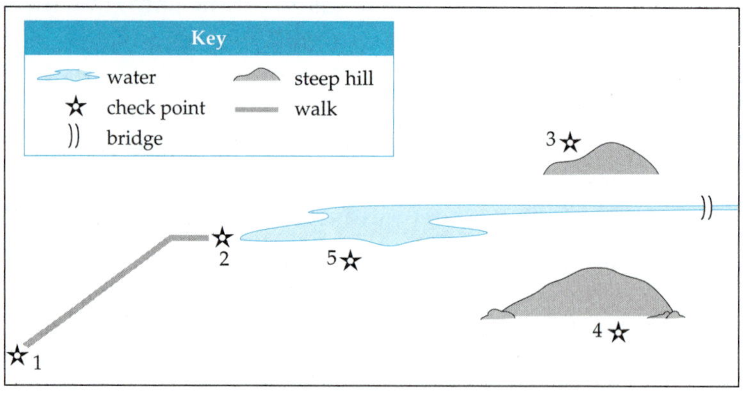

FIGURE 6.14.
Map of Rockville Park.

Figure 6.14 is a map of Rockville Park, where the orienteering competition is to be held. As the first stage, the race organizers have decided to start the contest at the entrance to the park, and to require contestants to use a walk that goes to a pond at the center of the park. After that, the course goes "cross-country" to the remaining checkpoints. The rules require contestants to avoid slopes of steep terrain and to cross streams only on bridges. Your class has entered a team, and they have asked you to develop from the map a set of instructions that they can follow, with directions expressed using the mathematics convention for measuring angles; i.e., 0° represents east. Use **Handout H6.2** which shows the map's proper scale, together with a ruler and a protractor to answer the following questions.

1. As a warm-up, think of drawing two points on a piece of paper and connecting them with a line segment. The line segment defines a distance (length) and an orientation (slope). Explain how to alter the drawing so that it also describes a specific direction, from one endpoint toward the other.

2. Now, on to the Rockville Course. First, consider the instructions needed to get your team from the park entrance (at checkpoint 1), along the walk to checkpoint 2.

   a) Draw the desired directed route on your map. Then write specific instructions to direct the athletes from the starting point to the bend in the walkway.

   b) Complete the instructions to get the athletes to checkpoint 2. Draw that route on your map as well.

   c) Had your team been able to go *directly* to checkpoint 2 instead of having to go along the walk, what one instruction would have been needed? Draw that directed route on your map. Describe the relationship between the routes in parts (a) and (b), and this one.

   d) At this point in the competition, how far east (over) and how far north (up) has your team traveled?

   e) The distances in parts (a) and (b) don't add to give either the distance in part (c) or the total distance in part (d). Explain why all these distances differ.

3. Draw on your map the route you would recommend to your team for completing the course. Then write a set of instructions for each of the stages; keep the number of instructions to a minimum. If more than one directed segment is needed for any stage, identify the condition causing it.

   a) Stage 2: Travel from checkpoint 2 to checkpoint 3.

   b) Stage 3: Travel from checkpoint 3 to checkpoint 4.

   c) Stage 4: Travel from checkpoint 4 to checkpoint 5.

4. Consider the cumulative effect of these instructions on the competitors.

   a) If you combined all of the instructions you've written for stages 2, 3, and 4, what is the result? That is, where is your team with respect to checkpoint 2?

   b) If you combined all of the instructions you've written for all 4 of the stages (Items 2 and 3), where is your team with respect to the starting point?

5. After finishing the orienteering course, the competitors have to get back to their cars (located at the park entrance) in order to go home. What one instruction would return them from the finish (checkpoint 5) to the starting point (checkpoint 1)?

VECTORS

Many physical quantities, such as height and weight, may be specified by a single number (with units of measure); the only concern is "How much...." Other quantities, like displacement (a distance in a particular direction), velocity, and force, are directed.

Force from firing engine

FIGURE 6.15. The effect of a force depends on the amount of the force and the direction at which it is applied.

Meteorologists might use displacement to describe the movement of a storm over the last five minutes. Its velocity specifies both how fast it is moving (its speed) and in which direction. Engineers must specify propulsion forces carefully in controlling the motion of spacecraft. (See **Figure 6.15**). In fact, if you think about it, many quantities you use informally include "direction" information.

Vectors are the mathematical tool needed when both amount and direction are important. You used them informally in Activity 6.3. It is customary to represent vector quantities geometrically using arrows. The tip (head) of the arrow points in the specified direction, and the length of the arrow represents the amount, or **magnitude**, of the quantity, according to some scale. **Figure 6.16** shows two wind velocities, the first one at 10 mph in the direction of 60 degrees north of east, and the second one at 15 mph blowing due east. (However, note that weather forecasters usually specify the direction *from* which the wind blows rather than the direction *toward* which it blows.)

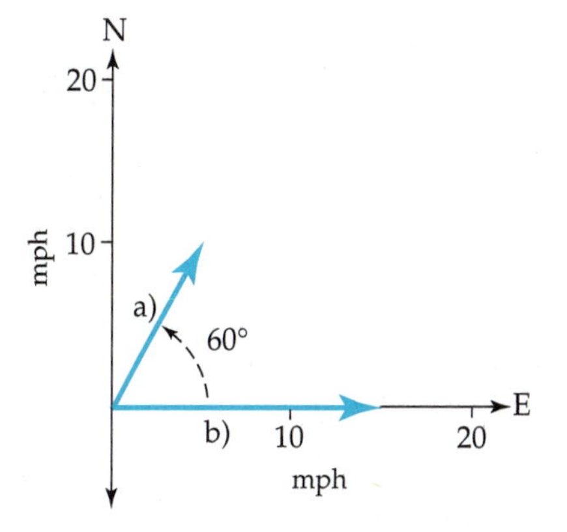

FIGURE 6.16.
A wind blowing (a) 10 mph at 60° N of E, and (b) due east at 15 mph.

Formally, a **vector** is defined as a directed line segment. Two vectors are defined to be equal if they have the same magnitude and same direction. Thus, two vectors do not have to be in the same place to be equal. **Figure 6.17** illustrates this idea by showing three equal vectors.

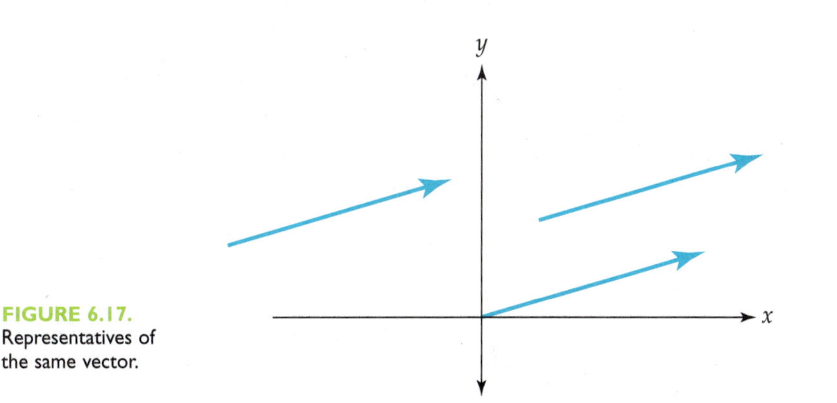

FIGURE 6.17.
Representatives of the same vector.

This text will denote a vector with a boldface letter, such as **v** (read as vector v). The vector representing the directed line segment from point A to point B is written as $\overrightarrow{AB}$ (read as "vector *AB*"). The magnitude of vector **v** is written as $|\mathbf{v}|$.

TAKE NOTE

Since boldface is difficult with pencil and paper, hand-written vectors usually use the arrow overhead notation. Thus, **v** would be written by hand as $\vec{v}$.

THE GEOMETRY OF VECTOR ARITHMETIC

Long ago you learned how to do arithmetic with numbers. More recently you have developed arithmetic on functions. Similar operations are possible with vectors, too.

Scalars and Scalar Multiples

Suppose a vector **v** represents a wind velocity of 5 mph blowing toward the east. To represent a wind velocity of 10 mph, use an arrow in the same direction with a length twice as long. **Figure 6.18** illustrates this concept.

Scale: 1 inch = 5 mph

FIGURE 6.18.
A scalar multiple of a vector **v**, using the scalar 2.

In general, multiplying a vector by a positive number multiplies its length by that number. Multiplying a vector by a negative number reverses the vector's direction and multiplies the length by the number's absolute value. Thus, if c is a non-zero number and **v** is a vector, the direction of the vector c**v** agrees with that of **v** if c is positive, and is opposite to that of **v** if c is negative. Since the number c acts like a *scaling factor*, it is called a **scalar**, and c**v** is called a **scalar multiple** of **v.** If c is zero, then c**v** is the zero vector, **0**. The vector **0** is the only vector with no direction.

Geometrically, scalar multiplication of vectors produces expansions, contractions, or reflections. (See **Figure 6.19.**) If $c > 1$, c**v** is an **expansion**. If $0 < c < 1$, then c**v** is a **contraction**. If $c = -1$, then c**v** is a **reflection**. If $c < 0$, write $c\mathbf{v} = -1|c|\mathbf{v}$, a composition of two scalar multiplications.

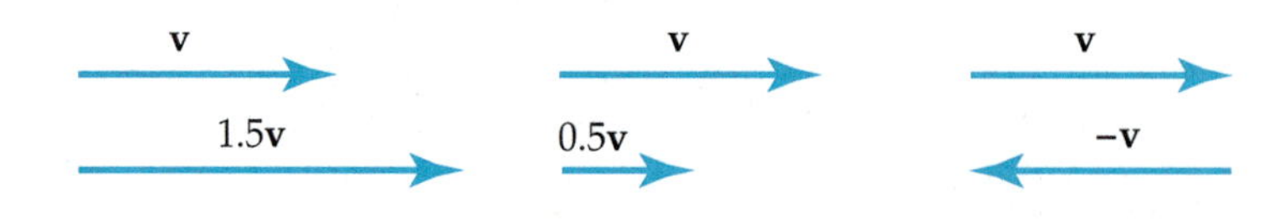

FIGURE 6.19.
Scalar multiplication can expand, contract, or reflect a vector.

Vector Addition

Each time you combined two or more orienteering instructions in Activity 6.3, you added vectors. The sum, or **resultant**, is the one vector that gives the same result as the successive individual ones. The **vector sum v + w** is found by drawing vector **w** so that its tail is at the tip of vector **v.** The resultant then has its tail at the tail of **v** and its head at the head of **w.** (See **Figure 6.20.**)

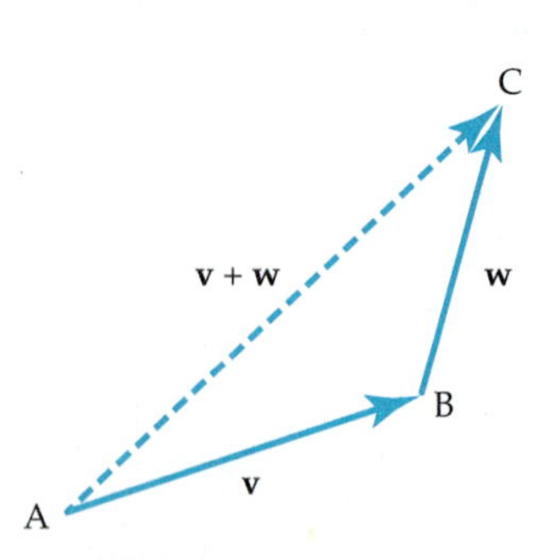

FIGURE 6.20.
Two vectors added together, by head-to-tail method.

EXAMPLE 4

© Susan Van Etten

Without any wind, a rescue helicopter's velocity is 50 mph. The helicopter pilot heads the chopper due east. A wind is blowing toward the north at 10 mph. Describe the actual motion of the helicopter.

SOLUTION:

Velocities are vector quantities. The actual velocity is the vector sum of the two given velocity vectors. See **Figure 6.21**.

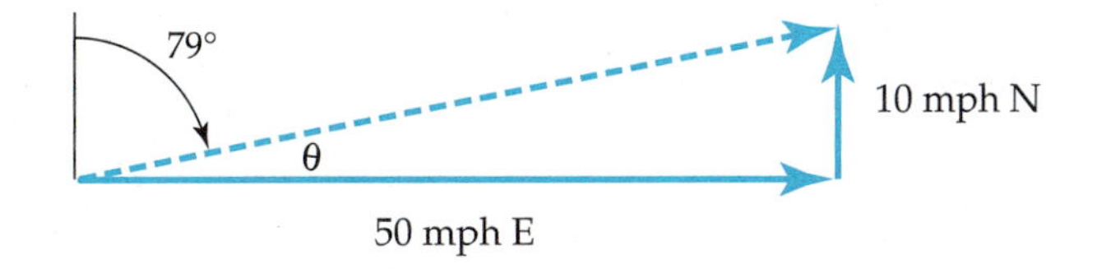

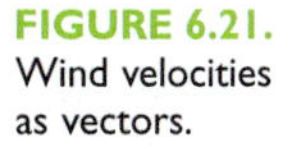

FIGURE 6.21. Wind velocities as vectors.

By measuring, the resultant velocity is approximately 51 mph and the angle of flight is approximately 11° north of due east. Since navigators measure direction clockwise from north, the helicopter pilot is flying on a heading of $90° - 11° = 79°$. Notice that a relatively strong wind changed the speed (magnitude) only slightly, but resulted in a significant change in direction.

Using the Pythagorean formula, the magnitude is $\sqrt{50^2 + 10^2} \approx 51$.

Also, the tangent of angle θ is $10/50 = 1/5$. Solving for θ, $\theta = \tan^{-1}(0.2) \approx 11.31°$.

An equivalent way to perform vector addition, called the **parallelogram method**, is useful when the two vectors are **concurrent**, or drawn so they start at the same point. Complete the parallelogram having the two vectors as adjacent sides. The vector sum is the diagonal of that parallelogram starting at the common point of the two original vectors. (See **Figure 6.22.**)

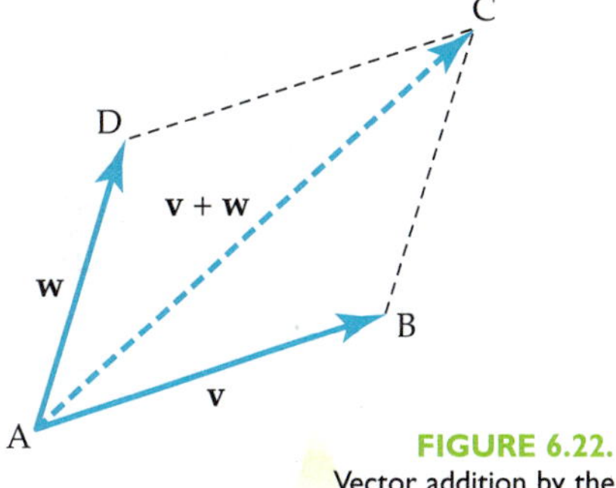

FIGURE 6.22. Vector addition by the parallelogram method.

The head-to-tail and parallelogram methods are equivalent, since $\overrightarrow{AD}$ and $\overrightarrow{BC}$ are different representatives of **w**. The parallelogram method also reveals that vector addition is a commutative operation; in other words, **v** + **w** = **w** + **v** = $\overrightarrow{AC}$.

Vector Subtraction

The subtraction **v** – **w** is defined as **v** + (–1)**w**. Thus, to perform the vector subtraction **v** – **w** when the vectors are drawn head-to-tail, add the opposite of **w** to **v**, as shown in **Figure 6.23**.

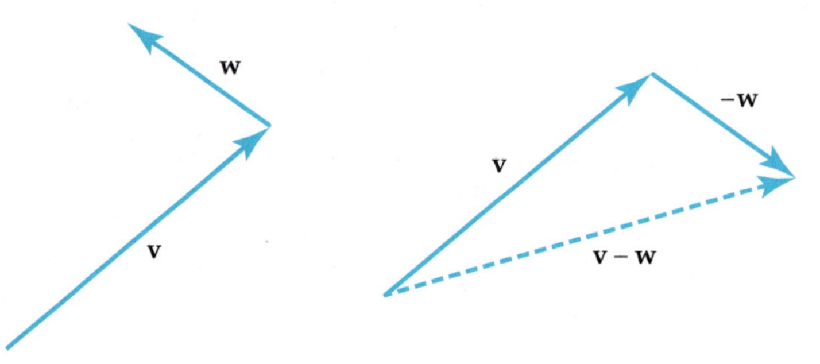

FIGURE 6.23.
Vector subtraction **v** – **w** with the head-to-tail method.

That is, draw –**w** from the tip of **v**, and complete the addition using the head-to-tail method. Alternately, if the vectors are drawn concurrent (so they start at the same point), the diagonal of the parallelogram from the tip of **w** to the tip of **v** provides the answer. (See **Figure 6.24**.)

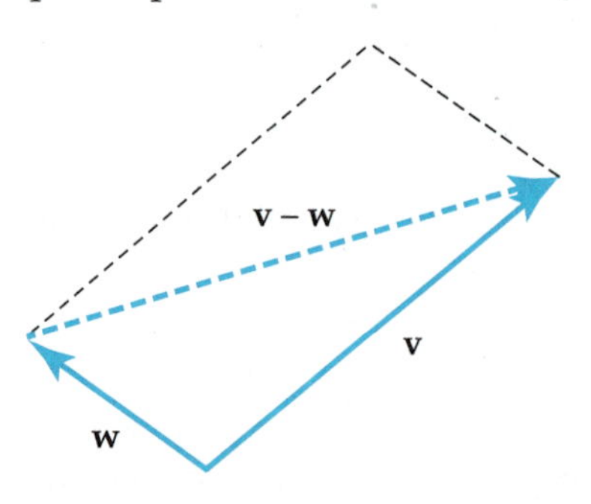

FIGURE 6.24.
Vector subtraction using the parallelogram method.

Thus, just as $a - b$ gives the directed distance "from b to a" on the number line, **v** – **w** gives the directed distance "from **w** to **v**" for concurrent vectors. In general, think of subtraction as "final minus initial."

EXAMPLE 5

An interesting challenge facing engineers is to determine the direction and magnitude of a propulsion force necessary to produce desired results in the trajectories of vehicles, such as the trajectories of space vehicles. Suppose at a particular moment, a vehicle is experiencing a force **F** = 1414 pounds in the direction shown in **Figure 6.25**. An engineer needs to apply a propulsion force so that the resultant force **r** acting on the vehicle is 1000 lbs in the direction of due east. How much propulsion force will be needed? In what direction should it be applied?

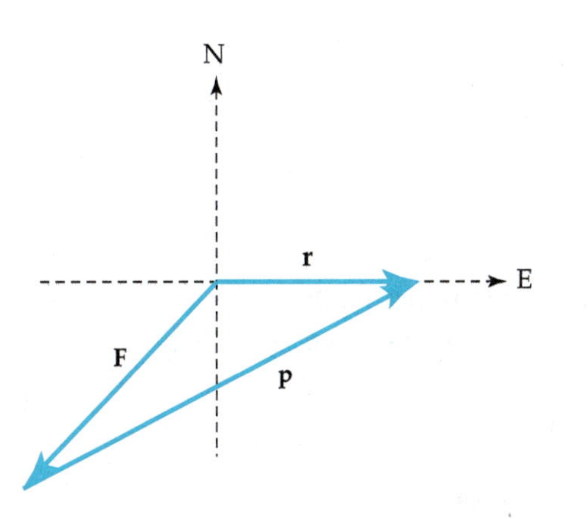

FIGURE 6.25.
A resistive force **F** acting on a vehicle with a desired resultant.

SOLUTION:

Forces are vector quantities. In this case, one of the individual forces (**F**) is known, and the vector sum is the net force. If **p** is the unknown propulsion force, then **F** + **p** = **r**. Therefore, **p** = **r** – **F**. In interpreting the drawing, an application of about 2240 lbs in the direction of 26.5° N of due E will have the desired effect.

Exercises 6.3

1. Consider vectors **a**, **b**, and **c** in **Figure 6.26**. Copy them onto a sheet of paper.

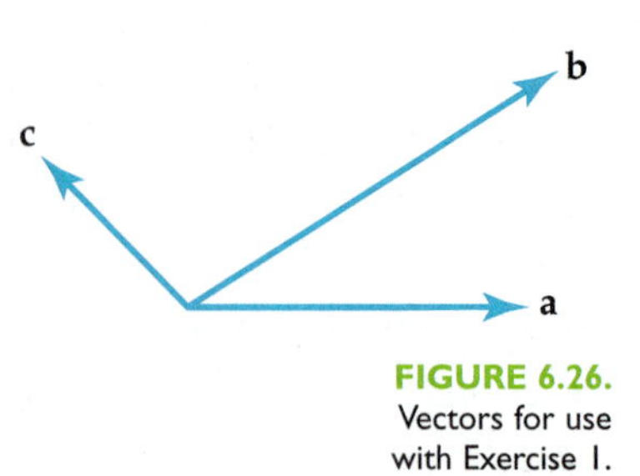

FIGURE 6.26. Vectors for use with Exercise 1.

   Using a geometric construction, draw a representative vector for:

   a) $\mathbf{a} + \mathbf{b}$

   b) $\mathbf{b} - \mathbf{c}$

   c) $\mathbf{a} - 2\mathbf{b}$

   d) $0.5\mathbf{a} - 1.5\mathbf{c}$

2. Vectors **u**, **v**, and **w** act as the sides of triangle *ABC*, as shown in **Figure 6.27**.

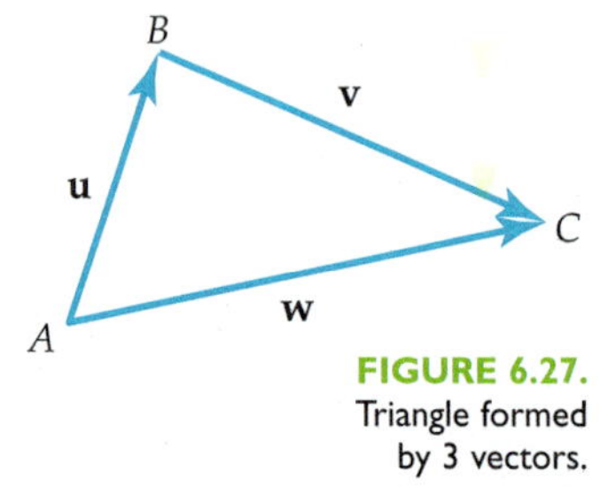

FIGURE 6.27. Triangle formed by 3 vectors.

   a) Express **w** in terms of **u** and **v**.

   b) Express **v** in terms of **u** and **w**.

3. The Parallelogram Method of Vector Addition (refer to Figure 6.22) demonstrates geometrically that $\mathbf{v} + \mathbf{w} = \mathbf{w} + \mathbf{v}$. In a similar manner, show that $\mathbf{v} - \mathbf{w} = -\mathbf{w} + \mathbf{v}$.

4. Sketch vector drawings to show that the following properties hold:

   a) Associative property for vector addition: $\mathbf{u} + (\mathbf{v} + \mathbf{w}) = (\mathbf{u} + \mathbf{v}) + \mathbf{w}$.

   b) Distributive property using multiplication by scalars and vector addition: $k(\mathbf{u} + \mathbf{v}) = k\mathbf{u} + k\mathbf{v}$.

5. Use a vector **p** to represent walking 4 miles due east, and a vector **q** to represent walking 2 miles due north. For each of the following, draw a representative vector to scale, and determine the resultant using the Pythagorean Theorem and the definition of $\tan^{-1}$.

   a) $2\mathbf{p} + 3\mathbf{q}$

   b) $3\mathbf{p} - 2\mathbf{q}$

Exercises 6.3

Corbis

6. A swimmer swims 4 mph in water without a current. She is attempting to swim across a river 100 yards wide with a current of 1.5 mph.

   a) If she heads perpendicular to the shore, estimate how far downstream she will be when she reaches the other side.

   b) How far will she travel?

   c) How long will it take her?

7. A fire department helicopter flies from its helicopter pad a distance of 50 km on a bearing of 30° (60° north of east) to drop water on a fire. Then it flies 40 km due southwest to a second fire. Place an *xy*-coordinate system so that the origin represents the helicopter pad, the *x*-axis points east, and the *y*-axis points north.

   a) At what point is the first fire located?

   b) At what point is the second fire located?

   c) What heading and distance must the pilot fly to return to the chopper pad?

8. Suppose a helicopter flying at 50 mph on a heading of 90° (due east) encounters a wind of 10 mph blowing due northeast (on a bearing of 45°). **Figure 6.28** represents the situation. Parts (a)–(e) illustrate how to find the resultant velocity (speed and direction) of the helicopter algebraically.

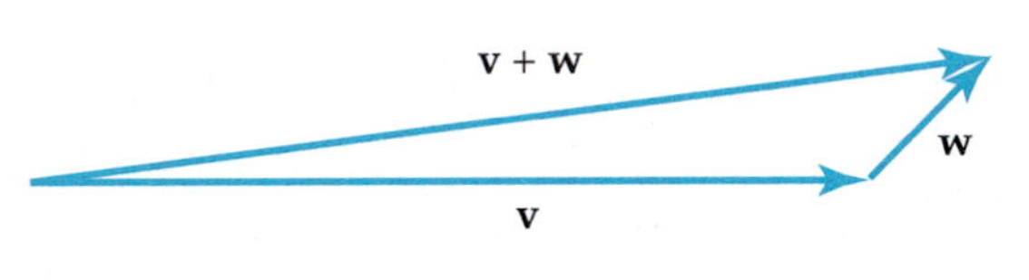

FIGURE 6.28. Vector diagram for Exercise 8.

   a) The wind velocity can be thought to be acting in two directions, one due east (which causes the speed to increase) and one due north (which deflects the motion into a new direction). Calculate how much of the wind velocity is acting due east. Explain your method.

   b) Calculate how much of the wind velocity is acting due north. Explain how you got this answer.

Exercises 6.3

c) Determine the resultant velocity in the easterly direction.

d) Knowing the total velocity in the east and north directions, use the Pythagorean Theorem and the definition of $\tan^{-1}$ to find the resultant velocity.

e) Verify your work by direct measurement in Figure 6.28.

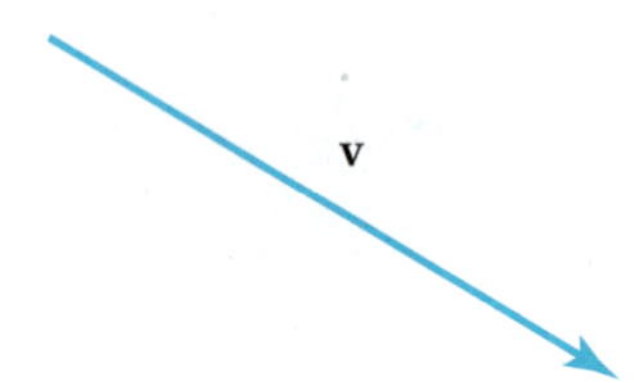

FIGURE 6.29.
Vector for use with Exercise 9(a).

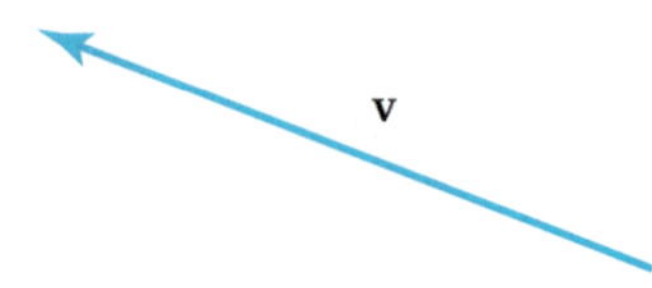

FIGURE 6.30.
Vector for use with Exercise 9(b).

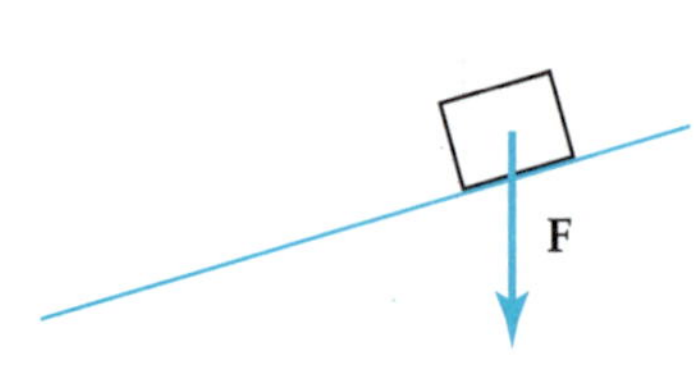

FIGURE 6.31.
Vector for use with Exercise 9(c).

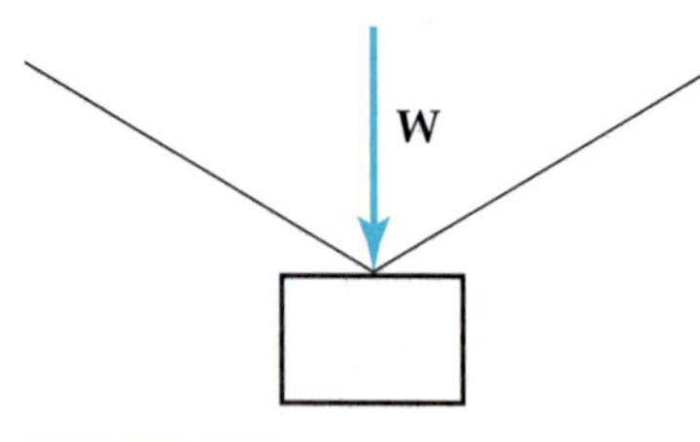

FIGURE 6.32.
Vector for use with Exercise 9(d).

9. A critical step in Exercise 8 was to resolve a vector into **components**, which is the reverse process to vector addition. In the case of the analysis of the wind, components were directed east and north, but that was out of convenience for that situation. The necessary condition is that the components combine by vector addition to produce the original vector. For each of the following vectors, draw the original vector and the specified components, then describe the length of each component as a multiple of the magnitude of the original vector, based on direct measurement.

   a) Form components of vector **v** in horizontal and vertical directions. (Refer to **Figure 6.29.**)

   b) Form components of vector **v** in horizontal and vertical directions. (See **Figure 6.30.**)

   c) Form components of vector **F**, one directed along the incline (parallel to the incline surface), and one directed perpendicularly *into* the incline. (See **Figure 6.31.**)

   d) Form two components of vector **W**, one directed along each of the wires supporting the weight of the object. (Note: The components will *not* be perpendicular to each other.) The angle formed by the two wires is 120°. (See **Figure 6.32.**)

   e) Look back at the work done in Activity 6.3. Which task asked you to resolve a vector into components?

10. Two tugboats exert forces **u** and **v** on a luxury liner. Force **u** has a magnitude of 3000 lb and acts in an azimuth direction of 300°, while force **v** has a magnitude of 2000 lb and acts in an azimuth direction of 240°. Apply a similar analysis to that done in Exercise 8, and determine the amount and direction of the resultant force on the luxury liner.

Exercises 6.3

11. Generalize the work done in Exercise 10. Consider the first vector **u** to have magnitude M_1, and act at an angle of A_1 (standard mathematical orientation), and let the second vector **v** have magnitude M_2 and act at an angle of A_2.

   a) Develop a formula that calculates the magnitude of the resultant vector.

   b) Develop a formula that calculates the direction of the resultant vector in standard mathematical orientation.

   c) Check the formulas you've developed by applying them to a situation in which you walk 8.66 mi at 45° N of E, then go 5 mi in the direction of 45° N of W. How far are you from the beginning, and in what direction?

12. In some parts of Activity 6.3, you combined more than two vectors.

   a) Describe how to find the resultant of three or more concurrent vectors.

   b) Make a copy of **Figure 6.33** and apply your method to find the resultant of the 3 vectors shown.

   c) Describe how to find the resultant when three or more vectors are drawn head-to-tail.

   d) Apply your method to find the resultant of the 3 vectors shown in **Figure 6.34**.

   e) Describe an analytic method similar to the process used in Exercises 10 and 11 to find the resultant when a mathematical description (magnitude and direction) of three or more vectors is provided.

   f) Apply your method to find the resultant for the following 3 displacement vectors: go 30°N of E for 5 miles, then head 10 miles on an azimuth of 315°, and finally go due south for 3 miles.

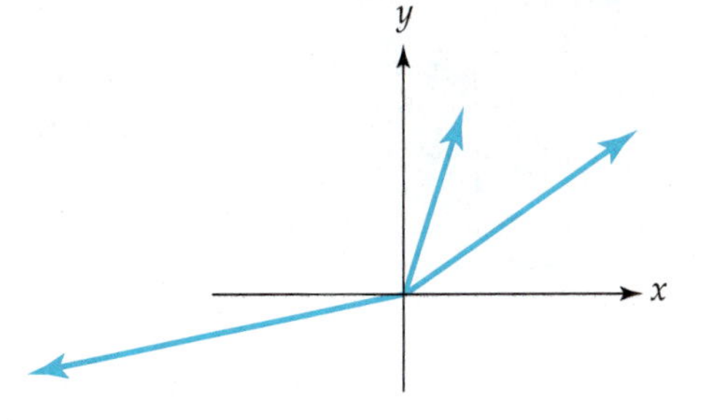

FIGURE 6.33.
Drawing for use with Exercise 12(b).

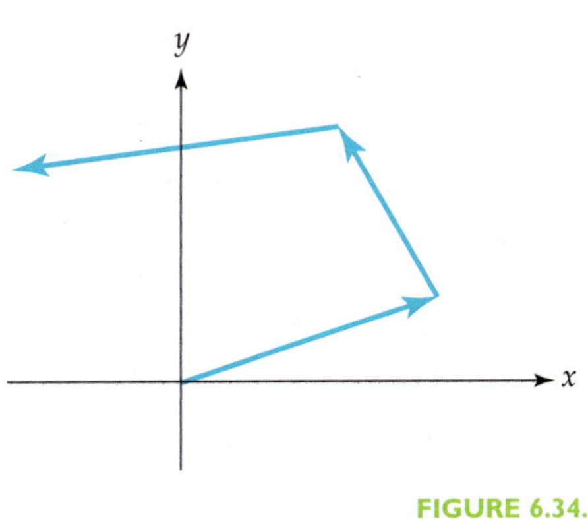

FIGURE 6.34.
Drawing for use with Exercise 12(d).

Exercises 6.3

13. Sometimes it's useful to find a vector, called the **equilibrant**, that produces a net sum of 0 when added to all the other vectors.

   a) What is the equilibrant of a velocity vector "2 mph directly downstream?" Interpret the equilibrant in terms of boating.

   b) What is the equilibrant of a displacement vector "12.5 miles in the direction of 228°?" Interpret the equilibrant in terms of hiking.

   c) Describe mathematically the relationship between a resultant and its equilibrant.

   d) Which question from Activity 2.3 asked you to determine an equilibrant?

© Susan Van Etten

14. **Investigation**: Surveyors verify property lines by doing a "traverse," which is a walk around the perimeter recording measurements at key locations. They use transits to measure angles and electronic sensors or measuring tapes for distances, and end up where they started—in other words, constructing real polygons using data in the form of vector information. In checking their work, they draw the various vectors, and if the resultant does not equal 0, an error in the measurements has been detected.

   a) **Table 6.3** contains data taken from a surveyor's traverse. Is there any error in the measurements? (Note: Actual measurements were rounded off for the purpose of this activity.)

   If they can't reproduce the data by repeating the traverse, they "absorb" the error into the measurements in equal amounts—in essence, adding 1/5 of the equilibrant to each of the vectors making up the polygon. For fairly precise results, however, it is better to work with the numbers than with the drawing to resolve the dilemma. In the rest of this investigation, you will work through the steps of such a calculation.

From	To	Azimuth	Distance
Pt. A	Pt. B	26°	285 ft
Pt. B	Pt. C	105°	610 ft
Pt. C	Pt. D	195°	720 ft
Pt. D	Pt. E	353°	203 ft
Pt. E	Pt. A	307°	647 ft

TABLE 6.3.
Traverse information for use with Exercise 14.

Exercises 6.3

b) Resolve each of the vectors forming the polygon into horizontal and vertical components. (Surveyors refer to horizontal components as latitudes and to vertical components as departures.) Organize your work into a table like **Table 6.4**.

Vector	Azimuth	Length	Latitude	Departure

TABLE 6.4. Sample table headings for Exercise 14(b).

c) Since the resultant is supposed to be **0**, the actual totals (latitude and departures, separately) are the total errors in each direction. Find the net latitude (total error in the horizontal direction) and net departure (total error in the vertical direction). Then find the average latitude and departure errors.

d) Use the average errors to find the "correction factor" for each measurement. That is, what correction should be added to each latitude and each departure to make them sum to 0?

e) Add two columns to your table from part (b) to represent the adjusted latitudes and departures. Complete those columns using the corrections you calculated in part (d).

f) Finally, use the adjusted latitudes and departures to compute adjusted values for the azimuths and lengths, and record the results in two more columns of your table.

LESSON 6.4

The Algebra of Vectors

Lesson 6.3 showed that vectors can be used to represent motion, velocity, and forces, and that complex questions can be answered using vector addition and subtraction, and multiplication of vectors by scalars. This lesson examines symbolic representations of vector operations.

Activity 6.4

Gridville is a simplified version of a large city with streets running parallel and perpendicular to one another. The major streets of Gridville are two-way streets one mile apart, represented by the lines of the grid. Buildings are represented by points and are located at the intersections of grid lines in order to identify their locations easily. Fire trucks and other emergency vehicles may only travel along grid lines. Diagonal movements are not allowed by ground transportation vehicles.

An earthquake has occurred in Gridville. Fire trucks and a helicopter are sent from the Fire Station at F to point A where a fire has erupted. After a preliminary examination, injured citizens are to be evacuated, either by air or land, to a regional medical facility located at point H. A reporter located at point R plans to interview the fire marshal at point F on her way to the emergency, and then interview the police chief at point P on her way to the hospital. **Figure 6.35** shows the location of the various buildings.

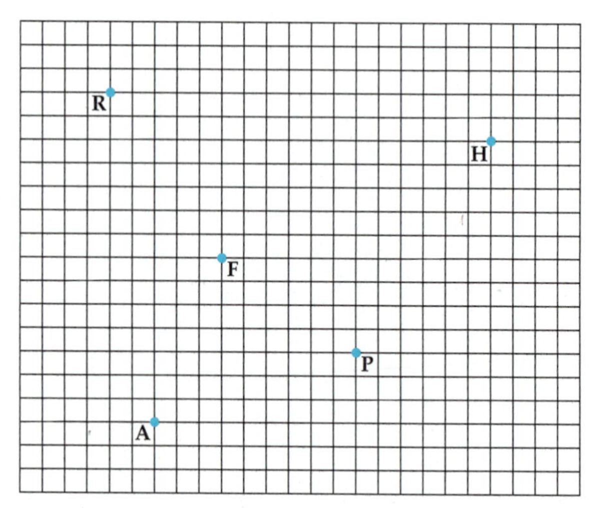

FIGURE 6.35.
A rectangular map of Gridville.

1. Introduce a rectangular coordinate system for your map, with the origin at the fire station.

   a) What are the coordinates of the location of the fire?

   b) What set of vectors describes a route to get the fire trucks from the station to the scene as quickly as possible?

   c) What one vector would get the helicopter from the fire station to the scene as quickly as possible?

   d) Which are more directly related to the meaning of location coordinates: fire truck directions or helicopter directions? Explain.

   e) Which are easier to provide: fire truck directions or helicopter directions? Explain.

2. The emergency rescue crews prepare to take the injured to the hospital.

   a) Write a set of vector directions for the fire trucks to get from point A to the hospital at point H.

   b) Find one vector instruction to get the helicopter pilot from point A to the hospital at point H. How did you determine that instruction?

   c) Explain how to determine the directions for the fire trucks from knowing *just* the coordinates for the fire (at point A) and the hospital.

   d) Explain how to determine the directions for the helicopter from knowing *just* the coordinates for the fire and the hospital.

3. During the interview, the reporter, who is from out-of-town, asks the fire marshal where the hospital is located.

   a) Give driving instructions from the fire station.

   b) Explain how to determine the hospital location, using just the coordinates of the fire location and directions from the fire to the hospital.

c) Can you determine the azimuth angle that describes the direction to the hospital from the fire station, from knowing *only* the azimuths of the helicopter directions to the emergency site from F and to the hospital from A? Explain.

d) Can you determine the distance from the fire station to the hospital from knowing *just* the lengths of the helicopter flights to the emergency site from F and to the hospital from A? Explain.

THE ALGEBRA OF VECTORS

The Standard Basis Vectors i and j

In Lesson 6.3, you learned to find the sum of two (or more) vectors geometrically. The equation $\mathbf{x} = \mathbf{v} + \mathbf{w}$ represents both the process of combining two vectors **v** and **w** into a single resultant, and **resolving** a vector **x** into two components. Much of the algebra of vectors is simplified if all vectors are first resolved into components that are parallel to the *x*- and *y*-axes.

The vector **i** is the basis vector in the positive *x*–direction, and is defined to be the directed line segment from (0, 0) to (1, 0). Similarly, vector **j** is the basis vector in the positive *y*–direction, and is defined to be the directed line segment from (0, 0) to (0, 1). Together, **i** and **j** are called the **standard basis**. By combining a stretching (or shrinking) of these basis vectors, *any* vector **v** may be written in the form $a\mathbf{i} + b\mathbf{j}$. The coefficients for the basis vectors tell you how they have been transformed. And, since **i** and **j** are represented concurrently, the resultant $a\mathbf{i} + b\mathbf{j}$ is represented by the segment from (0, 0) to (a, b).

Figure 6.36 illustrates the procedure for representing any vector as a **linear combination** of **basis vectors**.

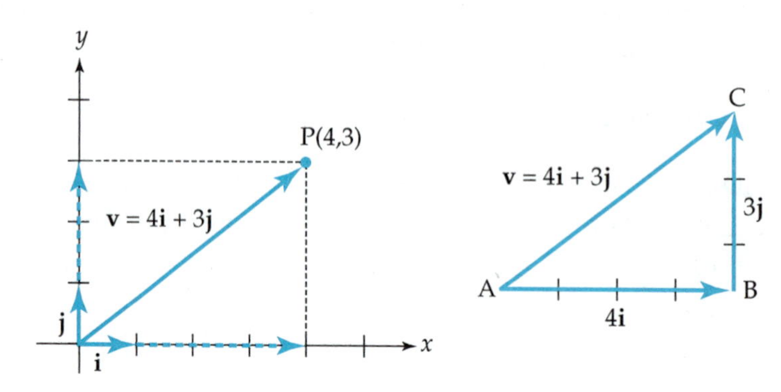

FIGURE 6.36. Vectors as linear combinations of the basis vectors **i** and **j**.

Any representation of a vector that starts at the origin is called a **position vector** because it points to a particular coordinate position. Thus, the representations described for **i**, **j**, and **v** above are position vectors. Note, however, that another representation for **v** is shown in Figure 6.36—one that doesn't start at the origin. Note that the basis vector **i** is not the complex number i studied previously. A boldface '**i**' indicates the basis vector.

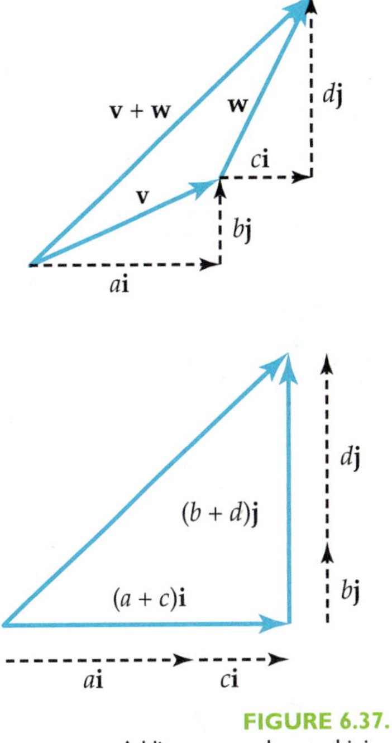

FIGURE 6.37.
Adding vectors by combining corresponding components.

Addition and Subtraction

Vectors are added algebraically by adding their corresponding components, as shown in **Figure 6.37**. If $\mathbf{v} = a\mathbf{i} + b\mathbf{j}$ and $\mathbf{w} = c\mathbf{i} + d\mathbf{j}$, then $\mathbf{v} + \mathbf{w} = (a\mathbf{i} + b\mathbf{j}) + (c\mathbf{i} + d\mathbf{j}) = (a + c)\mathbf{i} + (b + d)\mathbf{j}$.

Similarly, if $\mathbf{v} = a\mathbf{i} + b\mathbf{j}$ and $\mathbf{w} = c\mathbf{i} + d\mathbf{j}$, then subtraction is accomplished by subtracting corresponding components:
$\mathbf{v} - \mathbf{w} = (a\mathbf{i} + b\mathbf{j}) - (c\mathbf{i} + d\mathbf{j}) = (a - c)\mathbf{i} + (b - d)\mathbf{j}$.

Example 6

The helicopter pad in Gridville is located at O(0, 0). An emergency has occurred at point A(–3, –7). A regional hospital is located at B(12, 5). The helicopter pilot must fly to A, pick up evacuees, then fly from A to B. Express vectors $\overrightarrow{OA}$ and $\overrightarrow{AB}$ in terms of the basis vectors **i** and **j**.

Solution:

From **Figure 6.38**,

$\overrightarrow{OA} = -3\mathbf{i} - 7\mathbf{j}$, $\overrightarrow{OB} = 12\mathbf{i} + 5\mathbf{j}$, and $\overrightarrow{OA} + \overrightarrow{AB} = \overrightarrow{OB}$.

Thus, $\overrightarrow{AB} = \overrightarrow{OB} - \overrightarrow{OA}$, so $\overrightarrow{AB} = (12\mathbf{i} + 5\mathbf{j}) - (-3\mathbf{i} - 7\mathbf{j}) = 15\mathbf{i} + 12\mathbf{j}$.

The vector $\overrightarrow{AB} = 15\mathbf{i} + 12\mathbf{j}$ represents the path the pilot must fly from point A to point B.

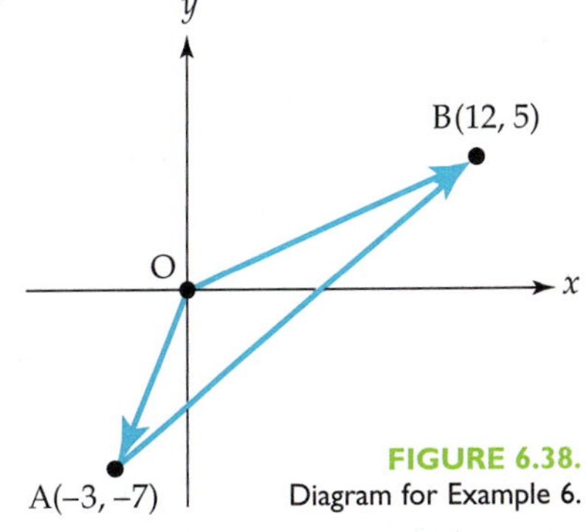

FIGURE 6.38.
Diagram for Example 6.

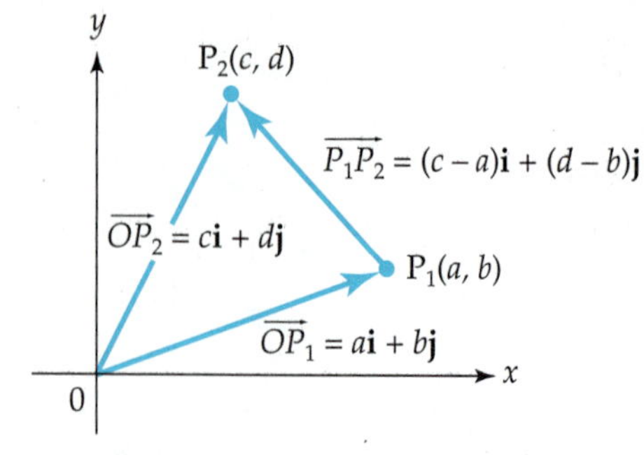

FIGURE 6.39.
Expressing a vector from one point to another.

In Example 6, it was necessary to find the vector from one point to another. Basis vectors make such computations easy. If P_1 has coordinates (a, b), and P_2 has coordinates(c, d), then vector $\overrightarrow{P_1P_2}$ records the directed changes, Δx and Δy, in the respective components. See **Figure 6.39**.
Since $\overrightarrow{OP_1} + \overrightarrow{P_1P_2} = \overrightarrow{OP_2}$, then $\overrightarrow{P_1P_2} = \overrightarrow{OP_2} - \overrightarrow{OP_1}$.

From the given basis representations, $\overrightarrow{OP_1} = a\mathbf{i} + b\mathbf{j}$ and $\overrightarrow{OP_2} = c\mathbf{i} + d\mathbf{j}$.
Substitution gives $\overrightarrow{P_1P_2} = \overrightarrow{OP_2} - \overrightarrow{OP_1} = (c\mathbf{i} + d\mathbf{j}) - (a\mathbf{i} + b\mathbf{j}) = (c - a)\mathbf{i} + (d - b)\mathbf{j}$.
In summary, the vector from $P_1(a, b)$ to $P_2(c, d)$ is:
$\overrightarrow{P_1P_2} = (c - a)\mathbf{i} + (d - b)\mathbf{j}$. Thus, as was the case with geometric methods, again final – initial produces the desired result.

Unit Vectors

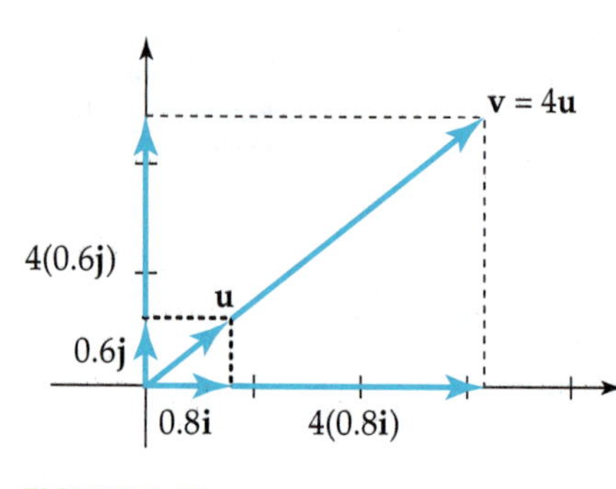

FIGURE 6.40.
Vector as a scalar multiple of its unit vector.

The vectors **i** and **j** are examples of **unit vectors**, since they have length 1. You have seen how easy it is to build other vectors using the basis vectors **i** and **j**. Sometimes, it is more convenient to think of a vector **v** as a scalar multiple of a unit vector in the same direction as **v**. The traditional label for such a generic unit vector is the name **u**.

Figure 6.40 shows how a unit vector $\mathbf{u} = 0.8\mathbf{i} + 0.6\mathbf{j}$ can be used to form a vector **v** that is four times as long.

Verify that **u** is indeed a unit vector: $|\mathbf{u}| = \sqrt{0.8^2 + 0.6^2} = 1$. Stretching both components of **u** by a single scalar produces similar triangles; if each component of **v** is four times longer than the corresponding component of **u**, then **v** is four times longer than **u**. Again check: $|\mathbf{v}| = \sqrt{3.2^2 + 2.4^2} = 4$.

The reverse of this process can be used to find a unit vector **u** in the same direction as vector **v**. For example, if $\mathbf{v} = 5\mathbf{i} - 12\mathbf{j}$, then shrinking it to unit length means making it 1/13 as long, since $|\mathbf{v}| = \sqrt{5^2 + (-12)^2} = 13$. That contraction is accomplished by the scalar multiplication, $\mathbf{u} = \frac{1}{13}\mathbf{v} = \frac{5}{13}\mathbf{i} - \frac{12}{13}\mathbf{j}$.

EXAMPLE 7

Let θ represent the angle between vector **u** and the positive x-axis in Figure 6.40. Resolve **v** into its components, expressed in terms of the magnitude (4) and direction (θ) of **v**.

SOLUTION:

$\mathbf{v} = 3.2\mathbf{i} + 2.4\mathbf{j} = 4\cos\theta\mathbf{i} + 4\sin\theta\mathbf{j}$. The notation is similar to that for complex numbers in polar form. Likewise, magnitude and direction angle (4 and θ) may remind you of the polar coordinates for a point in the plane.

PEOPLE AND MATH

If your future career involves piloting a boat, planning wilderness excursions, or even plotting the course of a space flight, you will be using the mathematics of coordinates and vectors. All of these jobs, and many others, use the process of **navigation**—determining the position of a point or object in relation to some exterior reference system. Maps, compasses, protractors, and other technology such as sonar devices all aid in the navigation process. In addition, sophisticated radio-navigation satellite systems such as the Navstar Global Positioning System (GPS) enable rescue services, such as the Coast Guard, to locate ships in distress more easily. The rescue services use transmissions from satellites to calculate reference points. Originally developed for wartime use, GPS permits land, sea, and airborne users to determine their positions on the earth by measuring their distances and directions from satellites in space.

United States Coast Guard

Exercises 6.4

1. Describe $\frac{\mathbf{v}}{|\mathbf{v}|}$ relative to the given vector $\mathbf{v} = a\mathbf{i} + b\mathbf{j}$.

2. For each of the following, write the vector $\overrightarrow{AB}$ as a linear combination of the standard basis vectors.

   a) A(–3, 4) and B(2, 8)

   b) $\overrightarrow{OA} = -2\mathbf{i} - 5\mathbf{j}$ and $\overrightarrow{OB} = -3\mathbf{i} + 2\mathbf{j}$

   c) $\overrightarrow{AB}$ has length 6 and a direction angle of 20° north of due west.

3. For each of the following, write the vectors $\mathbf{v} + \mathbf{w}$ and $\mathbf{v} - \mathbf{w}$ in terms of the standard basis.

   a) $\mathbf{v} = 5\mathbf{i} - 2\mathbf{j}$ and $\mathbf{w} = 4\mathbf{i} + 5\mathbf{j}$

   b) $\mathbf{v} = \overrightarrow{BA}$ and $\mathbf{w} = \overrightarrow{BC}$, with coordinates A(–2, 5), B(3, 8) and C(4, –1)

   c) $\mathbf{v}$ has length 4 and a direction angle of 120° (standard mathematical orientation), and $\mathbf{w}$ has length 2 and a direction angle of 330°.

4. For each of the following, evaluate $|\mathbf{v}|$.

   a) $\mathbf{v} = -3\mathbf{i} + 7\mathbf{j}$

   b) $\mathbf{v} = \overrightarrow{AB}$, with coordinates A(4, –5) and B(–2, 3)

   c) $\mathbf{v} = 6\mathbf{u}$, where $\mathbf{u} = 0.5\mathbf{i} + \frac{\sqrt{3}}{2}\mathbf{j}$

   d) $\mathbf{v} = \mathbf{0} = 0\mathbf{i} + 0\mathbf{j}$

5. For each of the following, find a unit vector in the same direction as the vector $\mathbf{v}$.

   a) $\mathbf{v} = -5\mathbf{i} + 3\mathbf{j}$

   b) $\mathbf{v} = \overrightarrow{AB}$ with coordinates A(2, –8) and B(–5, 4)

   c) $\mathbf{v}$ has length 6 and a direction angle of 225°

Exercises 6.4

6. Represent each of the following vectors in terms of its magnitude and direction.

   a) $\mathbf{v} = 2\mathbf{i} - 3\mathbf{j}$

   b) $\mathbf{v} = \overrightarrow{AB}$ with coordinates A(–3, –4) and B(–5, 2)

   c) $\mathbf{v} = 6\mathbf{u}$, where $\mathbf{u} = -0.6\mathbf{i} + 0.8\mathbf{j}$

7. A helicopter flying 50 mph with a heading azimuth of 315° encounters a wind of 10 mph, blowing in the direction of 30° north of east. The wind has the effect of blowing the helicopter slightly off course.

   a) Resolve the two vectors into linear combinations of the standard basis vectors **i** and **j**.

   b) Find the resultant vector, and interpret it in terms of its magnitude and direction.

   c) The pilot of the helicopter would like to maintain a *resultant* velocity of 50 mph along an azimuth of 315°. Taking into consideration the wind velocity, what air speed and heading should she set?

Trigonometry offers more direct ways to find the magnitude and direction of a resultant when two vectors do not act at right angles. Exercises 8 focuses on magnitude; Exercises 9 and 10 develop methods to determine direction angles. Subsequent exercises apply these results.

8. Consider the vectors **v** having length 20 and direction 0°, and **w** having length 10 and direction 30° (see **Figure 6.41**).

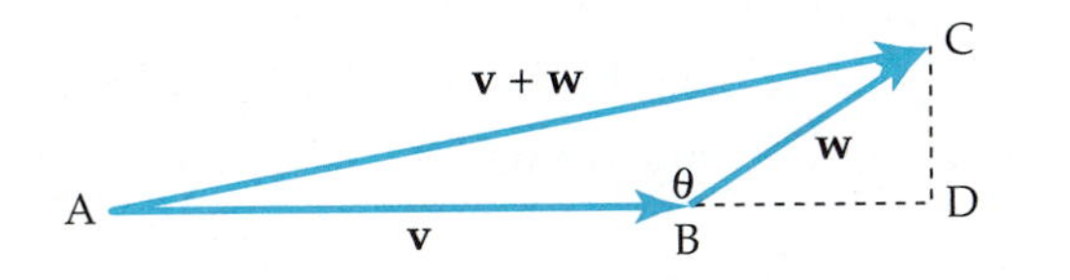

FIGURE 6.41. Representation of situation in Exercise 8.

   a) Use the **Law of Cosines** to find the value of $|\mathbf{v} + \mathbf{w}|$.

   b) Write a general formula for finding $|\mathbf{v} + \mathbf{w}|$ from $|\mathbf{v}|$, $|\mathbf{w}|$, and θ.

   c) Revisit Exercise 7(b) and apply the Law of Cosines to find the resultant speed. Explain how you found the measure of the needed angle.

Exercises 6.4

9. An interesting algebraic simplification develops when you apply the Law of Cosines to vectors that are expressed in terms of scalar combinations of basis vectors. To simplify things, **Figure 6.42** has two concurrent vectors with their tails at the origin.

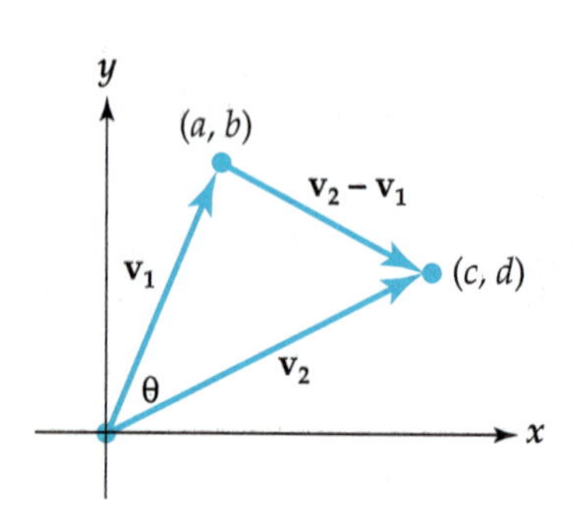

FIGURE 6.42.
Vector diagram for Exercise 9.

   a) Write the vectors $\mathbf{v}_1$, $\mathbf{v}_2$, and $\mathbf{v}_2 - \mathbf{v}_1$ as a scalar combination of basis vectors.

   b) Use the Pythagorean formula to write expressions for each of: $|\mathbf{v}_1|$, $|\mathbf{v}_2|$, and $|\mathbf{v}_2 - \mathbf{v}_1|$.

   c) Substitute these values into the Law of Cosines, treating $\mathbf{v}_1$ as side a, $\mathbf{v}_2$ as side b, and θ as angle C. Simplify your results as much as possible.

   d) If vector $\mathbf{v} = x_1\mathbf{i} + y_1\mathbf{j}$ and vector $\mathbf{w} = x_2\mathbf{i} + y_2\mathbf{j}$, then $\mathbf{v} \bullet \mathbf{w}$, read as "**v** dot **w**," is defined to be $x_1x_2 + y_1y_2$. This is called the **scalar product**, or **dot product** of **v** and **w**. Thus the expression $ac + bd$ that appears in your work for part (c) is the dot product of $\mathbf{v}_1$ and $\mathbf{v}_2$. Substituting into the expression you obtained in part (c), the Law of Cosines can be rewritten as: $\mathbf{v} \bullet \mathbf{w} = |\mathbf{v}||\mathbf{w}|\cos\theta$. Solve this expression for $\cos\theta$.

   e) Use your answer to part (d) to verify from their **i-j** component forms that the angle formed by the two given vectors in Exercise 7(b) is 105°.

10. The **Law of Sines** is useful for finding the direction angle of a resultant after the Law of Cosines has been used to find its magnitude.

B
63°
12
C
38°
A

FIGURE 6.43.
Diagram for use with Exercise 10(a).

   a) Recall from Chapter 5 that the Law of Sines is especially useful when solving a triangle in which two angles and one side are known. Show how to use the Law of Sines to find the remaining parts of the triangle in **Figure 6.43**.

   b) Explain how to find the direction angle for the resultant $\mathbf{v} + \mathbf{w}$, where vector **v** has magnitude 25 and direction angle of 130°, and **w** has magnitude 10 and direction angle of 55°.

Exercises 6.4

11. A fire department helicopter flies from its helicopter pad a distance of 50 km on an azimuth of 30° (60° north of east) to collect water from a pond. Then it flies 40 km due southwest to drop the water on a fire. Work with an xy-coordinate system with the helicopter pad at the origin, the x-axis pointing east, and the y-axis pointing north.

    a) Find the coordinates of the pond.

    b) At what point is the fire located?

    c) What heading and distance must the pilot fly to return to the helicopter pad?

Corbis

12. Two tugboats exert forces **w** and **v** on a luxury liner. Determine the resulting force on the luxury liner if **w** has a magnitude of 3000 newtons and acts in the direction of 150°, and **v** has a magnitude of 2000 newtons and acts in the direction of 210°. (Measure angles using the mathematical convention.)

13. A tow truck is attempting to retrieve a car stuck in the mud. The truck must operate from firm ground that is above and in front of the car. Past experience has taught that a horizontal force of about 1400 pounds will cause the truck to roll forward. A vertical force of 1800 pounds will lift the front wheels. A winch can exert 2000 pounds at an angle of 25° from the horizontal. Is the force sufficient to roll the truck forward? Lift the front wheels?

14. In Example 4, a helicopter heading east at 50 mph encounters a 10 mph wind that is blowing toward due north. Describe the resultant velocity of the helicopter as:

    a) a linear combination of basis vectors

    b) a scalar multiple of a unit vector expressed in terms of the direction angle

    c) a scalar multiple of a unit vector expressed in terms of the standard basis components in part (a).

15. **Investigation.** Vectors provide an alternative way to develop geometric relationships. Rather than using axioms (or theorems proven from those axioms), you use definitions and properties of vectors to develop similar arguments.

    Use vectors to prove that the midpoints of a quadrilateral form a parallelogram.

LESSON 6.5

Vector and Parametric Equations in Two Dimensions

You have examined vector operations both geometrically and symbolically. This lesson extends that symbolic work to vector equations.

Activity 6.5

Part 1: Exploration

NOAA

It is forest fire season, and helicopters are to be used to search for outbreaks of fire after an electrical storm. Two helipads will be used, each with five helicopters. One of the helipads is located at the origin O, while the other is at the point P(8, 10).

1. Each of the helicopters launched from the origin travel 2 miles in specified directions: Al — due east, Maria — due northeast, Carl — due north, Diego — due northwest, and Erica — due west. Use position vectors to describe each of their locations after traveling 2 miles.

2. The helicopters from the second helipad are also sent out with predetermined search patterns: Fran—due southeast, George—due south, Hank—due southwest, Katrina—due west, and Jack—along azimuth 330°. Use position vectors to describe each of their locations after traveling 2 miles.

3. Examine helicopter Diego's movement in more detail. His speed is 60 mph, and the traffic controller at helipad O wishes to track his position over time.

   a) Determine Diego's position vectors and locations for the first 5 minutes. Record your work in a copy of **Table 6.5.**

b) Describe the graph that is formed by successive locations as time increases. Explain how you can tell the direction of motion from examining the various position vectors in your table.

t (min)	Position Vector	Location
0		
1		
2		
3		
4		
5		

TABLE 6.5. Table for Item 3.

c) Write the velocity vector for Diego, using miles per minute. How is the velocity vector visible in your table from part (a)?

d) Write a **vector equation** for this situation—one that describes the various position vectors over time. Also, write an x-y equation that describes Diego's locations.

e) Rewrite your answer for the vector equation in part (d) as two component equations, with each equation acting in the direction of one of the standard basis vectors.

f) In the helicopter context, $t \geq 0$ is a reasonable domain. What would be the effect on the graph if the domain were extended to $-\infty < t < +\infty$?

4. Examine Jack's helicopter movement in more detail. He also travels at 60 mph, and the traffic controller at helipad O also wishes to track his position over time.

a) Determine Jack's position vectors and locations for the first 5 minutes; record your work in a copy of **Table 6.6**.

t (min)	Position Vector	Location
0		
1		
2		
3		
4		
5		

TABLE 6.6. Table for Item 4.

b) Describe the graph of these locations over time. From the table of position vectors, how can you determine Jack's velocity vector (using miles per minute)?

c) Again, write a vector equation that describes the position vectors in this situation, and an x-y equation that describes locations.

d) Rewrite your answer for the vector equation in part (c) as two component equations, one for each standard basis direction.

Part 2: Investigation

A vector equation can be separated into two parts that specify the x and y locations separately. For example, for the vector equation $\mathbf{r} = t(3\mathbf{i} - 5\mathbf{j})$, the equations $x = 3t$ and $y = -5t$ specify the x and y locations. The equations $x = 3t$ and $y = -5t$ are called parametric equations of the line.

Rewrite each of the following vector equations in parametric form, then use your calculator's parametric graphing mode to obtain a graph. For each equation, sketch the graph, indicate the WINDOW needed, and discuss the relationship between the equation and the features of the graph.

1. $\mathbf{r} = t(2\mathbf{i} + 3\mathbf{j}),\ 0 \le t \le 5$
2. $\mathbf{r} = \mathbf{i} + t(-2\mathbf{i} + 4\mathbf{j}),\ 0 \le t \le 5$
3. $\mathbf{r} = -2\mathbf{j} + t(2\mathbf{i} + 3\mathbf{j}),\ -6 \le t \le 6$
4. $\mathbf{r} = -4\mathbf{i} + 2\mathbf{j} + t(2\mathbf{i} + 3\mathbf{j}),\ 0 \le t \le 5$

POSITION VECTORS

In the previous two lessons you used vectors to represent quantities with magnitude and direction, regardless of their locations in space. However, many applications ask questions that emphasize location as well. In predicting whether a drive from a golf tee lands on the green, a fly ball clears the fence for a home run, or a hurricane will make landfall in a metropolitan area, one needs to specify a location in order to obtain a solution. In order for a satellite to escape the earth's atmosphere and be placed in a stationary orbit, engineers determining velocity and direction for the rocket launch must include a location as an initial condition.

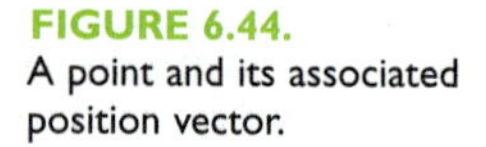

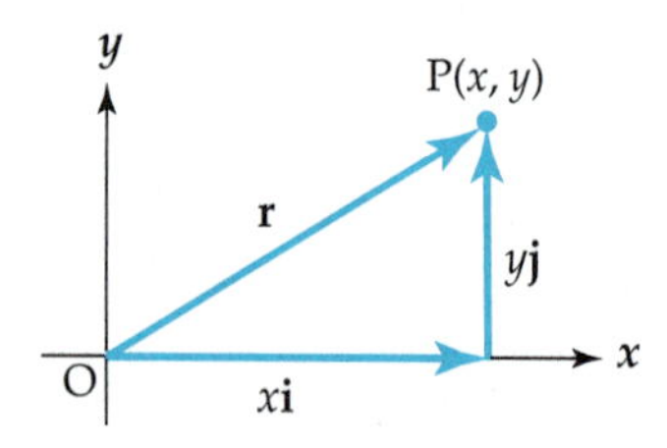

FIGURE 6.44. A point and its associated position vector.

To work with situations like the ones described, vectors are associated with particular positions. Recall that a **position vector** is defined as a directed line segment beginning at the origin. Thus, in **Figure 6.44**, vector $\mathbf{r} = \overrightarrow{OP} = x\mathbf{i} + y\mathbf{j}$ is the position vector to point $P(x, y)$.

VECTOR EQUATION FOR A LINE

In Activity 6.5, you tracked the paths of two particular helicopters over time. In both cases, you needed to know the flight path, which was expressed as a directional vector. As a function of time, that path traced out a linear graph, since the direction never changed. In general, any line can be specified by indicating one position on the line and the direction along which that line lies. One familiar equation for specifying this information is the slope-intercept form. Vectors are also convenient tools for specifying each of these features. In **Figure 6.45** the position vector **p** locates the point P(–3, 2) on the line, and the direction vector **v** = 2**i** + 1**j** determines the line's direction.

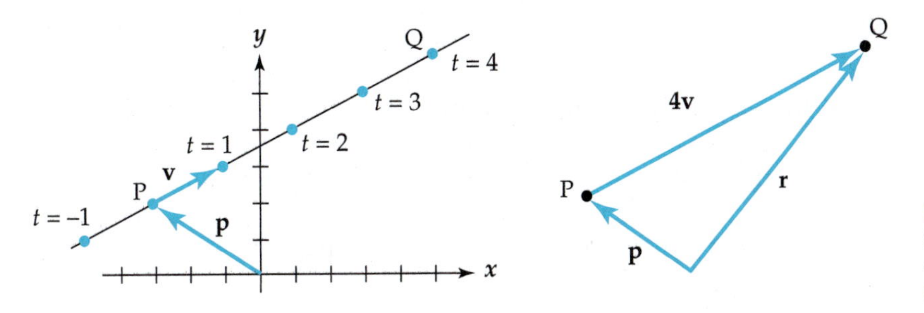

FIGURE 6.45.
Scalar multiples of a direction vector locating positions along a line.

For each scalar, t, the vector $t\mathbf{v}$ "stretches" the direction vector **v**. Maintaining its tail at point P, the head of the scalar multiple $t\mathbf{v}$ "points" to another location on the line. The position vector to that location is the sum of the two vectors **p** and $t\mathbf{v}$. Note that the vector $t\mathbf{v}$ defines Δx and Δy from P to the new point. For example, Figure 6.46 shows the position vector **r** for point Q as the position vector **p** plus the vector 4**v**. Every point on this line has a position vector of the form $\mathbf{r} = \mathbf{p} + t\mathbf{v}$. Thus, this is called the vector equation for this line.

In general, the vector equation for a line having direction vector $\mathbf{v} = c\mathbf{i} + d\mathbf{j}$, and going through a point P(a, b) may be written as:

$$\mathbf{r} = \mathbf{p} + t\mathbf{v} = (a\mathbf{i} + b\mathbf{j}) + t(c\mathbf{i} + d\mathbf{j}) = (a + ct)\mathbf{i} + (b + dt)\mathbf{j}, \text{ with } -\infty < t < +\infty.$$

Note that the independent variable is t, a scalar, and the dependent variable is **r**, a position vector. At times you may need to restrict the scalars, t, that may be used. Restricting scalars to a fixed interval $a \le t \le b$ produces a line segment, whereas using all possible scalar values $-\infty < t < +\infty$ generates a continuous line in both directions. Unless otherwise noted, assume that all real values of t are permitted.

EXAMPLE 8

Find a different vector equation for the line shown in Figure 6.45.

SOLUTION:

The point (–1, 3) is also on the line. Therefore,
$\mathbf{r} = (-1\mathbf{i} + 3\mathbf{j}) + t\cdot(2\mathbf{i} + 1\mathbf{j}) = (-1 + 2t)\mathbf{i} + (3 + 1t)\mathbf{j}$,
with $-\infty < t < +\infty$, is also a valid equation.

DISCUSSION/REFLECTION

1. Adjust the vector equation in Example 8 to describe:

   a) a ray that starts at point P and goes through Q.

   b) a line segment that starts at the point Q and ends at coordinates (–5, 1).

2. Write the general form for the vector equation of a line through the origin.

The components of a position vector may be thought of as specifying x and y locations separately. Thus, vector equations are closely related to parametric equations. In fact, the vector equation
$\mathbf{r} = a\mathbf{i} + b\mathbf{j} + t(c\mathbf{i} + d\mathbf{j})$ is equivalent to the parametric form:

$x = a + ct$ and $y = b + dt$, where $-\infty < t < +\infty$.

EXAMPLE 9

Write the vector equation $\mathbf{r} = 8\mathbf{i} + 9\mathbf{j} + t(2\mathbf{i} + 5\mathbf{j})$, with $-\infty < t < +\infty$, in both parametric form and in x-y form (no t's).

SOLUTION:

$\mathbf{r} = 8\mathbf{i} + 9\mathbf{j} + t(2\mathbf{i} + 5\mathbf{j}) = (8 + 2t)\mathbf{i} + (9 + 5t)\mathbf{j}$.

So, the parametric equations for this line are:

$x = 8 + 2t$

$y = 9 + 5t.$

Since $t(2\mathbf{i} + 5\mathbf{j})$ represents Δx and Δy from (8, 9), then the slope must be $(5t)/(2t)$, or 5/2. Thus the equation in point-slope form is:

$y - 9 = \frac{5}{2}(x - 8).$

Note that the control numbers for the line are *clearly* visible in the vector, parametric, and point–slope forms.

VECTOR EQUATIONS OF TRAJECTORIES

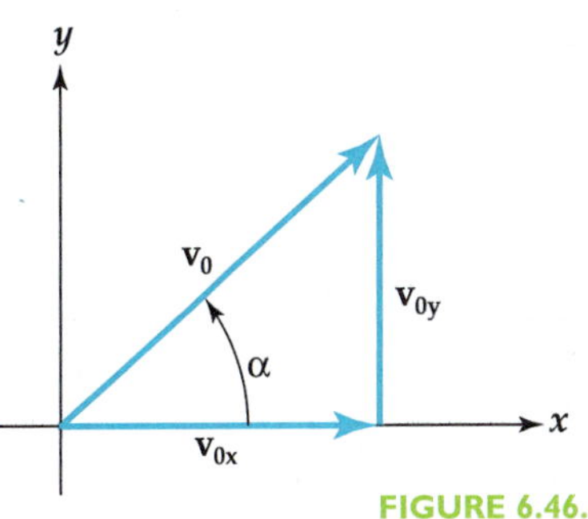

FIGURE 6.46.
A trajectory launched with velocity magnitude |v0| and angle a.

In practice, vector equations are the models of choice for many applications, including ballistics, weather tracking, and animation. To give you a flavor for constructing models using vectors, equations will be introduced to model trajectories. Suppose a projectile (such as a golf ball or baseball) is launched with a velocity vector $\mathbf{v_0}$ with magnitude $|\mathbf{v_0}|$ at a launch angle α, as shown in **Figure 6.46**.

The initial velocity vector $\mathbf{v_0}$ can be resolved into components $\mathbf{v_{0x}}$ and $\mathbf{v_{0y}}$, where $\mathbf{v_{0x}} = (\cos\alpha)|\mathbf{v_0}|\mathbf{i}$ and $\mathbf{v_{0y}} = (\sin\alpha)|\mathbf{v_0}|\mathbf{j}$. A vector equation describing position vectors $\mathbf{r}$ can model the trajectory at any time t after the initial launch. If the velocity remained constant (no forces of propulsion or resistance) the motion would be linear. Therefore, multiplying the constant velocity by the elapsed time t gives the position vector $\mathbf{r}$,

$\mathbf{r} = x\mathbf{i} + y\mathbf{j} = (|\mathbf{v_{0x}}|t)\mathbf{i} + (|\mathbf{v_{0y}}|t)\mathbf{j}$, with $t \geq 0$.

© Susan Van Etten

Even though air resistance and other forces affect the trajectory, suppose for the moment that the only force acting on the projectile after launch is gravity, which affects only the vertical position (since it decreases the upward velocity). Since the projectile may not have been launched from the origin, the general equation describing its location is:

$\mathbf{r} = x\mathbf{i} + y\mathbf{j} = (x_0 + |\mathbf{v_{0x}}|t)\mathbf{i} + (y_0 + |\mathbf{v_{0y}}|t - 0.5gt^2)\mathbf{j}$, with $t \geq 0$,

where g represents the acceleration due to gravity, the point $P(x_0, y_0)$ is the location from which the projectile is launched, and $t \geq 0$ measures the time after launching. It looks bad, but is similar in form to the vector equation for a line, with an extra term to describe how the vertical component strays from the line. If t is measured in seconds, distance in feet, and velocity in feet per second, the model becomes:

$$\mathbf{r} = x\mathbf{i} + y\mathbf{j} = (x_0 + |\mathbf{v_0}|\cos(\alpha)t)\mathbf{i} + (y_0 + |\mathbf{v_0}|\sin(\alpha)t - 16t^2)\mathbf{j}, \text{ with } t \geq 0,$$

where the launch point is $P(x_0, y_0)$, and initial velocity vector is $\mathbf{v_0}$, having direction angle α, assuming that gravity is the only force acting after launch.

EXAMPLE 10

Tiger Woods looks at a green that extends from 125 yards (375 feet) away to 140 yards (420 feet) away. Directly between him and the pin is a tall pine tree, which his caddie tells him is 50 yards (150 feet) from the ball. The caddie, Fluff Cowan, estimates the height of the tree at 50 feet. Tiger, a former Stanford student, carefully selects a golf club and golf ball which, coupled with his finely honed swing, will result in an initial velocity of 120 feet per second (about 82 mph). If he hits the ball perfectly, it will depart with a launch angle of $\alpha = 30°$. (Now you know what Tiger and Fluff are talking about between shots!) Does the ball hit the green? Does it clear the tree?

SOLUTION:

Refer to **Figure 6.47**. The frame of reference for the problem is a choice determined by the modeler. A convenient choice places the origin at the ball. That makes $x_0 = 0$ and $y_0 = 0$.

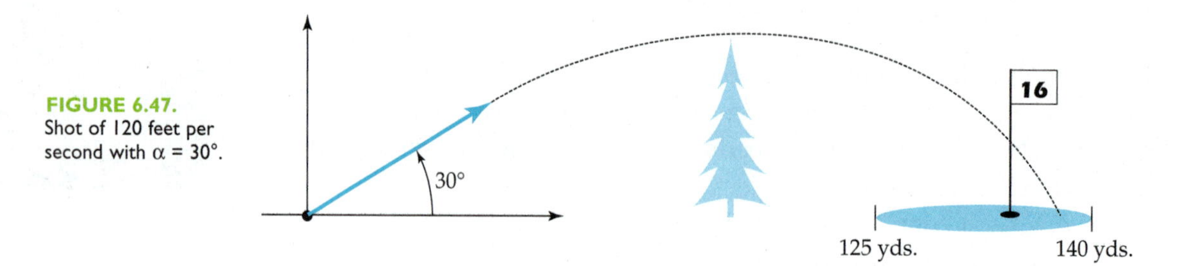

FIGURE 6.47. Shot of 120 feet per second with $\alpha = 30°$.

With $|\mathbf{v}_0| = 120$ and $\alpha = 30°$, substitute into the trajectory model equation:

$\mathbf{r} = x\mathbf{i} + y\mathbf{j} = (120 \cos 30°)t\mathbf{i} + [(120 \sin 30°)t - 16t^2]\mathbf{j},\ t \geq 0.$

Simplifying and approximating gives:

$\mathbf{r} = x\mathbf{i} + y\mathbf{j} \approx 103.9t\mathbf{i} + (60t - 16t^2)\mathbf{j},\ t \geq 0.$

Figure 6.48 shows the graph of this equation. Use TRACE to estimate where the ball hits the ground, and what the height of the trajectory is when $x = 150$. For example, when $t = 3.7$ sec, $x = 384.43$ ft and $y = 2.96$ ft. Therefore, one can expect that the shot travels about 385 feet.

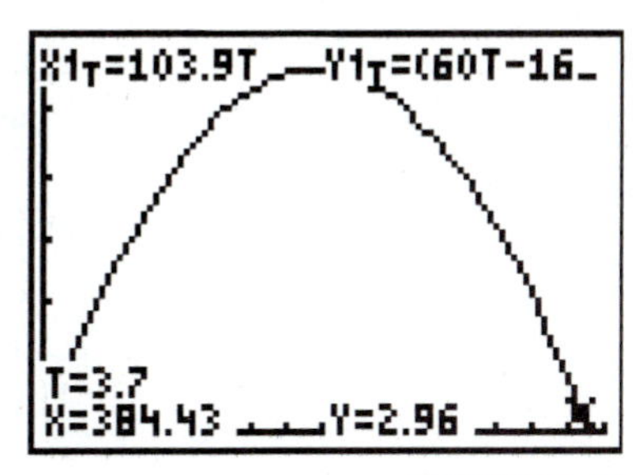

FIGURE 6.48. Graph of $\mathbf{r} = x\mathbf{i} + y\mathbf{j} \approx 103.9t\mathbf{i} + (60t - 16t^2)\mathbf{j},\ t \geq 0.$

Algebra is required to verify this approximation. The ball hits the ground when $y = 0$ for $t > 0$. Substituting $y = 0$ gives:

$y = 60t - 16t^2 = t(60 - 16t) = 0,$

which has two roots: $t = 0$ and $t = 3.75$ seconds. The first root corresponds to the launch position, while the second corresponds to the landing point. Now determine x when $t = 3.75$ seconds:

$x \approx 103.9t = 103.9(3.75) = 389$ feet.

So the ball lands about 130 yards away; Tiger will be on the green if he clears the tree.

To determine whether it clears the tree (150 feet away), first find t when $x = 150$:

$150 \approx 103.9t$, so $t \approx 150/103.9 = 1.44$ seconds.

Next, find the height of the ball when $t = 1.44$ seconds:

$y = 60t - 16t^2 = 60(1.44) - 16(1.44)^2 \approx 53.4$ feet.

Since the tree is estimated to be 50 feet high, Tiger needs to hope his caddie is right *and* hit the ball almost perfectly!

Exercises 6.5

1. Consider again the form for the vector equation for a line and the manner in which it conveys information about that line.

   a) In what way is a vector equation related to constructing a number line?

   b) Compare the vector form for lines to the parametric and point-slope forms of lines.

   c) What information is more easily obtained from a vector equation than from the general equation $Ax + By + C = 0$ for a line?

2. The two vector equations $\mathbf{r}_1 = -18\mathbf{i} + 13\mathbf{j} + t(2\mathbf{i} - \mathbf{j})$, $0 \le t \le 10$, and $\mathbf{r}_2 = 2\mathbf{i} + 3\mathbf{j} + t(-2\mathbf{i} + \mathbf{j})$, $0 \le t \le 10$, describe the same path of motion.

   a) Identify the difference between the motions described by the two equations.

   b) Adjust $\mathbf{r}_1$ to make a new equation $\mathbf{r}_3$ that describes the exact same motion as $\mathbf{r}_2$, but over a different time interval.

   c) Adjust $\mathbf{r}_2$ to make a new equation $\mathbf{r}_4$ that describes the same motion as $\mathbf{r}_1$, but over a different time interval.

3. There are many ways to express the vector equation for a particular line. **Table 6.7** contains values corresponding to the vector equation $\mathbf{r} = (1 + 3t)\mathbf{i} + (3 + 2t)\mathbf{j}$, $-\infty < t < +\infty$.

t	x	y
0	1	3
1	4	5
2	7	7
3	10	9
4	13	11

TABLE 6.7.
Values for the vector equation in Exercise 3.

Tables 6.8–6.10 come from different vector equations that describe the same line. Determine the vector equations they represent, and compare those equations to the original equation.

Exercises 6.5

a) See Table 6.8.

t	x	y
0	1	3
1	10	9
2	19	15
3	28	21
4	37	27

TABLE 6.8.
For Exercise 3(a).

b) See Table 6.9.

t	x	y
0	1	3
1	$1 + 3/\sqrt{13} \approx 1.83$	$3 + 2/\sqrt{13} \approx 3.55$
2	$1 + 6/\sqrt{13} \approx 2.66$	$3 + 4/\sqrt{13} \approx 4.11$
3	$1 + 9/\sqrt{13} \approx 3.50$	$3 + 6/\sqrt{13} \approx 4.66$
4	$1 + 12/\sqrt{13} \approx 4.33$	$3 + 8/\sqrt{13} \approx 5.22$

TABLE 6.9.
For Exercise 3(b).

c) See Table 6.10.

t	x	y
0	–8	–3
1	–5	–1
2	–2	1
3	1	3
4	4	5

TABLE 6.10.
For Exercise 3(c).

4. Change each of the following equations for a line into vector form.

a) $y = 1.5x$

b) $x = -4t,\ y = 2t$

c) $3x + 5y = 14$

d) $x = -3 - 2t,\ y = -5 + 3t$

Exercises 6.5

5. Change each of the following equations for a line into parametric form.

   a) $y = \frac{-3}{4}x$

   b) $\mathbf{r} = -2t\mathbf{i} + 3t\mathbf{j}$

   c) $\mathbf{r} = (-2 + 5t)\mathbf{i} + (4 - 2t)\mathbf{j}$

   d) $y - 4 = \frac{2}{3}(x + 1)$

6. Change the following equations for a line into general form, $Ax + By + C = 0$.

   a) $\mathbf{r} = -3t\mathbf{i} + 4t\mathbf{j}$

   b) $x = 0.5t$, $y = -0.25t$

   c) $x = 3 - 2t$, $y = -3 + t$

   d) $\mathbf{r} = (-1 + 3t)\mathbf{i} + (2 - 5t)\mathbf{j}$

7. Find vector equations for the lines containing the given point that are parallel and perpendicular to the given line. For each pair of lines, find the dot product of the respective direction vectors and describe your observations.

   a) $\mathbf{r} = (-2 + t)\mathbf{i} + (7 - 3t)\mathbf{j}$; $(4, -7)$

   b) $x = 2 - 3t$, $y = 5 + 2t$; $P(-2, 6)$

8. Consider the line described by the vector equation $\mathbf{r} = (-3 + 4t)\mathbf{i} + (2 - 3t)\mathbf{j}$

   a) What is the slope of the line?

   b) What angle does the line form with the positive x–axis? Explain.

   c) Write parametric equations for this line. Solve each of those equations for t, and set the two expressions equal to each other but do not simplify any further. With respect to the original vector equation, interpret the numbers appearing in this new equation.

   d) Interpret your equation in part (c) as a transformation (scale change and translation) of the tool kit equation $y = x$.

The form $\frac{x-h}{c} = \frac{y-k}{d}$ is called the **symmetric form**. The control numbers c and d are called the **direction numbers** for the line.

9. The general vector equation for a line is
   $\mathbf{r} = \mathbf{p} + t\mathbf{v} = (a + ct)\mathbf{i} + (b + dt)\mathbf{j}, -\infty < t < +\infty$.

   a) Write a formula for finding the angle that the line forms with the positive x–axis.

   b) Rewrite the equation for the line in symmetric form.

   c) Determine the symmetric form for a line passing through (a, b) and making an angle of θ with the positive x–axis.

© Susan Van Etten

10. A computer programmer wishes to animate two dots in straight-line paths across the screen in such a way that a collision takes place. Assume the lower left-hand corner is the origin, and coordinates are referenced as points in Quadrant I of the Cartesian plane. The programmer uses the vector equations $\mathbf{r}_1 = (3t)\mathbf{i} + (20 + t)\mathbf{j}$ and $\mathbf{r}_2 = (15 + 2t)\mathbf{i} + (3t)\mathbf{j}$ for the animation.

    a) Does a collision take place? Explain.

    b) Adjust the equations so that a collision takes place at $t = 12$. Explain your method.

    c) Adjust the equations so that a collision takes place at coordinates (48, 36). Explain your method.

11. In an attempt to avoid capture, a ship at location (30, 30) trying to smuggle drugs takes off on a heading of 180° (mathematical reference system) at 20 miles/hour. As shown in **Figure 6.49**, Coast Guard cutters at locations (30, 60) and (30, 0) take up the chase. In an effort to trap the smugglers, ship C_1 travels on a heading of 210° at 30 miles/hour, while C_2 heads 120° at 10 miles/hour.

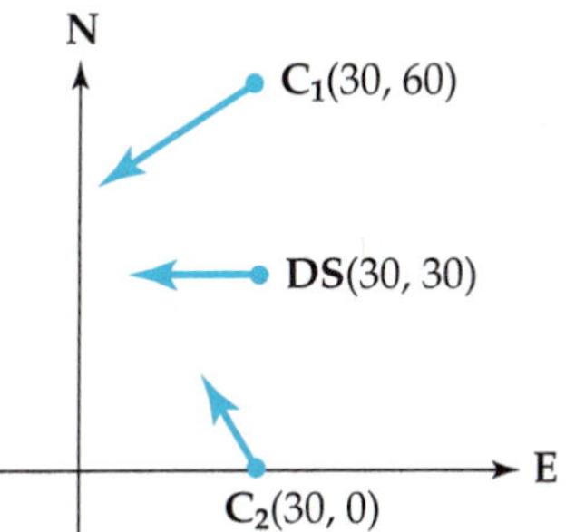

FIGURE 6.49.
Initial positions and directions for Exercise 11.

    a) If the smugglers simply continue on their path, will Coast Guard cutter C_1 be able to intercept the escape path in time?

    b) The smugglers realize that they can't simply outrun the Coast Guard, so after 1 hour, they change direction and head due south. Coast Guard cutter C_2 reacts by changing its direction to heading 240°. Will that Coast Guard ship catch the smugglers?

Exercises 6.5

12. Refer back to Tiger Woods' shot in Example 10. Keep the swing speed 120 feet/second.

    a) Suppose the launch angle is changed to 35°. Does the ball hit the green?

    b) At a launch angle of 35° does the shot clear the tree?

    c) Consider all possible launch angles. Find *all* angles whose trajectories clear the tree.

    d) What range of launch angles result in trajectories that both clear the tree *and* land on the green?

    e) If you want to give Tiger the largest margin of error on the launch angle for the shot, would it be better to ease up a little on the swing velocity or keep it at 120 feet/second? Explain.

13. How far can Mark McGwire hit a baseball if he hits it perfectly? Husson Furst, a famous baseball fan who collected a lot of data from his bleacher seat in Busch Stadium, estimates that if McGwire hits a fastball perfectly, the initial velocity will be about 100 mph. Assume that the only force acting on the ball after it is hit is gravity.

    a) If McGwire hits the fastball perfectly with a launch angle of 30°, how far will it go?

    b) Suppose he hits it to straightaway center field from an initial height of 3 feet, the wall is 440 feet away, and the top of the fence at that point is 15 feet high. Does the 30° launch angle give McGwire another home run?

14. Suppose that Mark McGwire hits it well, but the launch angle is slightly lower than in Exercise 13.

    a) Repeat the analysis from Exercise 13 with trajectory angles of 28° and 25°. Will those line drives become home runs?

    b) With an initial velocity of 100 mph, a launch angle of 21° and a launch angle of 69° will both produce the same distance for the ball—over 450 feet. Compare their actual trajectories. Will they produce similar results for a slugger like Mark McGwire?

    c) What angle will produce the *greatest* distance for a fly ball?

Exercises 6.5

15. Suppose McGwire hits the ball so that the initial velocity is less than 100 mph.

    a) Repeat the analysis from Exercise 13(a), with a launch angle of 30° and initial speeds of 90 mph and 75 mph.

    b) A promotional scheme at some stadiums is to have scoreboards announce "The Tale of the Tape," a computer estimate of how far a home run was hit, no matter where it lands. What are some of the problems encountered in determining such a measure? Suggest ways to calculate the distance a home run is hit.

© Susan Van Etten

16. **Investigation.** Based on the work done in Exercises 13–15, would you advise a track coach who works with the athletes in the shot put event to concentrate on form or on strength? Explain.

17. Rewrite each of the following vector equations in parametric mode. Then use your calculator to obtain its graph. Describe each graph, and explain its shape if possible. Note: Use square units in your graphing window, and radian mode.

    a) $\mathbf{r} = (4\cos t)\mathbf{i} + (4\sin t)\mathbf{j},\ 0 \le t \le 2\pi$

    b) $\mathbf{r} = (2\cos t)\mathbf{i} + (5\sin t)\mathbf{j},\ 0 \le t \le 2\pi$

    c) $\mathbf{r} = 3\mathbf{i} + 5\mathbf{j} + (2\cos t)\mathbf{i} + (2\sin t)\mathbf{j},\ 0 \le t \le 2\pi$

LESSON 6.6

Vector Equations in Three Dimensions

This chapter has examined a variety of coordinate systems and methods for solving problems dealing with motion in the plane. However, the ability to model the real world requires the capacity to describe location and motion in three dimensions. This lesson extends coordinates and vectors to three dimensions.

Activity 6.6

Part 1: Three-Dimensional Graphs in Isometric Perspective

The usual convention for representing three-dimensional coordinate systems in mathematics is to think of the x- and y-axes as defining the floor, with the z-axis pointing straight up. You may encounter other representations in physics or other courses.

The easiest way to begin building a sense of three-dimensional drawings and graphs is with the use of isometric dot paper. **Figure 6.50** illustrates the process.

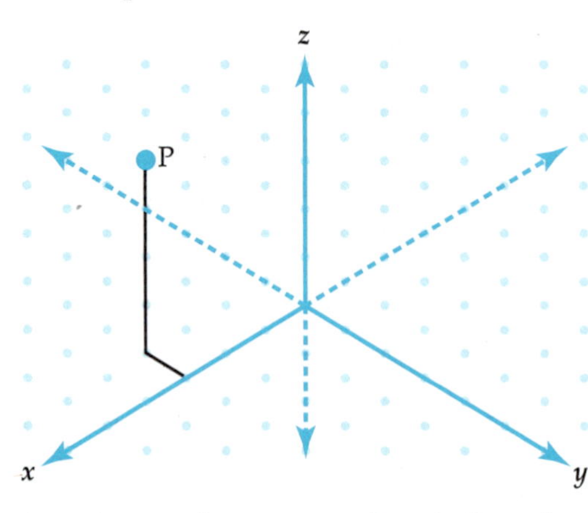

FIGURE 6.50. Three-dimensional coordinate system on isometric dot paper

Drawing in this system is easy because each axis lies along a row of dots, making points easy to interpret as locations in space. For example, Figure 6.50 shows the point P(3, –1, 4) as three dots along the positive x-axis (drawn solid), one unit parallel to the negative y-axis direction, and four units parallel to the positive z-axis direction.

In three-dimensional settings, it is customary to draw lines and segments that lie behind coordinate planes (or parts of objects) as dotted, or lighter.

1. Copy the coordinate axes shown in Figure 6.50, or use one of the grids provided in **Handout H6.4**.

   a) Draw the points contained in **Table 6.11** on a single coordinate system. Describe the shape produced by connecting the pairs AB, AC, AE, BD, BF, CD, CG, DH, EF, EG, FH, and GH.

Label	x	y	z	P(x, y, z)
A	1	–3	–1	(1, –3, –1)
B	1	–3	2	(1, –3, 2)
C	1	2	–1	(1, 2, –1)
D	1	2	2	(1, 2, 2)
E	–2	–3	–1	(–2, –3, –1)
F	–2	–3	2	(–2, –3, 2)
G	–2	2	–1	(–2, 2, –1)
H	–2	2	2	(–2, 2, 2)

TABLE 6.11. Three-dimensional points (isometric perspective)

   b) How might you identify the shape just from the values given in Table 6.11?

   c) What difficulties did you encounter in locating the points on an isometric grid? Can you visualize precisely the three-dimensional object whose outline is determined by the points?

2. Make another sketch of the coordinate axes shown in Figure 6.50, or use a second grid from your copy of Handout H6.4.

   a) Draw a cube with one corner at the origin, O(0, 0, 0), and the opposite corner being at P(1, 1, 1).

   b) Comment on the appearance of the two points O and P.

   c) Describe how you can use your answer from part (b) to determine the direction from which the viewer observes the cube. Explain.

3. Examine the two drawings you made in Items 1 and 2. Each figure is composed of rectangular faces, which are portions of planes.

a) How are these planar shapes represented in your drawings?

b) Use that fact to sketch a portion of the plane $y = 3$ on a new coordinate grid. (Either draw one last sketch of the coordinate axes shown in Figure 6.50, or use the third grid from your copy of Handout 6.4.) You may wish to recheck Table 4.11 if you have trouble getting started.

c) On the same grid, draw a portion of the plane $z = 2$.

d) Describe the intersection of the plane $y = 3$ and $z = 2$.

e) **Challenge**: What is the equation of that intersection?

Part 2: Three-Dimensional Graphs with Orthogonal Representation

Another way to extend the two-dimensional rectangular coordinate system with which you are familiar is shown in **Figure 6.51**. The positive direction for each axis is the labeled side.

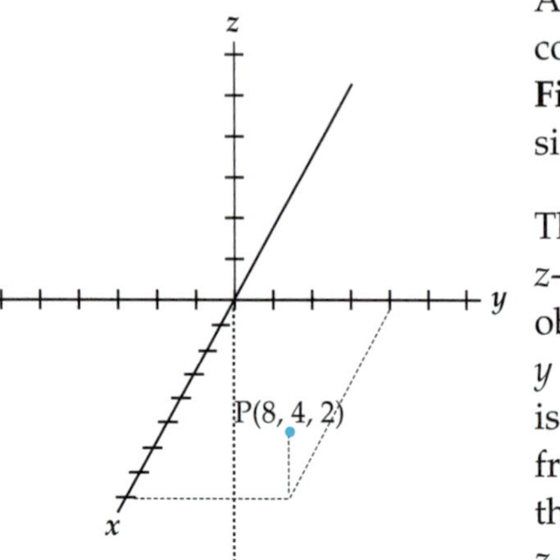

FIGURE 6.51.
Three-dimensional coordinate system and graph of point P.

The idea is that the *yz*-plane (containing the *y*-axis and the *z*-axis) is vertical, with the *x*-axis pointing forward. The observer is above the *xy*-plane, and slightly on the positive *y* side of the *xz*-plane. Visualizing space from a flat drawing is made easier by drawing the *x*-axis at an angle of about 30° from the *z*-axis (see Figure 6.51), and adjusting the length of the *x*-scale to about 70% of the scale length on the *y*- and *z*-axes. Drawing points is assisted by visualizing (or drawing in lightly) a portion of the box whose dimensions match the coordinates of the point to be drawn. Again, see Figure 6.51.

1. Copy the coordinate axes shown in Figure 6.51, or use one of the grids on **Handout H6.5**, and draw the points listed in **Table 6.12** on a single graph.

Label	x	y	z	P(x, y, z)
A	4	0	0	(4, 0, 0)
B	2	2	1	(2, 2, 1)
C	0	4	2	(0, 4, 2)
D	–2	2	3	(–2, 2, 3)
E	–4	0	4	(–4, 0, 4)
F	–2	–2	5	(–2, –2, 5)
G	0	–4	6	(0, –4, 6)

TABLE 6.12.
Three-dimensional points

2. Connect successive points (ABCDEFG) with line segments.

   a) Based on the table values for the three variables x, y, and z, are the points that look collinear really collinear? Explain.

   b) Describe the figure as it would appear from directly overhead.

   c) Describe what an observer standing at the origin would see watching an object traveling along the path.

3. Consider the points as images of an object moving in time. Write equations to describe that motion. Explain.

THREE-DIMENSIONAL COORDINATES

In previous lessons, vectors were used to represent motion, velocity, forces, and other quantities with magnitude and direction in two dimensions. For example, in the case of a Gridville rescue helicopter, only the (x, y) coordinates of the helicopter's position were considered. More typically, the helicopter would take off and land at distinct elevations, and would certainly fly at different altitudes. You also modeled trajectories of baseballs and golf balls, but you assumed that the ball was hit directly to a target without slice or hook. This assumption seems less than realistic!

FIGURE 6.52. Orbiter.

For yet another application, consider the photo of the Orbiter shown in **Figure 6.52**. How does one conveniently locate necessary points in the cargo bay? Can this be done in a manner that makes computing distances fairly easy?

Mathematical descriptions must be extended to three dimensions in order to answer questions of this sort. Conceptually, adding a z-axis to create 3-D just requires a line through the origin, perpendicular to the x- and y-axes. Drawing a two-dimensional coordinate system on your paper, setting it on your desk, and placing your pencil straight up at the origin illustrates such a system. The difficulty arises in *drawing* representative graphs of three-dimensional situations on two-dimensional surfaces such as paper (or computer screens). The use of perspective assists in both the process of making the drawing and in visualizing a situation from a sketch back into the real world. See **Figure 6.53**.

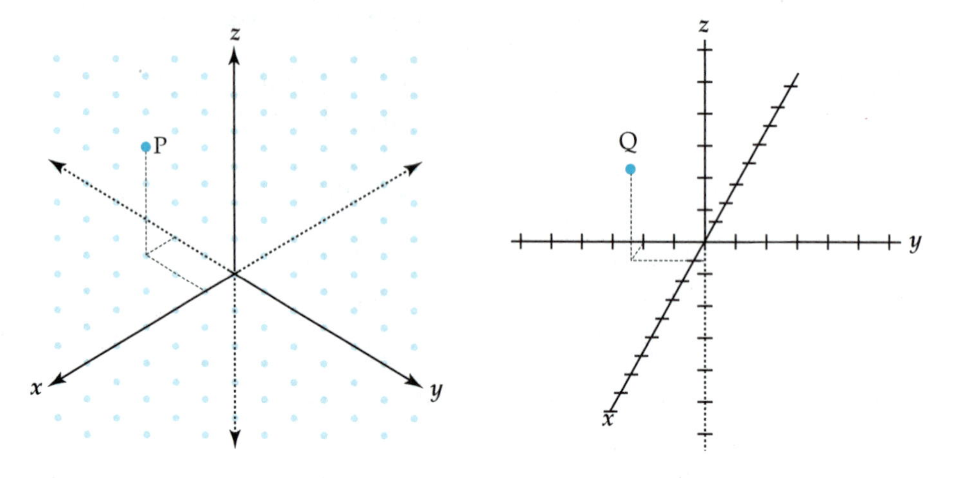

FIGURE 6.53. Isometric and orthogonal perspectives of three dimensions.

Isometric drawings are done on a lattice of dots constructed by making the angle between the positive x- and y-axes 120°. Squares are represented as rhombuses, and the third dimension is created with a z-axis at the same 120° angle to the other two axes. Points are located by going along one axis a certain number of units, then going parallel to the other two axes the appropriate distances. Specifying the coordinates of point P as (1, –2, 3) conveys information about how far (and in which direction) to go, but the actual displacements can be done in any order. The perspective of the viewer is from a point above Quadrant I of the x-y plane, looking toward the origin.

Orthogonal drawings use a vertical z-axis and a horizontal y-axis. That forces the x-axis to come directly out of the page (in reality); that illusion is created by drawing the x-axis at an angle and reducing the scaling on that axis. Point Q in Figure 6.53 also has coordinates (1, –2, 3) for comparison with P. Drawing the "frame" assists in spatial visualization. Again, the perspective in orthogonal drawings is from above Quadrant I of the x-y plane, but closer to the x-axis, and looking toward the origin.

Activity 6.7

Gridville's government office building is, as you might guess, laid out following a strict rectangular pattern. Employees there call it Girderville. The ground floor of the office building is a square 1000 feet by 1000 feet. Along each wall, and every 100 feet there is a hallway. There are 10 floors above the ground floor, each being 10 feet high. (The ground floor pattern is repeated on each floor.) At each intersection of a hallway with either another hallway or an exterior wall, there is a staircase (refer to **Figure 6.54**.)

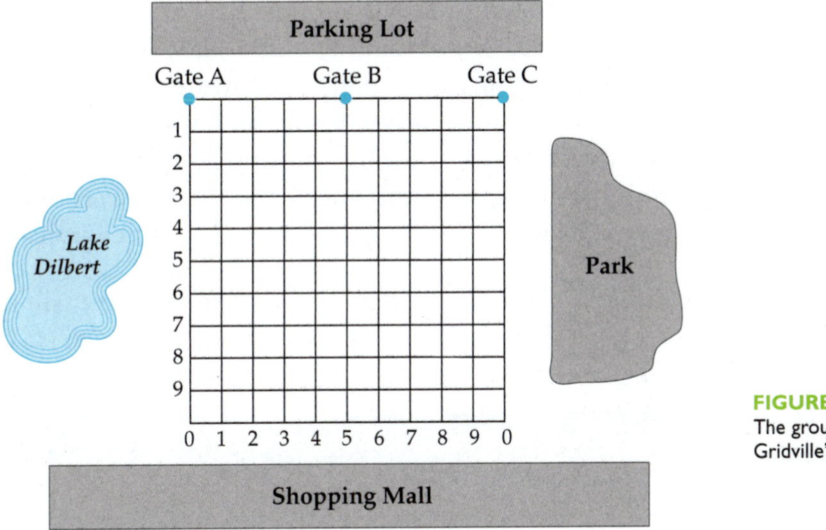

FIGURE 6.54.
The ground floor of Gridville's office building.

Sally, Jose, and Jefferson are friends who have just begun working in the building. Sally parks in the parking lot, enters Gate A, and walks down the hallway toward the shopping mall. At the 5th hallway, she makes a left and walks 3 hallways toward the Park. She then climbs 7 flights of stairs to get to her cubicle. Jose enters Gate B. He walks down the hallway toward the shopping mall. At the 7th hallway, he makes a left and walks 2 hallways toward the Park. He then climbs 4 flights of stairs to get to his office. Jefferson enters Gate C. He walks along the wall toward Lake Dilbert, turning left at the second hallway. He continues 6 hallways toward the shopping mall, then climbs 3 flights of stairs to get to his workspace.

1. Write directions for getting from:

   a) Sally's office to Jose's office

   b) Jose's office to Sally's office

   c) Jefferson's office to Sally's office

2. Sally, Jose, and Jefferson each has a cordless phone that is effective 500 feet from its base, which is located in their respective office. They plan to have a meeting in one of the offices, but would like to be able to monitor calls using their cordless phones. In whose office should they meet, and whose cordless phones will work there?

Sally, Jose, and Jefferson have found service centers on their floors with copiers, faxes, and most importantly, coffee. Each person would like to know which center is closest. From her office, Sally goes 2 hallways toward the shopping mall, then 4 hallways toward the Park, and arrives at Service Center A. From his office, Jose goes toward Lake Dilbert 2 hallways, then 2 hallways toward the Parking Lot, and arrives at Service Center B. From his office, Jefferson goes 2 hallways toward the Park, then 2 hallways toward the Parking Lot, and arrives at Service Center C.

3. Copy **Table 6.13**, and complete it by computing the walking distances from Sally's, Jose's, and Jefferson's offices to each of the three service centers.

TABLE 6.13. Walking distances to three service centers.

Office	Service Center A	Service Center B	Service Center C
Sally's			
Jose's			
Jefferson's			

VECTORS IN SPACE

Previous lessons used directed line segments to represent quantities possessing both magnitude and direction. The development of vectors easily extends to three dimensions.

Standard Basis Vectors

Adding an additional axis creates the need for a basis vector in the z-axis direction. The vectors **i** and **j** still represent the x- and y-axis directions in three dimensions, defined as directed line segments from the origin (0, 0, 0) to the points (1, 0, 0) and (0, 1, 0), respectively. The basis vector in the positive z–direction is the vector **k**, defined to be the directed line segment from the point (0, 0, 0) to the point (0, 0, 1). Scalar multiples of these basis vectors produce components parallel to either the positive or negative direction of the various axes. Geometrically, vector addition is defined as it was in two dimensions, and any vector can be resolved into linear combinations of the standard basis vectors (refer to **Figure 6.55**).

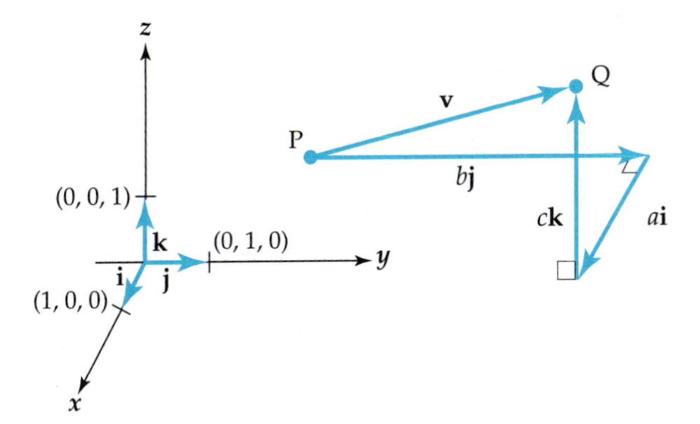

FIGURE 6.55.
Vector $\mathbf{v} = \overrightarrow{PQ}$ resolved into a linear combination of standard basis vectors.

Stretching the basis vectors **i** , **j** and **k** by scalars a, b, and c, and drawing the resulting vectors head-to-tail (in any order) produces vector
$\mathbf{v} = \overrightarrow{PQ} = a\mathbf{i} + b\mathbf{j} + c\mathbf{k}$.

The Position Vector **r**
$\mathbf{r} = \overrightarrow{OP} = x\mathbf{i} + y\mathbf{j} + z\mathbf{k}$

Position Vectors

The **position vector r** from the origin O to a typical point P(x, y, z) is defined in a manner analogous to the two-dimensional case (see **Figure 6.56**):

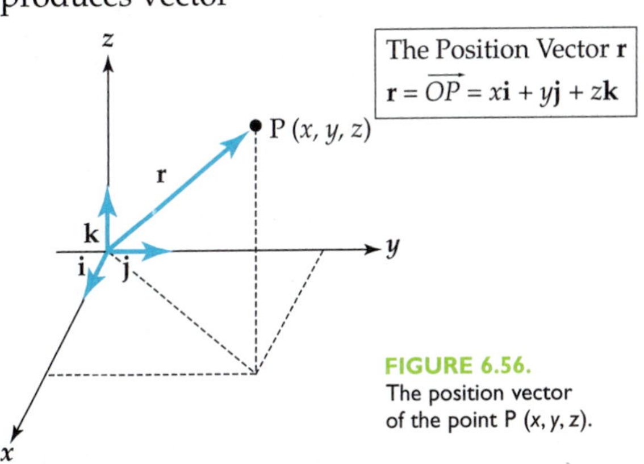

FIGURE 6.56.
The position vector of the point P (x, y, z).

Addition and Subtraction

Vectors are added algebraically by adding their corresponding standard basis components. If $\mathbf{v} = a\mathbf{i} + b\mathbf{j} + c\mathbf{k}$ and $\mathbf{w} = d\mathbf{i} + e\mathbf{j} + f\mathbf{k}$, then
$\mathbf{v} + \mathbf{w} = (a\mathbf{i} + b\mathbf{j} + c\mathbf{k}) + (d\mathbf{i} + e\mathbf{j} + f\mathbf{k}) = (a + d)\mathbf{i} + (b + e)\mathbf{j} + (c + f)\mathbf{k}$.

Similarly, subtraction of three-dimensional vectors may be accomplished component-wise. If $\mathbf{v} = a\mathbf{i} + b\mathbf{j} + c\mathbf{k}$, and $\mathbf{w} = d\mathbf{i} + e\mathbf{j} + f\mathbf{k}$, then
$\mathbf{v} - \mathbf{w} = (a\mathbf{i} + b\mathbf{j} + c\mathbf{k}) - (d\mathbf{i} + e\mathbf{j} + f\mathbf{k}) = (a - d)\mathbf{i} + (b - e)\mathbf{j} + (c - f)\mathbf{k}$.

Directed Line Segment Between Two Points

Just as in two dimensions, the subtraction, final – initial, yields the vector connecting two points. To see this formally, consider points $P_1(a, b, c)$ and $P_2(d, e, f)$, as shown in **Figure 6.57**.

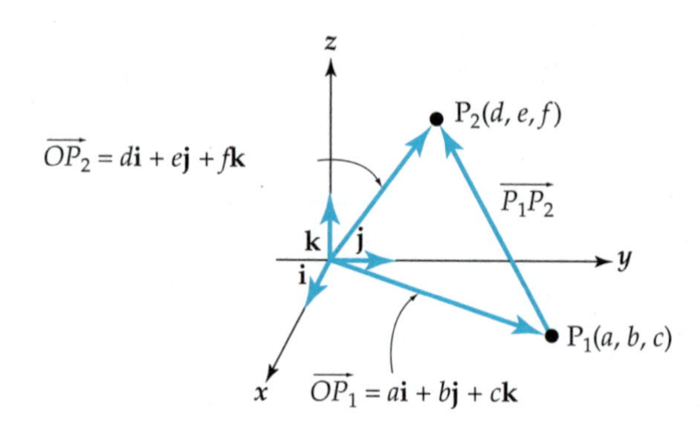

FIGURE 6.57. Vector from Point P_1 to Point P_2.

The position vector to $P_1(a, b, c)$ is $\overrightarrow{OP_1} = a\mathbf{i} + b\mathbf{j} + c\mathbf{k}$. Similarly, the position vector to $P_2(d, e, f)$ is $\overrightarrow{OP_2} = d\mathbf{i} + e\mathbf{j} + f\mathbf{k}$.

Since $\overrightarrow{OP_1} + \overrightarrow{P_1P_2} = \overrightarrow{OP_2}$, subtract to get the vector $\overrightarrow{P_1P_2}$:
$\overrightarrow{P_1P_2} = \overrightarrow{OP_2} - \overrightarrow{OP_1}$.

Substituting,
$\overrightarrow{P_1P_2} = \overrightarrow{OP_2} - \overrightarrow{OP_1} = (d\mathbf{i} + e\mathbf{j} + f\mathbf{k}) - (a\mathbf{i} + b\mathbf{j} + c\mathbf{k}) =$
$(d - a)\mathbf{i} + (e - b)\mathbf{j} + (f - c)\mathbf{k}$.

Thus, the vector from $P_1(a, b, c)$ to $P_2(d, e, f)$ is
$\overrightarrow{P_1P_2} = (d - a)\mathbf{i} + (e - b)\mathbf{j} + (f - c)\mathbf{k}$.

Magnitude of a Vector

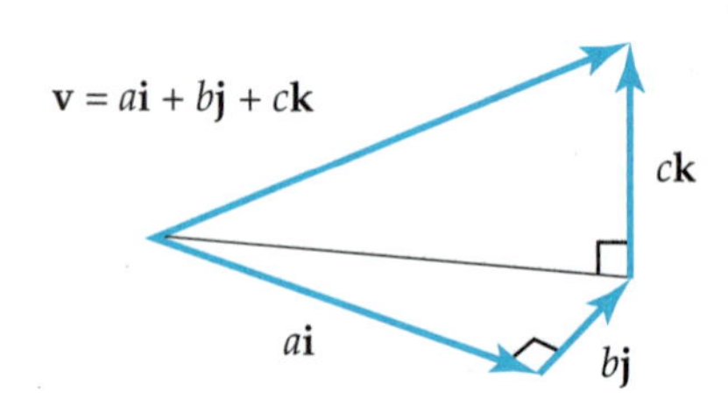

FIGURE 6.58. The magnitude of vector $\mathbf{v} = a\mathbf{i} + b\mathbf{j} + c\mathbf{k}$.

The magnitude of a 3-D vector may be determined from its components much as was done in two dimensions. Consider the vector $\mathbf{v} = a\mathbf{i} + b\mathbf{j} + c\mathbf{k}$ shown in **Figure 6.58.** By twice applying the Pythagorean theorem in the right triangles shown, the magnitude of vector $\mathbf{v} = a\mathbf{i} + b\mathbf{j} + c\mathbf{k}$ is $|\mathbf{v}| = \sqrt{|a|^2 + |b|^2 + |c|^2} = \sqrt{a^2 + b^2 + c^2}$. Remember, the symbol $|\mathbf{v}|$ denotes the magnitude of $\mathbf{v}$. In the exercises, you will have the opportunity to complete the details to show that $|\mathbf{v}| = \sqrt{a^2 + b^2 + c^2}$, where $\mathbf{v} = a\mathbf{i} + b\mathbf{j} + c\mathbf{k}$.

EXAMPLE 11

A helipad is located at O(0, 0, 0). An emergency has occurred on a mountain at point A(3, 5, 0.5). A Regional Hospital is located at B(2, 7, 0.25). The helicopter pilot must fly to A, pick up evacuees, then fly from A to B (see **Figure 6.59**). Express vectors $\overrightarrow{OA}$ and $\overrightarrow{AB}$ in terms of the basis vectors **i**, **j** and **k**. How far does the pilot have to travel from A to B?

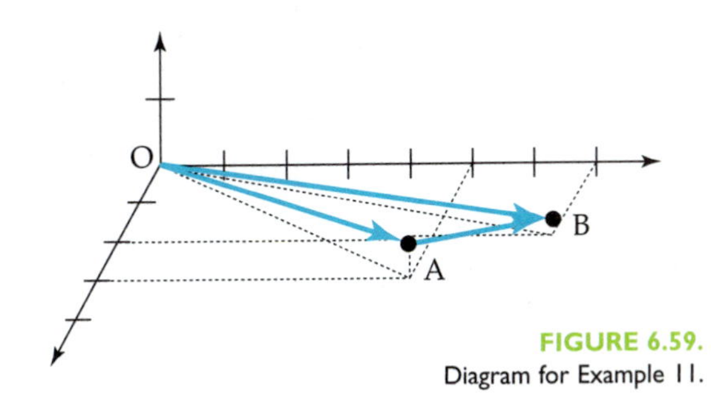

FIGURE 6.59.
Diagram for Example 11.

SOLUTION:

Thinking final – initial, vector $\overrightarrow{AB}$ is $\overrightarrow{OB} - \overrightarrow{OA}$.
Since $\overrightarrow{OA} = 3\mathbf{i} + 5\mathbf{j} + 0.5\mathbf{k}$ and $\overrightarrow{OB} = 2\mathbf{i} + 7\mathbf{j} + 0.25\mathbf{k}$, then
$\overrightarrow{AB} = (2\mathbf{i} + 7\mathbf{j} + 0.25\mathbf{k}) - (3\mathbf{i} + 5\mathbf{j} + 0.5\mathbf{k}) = -\mathbf{i} + 2\mathbf{j} - 0.25\mathbf{k}$.

The vector $\overrightarrow{AB} = -\mathbf{i} + 2\mathbf{j} - 0.25\mathbf{k}$ represents the magnitude and direction the helicopter pilot must fly from point A to point B. To determine how far the pilot must fly, use the formula for finding magnitude:

$|\overrightarrow{AB}| = \sqrt{(-1)^2 + (2)^2 + (-0.25)^2} = 2.25$, so the pilot must fly 2.25 miles.

Vector Equation of a Line

Just as in two dimensions, a position vector and a direction vector determine a unique line. The position vector **p** locates a point P_0 on the line, and the direction vector **v** selects the particular line from all possible lines through that point. Adding scalar multiples of **v** to **p** locates other points P on the line; the equation describing the position vectors for *all* points on the line has the form $\mathbf{r} = \mathbf{p} + t\mathbf{v}$, with $-\infty < t < +\infty$. The scalar, t, is the independent variable, and the position vector **r** is the dependent variable, "pointing to" points lying on the line. This equation is called the vector equation for a line.

In general, if the tip of the position vector $\mathbf{p} = x_0\mathbf{i} + y_0\mathbf{j} + z_0\mathbf{k}$ lies on a line L with direction vector $\mathbf{v} = a\mathbf{i} + b\mathbf{j} + c\mathbf{k}$, then the vector equation describing the set of all points P on the line is

$\mathbf{r} = \mathbf{p} + t\mathbf{v}$, or

$\overrightarrow{OP} = (x_0\mathbf{i} + y_0\mathbf{j} + z_0\mathbf{k}) + t(a\mathbf{i} + b\mathbf{j} + c\mathbf{k})$, where $-\infty < t < +\infty$.

EXAMPLE 12

In Example 11, a helicopter must fly patients from the mountain side at point A(3, 5, 0.5) to the regional hospital at B(2, 7, 0.25). Write a vector equation for the line through points A and B. Then restrict the domain to describe the path (segment) the helicopter pilot takes. (Refer back to Figure 6.59).

SOLUTION:

Let $P_0(x_0, y_0, z_0)$ be the point A(3, 5, 0.5). So, a position vector to the line is $\mathbf{p} = 3\mathbf{i} + 5\mathbf{j} + 0.5\mathbf{k}$. From Example 11, the vector from A to B is $\overrightarrow{AB} = -\mathbf{i} + 2\mathbf{j} - 0.25\mathbf{k}$, and is a direction vector for the line. Therefore, an equation for the line is

$\overrightarrow{OP} = (3\mathbf{i} + 5\mathbf{j} + 0.5\mathbf{k}) + t(-\mathbf{i} + 2\mathbf{j} - 0.25\mathbf{k})$, for $-\infty < t < +\infty$.

By the construction, $t = 0$ gives point A and $t = 1$ gives point B. Therefore, the path for the helicopter is described by parameter values in the interval $t = [0, 1]$. So, the path of the helicopter is given by $\overrightarrow{OP} = (3\mathbf{i} + 5\mathbf{j} + 0.5\mathbf{k}) + t(-\mathbf{i} + 2\mathbf{j} - 0.25\mathbf{k})$, for $0 \le t \le 1$.

The parameter t does not necessarily represent time. In Example 12, for convenience $t = 0$ was chosen to correspond to point A and $t = 1$ to correspond to point B, independent of the actual distance or flying time between points A and B. Thus t may be thought of as the fraction of the trip that has been completed, or as distance in 2.25-mile units.

EXAMPLE 13

Suppose the helicopter of Example 12 flies at a rate of 1 mile per minute. Write a vector equation for the line segment connecting points A and B that gives the position of the helicopter after m minutes.

SOLUTION:

From Example 11, the distance between points A and B is 2.25 miles. Thus, in the equation $\overrightarrow{OP} = (3\mathbf{i} + 5\mathbf{j} + 0.5\mathbf{k}) + t(-\mathbf{i} + 2\mathbf{j} - 0.25\mathbf{k})$, the direction vector has magnitude 2.25. The velocity vector has the same direction, but has magnitude 1 (miles per minute). Therefore, replace $-\mathbf{i} + 2\mathbf{j} - 0.25\mathbf{k}$ with the unit vector in the same direction, namely, $\frac{1}{2.25}(-\mathbf{i} + 2\mathbf{j} - 0.25\mathbf{k})$.

After substituting, the vector equation becomes:

$\overrightarrow{OP} = (3\mathbf{i} + 5\mathbf{j} + 0.5\mathbf{k}) + \frac{m}{2.25}(-\mathbf{i} + 2\mathbf{j} - 0.25\mathbf{k})$, for $0 \le m \le 2.25$, where m represents the minutes flown after leaving point A.

The methods used in Examples 11–13 extend to building models of the motion of satellites, storms, or "pens" in computer animation sequences.

As was true in two dimensions, the components of vector equations in three dimensions lead directly to parametric form. For example, in Example 12, $\overrightarrow{OP} = (3\mathbf{i} + 5\mathbf{j} + 0.5\mathbf{k}) + t(-\mathbf{i} + 2\mathbf{j} - 0.25\mathbf{k})$, for $-\infty < t < +\infty$, or $\overrightarrow{OP}$ $x\mathbf{i} + y\mathbf{j} + z\mathbf{k} = (3 - t)\mathbf{i} + (5 + 2t)\mathbf{j} + (0.5 - 0.25t)\mathbf{k}$, so the parametric equations of the line are:

$x = 3 - t$, $y = 5 + 2t$, $z = 0.5 - 0.25t$, for $-\infty < t < +\infty$.

Substituting $t = 0$ yields point A(3, 5, 0.5), and $t = 1$ gives point B(2, 7, 0.25).

Exercises 6.6

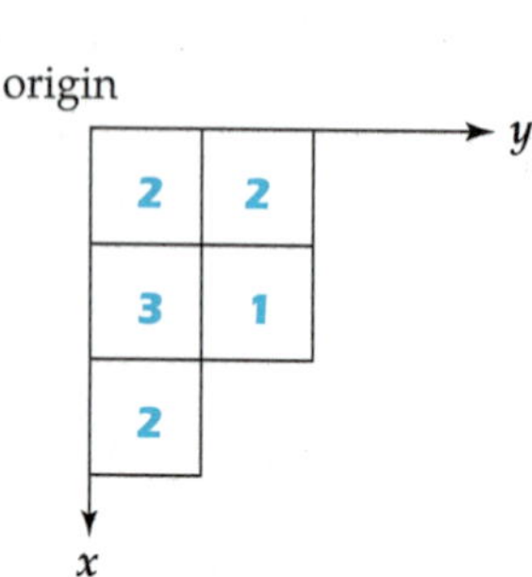

FIGURE 6.60.
Floor plan for a building structure.

1. **Figure 6.60** shows a floor plan, a view of a structure from directly above it. The numbers in each square represent the number of floors in that portion of the building. Make both an isometric drawing and an orthogonal drawing of the same building.

2. Show that $|\mathbf{v}| = \sqrt{a^2 + b^2 + c^2}$, where $\mathbf{v} = a\mathbf{i} + b\mathbf{j} + c\mathbf{k}$. If necessary, refer to Figure 6.58.

3. For each of the following situations, find the specified vector **v** and a unit vector **u** in the same direction.

   a) The direction vector for the line given by equation $\overrightarrow{OP} = (3 - 2t)\mathbf{i} + (5 + 3t)\mathbf{j} + (-4 - t)\mathbf{k}$.

   b) The vector from point A(7, 12, 8) to point B(–3, –5, 4).

   c) The position vector $\overrightarrow{OB}$, if $\overrightarrow{OA} = 11\mathbf{i} - 6\mathbf{j} + 3\mathbf{k}$ and $\overrightarrow{AB} = -5\mathbf{i} + 2\mathbf{j} - 5\mathbf{k}$.

4. a) Find a vector equation of the line through the point P(2, –5, 8) and in the direction defined by the vector $\mathbf{v} = 3\mathbf{i} + 5\mathbf{j} - 2\mathbf{k}$.

   b) Rewrite your equation in parametric form.

5. In this lesson, vectors in space were defined much as were vectors in the plane. Operations and properties also were seen to be very similar in both two and three dimensions. Let $\mathbf{v} = a\mathbf{i} + b\mathbf{j} + c\mathbf{k}$ and $\mathbf{w} = d\mathbf{i} + e\mathbf{j} + f\mathbf{k}$ be two vectors in space. Specify conditions on a, b, and c for which:

   a) $\mathbf{v} = \mathbf{0}$ (the zero vector)

   b) **v** and **w** are parallel

   c) **v** and **w** are perpendicular

6. a) Suggest a way to extend the definition of the dot product to three-dimensional vectors, and show that your definition holds for two–dimensional vectors.

   b) Use two vectors **v** and **w** that you *know* are perpendicular to each other, and check that your definition of dot product gives $\mathbf{v} \bullet \mathbf{w} = 0$.

Exercises 6.6

c) Describe two ways to determine the angle between two three-dimensional vectors.

7. Vector and parametric forms for equations of lines in three dimensions are closely related to their two-dimensional forms. However, not everything in 3-space behaves that nicely!

   a) In two dimensions, slope is the ratio of vertical change to horizontal change ($\Delta y/\Delta x$). In three dimensions, there are three changes. Discuss ways to convey the idea of slope for lines in three dimensions.

   b) Exercises 8 and 9 of Lesson 6.5 introduced the symmetric form for the equation of a line in two dimensions: $\frac{x-a}{c} = \frac{y-b}{d}$. Extend the symmetric form for a line to three dimensions. Write the symmetric form of the line in Exercise 4.

   c) In two dimensions, the general form for the equation of a line is $Ax + By + C = 0$. The extension of that expression in three dimensions is $Ax + By + Cz + D = 0$. Does this equation describe a line? If so, write the equation in vector, parametric, or symmetric form; if not, what *does* it describe?

8. If $\mathbf{v} = a\mathbf{i} + b\mathbf{j} + c\mathbf{k}$ and $\mathbf{w} = d\mathbf{i} + e\mathbf{j} + f\mathbf{k}$ are two vectors in space, the **cross product** of $\mathbf{v}$ and $\mathbf{w}$ is defined by:
   $\mathbf{v} \times \mathbf{w} = (bf - ce)\mathbf{i} + (cd - af)\mathbf{j} + (ae - bd)\mathbf{k}$.

   a) Find $\mathbf{v} \times \mathbf{w}$, if $\mathbf{v} = 2\mathbf{i} - 3\mathbf{j} + 2\mathbf{k}$ and $\mathbf{w} = -\mathbf{i} - 4\mathbf{j} + 3\mathbf{k}$.

   b) Use the definition of the cross product to show that $\mathbf{v} \times \mathbf{w} = -\mathbf{w} \times \mathbf{v}$.

   c) Verify for the vectors in part (a) that the cross product vector $\mathbf{v} \times \mathbf{w}$ is perpendicular to both $\mathbf{v}$ and $\mathbf{w}$.

   d) Verify for the vectors in part (a) that the magnitude of the cross product vector $\mathbf{v} \times \mathbf{w}$ is equal to the area of the parallelogram formed by vectors $\mathbf{v}$ and $\mathbf{w}$ when drawn as position vectors.

TAKE NOTE

The dot product of two vectors is a number; the cross product is another vector. Therefore, the dot product is sometimes called the scalar product, and the cross product called the vector product, of two vectors.

9. Graph the following vector equations on a single set of axes, for $0 \le t \le 1$: $\mathbf{v} = t\mathbf{j}$, $\mathbf{w} = \mathbf{j} + t(\mathbf{i} - \mathbf{j})$, $\mathbf{s} = \mathbf{i} + t\mathbf{j}$ and $\mathbf{r} = 2.5\mathbf{i} + 0.5\mathbf{j} + (0.5\cos 2\pi t\mathbf{i} + 0.5\sin 2\pi t\mathbf{j})$.

Exercises 6.6

10. Write a set of vector equations to trace the letters of the word HI.

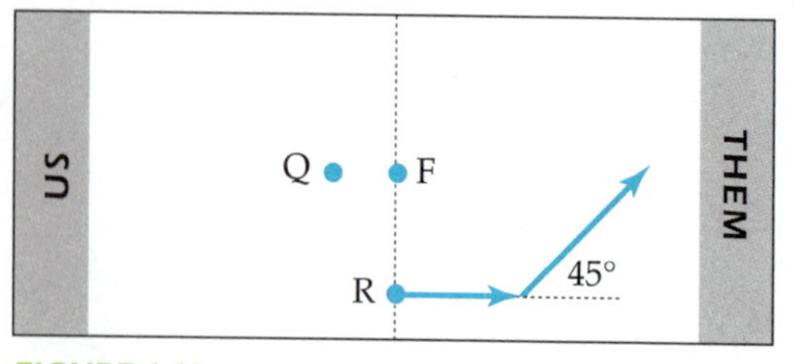

FIGURE 6.61.
Football Field for Exercise 11.

11. The local football team needs one score to win the game, but there is time for only one more play—a very special play that will work if executed perfectly. The play called is a "20-Post", in which the wide receiver (the fastest player on the field for *either* team) starts out on the sideline, runs down field and veers at a 45° angle toward the middle of the field, as shown in **Figure 6.61**.

The receiver lines up at point R, which is 10 yards away from the football, located at point F. When the ball is snapped, the quarterback drops back 7 yards from point F to point Q. He can count on 5 seconds of pass protection from his trusted blockers but must have thrown the ball by then. At the instant the ball is snapped, the receiver leaves point R running at 10 yards/sec, and makes the turn 20 yards down field. He needs at least 30 yards after that in order to outrun the defense and get open to receive the touchdown pass. Determine an initial velocity and release angle for the football from the quarterback, and the direction in which he must throw the ball for it to be caught for the winning touchdown. Also, write the trajectory of the football as a set of parametric equations.

Corbis

12. A crop duster is flying at a height of 500 feet along a heading of 60° N of E, with a ground speed of 40 mph. Its ground position is currently 2 miles away from the intended target for the day's spraying. The pilot is diving along a vertical parabolic path, maintaining the 40 mph ground speed, to release the dust at a height of 40 feet. See **Figure 6.62**. Write a set of three parametric equations to describe the trajectory.

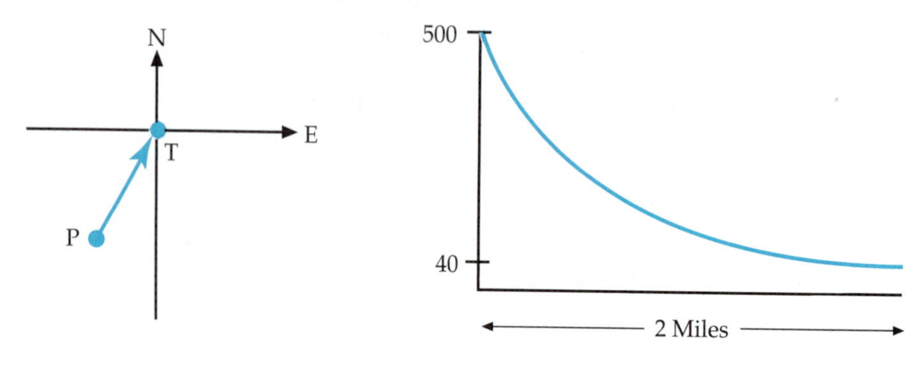

FIGURE 6.62.
The path flown by the crop duster.

Exercises 6.6

13. Michele Summer, a wildlife biologist on vacation in California, spots a hawk riding a thermal updraft in a lazy, circular arc. She notices that the bird rises about 30 feet each time it glides through one complete revolution without flapping its wings a single time. Assuming each revolution to be circular, she estimates the radius at 60 feet and the initial height of the bird at 300 feet (level with the top of a nearby redwood tree). She times the interval to complete each revolution at 20 seconds. She also estimates that the bird's location is approximately 400 feet east and 900 feet south of her present location. Using these figures, describe the path of the bird using a vector equation.

14. In practice, vector equations are the models of choice for many applications including ballistics, weather tracking, and animation, among many others. Suppose that Gridville is hosting an aircraft show and that one of the stunt planes is planning a circle-and-land maneuver. The current plan is described by the vector equation:

    $\mathbf{r}(t) = 3\cos(t)\mathbf{i} + \sin(t)\mathbf{j} + \left(\frac{1}{2} - \frac{1}{8}t\right)\mathbf{k},$

    where $\mathbf{r}$ is measured in miles, t in minutes from the beginning of the approach, and angles are measured in radians. The origin is located at the main viewing stand.

    a) When and where will the plane touch down?

    b) Describe the ground path of the plane.

    c) How close to the viewing stand will the plane be when it touches down?

15. a) What apparent angles are possible in viewing a real right angle from a point on or above its bisector? (This limits the possible ways that the x- and y-axes can appear in drawings of 3-dimensional figures!)

    b) In Part 1 of Activity 6.6, you drew a cube with sides of length 1 on isometric dot paper, and found that the top front corner was coincident with the bottom back corner (origin). Review Item 2(c) of Activity 6.6 Part 1. Use the equation $\cos(\theta) = \frac{v \bullet w}{|v||w|}$ and your definition of dot product, with $\mathbf{v}$ defined as the floor diagonal and $\mathbf{w}$ the line of sight. Determine the angle that the observer's line of sight makes with the x-y plane.

Exercises 6.6

c) Explain why drawing the positive direction of each axis at 120° angles on the page creates a unique viewing perspective.

d) If the x- and y-axes were drawn to make only a 45° angle on the page from the negative z-axis, the cube would still look like a cube. Make such a drawing keeping scale markings on all axes equal. How did the appearance change? Why?

e) Repeat part (d) for the situation in which the x- and y-axes make 75° angles on the page from the negative z-axis.

f) **Investigation.** Pretend you can grab the end of the positive x- and y-axes, and swing them toward each other (like bellow handles). How does changing the 120° angle affect the drawing of a cube?

16. When making orthogonal drawings in Part 2 of Activity 6.6, you were asked to reduce the scale on the x-axis to about 70% of that on the other two axes.

a) Draw a cube on such an orthogonal grid. How well does the drawing represent a cube?

b) Now, adjust the scaling on the x-axis to see its effect. Repeat part (a) twice more. On the first time, scale the x-axis exactly as the other two axes. On the second time, scale it to 50% of the other two axes. Which of the three drawings looks best?

17. **Investigation**. You have studied two 2-dimensional coordinate systems. Points in a Cartesian coordinate system require a pair of distances (x, y). In a polar coordinate system, descriptions require a distance and a direction (r, θ). In each system two pieces of information are required to locate a point uniquely, which is why the plane is called two-dimensional. In a similar manner, coordinates in space can be described by three distances (x, y, z); space is three-dimensional.

a) Explain why the coordinate system (latitude, longitude) can locate points uniquely on the Earth, which is three-dimensional.

b) Explain how astronomers can use a coordinate system (azimuth, altitude) to locate celestial objects in the sky.

Exercises 6.6

c) Research how the astronomical coordinate system (right ascension, declination) is defined. In what ways is this similar to the latitude, longitude system? In what ways is this similar to the azimuth, altitude system?

There are ways of describing the locations of points in space other than providing three directed distances along three perpendicular axes.

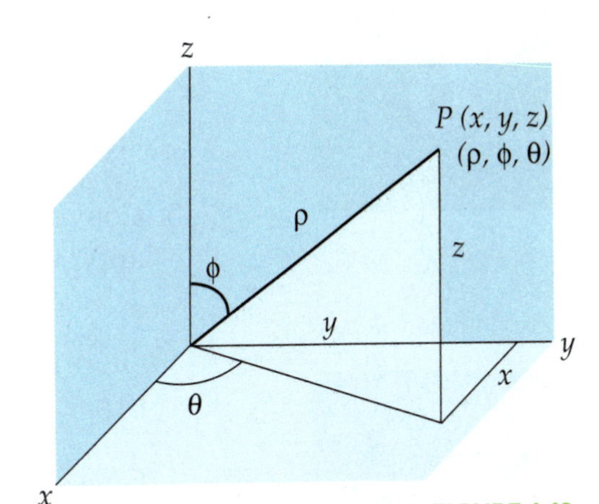

FIGURE 6.63. Definitions of spherical coordinate measurements.

d) One method, called **spherical coordinates**, specifies an angle θ, measured horizontally counterclockwise from the positive x-axis of a corresponding Cartesian system, an angle ϕ, measured vertically down from the positive z-axis, and a distance ρ, measured from the origin. See **Figure 6.63**. Compare this system to other systems you know.

e) Develop rules for transforming a Cartesian coordinate (x, y, z) into spherical coordinates. Also, develop rules for transforming spherical coordinates into Cartesian coordinates.

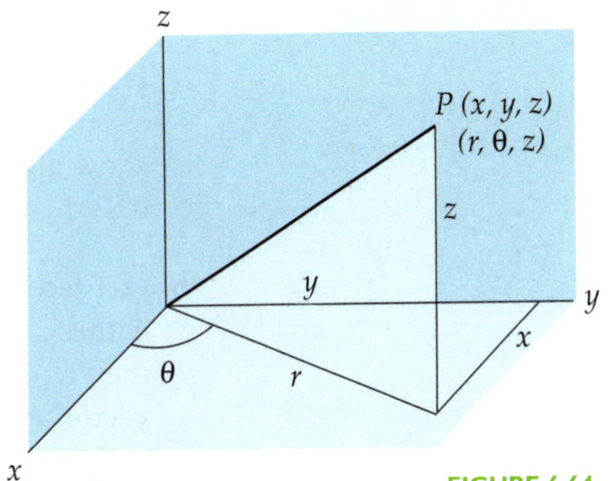

FIGURE 6.64. Definitions of cylindrical coordinate measurements.

f) Another 3-D system, called **cylindrical coordinates**, specifies an angle θ, measured counterclockwise horizontally from the x-axis of the corresponding Cartesian system, a distance r, measured in the horizontal plane to a point directly above or below the point being plotted, and a height z, measured parallel to the vertical z-axis. See **Figure 6.64**. Compare this system to others you know.

g) Develop rules for transforming a Cartesian coordinate (x, y, z) into cylindrical coordinates. Also, develop rules for transforming cylindrical coordinates into Cartesian coordinates.

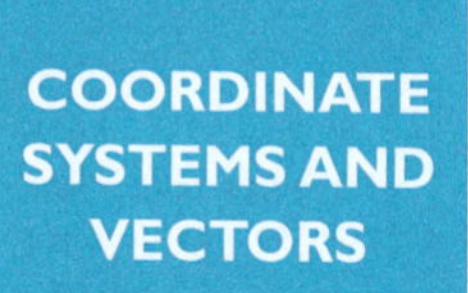

COORDINATE SYSTEMS AND VECTORS

Chapter 6 Review

1. Write a summary of the important mathematical ideas in this chapter.

2. Give three different polar coordinate pairs equivalent to the rectangular pair (3, –3). Include a pair for which r is negative and a pair for which θ is negative.

3. Convert each polar pair to rectangular coordinates. Round approximations to two decimal places.

   a) $(5, \frac{-\pi}{2})$

   b) (–6, 320°)

4. Give a general description of all polar coordinate pairs that represent the point with polar coordinates (3, 50°).

5. Starting at the origin, express each of the displacements shown in **Figure 6.65** along edges of a unit cube as a direction vector, and show algebraically that you end up where you started.

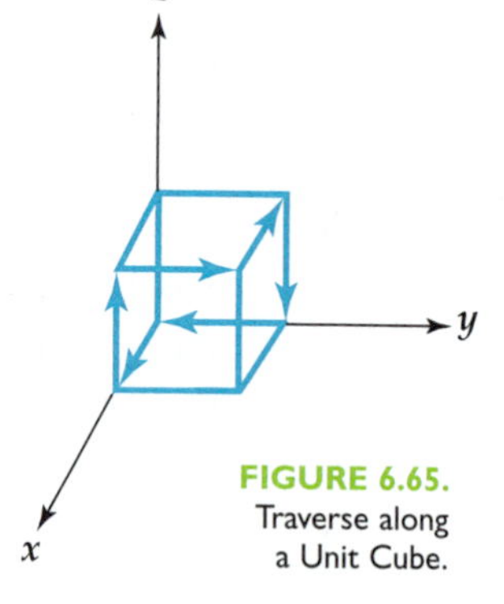

FIGURE 6.65. Traverse along a Unit Cube.

6. In a recent stretch, the San Francisco Giants won 7 out of 10 games at home, and 5 out of 8 games on the road. Altogether they won 12 of the 18 games during that time. Interpret the vector addition (7**i** + 3**j**) + (5**i** + 3**j**) = (12**i** + 6**j**) in the context of these facts.

7. Find the vertical and horizontal components of a navigational path 200 km in length with azimuth heading of 70°. Sketch the path and its components.

8. Sketch a graph of each of the following situations. Be sure to label axes appropriately.

   a) the Cartesian point (4, 3)

   b) the polar point (–2, π/4)

   c) the vector **v** = 4**i** –2**j**

d) the number $\mathbf{z} = \sqrt{2}$ cis 120°

e) the two-dimensional vector with magnitude 5 and direction angle 300°

f) the three-dimensional vector $\mathbf{v} = 2\mathbf{i} - 4\mathbf{j} + 3\mathbf{k}$

9. Express the line through the points P(6, 8) and Q(–2, 3) using the form specified. Where needed, assume the line goes *from* P *toward* Q.

a) slope-intercept form

b) point-slope form

c) vector form

d) parametric form

e) symmetric form

10. Perform the following operations:

a) Add the position vectors whose endpoints are the polar points P(5, π/6) and Q(2, π/3). What polar point is the endpoint of the resultant position vector?

b) Find $(2 - 3\mathbf{i})^5$, using De Moivre's Theorem.

c) Find all solutions to $3 - 4i = z^4$.

d) Simplify: 2(5 cis 315°)

e) Add, and write your answer in polar form: –2 cis 120° + 5 cis 210°.

f) Let **v** be the position vector to (–1, 1). Identify the terminal point of 3**v**.

11. Given points P(3, –2, 1), Q(2, 5, –1) and R(–3, –1, 4), find the following:

a) $\mathbf{v} = \overrightarrow{PQ}$

b) $\mathbf{w} = \overrightarrow{PR}$

c) $\mathbf{v} + \mathbf{w}$

d) $\mathbf{v} - \mathbf{w}$

e) $|\mathbf{v}|$

f) $|\mathbf{w}|$

g) $4\mathbf{w}$

h) $\mathbf{u}$ = unit vector in the direction of $\mathbf{v}$

i) $\mathbf{v} \bullet \mathbf{w}$

j) $\mathbf{v} \times \mathbf{w}$

k) the angle formed by the vectors **v** and **w** when drawn as position vectors

l) the distance QR using the Law of Cosines

12. Given two vectors $\mathbf{v} = a\mathbf{i} + b\mathbf{j}$ and $\mathbf{w} = c\mathbf{i} + d\mathbf{j}$, write the magnitude and direction of the vector sum $\mathbf{v} + \mathbf{w}$ in terms of:

a) the component scalars a, b, c, and d

b) the magnitudes and direction angles for **v** and **w**.

13. Develop a rule for performing the dot product operation on a pair of two–dimensional vectors when they are expressed in polar form.

14. Graph each of the following equations.

a) $r(\theta) = \theta$

b) $\mathbf{r}(t) = 2t\mathbf{i} + 3t\mathbf{j} - t\mathbf{k}$, with $-\infty < t < +\infty$

c) $\begin{cases} x = \sqrt{3}t \\ y = t \qquad \text{for } t \geq 0 \\ z = \sin(t) + 1 \end{cases}$

d) the limaçon $x^2 + y^2 = 2y + \sqrt{x^2 + y^2}$
(Hint: convert to polar form first.)

15. An air traffic controller is responsible for 2 planes whose paths will bring them near each other. The Cartesian coordinate system used by the airport has the origin at the control tower, with the positive x-axis due east and the positive y-axis due north. Distances are measured in miles, and time in minutes. The position vector to the first plane is $\mathbf{r}_1 = 5\mathbf{i} - 3\mathbf{j} + 2\mathbf{k}$; that plane's velocity vector is $\mathbf{v}_1 = -0.2\mathbf{i} + 0.5\mathbf{j} + 0.2\mathbf{k}$. The second plane currently has position vector $\mathbf{r}_2 = \mathbf{i} + 8\mathbf{j} - 3\mathbf{k}$, and velocity vector $\mathbf{v}_2 = 0.25\mathbf{i} - 0.7\mathbf{j} + 0.8\mathbf{k}$. An alarm will sound if the planes get within one mile of each other. Should the air traffic controller be concerned by the situation?

16. Given a three-dimensional position vector $\mathbf{v} = 3\mathbf{i} + 2\mathbf{j} - 4\mathbf{k}$, write the endpoint in Cartesian coordinates, spherical coordinates, and cylindrical coordinates.

17. What vector has length 8, makes a 45° angle with the positive z-axis, and has a projection on the x-y plane that makes a 20° angle with the positive x-axis?

Chapter 7

MATRICES

Corbis

CHAPTER INTRODUCTION

Journalist David Shenk became an enthusiastic Internet user soon after learning about it in 1991. As the 1990s progressed and the Internet grew, however, his enchantment turned to disillusion. By 1997 he had written *Data Smog*, a book highly critical of the Internet and the rapid growth of information for which it is partially responsible. Shenk believes there is so much information in the world today that it no longer contributes to the quality of life. Yet Shenk is also troubled by the quality of that information. "Information is not knowledge," he writes.

Indeed, the quantity of information is growing at an amazing rate. In the mid-1990s, there were estimates that the world's knowledge doubled every five years. By the late 1990s, estimates of the doubling period were about two years. How do we deal with such staggering amounts of information? How do we know that the information we obtain is accurate? The computer industry, which has contributed to the glut of information, is also helping to manage it and to answer these questions. Companies have been formed to develop and sell software which, with the participation of skilled professionals, searches data for useful information—a process called "data mining."

Once we determine that information is accurate, how do we organize it so it will be useful? The most common way to organize data is to place it in a rectangular configuration commonly called a table. To a mathematician, a table is a matrix. Mathematicians have developed an entire theory of matrices that helps people work with information.

As you explore matrices in this chapter, be aware that the processing of information with matrices sometimes takes surprising forms. Computer screens, for example, are really matrices: rows and columns of pixels, each represented inside the computer by one or more numbers that indicate the pixel's state. The theory of matrices facilitates the manipulation of pixels to create elaborate and colorful displays of information.

LESSON 7.1

Matrix Basics

To begin your work with matrices, it is imporant to understand a few basic terms and calculations. It's likely that you have seen at least some of these before. This lesson gives you the opportunity to review basic concepts, and to learn some new concepts as well.

Activity 7.1

A **matrix** is a rectangular configuration of data. Mathematicians usually enclose a matrix in brackets. A matrix has **rows**, which are horizontal, and **columns**, which are vertical (think of the columns that support a building). The number of rows and columns in a matrix are its **dimensions**; it is customary to state the number of rows first. Thus,

$$\begin{bmatrix} 5 & -6 & 4 \\ 2 & 1 & -7 \end{bmatrix}$$

is a 2 x 3 matrix. When matrices are used in a context, it is customary to write column labels above the matrix and row labels to its left to indicate the meaning of the data.

1. A retail store finds that it is better able to keep its stock of television sets current if it obtains them from several suppliers. For one brand of television it sells, the store uses two suppliers, A and B. It currently has in stock twelve 19″ sets, eight 27″ sets, and four 35″ sets from supplier A; and nine 19″ sets, six 27″ sets, and three 35″ sets from supplier B. Show two ways the company can store this information in a matrix. Give the dimensions of each matrix.

2. In order to prepare for an upcoming sales promotion on this brand of TV sets, the company orders additional stock. It orders five 19″ sets, three 27″ sets, and one 35″ set from supplier A; and four 19″ sets, four 27″ sets, and two 35″ sets from supplier B. Show two ways the order can be represented in a matrix. Then show how to use matrix addition to calculate the total inventory when the order is

received. Is it possible to add either matrix in your answer to 1 to either matrix in your answer to 2? Explain.

3. Matrices are multiplied in an unusual way. Each of the members of a row of the left matrix is multiplied by the corresponding member of a column of the second matrix; then the products are added to get a single entry in the product matrix. For example, to find the product of the matrices whose first row and first column are

$$\begin{bmatrix} 2 & 5 & -1 \\ & & \end{bmatrix}\begin{bmatrix} 3 & \\ 4 & \\ 7 & \end{bmatrix},$$

calculate 2 x 3 + 5 x 4 + –1 x 7 = 19. Since 19 was calculated by multiplying the first row of the left matrix by the first column of the right matrix, 19 is placed in the first row and the first column of the product matrix. This process is repeated until every row of the left matrix has been paired with every column of the right matrix.

The retail store pays the same wholesale price for a given set to both suppliers: $88 for a 19″ set, $157 for a 27″ set, and $292 for a 35″ set. Show two ways to store this information in a matrix. Then show ways in which matrix multiplication can be applied to each of these two, and to your answers to 2 to obtain information that is useful to the company. Interpret the data in each answer that you give.

4. Discuss the roles that labels play when matrix multiplication is used in a context. That is, how are the labels of the two matrices you multiplied in 3 related to the labels of the answer?

MATRIX BASICS

Matrices provide a common-sense way to organize data. Once data are in a matrix, operations can be performed on all of the matrix **members** (also called **elements** or **entries**) simultaneously. Although the time saved can be small when the matrices themselves are small, mathematical models that use matrices can have a million or more elements. Such applications of matrices result in considerable time savings.

Matrix Terminology and Notation

Matrices are often named with capital letters; in some cases the letter naming a matrix is enclosed in brackets. Thus A or [A] can designate a matrix. Elements of a matrix are named with the small-case letter that names the matrix, subscripted to indicate the element's position within the matrix. For example, $a_{2,3}$ refers to the element of matrix A in the second row and third column.

Care should be taken when referencing an element of a matrix or stating its dimensions: rows are always listed first. Thus, $a_{2,3} = 9$, the element of matrix A in the second row and the third column, but $a_{3,2} = 8$, the element in the third row and second column (see matrix in the right margin).

$$A = \begin{bmatrix} 5 & 4 & 1 \\ -2 & 7 & 9 \\ -3 & 8 & 0 \end{bmatrix}$$

When matrices occur in a context, it is good practice to label rows and columns accordingly. For example, the labels in the following matrix explain that the batteries are organized by type and size:

$$\begin{array}{r} \\ \\ \text{Heavy duty} \\ \text{Alkaline} \end{array} \begin{array}{c} \text{Battery Size} \\ \begin{array}{ccc} \text{AA} & \text{C} & \text{D} \end{array} \\ \begin{bmatrix} 6 & 8 & 3 \\ 9 & 4 & 5 \end{bmatrix} \end{array}.$$

Note, however, that the labels do not state the units of measure; that is, whether 6 is the number of AA heavy duty batteries or, say, the price. Units of measure usually do not appear in matrices, but are established by the context in which the matrices are used.

Matrices are often used to represent position vectors, which you studied in Chapter 6.

Example 1

Write a matrix to represent the vector $2\mathbf{i} + 3\mathbf{j} + 5\mathbf{k}$.

SOLUTION:

The vector can be represented by either the 1 x 3 matrix [2, 3, 5] or the 3 x 1 matrix $\begin{bmatrix} 2 \\ 3 \\ 5 \end{bmatrix}$.

Mathematicians often apply the term **vector** to one-dimensional matrices even when they are not used in an explicit vector situation.

```
NAMES MATH EDIT
1:[A]
2:[B]
3:[C]
4:[D]
5:[E]
6:[F]
7↓[G]
```

FIGURE 7.1.
The matrix screen of one type of graphing calculator.

If matrix calculations were done by hand, the power of matrices would be minimal. Much of the computational technology available today—graphing calculators and spreadsheets, for example—does matrix calculations. **Figure 7.1** shows the matrix screen on one type of graphing calculator. In this case, three sub-menus allow access to the names of all the calculator's matrices, various matrix operations, and the editing screen.

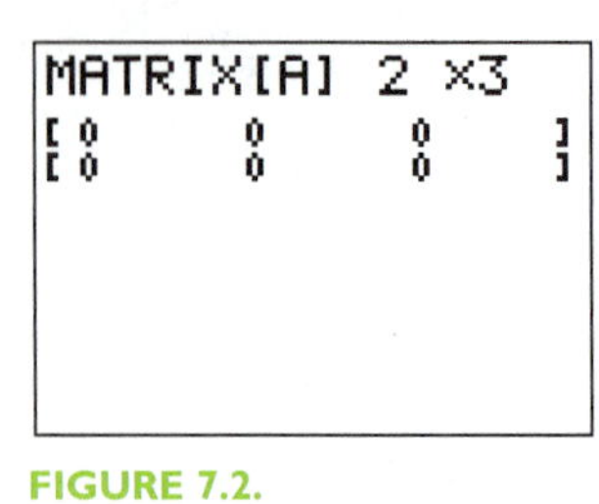

FIGURE 7.2.
Editing matrix A.

When this calculator's editing sub-menu is chosen and one of the matrices is selected, the user first enters the dimensions, then moves around the matrix to enter the elements (**Figure 7.2**). Movement is controlled by the calculator's cursor keys.

Entering a matrix on a spreadsheet is straightforward: type the elements into any convenient rectangular array of cells. Special commands are needed only when operations are being performed. Note that many spreadsheet manuals use the term **array** when discussing matrices.

Addition and Subtraction of Matrices

Addition and subtraction are perhaps the simplest of matrix operations because they are performed exactly as you would expect them to be performed. To add two matrices, add corresponding elements. For example, the matrix sum A + B has $a_{2,3} + b_{2,3}$ in the second row and third column (provided that both matrices have at least two rows and three columns).

Matrices cannot be added (or subtracted) unless they have the same dimensions. Furthermore, when matrices are used in a context, they should not be added or subtracted unless their labels match. For example, addition of the following matrices is mathematically possible, but it makes no sense.

$$\begin{array}{r} \\ \\ \text{Heavy duty} \\ \text{Alkaline} \end{array} \begin{array}{c} \text{Battery Size} \\ \begin{array}{ccc} \text{AA} & \text{C} & \text{D} \end{array} \\ \begin{bmatrix} 6 & 8 & 3 \\ 9 & 4 & 5 \end{bmatrix} \end{array} \qquad \begin{array}{r} \\ \\ \text{Heavy duty} \\ \text{Alkaline} \end{array} \begin{array}{c} \text{Battery Brand} \\ \begin{array}{ccc} \text{Duracell} & \text{Eveready} & \text{Ray - O - Vac} \end{array} \\ \begin{bmatrix} 12 & 23 & 15 \\ 18 & 17 & 22 \end{bmatrix} \end{array}$$

Calculator addition of matrices is accomplished by first entering data into two matrices with the same dimensions. Typically, the names of the matrices are obtained from the matrix menu and entered on the home screen. The calculator's regular addition key is used. **Figure 7.3** shows a calculator addition performed in two ways. In the second, the results are stored in a third matrix, which is automatically dimensioned properly.

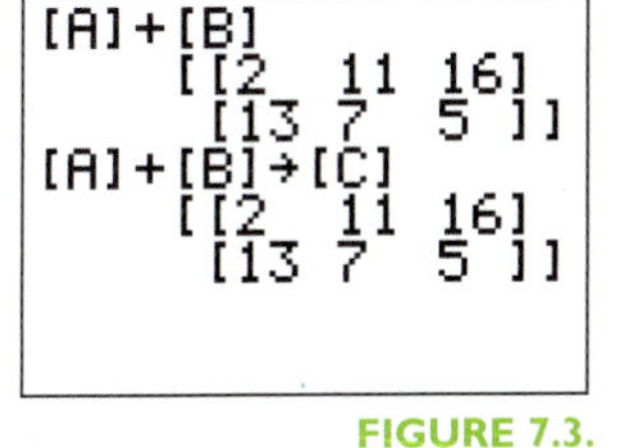

FIGURE 7.3.
Matrix addition on a calculator.

Matrix addition on a spreadsheet varies with the spreadsheet. The examples in this chapter use Microsoft Excel, one of the most popular spreadsheets. **Figure 7.4** shows the results of a matrix addition in Excel; the sum is stored in cells C5 through D6. First, the block C5:D6 is selected (by dragging the mouse); then an equal sign (=) is typed. Block A1:B2 is selected, a plus sign (+) typed, then block D1:E2 selected. To complete the calculation, the Enter key is pressed while a special key is held down. The special key is usually the Command key (Macintosh) or the Control-Shift key (PCs); check your spreadsheet manual for specifics. The completed formula is {=A1:B2+D1:E2}. (To indicate that this is a matrix formula, Excel inserts braces when the Enter key is pressed.)

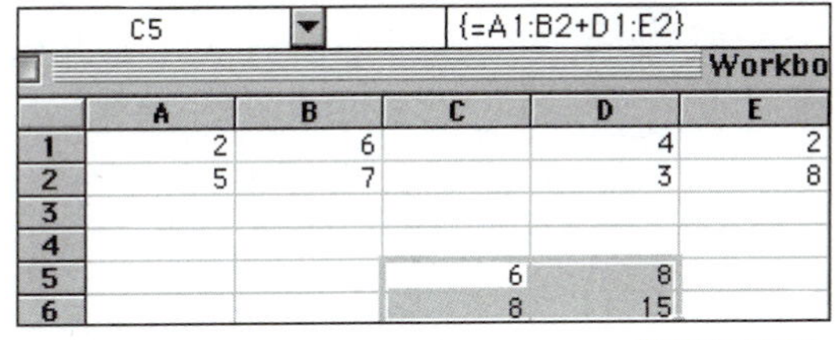

FIGURE 7.4.
Matrix addition in Excel.

In some applications of matrices, it is necessary to add the same value to every element of a matrix. On a calculator, this type of calculation is best done by first filling one matrix with a single value; most technology has special features for this purpose (see **Figure 7.5**). Once the matrix is filled with a constant, it can be added to any other matrix of the same dimensions.

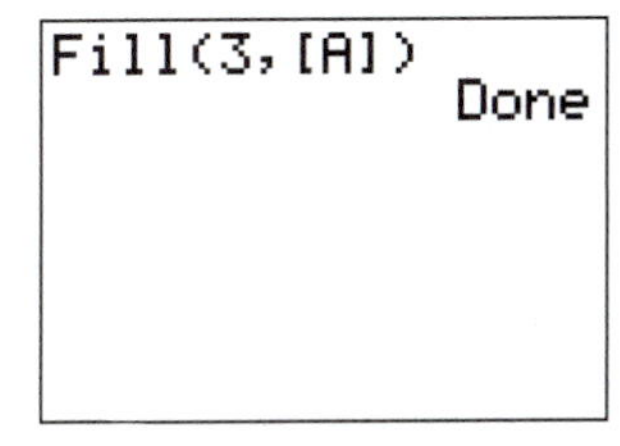

FIGURE 7.5.
Filling matrix [A] with 3s on a graphing calculator.

In Excel, a block can be filled with a single value by first selecting the block, typing an equal sign followed by the value, and pressing the Enter key while holding down the same key used to enter a matrix addition.

Multiplication of Matrices

There is more than one type of matrix multiplication. The simplest is **scalar multiplication**, in which every element of a matrix is multiplied by the same constant (often called a scalar). For example, to double every element of matrix A, type 2[A] on a typical calculator. To store the result as matrix B, type 2[A]→[B]; to replace matrix A with its double, type 2[A]→[A].

Scalar multiplication is done in Excel by first selecting the block of cells in which the result will be stored, typing =2*, selecting the block to be doubled, then pressing Enter while holding down the same key used to enter a matrix addition.

Another type of matrix multiplication is called simply **matrix multiplication.** It is performed in a row by column fashion. That is, the first element in a row of the first (left) matrix is multiplied by the first element of a column of the second (right) matrix. The second element of the same row of the first matrix is multiplied by the second element of the same column of the second matrix, and so forth. After the products are calculated, they are summed to obtain a single value.

EXAMPLE 2

Show the calculation of the entry determined by the indicated row and column, and place the result in the proper location of the product matrix.

$$\begin{matrix} & & \downarrow & \\ \rightarrow & \begin{bmatrix} 2 & 5 & -1 \\ & & \end{bmatrix} & \begin{bmatrix} 3 & \\ 4 & \\ 7 & \end{bmatrix} \end{matrix}$$

SOLUTION:

$$\begin{matrix} & & \downarrow & \\ \rightarrow & \begin{bmatrix} 2 & 5 & -1 \\ & & \end{bmatrix} & \begin{bmatrix} 3 & \\ 4 & \\ 7 & \end{bmatrix} \end{matrix} = \begin{bmatrix} 2(3)+5(4)-1(7) & \\ & \end{bmatrix} = \begin{bmatrix} 19 & \\ & \end{bmatrix}$$

Because of the way matrix multiplication is performed, the number of elements in a row of the first matrix must equal the number of elements in a column of the second matrix.

Some observations about matrix multiplication:

- In general, the entry in the nth row and mth column of the product is obtained from the nth row of the first (left) matrix and the mth column of the second (right) matrix.

- The order of multiplication matters. Usually, the product is different if the order of the two matrices is reversed and may, in fact, be undefined.

- The dimensions of the product can be determined from the dimensions of the matrices being multiplied. If an $m \times n$ matrix is multiplied by an $n \times p$ matrix, the result is an $m \times p$ matrix. The second dimension of the first matrix must equal the first dimension of the second matrix.

- When matrix multiplication is used in a context, matrix labels act in a manner similar to dimensions. For example, if the first matrix is battery type X battery size and the second is battery size X battery cost, then the product matrix is battery type X battery cost. Again, the second label of the first matrix must be the same as the first label of the second matrix.

Graphing calculators and spreadsheets can multiply two matrices if the dimensions are compatible. **Figure 7.6** shows the result of multiplying a 4 X 2 and a 2 X 3 matrix on a typical calculator. Note that no multiplication symbol is needed.

```
[A][B]
    [[26 29 24]
     [55 79 75]
     [13 40 47]
     [68 68 52]]
```

FIGURE 7.6.
Matrix multiplication on a graphing calculator.

The Excel spreadsheet uses a special function to perform matrix multiplication: select the block of cells to contain the result, type =MMULT(, select the first matrix, type a comma, select the second matrix, type a right parenthesis, then press Enter while holding down the same key used to perform matrix addition. **Figure 7.7** shows the product of the matrix in cells A1:B4 and the matrix in cells D1:F2 stored in cells C6:E9. The formula that generates the product is {=MMULT(A1:B4,D1:F2)} (again, Excel encloses the formula in braces when the Enter key is pressed).

C6 {=MMULT(A1:B4,D1:F2)}

Workbook1

	A	B	C	D	E	F
1	2	3		1	7	9
2	7	6		8	5	2
3	5	1				
4	4	8				
5						
6			26	29	24	
7			55	79	75	
8			13	40	47	
9			68	68	52	

FIGURE 7.7.
Matrix multiplication in Excel.

Exercises 7.1

1. A lumber store sells three kits for building decks onto houses. The lumber in each kit consists of 2 x 6 boards that serve as floor joists, 2 x 4 boards used primarily for railings, and 1 x 6 boards used for flooring. Kit A uses 18, 26, and 20 of each type, respectively; kit B uses 24, 38, and 30; kit C uses 32, 54, and 42. A 2 x 6 costs the store \$2.85; a 2 x 4 costs \$1.72; and a 1 x 6 costs \$1.96.

   a) Show how this information can be stored in a pair of matrices. Multiply the matrices and interpret the resulting information.

   b) The store receives an order from a builder for 6 deck A kits, 10 deck B kits, and 15 deck C kits. Show how to use matrix multiplication to calculate the amount of lumber needed to fill the order and the cost of the material to the store.

2. Refer back to the matrix addition spreadsheet in Figure 7.4. Explain how the 15 in cell D6 was obtained.

3. Refer back to the matrix multiplication spreadsheet in Figure 7.7. Explain how the 40 in cell D8 was obtained.

4. In each of the following pairs, tell whether it is possible to multiply matrix A and matrix B in either order. If it is possible, state the order in which the matrices can be multiplied and the dimensions of the product.

   a) $A = \begin{bmatrix} 5 & 1 \\ 4 & 2 \end{bmatrix}$; $B = \begin{bmatrix} 1 & 2 \\ 7 & 2 \\ 3 & 5 \end{bmatrix}$

   b) $A = \begin{bmatrix} 3 & 2 & 1 & 6 \end{bmatrix}$; $B = \begin{bmatrix} 2 & 8 \\ 9 & 1 \\ 5 & 4 \\ 3 & 6 \end{bmatrix}$

   c) $A = \begin{bmatrix} 4 & 3 & 6 \end{bmatrix}$; $B = \begin{bmatrix} 5 \\ 2 \\ 7 \end{bmatrix}$

Corbis

Exercises 7.1

5. A **square matrix** is a matrix in which the number of rows equals the number of columns. When is it possible to add two square matrices? To multiply them? Construct an example to support your answer.

6. When three matrices are multiplied, does it matter which two are multiplied first? That is, does it matter whether the product is calculated as (AB)C or A(BC)? Construct an example to support your conclusion.

7. A **multiplicative identity** matrix is a square matrix that does not change the values in any matrix of the same dimensions when the two are multiplied.

   a) Show that $\begin{bmatrix} 1 & 0 \\ 0 & 1 \end{bmatrix}$ is the 2 x 2 multiplicative identity and that order of multiplication does not matter when this matrix is multiplied by another 2 x 2 matrix.

   b) Find a 3 x 3 identity matrix.

   c) Describe an n x n identity matrix.

   d) Is there a multiplicative identity matrix for 2 x 1 matrices? If so, does it work equally well on the left and right? If not, explain why not.

   e) There are also identity matrices for matrix addition. Describe the properties of an additive identity matrix.

8. Exercise 7 examined multiplicative and additive identities for matrices. In general, an identity element for an operation is an object that, when used with the specified operation on any permitted object, always returns the original object. In Exercise 7, the objects were matrices and the operations were multiplication and addition. Identify the identity elements for the following object/operation pairs.

   a) real numbers with addition

   b) real numbers with multiplication

   c) functions with composition

Exercises 7.1

9. Simple secret codes are sometimes constructed by replacing each letter in a message with its numerical position in the alphabet (a blank space is 27), then adding a constant, multiplying by a constant, or both. The message is sent in numerical form. For example, the letter C is the third letter of the alphabet. If the coding method doubles and adds 5, then C is coded as $2 \times 3 + 5 = 11$.

   a) Show how ONE FINE DAY can be coded in one step by storing numerical values in a matrix with two rows, then multiplying by a scalar and adding a constant matrix. Use the double and add 5 method.

   b) This code suffers from a serious flaw: each occurrence of a given letter is coded with the same numerical value. In messages of even moderate length, statistical patterns emerge. Code crackers know, for example, that E is the most common letter in English. Applying statistics to such messages can enable unwanted parties to read them. Show that if the coding method multiplies by a square matrix such as $\begin{bmatrix} 2 & 3 \\ 1 & 5 \end{bmatrix}$, this flaw is avoided.

10. Calculator animation can be done with a technique similar to the one used for coding in Exercise 9. The coordinates of the pixels that describe an object are stored in a matrix of two rows (the first row contains x-coordinates, the second y-coordinates). Thus a point is represented as a 2×1 matrix. Repeatedly adding a constant matrix produces the illusion of motion along a linear path. Matrix multiplication can be used to rotate an object about the origin.

    a) Calculate the matrix product $\begin{bmatrix} \cos 30° & -\sin 30° \\ \sin 30° & \cos 30° \end{bmatrix} \begin{bmatrix} 5 \\ 2 \end{bmatrix}$. Plot on graph paper the point (5, 2) and the point produced by the product. Connect each point to the origin.

    b) Show analytically that the points are the same distance from the origin.

    c) Show that the angle between the segments connecting the two points to the origin is 30°. (Hint: Use trigonometry to find the angle between each segment and the x-axis.)

    d) Matrix multiplication can also be used to reflect an object through the x-axis, the y-axis, or the origin. Find a square matrix that produces each type of reflection.

Exercises 7.1

e) **Investigation.** As a project, write a calculator program that stores the coordinates of a polygon in a matrix, draws the polygon, then draws successive rotations of the polygon through increments of a specified angle.

PEOPLE AND MATH

Would you enjoy a job that paid you to play games in an arcade? If you create video games as a **multimedia computer programmer**, playing will actually be part of your job. After deciding on a concept such as a pinball game, the computer programmer plays to observe details of the movement, look, and sound of a real pinball machine. At the design stage, collaboration with graphic artists and audio engineers often takes place. The next step for the programmer is breaking the action down into small parts and writing corresponding code in programming language that will instruct the computer to animate a shape matrix to create realistic effects. The programmer then tests, debugs, and refines the program until the results on the computer screen matrix are exactly right.

11. The diagram in **Figure 7.8** shows the way a group of people ranked four candidates in an election. For example, 10 people ranked A first, B second, C third, and D fourth.

    a) Show a way to use matrix multiplication to find the point totals for candidates A, B, C, and D if a first-place ranking is worth 4 points, second 3 points, third 2 points, and fourth 1 point.

    b) Show how to store several different point systems in a single matrix and calculate the associated point totals with a single multiplication. You might, for example, feel that the difference between a first-place ranking and a second-place ranking should be more than one point, or that a last-place ranking should be worth nothing. Can you find a point system that changes the winner?

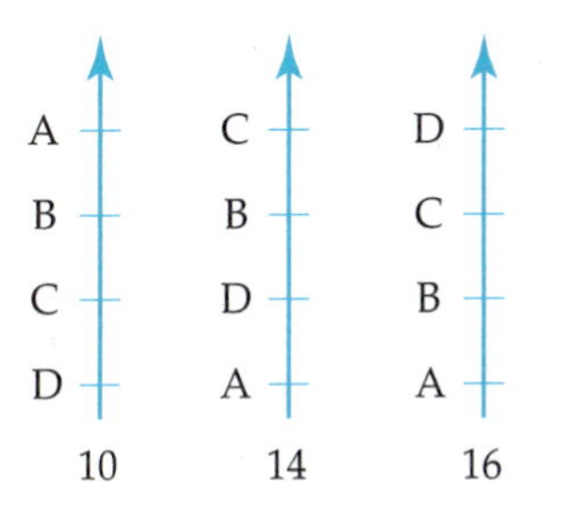

FIGURE 7.8.

12. A graph (network) is a collection of nodes (vertices) and edges connecting some or all of the nodes. For example, the graph in **Figure 7.9** has four vertices and three edges.

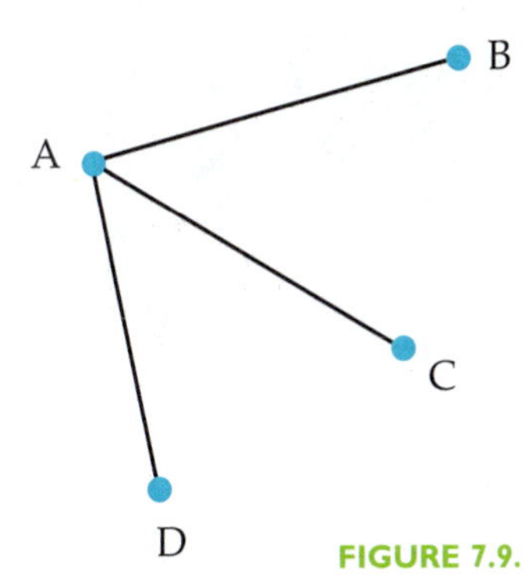

FIGURE 7.9.

Exercises 7.1

$$\begin{array}{c} \\ A \\ B \\ C \\ D \end{array}\begin{array}{c} \begin{array}{cccc} A & B & C & D \end{array} \\ \left[\begin{array}{cccc} 0 & 1 & & \\ 1 & & & \\ & & & \\ & & & \end{array}\right] \end{array}$$

a) Matrices are used to represent graphs. In a matrix representation, each vertex of the graph is assigned a row and column. A zero is recorded in the matrix if the row and column vertices do not have an edge between them; a one is recorded if they do have an edge between them. Such a matrix is called an **adjacency matrix**. Complete the adjacency matrix on the left for the graph in Figure 7.9.

b) A complete graph is one in which every pair of vertices has an edge. Sketch a complete graph with four vertices and give its adjacency matrix.

c) Useful information can be obtained from an adjacency matrix by raising it to a power. For example, if you square an adjacency matrix, the entries indicate the number of ways to travel between two vertices by using exactly two edges; if you cube the adjacency matrix, the entries indicate the number of ways to travel between two vertices using exactly three (not necessarily distinct) edges. Square the matrix you found in (b) and interpret the results.

d) Select one of the entries in the squared matrix. Examine the way in which that entry was calculated, and explain why the calculation produces the number of two-edge connections between the pair of vertices. Identify each of the paths counted by the selected entry.

13. An independent car rental company operates in a city in which it has three locations. A customer may rent a car at any location and drop it at the same location or at one of the other two. Company statistics show that of the cars rented at location A, 60% are returned to A, 30% to B, and 10% to C; of the cars rented at location B, 70% are returned to B, 10% to A, and 20% to C; of the cars rented at location C, 50% are returned to C, 30% to A, and 20% to B.

© Susan Van Etten

a) Show how to record this information in a square matrix in which the rows indicate the location in which a car is rented and the columns indicate the location to which it is returned. Use decimals to represent return rates.

b) On a given day, the company rents 40 cars at location A, 50 at location B, and 60 at location C. Show how the company can use matrix multiplication to estimate the number of cars that will be returned to each location. Explain why your calculation produces the correct information.

LESSON 7.2

The Multiplicative Inverse

Matrices make it possible to think of a collection of numbers as a single entity, and matrix operations allow calculations on entire groups of numbers in a single step. Just as numbers and their operations lead to algebra, matrices and their operations lead to matrix algebra. This lesson considers matrix equations and their solutions.

Activity 7.2

A (hypothetical) furniture store is planning three major sales events in the next three months. In these events, the store will feature three models of couches. The same three models will be featured in each event.

The store has run these three events for several years. During the first event, it usually sells about 45 of couch A, 30 of couch B, and 20 of couch C. During the second event, sales are about 35, 20, and 10 of each couch, respectively. In the third event, sales are about 20, 10, and 6 of each couch, respectively.

The company has set goals of \$35,500, \$23,500, and \$13,000 in revenues for the sales of these three couches during each of the respective events.

Your task in this activity is to explore pricing schemes that might bring the company close to its desired revenues. For example, do you think the company plans to use the same pricing scheme in each of the three sales?

Develop a matrix model with which you can use matrix multiplication to test various pricing schemes. Prepare a summary stating your conclusion, describing the model you used, and any assumptions on which the model rests.

THE MATRIX INVERSE

Arthur Cayley

Matrices were introduced in the 1850s by the British mathematician Arthur Cayley (1821–1895). Cayley's work included the development of an algebra of matrices, but his matrices were not received enthusiastically by all mathematicians. One reason for their less-than-warm reception was the strange way in which matrix multiplication was performed and the related fact that, in general, $AB \neq BA$. However, as the decades passed, the number of real-world situations for which matrix models proved useful grew. The development of computers in the twentieth century removed much of the tedium associated with matrix calculations, and thereby ensured a permanent place for matrix theory in the tool kit of the mathematical modeler.

In many applications of matrices, the modeler encounters matrix equations of the form $AX = B$, where A, X, and B are matrices and X represents an unknown matrix. Many such equations have a unique solution; some do not. Often, matrix A is a square matrix, and X and B are matrices of a single column. For example, this matrix equation has one unknown matrix (with two unknown entries):

$$\begin{bmatrix} 4 & 5 \\ 2 & 3 \end{bmatrix} \begin{bmatrix} x \\ y \end{bmatrix} = \begin{bmatrix} 1 \\ 6 \end{bmatrix}.$$

To see how matrix equations are solved, consider the matrix $A = \begin{bmatrix} 4 & 5 \\ 2 & 3 \end{bmatrix}$. For 2 x 2 matrices, the multiplicative identity matrix is $I = \begin{bmatrix} 1 & 0 \\ 0 & 1 \end{bmatrix}$. That is, $AI = IA = A$.

For matrix A, there is a special 2 x 2 matrix called the **multiplicative inverse** of A, which is often written A^{-1}. When A and A^{-1} are multiplied in either order, the result is the identity matrix. That is, $AA^{-1} = A^{-1}A = I$.

EXAMPLE 3

Show that the multiplicative inverse of $\begin{bmatrix} 4 & 5 \\ 2 & 3 \end{bmatrix}$ is $\begin{bmatrix} 1.5 & -2.5 \\ -1 & 2 \end{bmatrix}$.

SOLUTION:

$$\begin{bmatrix} 4 & 5 \\ 2 & 3 \end{bmatrix}\begin{bmatrix} 1.5 & -2.5 \\ -1 & 2 \end{bmatrix} = \begin{bmatrix} 1.5 & -2.5 \\ -1 & 2 \end{bmatrix}\begin{bmatrix} 4 & 5 \\ 2 & 3 \end{bmatrix} = \begin{bmatrix} 1 & 0 \\ 0 & 1 \end{bmatrix}.$$

Since $\begin{bmatrix} 1 & 0 \\ 0 & 1 \end{bmatrix}$ is the multiplicative identity element for 2 x 2 matrices,

$\begin{bmatrix} 1.5 & -2.5 \\ -1 & 2 \end{bmatrix}$ is the multiplicative inverse of $\begin{bmatrix} 4 & 5 \\ 2 & 3 \end{bmatrix}$.

Now, consider a matrix equation involving A: AX = B, in which X is an unknown 2 x 1 matrix.

Left-multiply both sides of the equation AX = B by A^{-1}: $A^{-1}(AX) = A^{-1}B$. This is equivalent to $(A^{-1}A)X = A^{-1}B$, which is the same as $IX = A^{-1}B$. But since I, the identity matrix for 2 x 2 matrices, is also the identity for a 2 x 1 matrix, the last equation is $X = A^{-1}B$. Thus, provided the multiplicative inverse of A can be found, the matrix equation AX = B can be solved.

EXAMPLE 4

Use multiplication by an inverse matrix to solve the matrix equation $\begin{bmatrix} 4 & 5 \\ 2 & 3 \end{bmatrix}\begin{bmatrix} x \\ y \end{bmatrix} = \begin{bmatrix} 1 \\ 6 \end{bmatrix}$.

SOLUTION:

$$\begin{bmatrix} 1.5 & -2.5 \\ -1 & 2 \end{bmatrix}\begin{bmatrix} 4 & 5 \\ 2 & 3 \end{bmatrix}\begin{bmatrix} x \\ y \end{bmatrix} = \begin{bmatrix} 1.5 & -2.5 \\ -1 & 2 \end{bmatrix}\begin{bmatrix} 1 \\ 6 \end{bmatrix}, \text{ so } \begin{bmatrix} 1 & 0 \\ 0 & 1 \end{bmatrix}\begin{bmatrix} x \\ y \end{bmatrix} = \begin{bmatrix} -13.5 \\ 1 \end{bmatrix}.$$

Therefore, $x = -13.5$; $y = 1$.

A few notes about the solution of matrix equations:

- The terms **multiplicative inverse** and **multiplicative identity** are often shortened to **inverse** and **identity**, respectively. Since additive identities and inverses also exist, the shorter terms should be used only if it is clear they do not refer to addition.

- Not all matrices have multiplicative inverses; therefore, not all matrix equations have solutions.

- Often the numerical entries in a matrix inverse are not whole numbers. The exact form of a matrix inverse can be found by using the determinant (a topic considered in Exercises 7.2). Although exact forms can be found by hand, the many calculations make the work tedious for all but the smallest matrices (i.e., 2 x 2).

FINDING INVERSES

To find the inverse of a particular matrix by hand, it is necessary to write a set of equations in which the variables represent the entries of the (unknown) inverse matrix, then create and solve equations describing the multiplication of the inverse and the original matrix.

EXAMPLE 5

Set up equations for finding the inverse of $\begin{bmatrix} 4 & 5 \\ 2 & 3 \end{bmatrix}$.

SOLUTION:

Represent the inverse as a matrix with unknown entries:
$\begin{bmatrix} a & b \\ c & d \end{bmatrix}$.

Write an equation describing the relationship between the original matrix, its inverse, and the identity matrix, then perform the multiplication:
$\begin{bmatrix} 4 & 5 \\ 2 & 3 \end{bmatrix}\begin{bmatrix} a & b \\ c & d \end{bmatrix} = \begin{bmatrix} 1 & 0 \\ 0 & 1 \end{bmatrix}$, so $\begin{bmatrix} 4a+5c & 4b+5d \\ 2a+3c & 2b+3d \end{bmatrix} = \begin{bmatrix} 1 & 0 \\ 0 & 1 \end{bmatrix}$.

Since the two matrices in this last equation are equal, corresponding entries must be equal. For example, $4a + 5c = 1$. There are a total of four such equations, two involving a and c, two involving b and d.

Therefore, finding the inverse requires solving two systems of equations in two variables: $\begin{cases} 4a + 5c = 1 \\ 2a + 3c = 0 \end{cases}$ and $\begin{cases} 4b + 5d = 0 \\ 2b + 3d = 1 \end{cases}$.

The solution of each system can be completed by symbolic methods.

Finding the multiplicative inverse of a square matrix requires finding 4, 9, 16, or more unknowns, so the process is tedious if done by hand. Therefore, in almost all realistic cases, calculators or computers are used to find inverses. Most graphing calculators can find the inverse of a square matrix (see **Figure 7.10**).

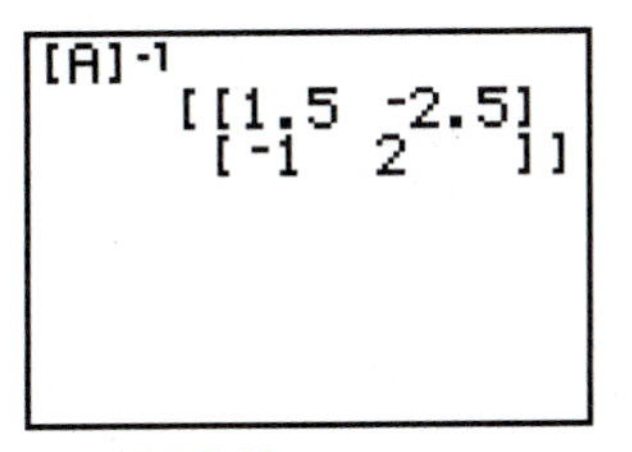

FIGURE 7.10.
Finding the inverse of [A] = $\begin{bmatrix} 4 & 5 \\ 2 & 3 \end{bmatrix}$ on one type of calculator.

In Excel, the matrix inverse command is MINVERSE. For example, to find the inverse of a 2 x 2 matrix contained in cells A1:B2, highlight a 2 x 2 block to contain the answer, type =MINVERSE(A1:B2), and press Enter while holding down the key used to execute matrix operations.

Exercises 7.2

1. Show that the matrix

$$\begin{bmatrix} -14 & 9 & 2 \\ 13 & -8 & -2 \\ -5 & 3 & 1 \end{bmatrix}$$

is the multiplicative inverse of the matrix

$$\begin{bmatrix} 2 & 3 & 2 \\ 3 & 4 & 2 \\ 1 & 3 & 5 \end{bmatrix}.$$

2. Show how to use the inverse of

$$\begin{bmatrix} 2 & 3 & 2 \\ 3 & 4 & 2 \\ 1 & 3 & 5 \end{bmatrix}$$

to solve the matrix equation

$$\begin{bmatrix} 2 & 3 & 2 \\ 3 & 4 & 2 \\ 1 & 3 & 5 \end{bmatrix}\begin{bmatrix} x \\ y \\ z \end{bmatrix} = \begin{bmatrix} 5 \\ 1 \\ 3 \end{bmatrix}.$$

3. The problem posed in Activity 7.2 can be solved by writing and solving a matrix equation. Show how to do this.

4. Group decisions are sometimes made by having the members of the group rank the choices and assigning points to the rankings. (See Exercise 11 of Lesson 7.1.)

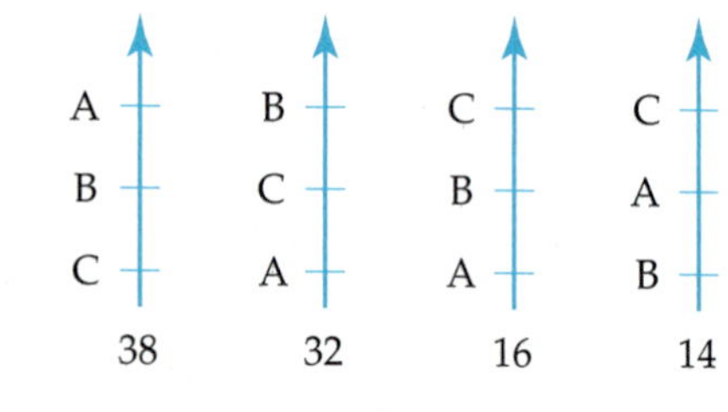

FIGURE 7.11.

a) The diagram in **Figure 7.11** represents the way the members of a group rank three alternatives. For example, 38 voters rank A first, B second, and C third. Use matrix multiplication to determine the point totals for A, B, and C if a first-place ranking is worth 3 points, a second-place ranking 2 points, and a third-place ranking 1 point.

b) Use a matrix equation to find a point system that results in A's winning. (Hint: pick point totals for A, B, and C, then find a point system that produces these totals.)

Exercises 7.2

5. A company manufactures three models of televisions at three different plants. At full capacity, plant 1 produces 200 of model A, 150 of model B, and 100 of model C in a day; plant 2 produces 300 of model A, 100 of model B, and 50 of model C; plant 3 produces 100 of each model. The company receives an order from a large retail chain for 8300 each of models A and B and 5400 of model C. Show how to use a matrix equation to determine the number of days to operate each plant to fill the order exactly.

6. Example 5 of this lesson described finding the inverse of $\begin{bmatrix} 4 & 5 \\ 2 & 3 \end{bmatrix}$ by symbolic means.

   a) Complete the example by solving the two systems of equations $\begin{cases} 4a+5c=1 \\ 2a+3c=0 \end{cases}$ and $\begin{cases} 4b+5d=0 \\ 2b+3d=1 \end{cases}$.

   b) Generalize the method used in Example 5. That is, find the inverse of the general 2 x 2 matrix $\begin{bmatrix} p & q \\ r & s \end{bmatrix}$.

   c) Compare the four members of the matrix you found in (b). Use them to describe a short cut for finding the inverse of a 2 x 2 matrix.

   d) Your solution in part (b) and the fact that division by 0 is undefined will help you understand why a matrix may not have an inverse. Explain.

7. The process of calculating the inverse of a square matrix involves a lot of computation, meaning that few people find inverses of matrices larger than 2 x 2 by hand. However, because the entries in an inverse are often not whole numbers and because some matrices do not have inverses, it is useful to understand part of the process. Basically, finding the inverse of a square matrix has two steps:

   - A new square matrix is derived from the first by performing certain calculations on the members of the first matrix.

   - Every member of this new matrix is divided by a constant that is also calculated from the original matrix. This constant is called the **determinant** of the original matrix. (Calculators that do matrix calculations have a determinant function; it is usually abbreviated det.)

Exercises 7.2

The determinant can be used to express an inverse in exact (non-decimal) form. Suppose, for example, that the determinant of a matrix is 2 and that the decimal form of the inverse of the matrix is $\begin{bmatrix} 2 & -1 \\ -2.5 & 1.5 \end{bmatrix}$.

Since the calculation of each member of the inverse includes division by 2, the inverse can be rewritten by multiplying each element by 2 and expressing the division by 2 as a scalar factor:
$\frac{1}{2}\begin{bmatrix} 4 & -2 \\ -5 & 3 \end{bmatrix} = \begin{bmatrix} 2 & -1 \\ \frac{-5}{2} & \frac{3}{2} \end{bmatrix}$.

This method—multiplying each entry of the inverse by the determinant then writing the division as a scalar factor—applies in general; that is, to any matrix inverse.

Since division by the determinant is the only division used in the calculation of the inverse, the process in this example can be used to find an exact form of the inverse of almost any square matrix. An exception occurs when the determinant is 0. Mathematicians have used the fact that division by 0 is undefined to prove that a square matrix does not have an inverse if its determinant is 0. A square matrix with determinant 0 is called a **singular matrix**.

a) Of what matrix is $\begin{bmatrix} 2 & -1 \\ -2.5 & 1.5 \end{bmatrix}$ the inverse? Explain your reasoning.

b) Refer back to your work in Exercise 6(b). In general, how do you think the determinant of a 2 x 2 matrix is calculated?

c) Find the exact form of the inverse of $\begin{bmatrix} 3 & 4 \\ -2 & 5 \end{bmatrix}$. Explain how you obtained your result.

d) Find the exact form of the inverse of $\begin{bmatrix} 4 & -5 & 3 \\ 8 & 2 & -5 \\ -7 & 1 & 6 \end{bmatrix}$. Explain your work.

e) Find the exact form of the inverse of $\begin{bmatrix} 4 & 3 & 1 \\ 8 & 6 & 2 \\ 3 & 5 & 7 \end{bmatrix}$. Explain your work.

Exercises 7.2

8. The operation of addition also has identity matrices and inverse matrices. An additive identity matrix is one that produces no change under addition. An additive inverse is one that produces the identity matrix under addition. Select one or two sample matrices and discuss the additive identity and additive inverse.

9. An important application of matrices is the solution of systems of equations, since solving a system of n equations with n unknowns is tedious if done by hand. For example, consider the system

$$\begin{cases} 2x - 3y + 4z = 7 \\ 3x + 2y - 3z = 6 \\ x - 5y + 2z = 2. \end{cases}$$

a) This system is equivalent to a matrix equation involving a single 3 x 3 matrix and two 3 x 1 matrices, one of which contains the variables x, y, and z, Show how to write this system of equations as a matrix equation.

b) Solve the system by solving the matrix equation you wrote in (a). Round answers to two decimal places.

c) The exact form of the solutions can be found by removing the determinant of the 3 x 3 matrix from its inverse. Show how to do this and give the exact solutions.

d) You may be wondering why the system you solved in this exercise couldn't be solved by using 1 x 3 matrices instead of 3 x 1 matrices. It can! Show how to solve the system by writing a matrix equation with two 1 x 3 matrices and one 3 x 3 matrix.

10. Solve each system. Give the solutions rounded to two decimal places and in exact form.

a) $\begin{cases} x + 2y - 3z = 7 \\ 2x + 4y - z = 2 \\ 2x + 3y + 5z = 9 \end{cases}$

b) $\begin{cases} 3a - b + 2c + d = 55 \\ a + 5b - 2c + 3d = 30 \\ 2a - b + 4c - 2d = 72 \\ -3a + 2b + c + 2d = 83 \end{cases}$

Exercises 7.2

11. Solve each system of two equations in two variables. When an equation has two variables, it can be represented geometrically by a line in a coordinate plane. After you have solved each system, graph the pair of lines and interpret your solution geometrically.

a) $\begin{cases} 3x + 2y = 7 \\ 5x + y = 18 \end{cases}$

b) $\begin{cases} 2x - 3y = 17 \\ -4x + 6y = 9 \end{cases}$

c) $\begin{cases} 2x - 3y = 17 \\ -4x + 6y = -34 \end{cases}$

LESSON 7.3

Systems of Equations in Three Variables

There is a direct connection between algebra and matrix algebra: systems of equations can be represented and solved as matrix equations. The simplest systems of equations have two equations and two variables. The focus of this lesson is systems of three equations in three variables, their matrix representations, and their geometric representations.

Activity 7.3

For several years you have used two-variable equations of the form $Ax + By = C$ as mathematical models. These equations are functions with linear graphs. The three-variable counterpart of such equations is $Ax + By + Cz = D$. Your task in this activity is to investigate the geometric representation of $Ax + By + Cz = D$ by applying your knowledge of three-dimensional coordinate systems from Chapter 4.

1. Begin by considering the equation $x + y + z = 5$. Use isometric dot paper to plot triples (x, y, z) that satisfy this equation. Plot enough triples to confirm or reject the assertion that this equation produces a line in three dimensions.

2. If you decide that the graph of $x + y + z = 5$ is not linear, plot enough additional pairs to establish the shape of its geometric representation. Identify the shape and support your conclusion.

3. Discuss all the possible configurations of two such geometric objects in three dimensions. That is, if two different equations of the form $Ax + By + Cz = D$ were graphed in three dimensions, what could be the geometric relationships between the two?

4. Answer Item 3 for three such objects.

5. Mathematicians sometimes use the word *linear* when discussing equations of the form $Ax + By + Cz = D$. Why do you think they do this?

LINES AND PLANES IN SPACE

x	y	z
0	0	4
0	1	3
0	2	2
0	3	1
0	4	0
1	0	3
1	1	2
1	2	1
1	3	0
2	0	2
2	1	1
2	2	0
3	0	1
3	1	0
4	0	0

TABLE 7.1.
Positive-integer solutions to $x + y + z = 4$.

Among the simplest three-variable equations are those of the form $Ax + By + Cz = D$, where A, B, C, and D are real numbers. (Note that since one or more of A, B, C, and D may be 0, equations such as $x = 3$ are in this group.)

As an example, consider the equation $x + y + z = 4$. One way to begin an analysis of this equation is by constructing a table of values. Since there is a fairly small set of non-negative integers that satisfy the equation, a table can display them all. **Table 7.1** shows all non-negative integer triples that satisfy $x + y + z = 4$.

When an equation has three variables, values are assigned to two of them, and the third is determined by the equation. Thus, one of the variables is a function of the other two. In mathematics, z is often expressed as a function of x and y: $z = f(x,y)$. The equation $x + y + z = 4$ can be written $z = 4 - x - y$ or $f(x,y) = 4 - x - y$.

Table 7.1 is relatively easy to generate with a spreadsheet. Since z is a function of x and y, the third column can be generated with a single formula. Portions of the first two columns can be generated with recursive formulas if the table is built in a systematic way such as the one shown.

The points can be plotted on isometric graph paper on which x-, y-, and z-axes have been sketched. Plotting can also be done with a three-dimensional graphing utility with scatter plot capability. **Figure 7.12** shows the results of plotting the triples in Table 7.1 with a utility.

Corbis

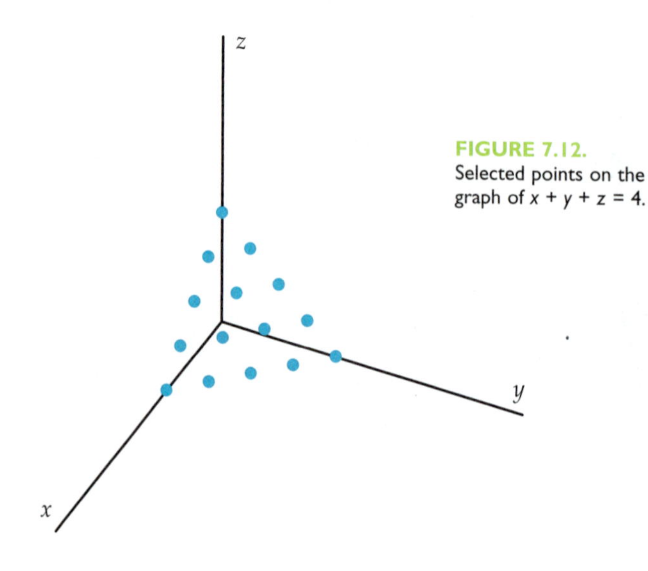

FIGURE 7.12.
Selected points on the graph of $x + y + z = 4$.

The points are part of a plane that is the geometric representation of $x + y + z = 4$. The plane intersects the coordinate system's reference planes in three lines. The intersections with the coordinate planes are called the **traces** of the plane and help in visualizing the graph as a plane. A portion of each trace can be seen by drawing the segments that connect the points that lie in each reference plane (see **Figure 7.13**), i.e., where some coordinate is 0.

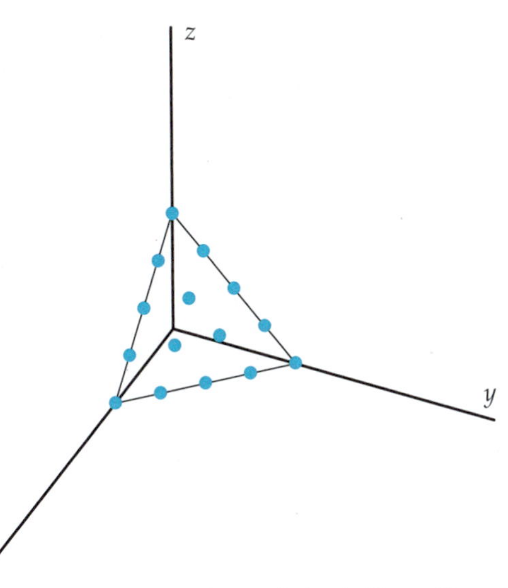

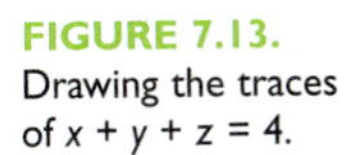
FIGURE 7.13. Drawing the traces of $x + y + z = 4$.

In general, the geometric representation of $Ax + By + Cz = D$ is a plane. However, mathematicians apply the term **linear** to such equations because they involve the first power of each variable. (In a similar manner, the term **quadratic** applies to all polynomials having maximum degree 2 among all its variables.)

A system of three distinct equations of this type is represented by three planes, and the way in which three distinct planes can be configured is an important consideration when solving a system of three equations in three variables. **Table 7.2** compares geometric results with algebraic. Note that the geometric result is not necessarily unique since there may be other configurations that produce the same algebraic result (see Question 1 of Exercises 7.3).

Sample Geometric Result	Algebraic Result
Three planes are parallel (**Figure 7.14**).	The system has no solution.
Three planes intersect in a point (**Figure 7.15**).	There is a single solution: an ordered triple (x, y, z).
Three planes intersect in a line (**Figure 7.16**).	There are infinitely many ordered triples, all of which represent points on the line of intersection.

TABLE 7.2.

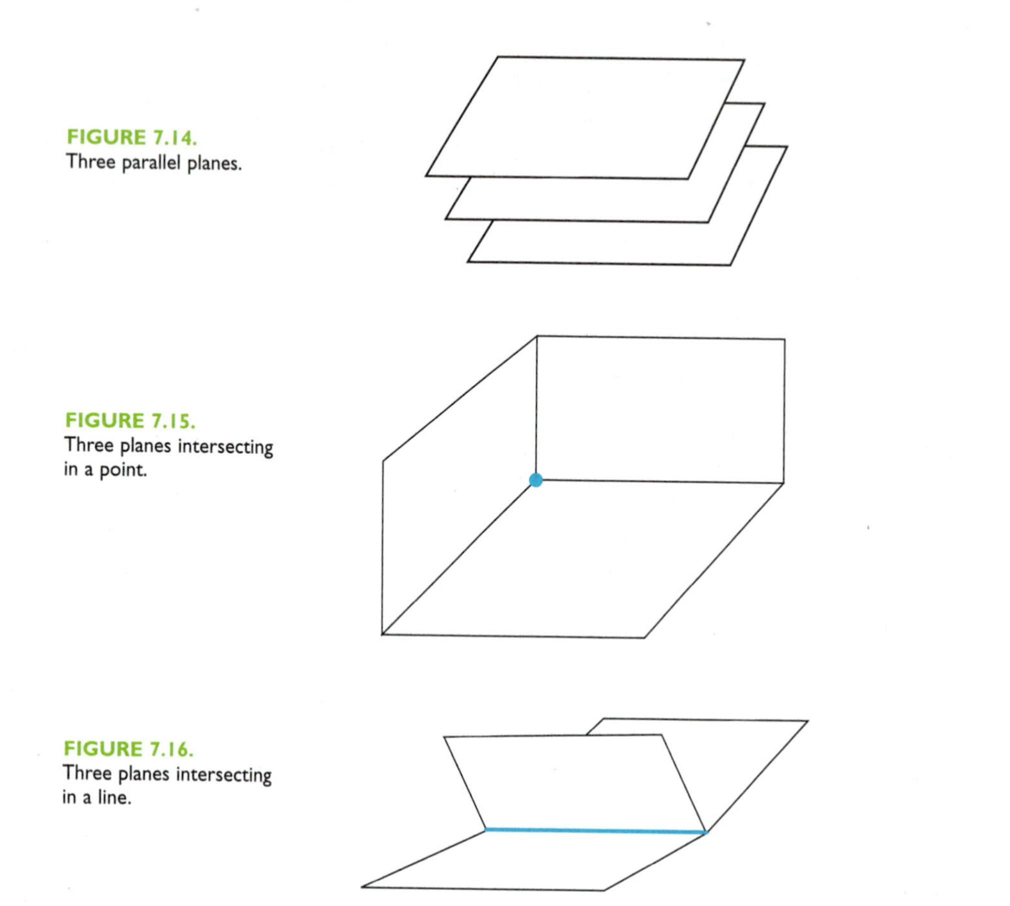

FIGURE 7.14.
Three parallel planes.

FIGURE 7.15.
Three planes intersecting in a point.

FIGURE 7.16.
Three planes intersecting in a line.

SYSTEMS OF THREE EQUATIONS IN THREE VARIABLES

When a system of three equations in three variables has a single solution, it can be found by symbolic means or by using technology to solve a related matrix equation. The symbolic process reduces the system to a system with two variables and two equations and then reduces this system to a single equation with a single variable. (See Example 2 in Lesson 3.2.)

Solutions Using Matrix Inverses

Because of the many multiplications and additions that are necessary to solve a three variable/three equation system by hand, mistakes are common to make. Perhaps the simplest way to find the solution is to represent the system as a matrix equation and use technology to solve the matrix equation.

Example 6

Solve the system $\begin{cases} 3x+2y-z=9 \\ x-4y+3z=-5 \\ 2x-2y-5z=7 \end{cases}$ using matrices.

Solution:

The system is equivalent to the matrix equation

$$\begin{bmatrix} 3 & 2 & -1 \\ 1 & -4 & 3 \\ 2 & -2 & -5 \end{bmatrix} \begin{bmatrix} x \\ y \\ z \end{bmatrix} = \begin{bmatrix} 9 \\ -5 \\ 7 \end{bmatrix},$$

which has the solution

$$\begin{bmatrix} x \\ y \\ z \end{bmatrix} = \begin{bmatrix} 3 & 2 & -1 \\ 1 & -4 & 3 \\ 2 & -2 & -5 \end{bmatrix}^{-1} \begin{bmatrix} 9 \\ -5 \\ 7 \end{bmatrix} = \begin{bmatrix} 2 \\ 1 \\ -1 \end{bmatrix}.$$

Thus, the system's solution is $x = 2$, $y = 1$, $z = -1$, which can be verified by substitution into each of the original three equations.

Solving Singular Systems

Not all systems of three equations in three variables have a single solution. In such cases, the related matrix equation will fail to have a solution because the 3 x 3 matrix used to find the solution has no inverse; the matrix is singular. In these situations, the matrix equation approach of Example 6 is useless, but another matrix technique gives useful information.

Consider the system $\begin{cases} 2x-9y-3z=-20 \\ x-2y-4z=0 \\ x-3y-3z=-4. \end{cases}$

If you attempt to solve the related matrix equation, you find that

$$\begin{bmatrix} 2 & -9 & -3 \\ 1 & -2 & -4 \\ 1 & -3 & -3 \end{bmatrix}$$

has no inverse. (Your calculator may call it a singular matrix.)

The system

$$\begin{cases} 2x - 9y - 3z = -20 \\ x - 2y - 4z = 0 \\ x - 3y - 3z = -4 \end{cases}$$

can be represented with the **augmented matrix**

$$\begin{bmatrix} 2 & -9 & -3 & -20 \\ 1 & -2 & -4 & 0 \\ 1 & -3 & -3 & -4 \end{bmatrix},$$

which has an additional column containing the equations' constant terms. (Your calculator may have a feature that allows you to augment an existing matrix; that is, to add one or more columns to it.) To see how this augmented matrix representation can provide new information about the system of equations, first consider a variation of the procedure used to solve systems symbolically.

A system of equations can be transformed into an equivalent system by adding a multiple of any equation to any other equation. When 3 x 3 systems are solved symbolically, the proper choice of multiples eliminates one of the variables from two of the equations, and so creates a system of two equations in two variables.

Example 7

Show how to add multiples of equations to each other to reduce the system $\begin{cases} 2x - 9y - 3z = -20 \\ x - 2y - 4z = 0 \\ x - 3y - 3z = -4 \end{cases}$ to two equations in two variables.

Sample Solution:

Transform $\begin{cases} 2x - 9y - 3z = -20 \\ x - 2y - 4z = 0 \\ x - 3y - 3z = -4 \end{cases}$ to $\begin{cases} 2x - 9y - 3z = -20 \\ x - 2y - 4z = 0 \\ -y + z = -4 \end{cases}$

by adding –1 times the second equation to the third equation, eliminating x from the third equation.

To complete the reduction to a 2 x 2 system, add –2 times the second equation to the first to obtain $\begin{cases} -5y + 5z = -20 \\ -y + z = -4 \end{cases}$.

You now have two equations in only two variables. This system and the original are equivalent—that is, any solution of one system is a solution of the other.

Skilled manipulation of a system can reduce it to an equivalent form that produces useful information about the system and its geometric representation. The matrix counterpart of this reduced form is known as **reduced row-echelon form**. The reduced row-echelon form is characterized by 0s below a diagonal through the first, second, and third entries of the first, second, and third row, respectively, and by either 1s or 0s along the diagonal:

$$\begin{bmatrix} 1 & \_ & \_ & \_ \\ 0 & 1 & \_ & \_ \\ 0 & 0 & 1 & \_ \end{bmatrix}.$$

Finding the reduced row-echelon form of an augmented matrix is tedious if done by hand. Fortunately, many calculators can find this form automatically.

Figure 7.17 shows how the reduced form of $[A] = \begin{bmatrix} 2 & -9 & -3 & -20 \\ 1 & -2 & -4 & 0 \\ 1 & -3 & -3 & -4 \end{bmatrix}$ is found on one type of calculator.

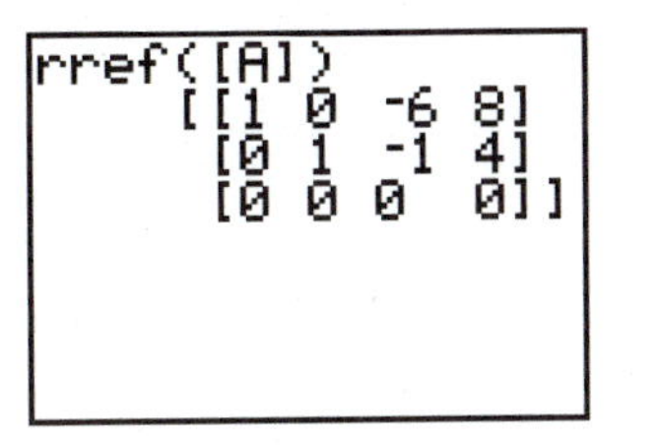

FIGURE 7.17. Selecting reduced row-echelon form from the calculator's matrix math menu (left) and the results (right).

The reduced row-echelon form of $\begin{bmatrix} 2 & -9 & -3 & -20 \\ 1 & -2 & -4 & 0 \\ 1 & -3 & -3 & -4 \end{bmatrix}$ is $\begin{bmatrix} 1 & 0 & -6 & 8 \\ 0 & 1 & -1 & 4 \\ 0 & 0 & 0 & 0 \end{bmatrix}$.

Since the first column of the matrix representation of a system corresponds to the variable x, the second column to y, the third to z, and the fourth to the constant, the algebraic counterpart of the reduced matrix is the system of three equations

$$\begin{cases} x - 6z = 8 \\ y - z = 4 \\ 0 = 0. \end{cases}$$

Of course, the last equation is always true. It can be ignored since it provides no information about x, y, or z.

The system can be modified slightly by writing x and y each in terms of z: $\begin{cases} x = 8 + 6z \\ y = 4 + z. \end{cases}$

Solutions of the original system can then be found by substituting values for z. For example, if $z = 2$, then $x = 8 + 6(2) = 20$, and $y = 4 + 2 = 6$. Thus, (20, 6, 2) is a solution of the original system, which can be verified by substitution into each of the three original equations. Since z can be chosen arbitrarily, an infinite number of solutions can be found for this system. The geometric interpretation of this result is considered in Exercises 7.3, which follow.

Exercises 7.3

1. Table 7.2 gives one geometric interpretation of a system of three equations in three variables with no solution. Discuss other geometric configurations of three planes in which there are no common intersections. Sketch each configuration.

2. Give examples from everyday life that resemble three planes with no points of intersection, one point of intersection, and many points of intersection.

3. Describe the location of the plane $2x - y = 6$. Make a sketch on isometric graph paper with enough sample points to visualize the plane. Include all traces.

4. a) Describe the location of the plane $z = 4$.

   b) What are the equations of the xy-, yz-, and xz-planes?

5. The equation of a line in two dimensions is similar to the equation of a plane in three dimensions. Because of the similarity, one way to think about planes in three dimensions is by drawing analogies with lines in two dimensions.

   a) Find an equation of a line that is parallel to $2x - y = 6$ in two dimensions. Explain the process you used.

   b) How might the process you used to find the equation of a parallel line in two dimensions be extended to the problem of finding a parallel plane in three dimensions? For example, find an equation of a plane parallel to $x + y + z = 5$.

   c) Discuss how the analogy you drew in (b) could be verified.

6. In two dimensions, when a point is not on a given line, you can draw only one line through the point that does not intersect the given line. The line that you draw must be parallel to the given line.

   a) Is this fact true of a point and a line in three dimensions? Explain.

   b) Is a similar fact true of points and planes in three dimensions? Explain.

Exercises 7.3

c) Is the point (3, –2, –1) on the plane $x - y + 2z = 7$? Explain.

d) Find an equation of each plane parallel to $x - y + 2z = 7$ that passes through the point (3, –2, –1).

7. The intercepts of linear functions in two variables are often important in modeling situations. Discuss each of the following intercepts of $2x - y - 3z = 6$.

a) The intersection with the x–axis.

b) The intersection with the y–axis.

c) The intersection with the z–axis.

d) The intersection with the xy–plane.

e) The intersection with the xz–plane.

f) The intersection with the yz–plane.

8. In this lesson's examples, the system

$$\begin{cases} 2x - 9y - 3z = -20 \\ x - 2y - 4z = 0 \\ x - 3y - 3z = -4 \end{cases}$$

was reduced to the equivalent system

$$\begin{cases} x - 6z = 8 \\ y - z = 4 \end{cases}$$

by reducing the augmented matrix

$$\begin{bmatrix} 2 & -9 & -3 & -20 \\ 1 & -2 & -4 & 0 \\ 1 & -3 & -3 & -4 \end{bmatrix}.$$

You may want to re-read the discussion of this system in the lesson before doing this exercise.

a) It is possible to tell by examining the three equations that the geometric representation of the system is three different planes, no two of which are parallel. Explain.

b) Explain why the three planes must intersect in a line.

Exercises 7.3

c) Does the discussion of this system in the lesson lead you to any conclusions about the algebraic representation of lines in space? Explain.

d) In Chapter 6, you found a vector equation and a set of parametric equations for a given line. Find both for the line of intersection for these planes.

9. Although this lesson's examples use systems of three equations in three variables, the solution processes can be applied to any system in which the number of equations equals the number of variables.

Consider the two-variable system $\begin{cases} 2x - 3y = 7 \\ 5x + 4y = 19. \end{cases}$

a) Show how to solve this system by solving a matrix equation.

b) Show how to solve this system by reducing an augmented matrix.

c) Give the system's solution in exact form. (You can solve the system by hand, use a conversion-to-fraction feature if your calculator has one, or apply the determinant procedure described in Lesson 7.2.)

d) Interpret the results of this exercise geometrically.

10. Suppose a particular system of three equations in three variables, x, y, and z is solved by finding a reduced row-echelon matrix. Compare the algebraic meanings of [0 0 0 1] and [0 0 1 0] as final rows of the reduced row-echelon matrix. One of these implies there is a single solution, the other implies there is no solution.

a) Which is which? Explain.

b) A system of equations includes the equations $2x + y - 3z = 11$ and $2x + y - 3z = 12$. Show how adding a multiple of one of the equations to the other justifies the answer you gave in (a).

Exercises 7.3

11. Solve each of the following systems by two methods: writing and solving a matrix equation; reducing an augmented matrix. Interpret the results geometrically. If the geometric interpretation is a line, give a vector and a parametric representation of the line.

a) $\begin{cases} 2x - 3y + z = 6 \\ x + 4y + 2z = 11 \\ 3x - y - z = -5 \end{cases}$

b) $\begin{cases} 5x - 2y - 4z = 7 \\ 3x + 6y + 3z = 5 \\ x - 2y - 2z = 13 \end{cases}$

c) $\begin{cases} 3x + y - 7z = -1 \\ x + y - z = 3 \\ 2x + y - 4z = 1 \end{cases}$

12. The techniques applied to systems of three equations in three variables can, in general, be applied to systems of n equations in n variables. However, geometric interpretations are difficult unless you have a thorough understanding of four or more spatial dimensions, which very few people do. Apply the techniques you have learned to the following systems. Determine a single solution, if it exists; if it does not, state whether there are no solutions or many.

a) $\begin{cases} 2a + 5b - 3c + 4d = 17 \\ a - 3b - 4c + d = -5 \\ 2a - b + 2c - 2d = 9 \\ 3a + 2b + 5c + 3d = -8 \end{cases}$

b) $\begin{cases} a + b + 3c + 3d = -2 \\ 3a + b + 7c + 13d = 4 \\ a + 2b + 5c - 2d = -1 \\ 2a + b + 5c + 8d = 2 \end{cases}$

13. The technique of reducing an augmented matrix can be applied to systems in which the number of variables and the number of equations are unequal. Apply the reduction technique to each of these systems and discuss the results.

a) $\begin{cases} 2a + 3b - 4c = 2 \\ a - b + c = 7 \end{cases}$

b) $\begin{cases} 2a + 3b - 4c = 2 \\ a - b + c = 7 \\ a + b = 5 \end{cases}$

c) $\begin{cases} 2a + 3b - 4c = 2 \\ a - b + c = 7 \\ a + b = 5 \\ b + c = 3 \end{cases}$

14. Systems of equations are useful in solving problems about resource allocation under certain constraints. As an example, consider a hypothetical agricultural situation. The problem is to determine the number of acres of each of three crops to be planted in 800 available acres of farm land. Per-acre cost of seed is \$45 for crop A, \$15 for crop B, and \$30 for crop C. Planting crop A requires 5 hours of labor per acre; planting crop B requires 5 hours of labor per acre; and planting crop C requires 4 hours of labor per acre. \$25,000 is budgeted for seed, and there are 3600 hours of labor available. Create either a matrix equation or a system of equations to model the situation in which all resources are fully used, and find all solutions.

© Susan Van Etten

15. Reconsider the situation in Exercise 14 if D, a fourth crop, has a seed cost of \$25 per acre and requires 4 hours of labor per acre to plant. (Information for the other three crops remains the same, as does the available money for seed and hours of labor.) Determine the number of acres of each crop to be planted.

CHAPTER 7

MATRICES

Chapter 7 Review

1. Summarize the important mathematical ideas of this chapter.

2. **Table 7.3** shows nutritional content per serving of various foods.

Food	Serving Size	Calories	Protein (grams)	Cholesterol (grams)	Carbohydrates (grams)	Fiber (grams)
Halibut, baked	3 oz	119	22.7	35	0	0
White rice, cooked	1 cup	267	4.8	0.1	59	0.21
Grapes, green	1 cup	102	1	0	27	0.70
Pinto beans, cooked	1 cup	235	14	0	44	5.2
Ground beef, lean	4 oz	298	28	99	0	0
Peanut butter, creamy	1 T	95	4.6	0	3	0.53
Chicken, light w/o skin, roasted	3.5 oz	173	30.9	85	0	0
Carrots, cooked	0.5 cup	35	0.9	0	8	1.2
Potato, baked w/skin	1 large	220	4.7	0	51	1.2

TABLE 7.3.

Source: Kirschman, Gayla J. and John D. Kirschman. 1996. *Nutrition Almanac* 4th ed. New York: McGraw Hill.

The following amounts of these foods are consumed in a week: baked halibut, 6 oz; white rice, 2 cups; green grapes, 2 cups; cooked pinto beans, 2 cups; ground beef, 12 oz; creamy peanut butter, 6 tablespoons; light chicken, 7 oz; cooked carrots, 2 cups; 3 large baked potatoes.

Show how all of this information can be stored in matrices and how matrix multiplication can be used to determine the nutritional content of the foods consumed in a week.

3. Show how to use matrix multiplication to rotate the triangle with vertices at (5, 2), (7, –1), and (4, –2) through an angle of 90° about (0, 0). Make a sketch showing the triangle and the rotation.

4. Show how to use matrix multiplication to find the number of ways to go from vertex A to vertex B in **Figure 7.18** using exactly three edges. Identify clearly in your matrix product which entry (or entries) count these routes. Then verify your matrix solution by listing each route.

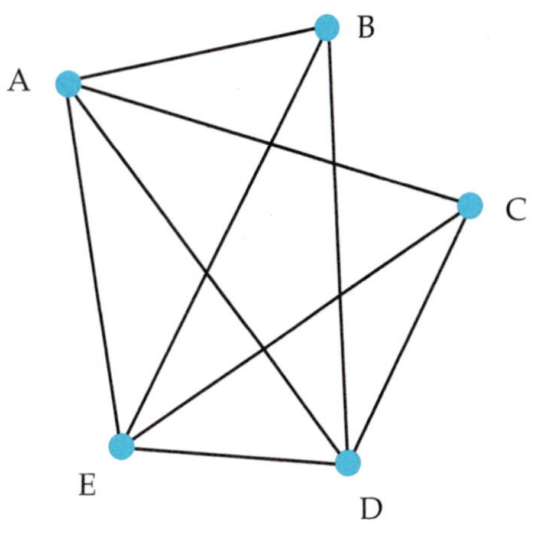

FIGURE 7.18.

5. Solve the following matrix equation. Show your work and round approximations to two decimal places.

$$\begin{bmatrix} 2 & -4 & -1 \\ -3 & 3 & 5 \\ 1 & 2 & 3 \end{bmatrix}\begin{bmatrix} x \\ y \\ z \end{bmatrix} = \begin{bmatrix} 3 \\ 4 \\ -1 \end{bmatrix}$$

6. Give an exact solution for the matrix equation in Exercise 5.

7. Solve the following system of equations. Give the solutions in exact form.

$$\begin{cases} 5a - 2b + c + d = 31 \\ 2a + 3b + 2c - 2c = -22 \\ 3a + 7b - 4c - 5d = 17 \\ 2a - b - c + 4d = -11 \end{cases}$$

8. Sketch the portion of the plane $2x + y + z = 6$ that has all coordinates positive. Discuss the traces in the xy-, yz-, and xz-planes, and intercepts with the x-, y-, and z-axes.

9. Show how the following system can be solved by two different methods. Give the answers in exact form and give a geometric interpretation of the result.

$$\begin{cases} x + 2y + 2z = 7 \\ x + y + 3z = 11 \\ 4x + y + 5z = 29 \end{cases}$$

10. Solve the following system and give a geometric interpretation of the results.

$$\begin{cases} 5x + 2y + z = 9 \\ 3x + y + z = 6 \\ x + 2y - 3z = -3 \end{cases}$$

11. In this chapter, you have seen that augmenting a matrix with a single column and finding the reduced row-echelon form of the augmented matrix provides useful information about a system of equations. It is also possible to augment a matrix by adding several columns. For example, the 3 x 3 matrix

$$\begin{bmatrix} 3 & 2 & 3 \\ 2 & 1 & 3 \\ 4 & 2 & 5 \end{bmatrix}$$

can be augmented with three additional columns representing the 3 x 3 identity matrix:

$$\begin{bmatrix} 3 & 2 & 3 & 1 & 0 & 0 \\ 2 & 1 & 3 & 0 & 1 & 0 \\ 4 & 2 & 5 & 0 & 0 & 1 \end{bmatrix}.$$

When this matrix is reduced, it provides useful information about the original matrix. Find the reduced row-echelon form of this matrix and interpret the results. (Hint: Write a square matrix from the last three columns of the result and multiply it by the original matrix.)

12. A retail appliance company uses three types of vehicles for deliveries: vans, small trucks, and large trucks. The capacities of these vehicles are shown in **Table 7.4**.

TABLE 7.4.

Type of appliance	Van	Small Truck	Large Truck
Range, dishwasher, washer, dryer	2	4	8
Refrigerator	0	1	3
Air conditioner, microwave oven	3	5	6

For example, the ideal full load for a small truck is four ranges, dishwashers, washers, or dryers, one refrigerator, and five air conditioners or microwave ovens.

The company has several of each type of vehicle.

On a given day, the company has 26 of the first group of appliances, 6 refrigerators, and 30 from the last group of appliances scheduled for delivery.

a) Can the delivery be accomplished with fully loaded vehicles? Explain.

b) Suggest other ways that the company could accomplish the delivery. Discuss what might be an optimal solution.

13. In Exercise 12, suppose a customer calls at the last minute and cancels a dishwasher order. How do you think the company should reschedule the vehicles?

Chapter 8

ANALYTIC GEOMETRY

Corbis

CHAPTER INTRODUCTION

Automated Guided Vehicles (AGVs) are considered by many to be the most flexible material-handling systems used by industry. As these vehicles move about within a factory, they require continuous control by either an on-board computer or one that is centralized in the factory. The computers that operate these vehicles must pay strict attention to location and distance in order for the AGV to pick up and drop off materials at specified locations. AGVs must be able to follow optimal routes that avoid collisions with other vehicles, equipment, and people.

Several different methods are used to guide AGVs. Fixed path guidance, such as a track on the factory floor, is probably the least expensive method, but it does not lend itself to layout changes easily. Free path guidance, on the other hand, is more flexible, but is probably more expensive because it requires programmed software.

Mathematics plays a large part in programming the computers that control these vehicles or other similar devices. In order for these machines to traverse some "optimal" path, the person programming the movements of the AGV must be familiar with the layout of the factory floor and have some way of mathematically modeling the desired path. With the help of **analytic geometry**, the study of geometric figures by algebraic representation, many situations in today's society, such as the automated guided vehicle example, can be examined, explained, and modeled.

Analytic Geometry and Loci

In previous chapters, you used algebra to help you solve problems and model real world events. In this lesson you will use algebra to study geometric figures, to deduce properties from these figures, and to solve problems that can be modeled geometrically. The study of geometric figures by algebraic representation is known as analytic geometry or **coordinate geometry**.

Activity 8.1

Draw two points, A and B, on a piece of dot paper or graph paper (see **Figure 8.1**).

• A

• B

FIGURE 8.1. Two points.

1. Use your ruler to help draw a point that is equidistant from both points. Label this new point C. Is C the only point that is equidistant from A and B?

2. Draw at least ten more points that are equidistant from A and B. Are there still more? How many more?

3. How might you describe the set of all points that are equidistant from the two given points A and B? Justify your answer.

A LOCUS OF POINTS

In mathematics, the word **locus** is used to define a set of points whose location satisfies a certain condition or conditions. The plural of locus is **loci** (*lo ci*). In Activity 6.1, you drew a subset of the locus of points that were equidistant from two given points. A drawing of the entire set would show the perpendicular bisector of the segment joining the two given points.

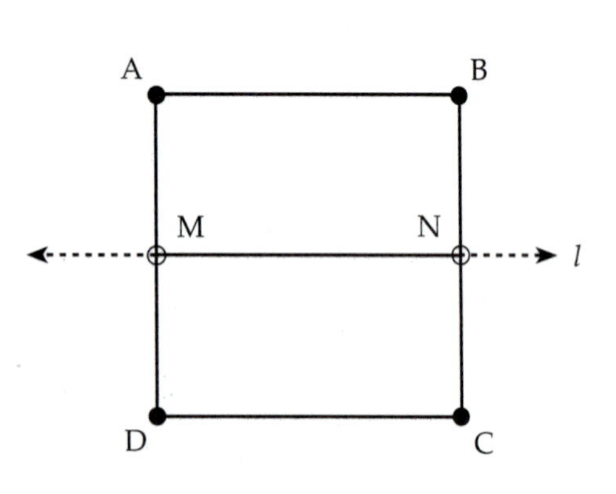

FIGURE 8.2.
Locus of points that satisfies more than one condition.

When a locus must satisfy more than one condition, each condition must be taken into account. For example, to find the locus of points in the interior of a given square, ABCD, and equidistant from one pair of opposite sides, AB and CD, you must determine the locus of points that satisfies both conditions. **Figure 8.2** shows the dotted line l, the locus of points equidistant from the pair of opposite sides. The points that satisfy the second condition are all points in the interior of the square. Hence, the desired locus is the points on that line that are also in the interior of the square. These are shown by segment MN, excluding the endpoints M and N.

ANALYTIC GEOMETRY

In determining the locus of points that meets certain conditions, it is often helpful to use analytic geometry. Analytic geometry is sometimes referred to as coordinate geometry, since the objects are described by coordinates in the plane or space.

As you may remember from previous courses, coordinate geometry was invented by Déscartes in the 17th century. He was the first to recognize the power of using numbers to identify locations of points, and it was through his efforts that the unification of the algebra and geometry of his time was achieved. Coordinate geometry provides one of the most important tools available to the mathematician. Without it, computer graphics would be impossible.

When working a geometry problem, it is sometimes easier and more useful to use a coordinate representation of the problem than to look at it purely geometrically. For example, using a coordinate description of Activity 8.1, it is possible to verify that the locus of points (x, y) that is equidistant from two given points, A and B, is a line perpendicular to the line segment AB (see **Figure 8.3**). And since every point on the line is equidistant from A and B, the line also bisects the segment AB. Hence, the locus must be the perpendicular bisector of segment AB (see Exercise 10).

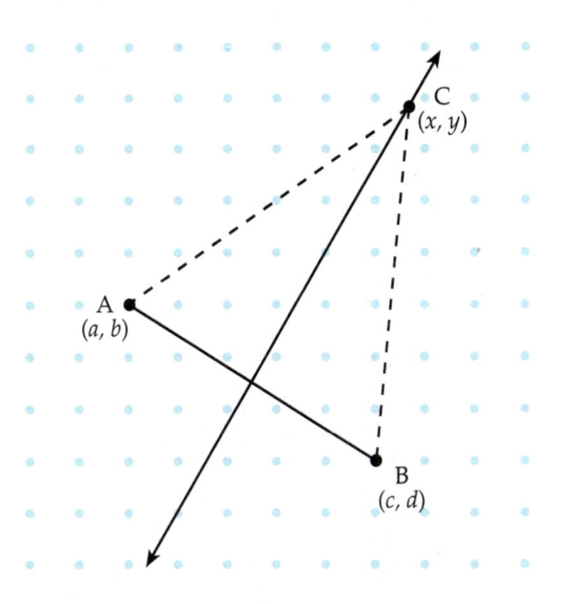

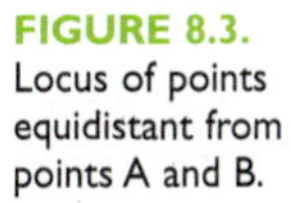

FIGURE 8.3.
Locus of points equidistant from points A and B.

REPRESENTING FIGURES WITH COORDINATE GEOMETRY

When examining a geometric problem analytically, you need to think about more than just the figure itself. The figure must first be placed on the coordinate plane in some convenient location. Locating a polygon with one vertex at the origin and one or more of its sides on the axes may help simplify the coordinates of the vertices (see **Figure 8.4**). Placing a figure so that it has one of the axes as a line of symmetry can also be helpful (see **Figure 8.5**).

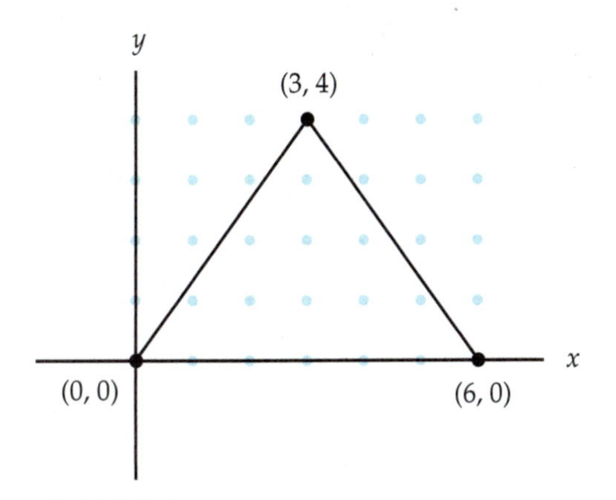

FIGURE 8.4.
Isosceles triangle with one vertex at the origin.

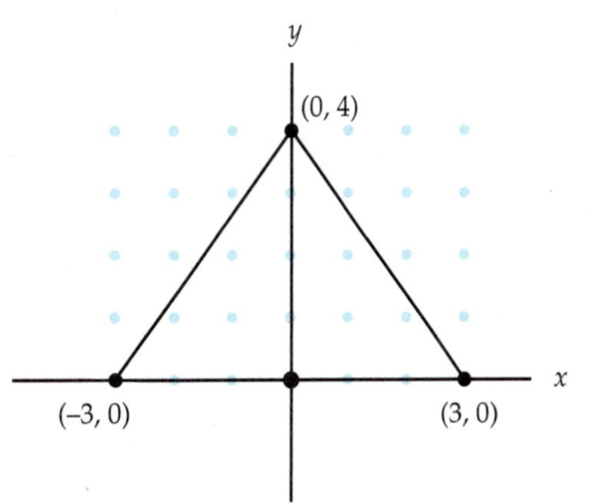

FIGURE 8.5.
Isosceles triangle with the y-axis as a line of symmetry.

After a geometric figure is located on the coordinate plane, it is possible to assign coordinates to the vertices. Algebraic equations can then be used to describe lines, and formulas can be used to find midpoints, lengths of segments, and slopes of lines.

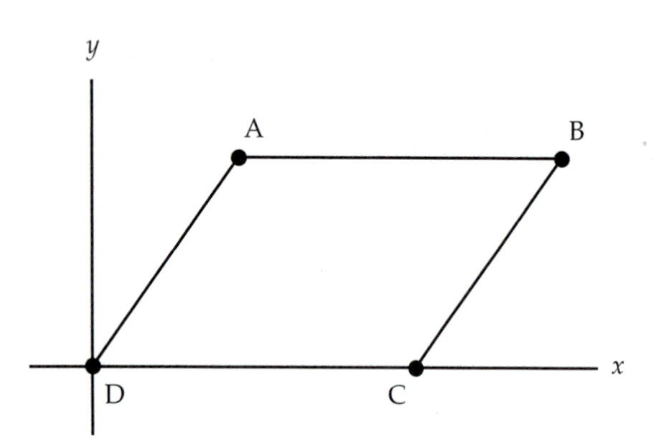

FIGURE 8.6.
Parallelogram ABCD.

EXAMPLE 1

Given parallelogram ABCD.

Prove that the opposite sides are equal in length.

SOLUTION:

Begin by placing parallelogram ABCD on the coordinate plane (see **Figure 8.6**).

By definition, a parallelogram is a quadrilateral with both pair of opposite sides parallel. If the figure is placed so that one vertex, D, is at the origin, and side DC on the x-axis (see **Figure 8.7**), then D = (0, 0) and assuming DC is a units in length, then vertex C = $(a, 0)$. If you also assume A = (b, c), then the slope of AD is c/b, which must equal the slope of BC. So B = $(a + b, c)$.

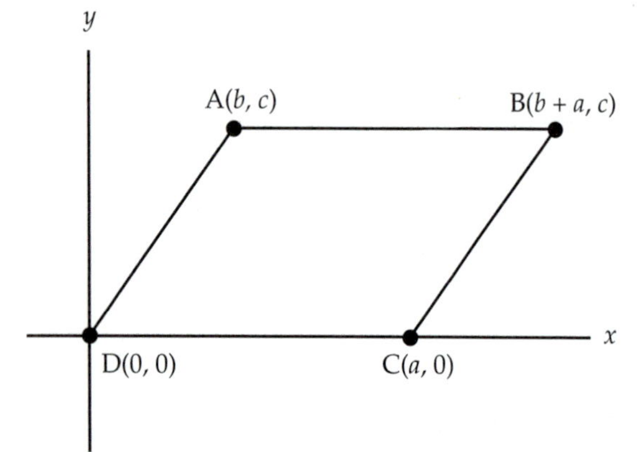

FIGURE 8.7.
Parallelogram with labeled vertices.

Once the coordinates of the vertices are determined, the distance formula can be used to find the lengths of all four sides and show that the opposite sides are equal in length.

It is given that DC = a

$$AB = \sqrt{((b+a)-b)^2 + (c-c)^2} = \sqrt{a^2 + 0^2} = a$$

$$DA = \sqrt{(b-0)^2 + (c-0)^2} = \sqrt{b^2 + c^2}$$

$$BC = \sqrt{((b+a)-a)^2 + (c-0)^2} = \sqrt{b^2 + c^2}$$

HELPFUL INFORMATION

As you saw in the previous example, the distance formula is helpful in finding the distance between two points once the coordinates of the points have been determined. In your work in previous courses, you used many formulas such as the distance formula and midpoint formula, and you also found equations of lines when information about the lines was provided. As your studies in analytic geometry proceed, the following definitions, theorems, and formulas will be useful:

Slope of a line:

The slope of a line through two points (x_1, y_1) and (x_2, y_2) is $\frac{y_2 - y_1}{x_2 - x_1}$.

Parallel and perpendicular lines:

Two distinct, nonvertical lines are parallel if and only if the lines have the same slope.

Two nonvertical lines are perpendicular if and only if the product of their slopes is –1.

Distance Formula:

In the coordinate plane, the distance between two points (x_1, y_1) and (x_2, y_2) is $\sqrt{(x_2 - x_1)^2 + (y_2 - y_1)^2}$.

Midpoint Formula:

In the coordinate plane, if a segment AB has endpoints (a, b) and (c, d) respectively, the midpoint of segment AB is $\left(\frac{a+c}{2}, \frac{b+d}{2}\right)$.

Equation of a line when given

a) the slope m, and the y-intercept $(0, b)$: $y = mx + b$

b) a point (h, k) on the line and the slope m: $y - k = m(x - h)$ or $y = k + m(x - h)$

Exercises 8.1

1. a) Find an equation of the line AB that contains the points A(5, 2) and B(–3, 8).

   b) Find the midpoint P of AB.

   c) Show that P is a point on the line AB.

2. A and B are the endpoints of the base of the isosceles triangle ABC.

   a) If A = (–1, 8) and B = (7, 4), give coordinates for vertex C.

   b) Use the distance formula to show that your ΔABC is indeed isosceles.

   c) Check with others in your class. Does everyone have the same vertex for C? If not, explain how this can happen.

3. Rectangle ABCD has vertices A(4, 8), B(8, 2), and C(5, 0).

   a) Find the coordinates of vertex D. Explain how you found the fourth vertex.

   b) Verify that your quadrilateral is a rectangle.

4. Prove that ΔABC, with A(3, 9), B(4, 0), and C(–1, 4), is a right triangle (see **Figure 8.8**).

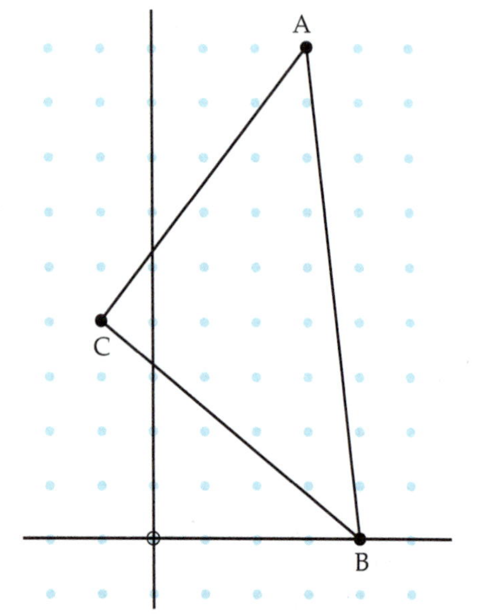

FIGURE 8.8.

5. If the lengths of the sides of a triangle are fixed, the triangle is rigid. That is, its size and shape cannot change. In fact, triangles are the only polygons that possess this property. Due to their rigid nature, triangles play a very important part in the construction industry. Just look around you; there are many instances of triangles being used to increase the strength of a structure. One example with which you may be familiar is the use of building trusses to create the framework for roofs of houses.

Corbis

Exercises 8.1

The truss shown in **Figure 8.9** is a 1:4 truss, which means that for every four units of horizontal distance of construction material, the truss rises or falls one unit.

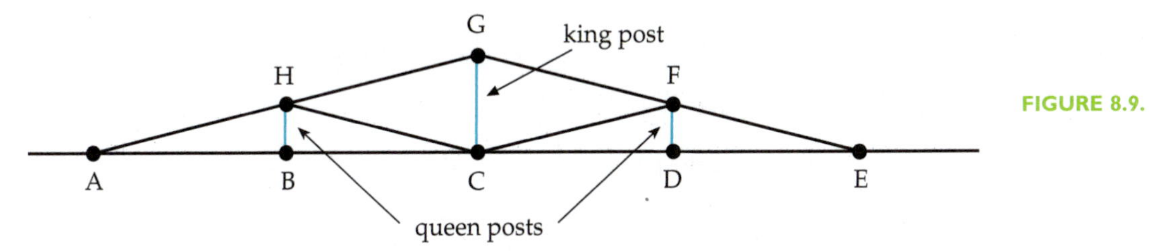

FIGURE 8.9.

a) Assume the truss is 16 feet long, how tall is the king post?

b) Use Figure 8.9 along with the following information to draw the truss on a rectangular coordinate system and label the vertices. The truss is 1:4. The length of AB is b units. C is the midpoint of AE. B and D are midpoints of AC and CE, respectively. Triangle AGE is isosceles. $HB \perp AE$ and $FD \perp AE$.

c) Use your coordinatized figure to find the lengths of the boards AE, AG, GE, BH, CG, DF, CH, and CF in terms of b.

d) In building a house, a truss is needed with overall length 12 meters. Use your information to find the lengths of boards AG, GE, BH, CG, DF, CH, and CF to the nearest tenth of a meter.

e) At what angle do AE and AG meet in part (d)?

6. An automated guided vehicle needs to maneuver from its home base P_1 to another location P_2 (see **Figure 8.10**). In order for it to get from P_1 to P_2, it must move between two parallel banks of shelves. Since it is very important that the robot not touch either shelf, describe the best path for the robot to take.

P_2

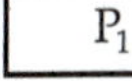

FIGURE 8.10.

Exercises 8.1

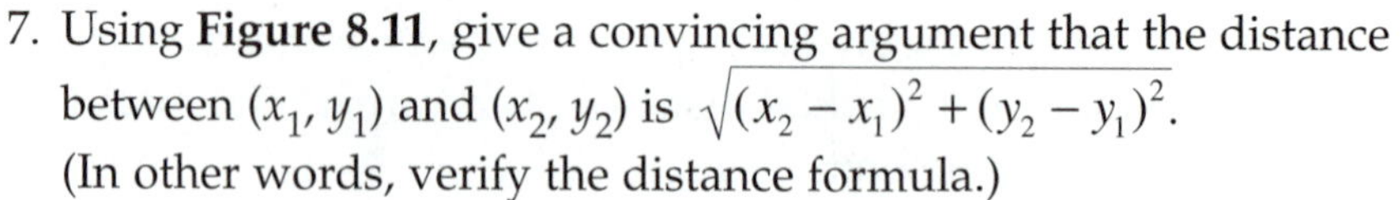

7. Using **Figure 8.11**, give a convincing argument that the distance between (x_1, y_1) and (x_2, y_2) is $\sqrt{(x_2 - x_1)^2 + (y_2 - y_1)^2}$. (In other words, verify the distance formula.)

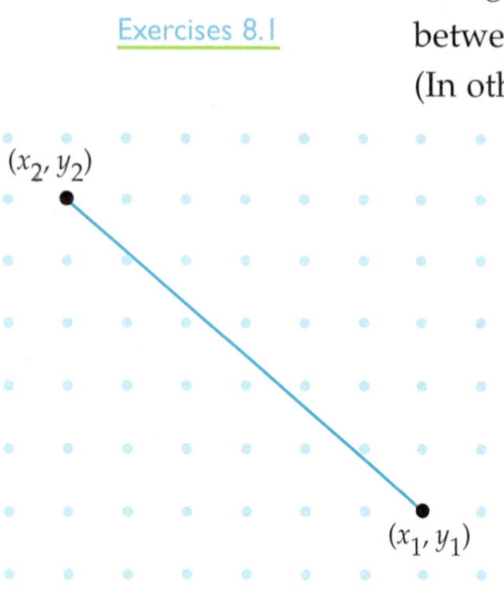

FIGURE 8.11.

8. Prove that M(2, 1) is the midpoint of AB if A = (–2, –4) and B = (6, 6) by showing

   a) that MA = MB, and

   b) that (2, 1) lies on the line segment AB.

   c) To prove that M is the midpoint of AB, why is it not sufficient only to show that MA = MB?

9. It is very important for archaeologists to keep precise records of locations of discoveries at archaeological sites. To do so, site plans are keyed to some carefully selected point such as a survey beacon or landmark that appears on a large-scale map. From this point, a grid of squares made of rope or heavy string can be laid out over the area. The grid provides a system of coordinates, which is vital for record keeping. Many systems are set up so that one axis points north.

   Assume the origin of the grid in **Figure 8.12** is in the lower left corner.

   a) What are the approximate coordinates of the tip of the tail of the dinosaur?

   b) What are the coordinates of the tip of its head?

   c) If the scale in Figure 8.12 is 1 linear unit = 1 meter, approximately how far is it from the tip of the dinosaur's tail to the tip of its nose?

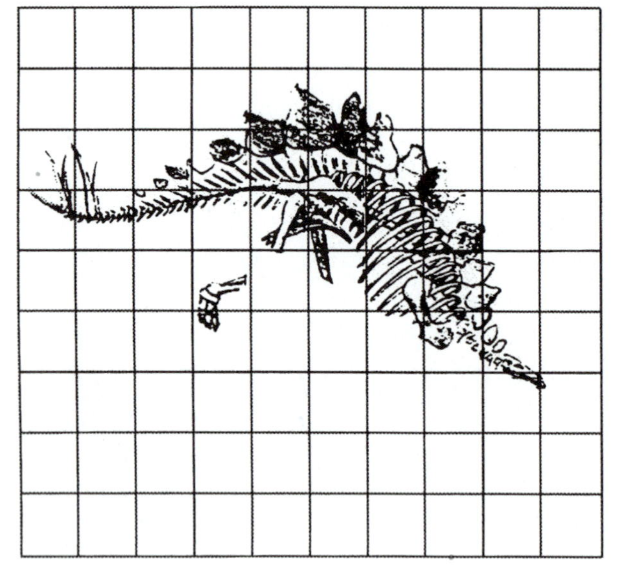

FIGURE 8.12.

PEOPLE AND MATH

If you are interested in statistics and computer science and are fascinated with the mysteries of the past, a career as an **archaeologist** might be for you. Archaeologists spend much of their time searching for areas likely to be rich in artifacts. Once they select such an area, a rectangular quadrant with an accurate grid system is constructed and only then can the digging and reclamation of artifacts proceed.

As artifacts are discovered, the location of each within the grid is carefully documented. The chemical composition of all objects unearthed at the dig becomes part of the huge amount of data needing to be analyzed. Even the amount and types of pollen embedded in layers of sediment become tools for recreating the past by providing clues about climatic and environmental changes. Once they gather all of the data, archaeologists can use computer programs to analyze what the data reveal about the past.

Exercises 8.1

10. In Activity 8.1, you drew a subset of the locus of points equidistant from two given points A and B. In this exercise, you will use analytic geometry to represent the desired locus with an equation and prove that it is the perpendicular bisector of the segment joining A and B.

    Figure 8.13 shows points A and B placed on the coordinate plane with A = (a, b) and B = (c, d). Let the point C represent any point in the desired set of points (the locus) that is equidistant from A and B. Since C is a variable point, its coordinates can be represented with variables such as (x, y).

FIGURE 8.13.

    a) Use the distance formula to find the distances from A to C and from B to C.

    b) Use the fact that the distance from A to C must equal the distance from B to C to find an equation for the locus of points that is equidistant from A and B. (Hint: Don't forget that x and y are the variables in the equation and that a, b, c, and d are constants.)

    c) What kind of equation is the equation you found in (b)?

    d) What is the slope of the equation from (b)?

    e) Find the slope of AB.

Exercises 8.1

f) What can you say about the locus of points and the segment AB? Explain.

g) Explain why you can say that the locus is the perpendicular bisector.

11. The distance from point A to (2, 4) is 5.

a) Could point A be (–2, 1)? Why or why not?

b) Name three other possible locations for A.

12. Sketch and describe each locus.

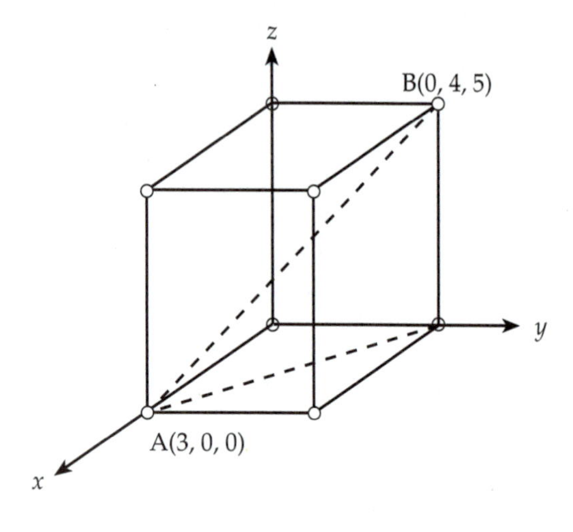

FIGURE 8.14.

a) In a plane, points equidistant from two given parallel lines.

b) In a plane, points equidistant from the sides of an angle.

c) In a plane, points that are a given distance d from a given line l.

d) In a plane, points that are equidistant from two intersecting lines.

e) In a plane, points that are equidistant from two intersecting lines and a given distance d from the point of intersection of those lines.

f) In a plane, points a fixed distance r from a given point.

13. Use coordinate geometry to show that the medians of triangle ΔABC are concurrent (intersect in a single point). A = (1, 2), B = (7, 2), and C = (5, 6). Give the coordinates of the point of concurrency.

14. The distance between two points can be found in space just as it can be found in the plane. Use **Figure 8.14** to help you find the distance between A(3, 0, 0) and B (0, 4, 5).

15. The distance formula for a two-dimensional coordinate system can be extended to a three-dimensional coordinate system. Use Figure 6.14 in Exercise 14 to help you complete the following statement: In three dimensions, the distance between two points (x_1, y_1, z_1) and (x_2, y_2, z_2) is ____________________.

Modeling with Circles

From your work in previous chapters, you are familiar with equations for different functions such as linear, exponential, sine, and polynomial. In this lesson, you will explore a familiar figure—the circle. You will then apply what you know about circles and the distance formula to find an equation for any given circle.

Activity 8.2

1. Choose a point C on a sheet of paper. Use your ruler or other measuring device to locate at least 10 points that are exactly 4 cm from C.

2. Describe the locus of points that is 4 cm from C.

3. Consider a circle with center at (0, 0) and a radius of 3 units. Use coordinate geometry to find an equation of the locus of points (see **Figure 8.15**).

FIGURE 8.15.

4. Find an equation for a circle with its center at the origin and radius equal to r units.

5. Now consider translating the circle so that its center is no longer at the origin. Find an equation of the circle whose center is at (2, 3) and radius = 4 units.

6. For a final generalization, find the equation of a circle with a center (h, k) and a radius of r.

CIRCLES

Fascination with circles goes back beyond recorded history. The first recorded study of the circle dates back to around 650 B.C., when proofs of several theorems are attributed to Thales. Even Euclid in Book III of the *Elements* wrote about the properties of circles.

The study of pi (π) also predates recorded history. The early values of π were probably found by measurement. The first recorded values can be found in the Egyptian Rhind Papyrus, a collection of mathematical examples copied by a scribe named Ahmes around 1650 B.C. The first theoretical calculation of $\frac{223}{71} < \pi < \frac{22}{7}$ is attributed to Archimedes (287–212 B.C.).

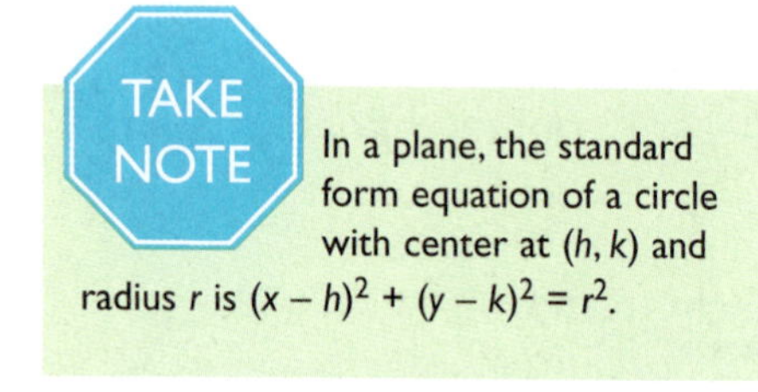

Since elementary school you've studied circles and used the traditional definition that a circle is a set of points in a plane a given distance from a point called the center. In Activity 8.2, you used your definition, started with a specific case, noted relationships, and then made generalizations to help you discover the equation for any circle in a plane.

EXAMPLE 2

Write an equation of a circle with center (–3, 1) and radius 6.

SOLUTION:

If the center and the radius of a circle are known, an equation for the circle can be found by substituting the coordinates of the center for h and k, and the length of the radius for r into the form $(x - h)^2 + (y - k)^2 = r^2$. Thus, the equation of the circle in standard form is $(x - (-3))^2 + (y - 1)^2 = 6^2$ or $(x + 3)^2 + (y - 1)^2 = 36$.

EXAMPLE 3

Find the center and the radius of the circle whose equation is $(x - 4)^2 + (y + 3) = 12$.

SOLUTION:

Since the equation for the circle is given in standard form, its center and radius are easily identified. Its center is $(4, -3)$ and the radius is $\sqrt{12}$ or $2\sqrt{3}$.

Finding the center and the radius is not difficult if the equation for the circle is given to you in standard form $(x - h)^2 + (y - k)^2 = r^2$. But what do you do if the equation is not in that form? For example, consider the equation $x^2 - 2x + y^2 + 4y - 3 = 0$. Could this be an equation for a circle? If so, what are the center and radius of the circle?

In order to put this equation into the form you recognize, you need to recall a process called "completing the square." To complete the squares, you must add values to the left side of the equation that will make $x^2 - 2x$ and $y^2 + 4y$ perfect squares. Add 1 to the expression $x^2 - 2x$, add 4 to $y^2 + 4y$ and don't forget to add 5 to the right side of the equation, since you added a total of 5 to the left.

You now have $x^2 - 2x + 1 + y^2 + 4y + 4 = 3 + 5$, which can be rewritten as $(x - 1)^2 + (y + 2)^2 = 8$. Now you can recognize that this is the equation for a circle whose center is $(1, -2)$ and radius is $\sqrt{8}$.

Exercises 8.2

1. Write an equation for each of the following circles:

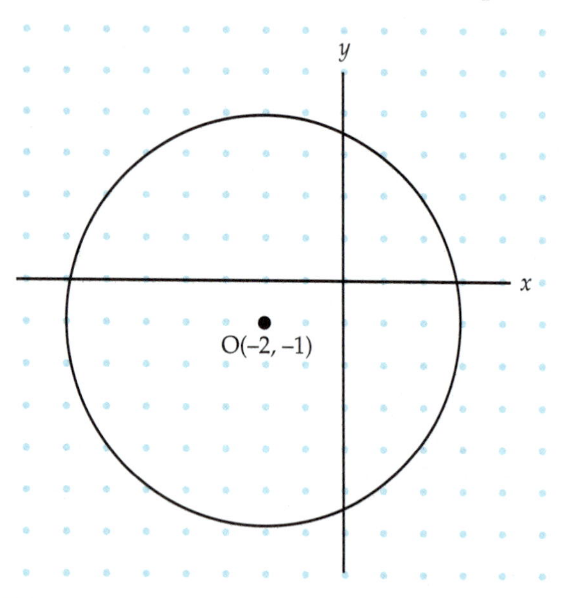

FIGURE 8.16.

a) Center (2, –5) and radius 7.

b) See **Figure 8.16.**

c) Endpoints of the diameter are (2, –2) and (0, 4).

d) Center (–2, 3) and contains the point (1, 5).

2. Find the center and radius of each circle:

a) $x^2 + (y + 4)^2 = 17$

b) $(x - a)^2 + (y - b)^2 = 100$

c) $x^2 + 3x + y^2 - 6y + 1 = 0$

3. Use the function grapher on your graphing calculator to graph the circle $x^2 + y^2 = 25$.

a) What do you have to do to get the equation into a form the calculator can understand?

b) Does the graph look like a circle?

c) Use the function grapher on your graphing calculator to graph $(x + 3)^2 + (y - 4)^2 = 25$.

4. a) Does the point (7, –1) lie on the circle $(x - 3)^2 + (y + 4)^2 = 25$? Explain how you know.

b) With respect to the circle in part (a), where does the point (5, 1) lie? Explain how you know.

c) When would a point lie inside the circle?

5. a) Sketch as many circles as you can that are tangent to both the x- and y-axes and have a radius of 5 units.

b) Give an equation for each of the circles sketched in (a).

Exercises 8.2

6. Describe and sketch the locus of points defined by each of the following:

a) $(x - 3)^2 + y^2 \leq 9$

b) $(x + 2)^2 + (y - 1)^2 > 10$

7. Does the equation in parts (a)–(e) represent a circle? Explain why or why not.

a) $x^2 + y^2 = 16$

b) $x^2 + 5y^2 = 16$

c) $3(x - 7)^2 + 4(y - 1)^2 = 50$

d) $2(x - 3)^2 + 2(y + 1)^2 = 36$

e) $x^2 + 3 = 21 + y^2$

8. How can you tell by looking at an equation if it is an equation for a circle?

9. In parts (a)–(c), translate the graph of the equation two units to the right and three units up. Write an equation for the new graph.

a) $y = x^3$

b) $y = |x|$

c) $x^2 + y^2 = 9$

10. In geometry, it is known that the tangent to a circle at a given point P is perpendicular to the radius drawn to P. Find an equation of the line AB that is tangent to the circle at point P in **Figure 8.17**.

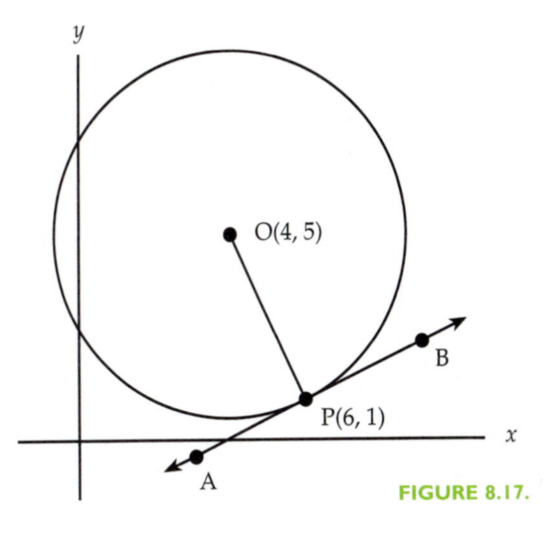

FIGURE 8.17.

Exercises 8.2

11. Often, people have something that is circular in shape and need to find the center. For example, a gardener may want to plant a tree or place a fountain exactly in the center of a circular garden. In order to find a way of locating the center of a circle, a bit of discovery is needed on your part.

© Susan Van Etten

a) Draw a circle on a piece of paper by tracing around a circular object such as a jar lid. Cut it out and by folding the cut-out circle, find its center. Explain what you did and how you know the folds determined the center.

b) Place a point P on your paper. Use a compass, string, or drawing utility such as Geometer's Sketchpad or Cabri to draw a circle with center at P and radius approximately 2 inches. Draw a **chord** AB on your circle. (Remember that a **chord** of a circle is a line segment joining two points on a circle.) Use the drawing utility, Plexiglas mirror, paper folding, or compass to construct the perpendicular bisector of AB. What do you notice about the perpendicular bisector of AB? Draw another chord CD and construct its perpendicular bisector. What can you say about the intersection of the two perpendicular bisectors? Explain why is this so.

c) Suppose an archaeologist finds a shard (a broken piece) that appears to be a portion of a round ceramic plate and wants to know the diameter of the original object. Use a circular object to draw an arc (portion of a circle) on your paper to represent the archaeologist's shard (see **Figure 8.18**).

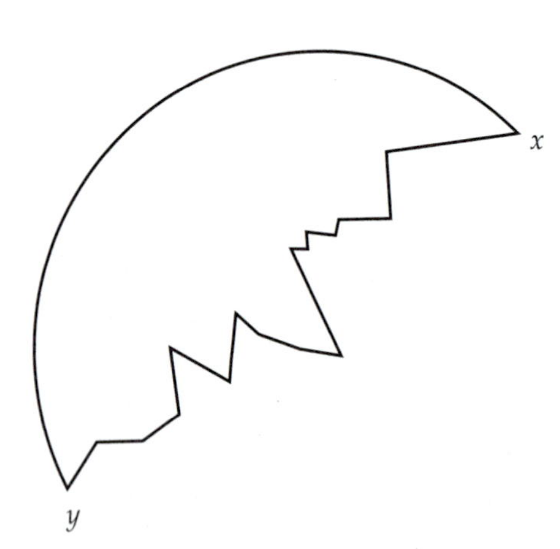

FIGURE 8.18.

Determine the center of the object. Use your constructed center to determine the diameter of the original object. You may want to test your answer against the original object.

Write a short paragraph explaining the procedure you used to find the desired diameter.

Exercises 8.2

FINDING CENTERS OF ANCIENT CIRCLES

In 1980, archaeologists began excavation of an ancient stadium in the Greek city of Corinth. While excavating the ruins, a curved starting line for a racecourse dating back to 500 B.C. was uncovered. Archaeologists wanted to find the center and radius of the circular arc so they could determine how the racetrack was laid out. They also wanted to determine the units of measurement used by the Corinthians at that time.

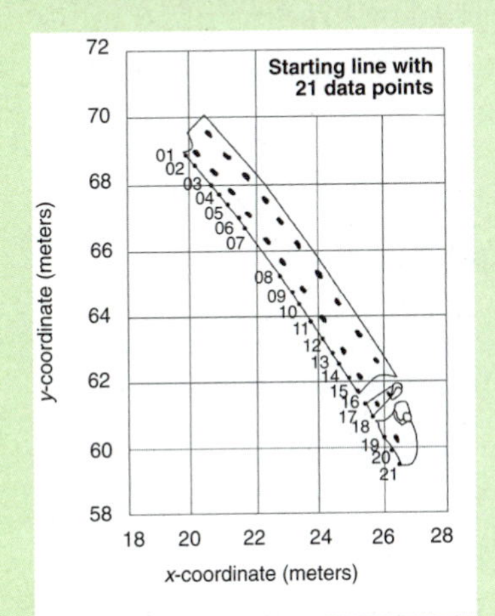

Archaeologists surveyed and located 21 points on the starting line. These points were given x- and y-coordinates with respect to an arbitrary coordinate system. **Table 8.1** gives the coordinates of the 21 points.

After measurements were made, mathematicians Chris Rorres and David Gilman Romano noted that "it's not possible to fit a circle exactly through these 21 points due to many cumulative errors. These include errors made by the original surveyors laying out the starting line; errors made by the stone cutter and builders of the starting line; shifting of the starting line over two and a half millennia; errors locating points to measure due to erosion and other damage along the edge; and errors in the measuring instruments."

Archeologists chose to find the center based on the fact that three points determine a circle. They calculated the exact circles that passed through each of 680 triplets of data points, then averaged the x- and y-coordinates of the triplet circles to determine the x- and y-coordinates of the center of the final circle.

Rorres and Romano used a different approach. They found a circle for which the sum of the squares of the distances of the data points to the circle was a minimum.

The two methods yielded different results. Rorres and Romano mathematically support their solution to the problem and say that the three-points method overestimates the true radius.

Data Point	x-coordinate (meters)	y-coordinate (meters)
01	19.880	68.874
02	20.159	68.564
03	20.676	67.954
04	20.919	67.676
05	21.171	67.379
06	21.498	66.978
07	21.735	66.692
08	22.810	65.226
09	23.125	64.758
10	23.375	64.385
11	23.744	63.860
12	24.076	63.359
13	24.361	62.908
14	24.597	62.562
15	24.888	62.074
16	25.375	61.292
17	25.166	61.639
18	25.601	60.923
19	25.979	60.277
20	26.180	59.926
21	26.142	59.524

TABLE 8.1. Coordinates of 21 points on the stadium starting line.

Exercises 8.2

12. From Table 8.1, choose any three of the 21 points located by the archaeologists in the Corinthian stadium excavation and plot the points on a coordinate plane. Determine the center and the radius of the circle that passes through your three points.

13. Sketch and describe the locus of points in a plane that are the vertices of the right angles in the right triangles with a given hypotenuse AB.

© Susan Van Etten

14. Radio station KXEM claims to have a broadcast radius of 50 miles. Angelo's house is located on the edge of the broadcast range, 40 miles east and 30 miles south of the transmitter.

a) If Angelo travels due north from his house, how many miles can he travel before he loses the KXEM radio signal?

b) A straight road passes through two small towns in the KXEM broadcast area. One town is located 10 miles east and 40 miles north of the transmitter. The other is 10 miles west and 20 miles north of the transmitter. At what locations will a car traveling on this road run out of the broadcast area?

c) Cecile lives 20 miles from Angelo's house. Like Angelo, she lives on the edge of the broadcast range. Give all possible locations for Cecile's house in relation to the radio station transmitter.

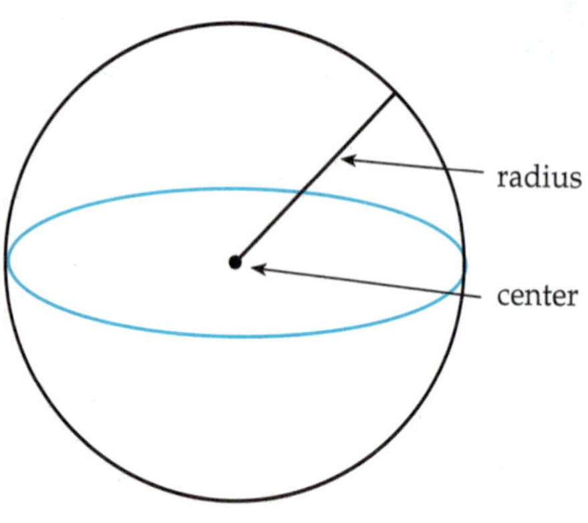

FIGURE 8.19.

15. A sphere (see **Figure 8.19**) can be defined as a set of points in space a given distance from a point called the center of the sphere. The given distance is called the radius of the sphere.

a) Find the equation for a sphere with center at (0, 0, 0) and radius of r. (Hint: Remember how to find the distance between two points in space.)

b) Find the equation for a sphere with center (2, –1, 4) and radius 3.

c) Find the equation for a sphere with center at (1, 2, 3) and that contains the point (2, –1, 5).

Exercises 8.2

16. Throughout this course, you've used coordinate systems to determine locations of points in the plane. Determining locations on a sphere such as our planet Earth is another matter. The most common way of locating a point on a sphere such as Earth is to give its longitude and latitude expressed in degrees, minutes, and seconds (see **Figure 8.20**). Latitude runs from 0° at the equator to 90°N or 90°S at the poles. The lines of latitude run east and west and are called **parallels**. Lines of longitude are called **meridians** and run north and south, intersecting at the poles. Longitude runs from 0° at the prime meridian to 180° east or west on the other side of the globe.

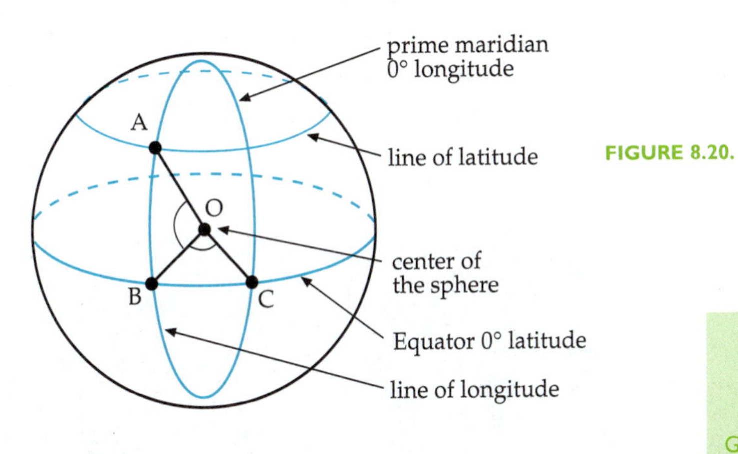

FIGURE 8.20.

If the measure of ∠COB = 35°, then B has a coordinate of 35° west longitude. If the measure of ∠AOB = 45°, then A has a coordinate of 45° north latitude. Hence, A's location can be described as N45° W35°.

Global Positioning Systems (GPS) are satellite radio positioning systems that provide information to GPS users anywhere on or near the surface of Earth. Since the advent of GPS technology, the need to understand some spherical geometry has become even more important. Civilians can now purchase handheld receivers, some of which cost less than $100. These receivers can relay to the user his or her position coordinates, and velocity and time information.

GPS ACCURACY

For national security reasons the Department of Defense can purposely introduce error into GPS signals. Due to this error, GPS systems typically have a margin of error of approximately 15 meters. For systems that need to be more accurate, it is possible to reduce this error to approximately 1 meter by using an extension of the GPS system known as Differential GPS (DGPS), which is a land-based radio system that transmits position corrections to GPS receivers.

Exercises 8.2

One formula that can be used to find the distance between two points on a sphere when the latitude and longitude of the two points are known is as follows:

Distance = $R \times a$, where R is the radius of Earth ($\approx$ 4000 miles)

and

$a = \cos^{-1}(\cos(\text{latA}) \times \cos(\text{latB}) \times \cos(\text{lonB} - \text{lonA}) + \sin(\text{latA}) \times \sin(\text{latB}))$ expressed in radians.

a) Katy Trail State Park, a 200-mile biking and hiking trail, parallels the Missouri River as it runs east and west across the middle of Missouri. Bikers and hikers often find a GPS receiver helpful and convenient in locating their position on the trail. Suppose you begin your hike at Frontier Park in St. Charles (N38° 48.403′ W90° 29.075′) and end it at Jungs Station Rd. (N38° 43.451′ W90° 32.822′). Use the above formula to find out how far you are from your starting point.

b) Without using the above formula (and assuming the radius of Earth is approximately 4000 miles), find how far north of the equator St. Louis is if the city is located at 38.6° latitude north. Describe how you solved the problem.

Modeling with Parabolas

Curves formed by cutting cones in certain ways have been of interest for over two thousand years, but it wasn't until about 220 B.C. that Appollonius of Perga began the formal study of conics and their properties. During his lifetime he wrote eight books on conics and gave ellipses, parabolas, and hyperbolas their names.

Conic, an abbreviation for **conic section**, comes from the word cone. **Figure 8.21** shows a right circular cone formed by two intersecting lines when one line is rotated about the other. The fixed line is called the **axis** of the cone, the other line is called the **generator**, and the point where the two lines intersect is called the **vertex**. The cone consists of two parts that are called **nappes**.

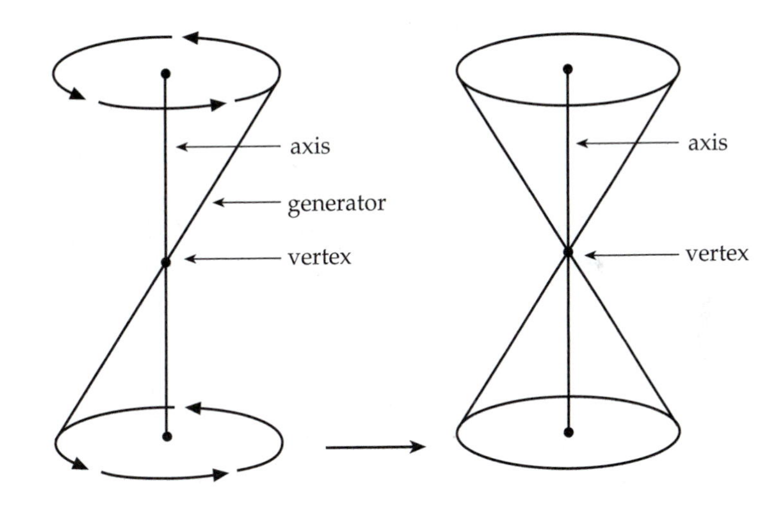

FIGURE 8.21.
Generating a cone.

Conics are curves that come from the intersection of a right circular cone and a plane. The study will focus on the conics formed when the plane does not pass through the vertex (see **Figure 8.22**). If the plane is perpendicular to the axis of the cone the curve is a **circle** (see (a)). If the plane is oblique to the axis and intersects all edges of one nappe of the cone, the curve is an **ellipse** (see (b)). A **parabola** is formed when the plane is tilted so that it is parallel to one generation of the cone thus intersecting

only one nappe of the cone (see (c)). If the plane intersects both nappes, a **hyperbola** is formed (see (d)).

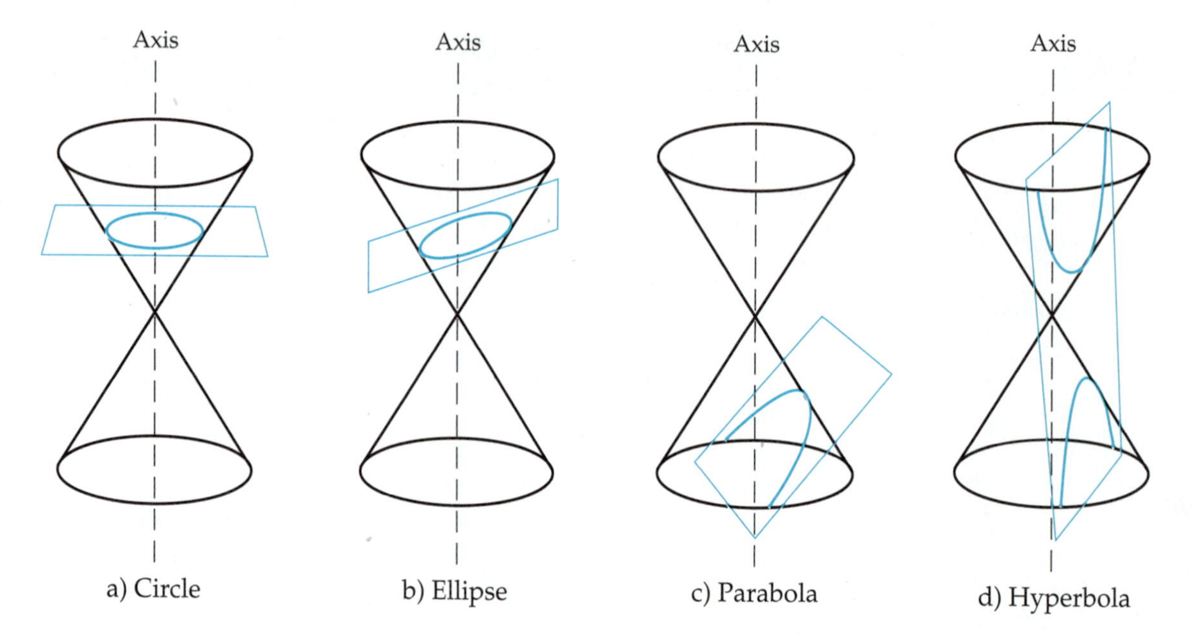

FIGURE 8.22.

Activity 8.3

As the lead engineer for the robotics division of a large manufacturing firm, you have been asked to develop the guidance software for an AGV (automated guided vehicle) located on the manufacturing floor. Your current area of interest is the region of floor on which the AGV must move between a large bank of equipment located along a wall of the building and a delicate piece of machinery located near the center of the floor.

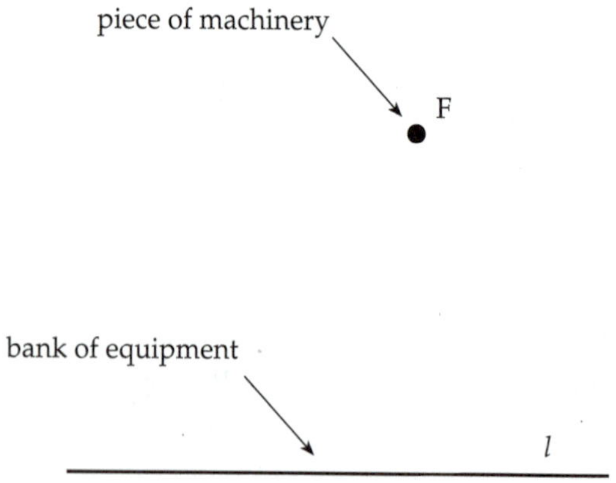

piece of machinery

F

bank of equipment

l

Your job is to find an optimal path for the vehicle that can be modeled mathematically. The AGV's path must be one that stays as far away from both the wall (l) and the piece of machinery (F) as possible. See **Figure 8.23**.

FIGURE 8.23.
Map of the manufacturing area.

Part 1

a) Use Handout 8.1, which shows a line l and a point F that is not on l. Determine the locus of points that is equidistant from the given line and the given point. Materials that might be of help to you are a ruler, a compass, wax paper or patty paper (for paper folding), or a geometric drawing utility such as Geometer's Sketchpad. (Hint: Recall that the distance from a point to a line is the length of the perpendicular from the point to the line.)

b) Describe the locus of points you located.

Part 2

Now that you have an idea of how the locus looks, it's time to try to find an algebraic model that fits the situation.

a) One way to begin is to use your knowledge of coordinate geometry. For example, overlay a coordinate system on the factory floor in such a way that the piece of machinery is located at (0, 1) and the wall is modeled by the equation $y = -1$ (see **Figure 8.24**). Let $P(x, y)$ represent every point whose distance from the point (0, 1) is equal to the distance from the line whose equation is $y = -1$, then use the distance formula to find the equation of the desired locus.

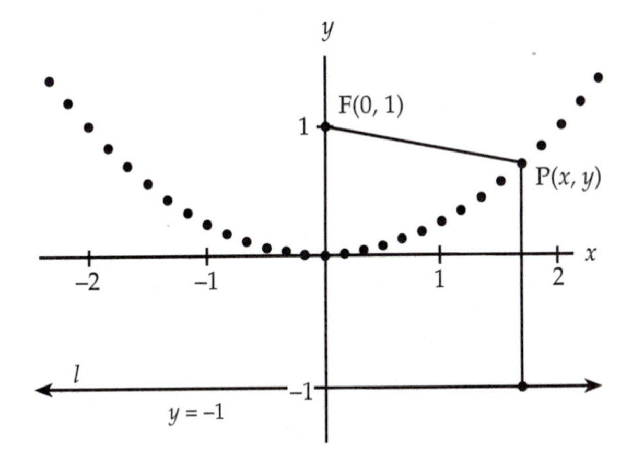

FIGURE 8.24. Locus of points equidistant from l and F.

b) The equation found in part (a) is the equation for what kind of curve? Explain how you know.

c) Generalize the problem by placing the piece of machinery at the point $(0, p)$ and letting the wall be represented by the equation $y = -p$. Find the equation for the desired locus.

PARABOLAS

TAKE NOTE If a parabola has a vertex at (0, 0), a focus at $(0, p)$ and a directrix $y = -p$, the standard form equation of the parabola is $x^2 = 4py$. Note that p is a directed distance from the vertex to the focus. When $p > 0$, the parabola opens upward, and when $p < 0$, it opens downward.

From your explorations in Activity 8.3, you now know that a **parabola** can be defined geometrically as the locus of points in a plane that is equidistant from a given line and a given point not on the line. Mathematicians call the given line the **directrix** of the parabola, and call the given point the **focus**.

Figure 8.25 shows a parabola, its directrix and its focus. The line through the focus, perpendicular to the directrix, is called the **axis of symmetry**. The point where the axis of symmetry intersects the parabola is called the **vertex**.

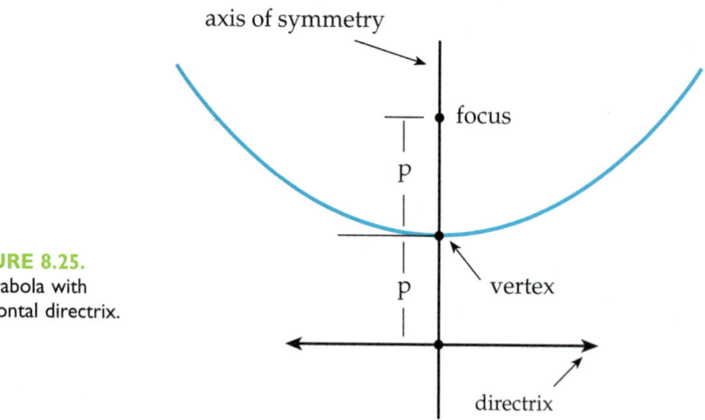

FIGURE 8.25. A parabola with horizontal directrix.

EXAMPLE 4

Find equations for the directrix and the parabola if the focus is the point (0, –5) and the vertex is the origin. Graph the function.

FIGURE 8.26.

SOLUTION:

First sketch the information given (see **Figure 8.26**).

From the given information and the definition of a parabola, you know that the directrix is the horizontal line $y = 5$ and that $p = -5$. Substituting into the form $x^2 = 4py$, an equation for the parabola is $x^2 = -20y$. When graphing the function, it is helpful to locate a point or two on the parabola (see **Figure 8.27**). Keeping in mind the locus definition for a parabola, you know that when $y = -5$, $x = \pm 10$. Why?

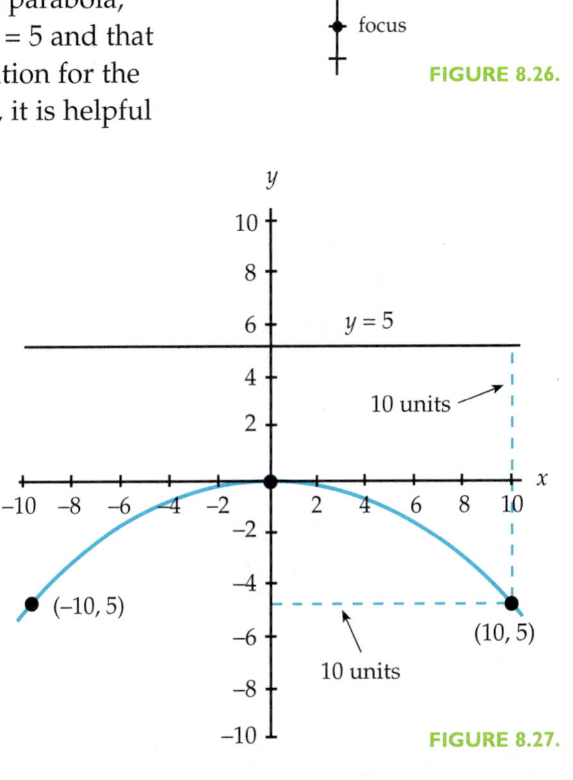

FIGURE 8.27.

When a parabola with a vertex at the origin and the y-axis as its axis of symmetry is shifted h units horizontally and k units vertically, the result is a parabola with vertex at (h, k) and an axis of symmetry parallel to the y-axis as shown in **Figure 8.28**.

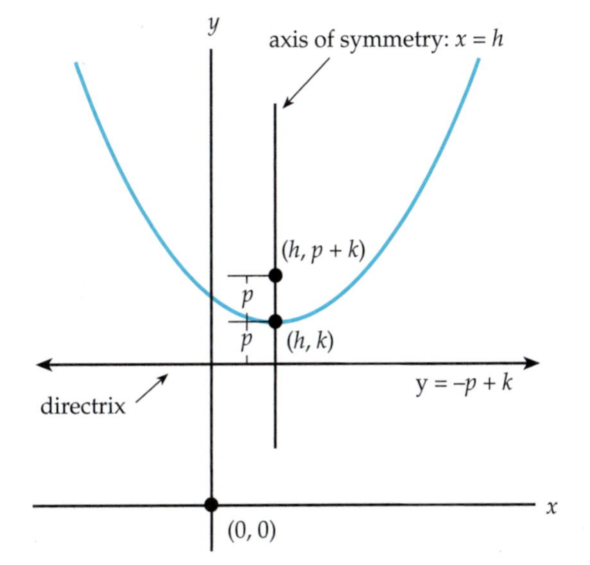

FIGURE 8.28.

TAKE NOTE

If a parabola has a vertex at (h, k), a focus at $(h, k + p)$, and a directrix $y = -p + k$, the standard form equation of the parabola is $(x - h)^2 = 4p(y - k)$. If $p > 0$, the parabola opens upward, and if $p < 0$, it opens downward.

EXAMPLE 5

Find an equation of the parabola with focus at (5, 3) and directrix $y = -5$.

SOLUTION:

From the given focus and directrix, you know the vertex is (5, –1) and that $p = 4$. Using the form $(x-h)^2 = 4p(y-k)$, the desired equation is $(x-5)^2 = 16(y+1)$.

© *Susan Van Etten*

© *Susan Van Etten*

PARABOLIC POPULARITY

Parabolas are everywhere—even in fountains. According to mathematician and author Lee Whitt in *The UMAP Journal* (Fall 1982), "Over one billion parabolas have been sold. The biggest seller is the auto headlight, the smallest seller is the huge 200-inch Hale telescope mirror. Popular models can be found in bridge designs, roof construction, fluid dynamics, and even starring in some movies (including *Papillion* with Steve McQueen, and *The Man with the Golden Gun* with Sean Connery)." The advent of satellite dishes for home television viewing has increased their popularity. Parabolas are BIG business.

The main cable of a suspension bridge appears to be a parabola. If this cable evenly supports a uniform horizontal load by infinitely many vertical cables, which are massless and the gravity field is parallel, the cable *is* a parabola. Since this ideal situation doesn't exist, the cable is only approximately parabolic.

REFLECTION PROPERTY

In addition to being used in uniform load-bearing structures and in modeling projectile motion, the parabola has a reflection (or focal) property that is incredibly useful.

When a parabola is rotated around its axis of symmetry, a three-dimensional surface called a **paraboloid of revolution** is formed. Headlights of a car and satellite dishes are two examples of paraboloids. The reflection property of a parabola can be demonstrated by using a paraboloid with a mirrored surface. If a light is placed at the focus of the paraboloid, all rays from the light will reflect off the mirror in lines parallel to the axis of symmetry (see **Figure 8.29**).

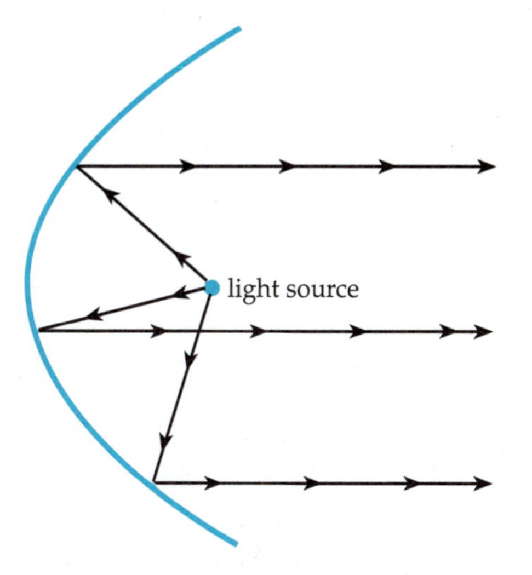

FIGURE 8.29.
Light rays reflecting from a source at the focus of the parabola.

Flashlights, search lights, and headlights of cars are just a few of the items that illustrate this property. The high and low beams of certain lights are achieved by placing one light source at the focus for the high beam and the other source to the left and above the focus for the low beam. Hence, the low beam reflects the light down and to the right, away from on-coming cars. Since this property applies to sound as well as light, parabolic reflectors are used in some horn and siren systems.

This reflection property, unique to parabolas also works in reverse, as shown in **Figure 8.30**. If signals come to a parabolic receiver from a distant source and are essentially parallel, they are reflected by the surface to a single point at the focus.

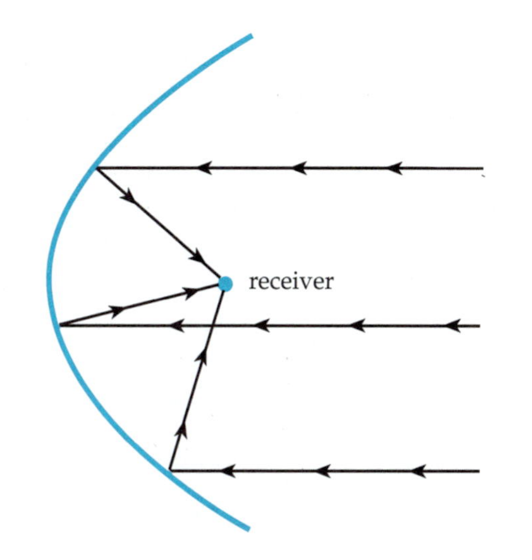

FIGURE 8.30.
Distant signals reflected to the focus of the parabola.

Applications of this property can be found in communications systems, electronic surveillance systems, radar systems, reflector-microphones found on the sidelines of football games, telescopes, and even in some solar furnaces. For example, one such furnace located atop a mountain in France has a 31-foot diameter mirror and is capable of creating temperatures as high as 5400°F at its focus.

Some telescopes use parabolic reflectors to collect radiation from space and concentrate it at the focus where it can be photographed or reflected to an eyepiece through the use of auxiliary mirrors. These receivers vary widely in size. The main data-acquisition station for deep-space probes, located at the Goldstone Tracking Station in California, has two parabolic dish antennae that measure 85 feet and 210 feet in diameter.

Arecibo Observatory, Reference #12

Some very large fixed radio telescopes have been built by carving huge parabolic bowls out of the ground and lining them with reflecting metal mesh. The world's largest radio telescope, a 1000-foot diameter "bowl" at Arecibo, Puerto Rico, is one of the best known installations of this type. Unfortunately, for our purposes, it is a spherical (not parabolic) reflector, but the enormity of its reflecting surface makes it worth mentioning here. When repairs are made on it, workers have to move around on snow skis.

A fun, interesting exploration that demonstrates both the receiving and sending versions of the reflection property can be found in some large science centers. This demonstration consists of two parabolic reflectors placed several feet apart (see **Figure 8.31**). If a person who is facing one of the reflectors stands at the focus of that reflector and whispers something, the person standing at the focus of the other reflector can clearly hear what was said.

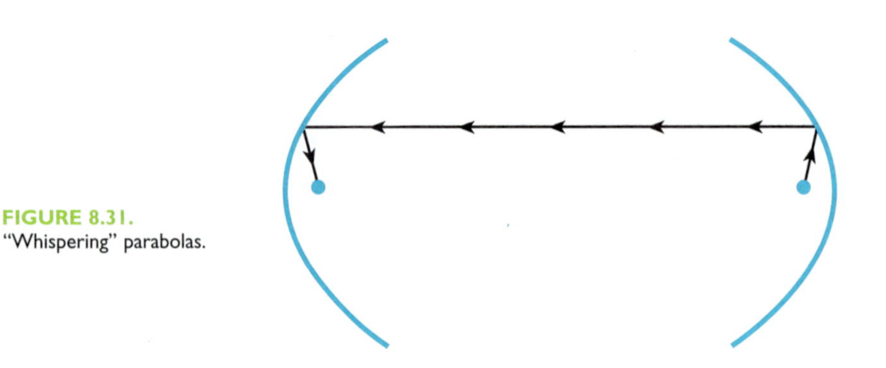

FIGURE 8.31.
"Whispering" parabolas.

The final application discussed here is one with which most people are familiar. When you cup your hands around your mouth to yell, the ideal shape is, you guessed it, a parabola! Likewise, when you cup your hands around your ears to hear a distant sound more clearly, the ideal shape is again a parabola.

It must seem that the number of applications of the parabola is endless. In reality, that is not quite true, but there are many more applications than what are discussed here. Now it's up to you to continue the search.

Exercises 8.3

1. For this exercise you will need a compass and straight edge, a geometric drawing utility such as Geometer's Sketchpad or Cabri, or a Plexiglas mirror such as a Mira.

   a) Use your method of choice and follow the steps below to construct points on the parabola with a given directrix and focus.

      Step 1: Construct a horizontal line (l) and a point (F) not on the line.

      Step 2: Place a point G on line l and construct segment FG.

      Step 3: Construct the perpendicular bisector of segment FG.

      Step 4: Construct a line perpendicular to line l through point G.

      Step 5: Locate the point of intersection of the lines from steps 3 and 4. Label the point P.

      Step 6: (If you are using a drawing utility) Trace the path of P as you move point G along line l.

      Step 6: (If you are not using a drawing utility) Repeat steps 2–5 at least three more times so that you can identify more points P on your parabola.

   b) Write a convincing argument that point P is a point on a parabola with line l as the directrix and point F as the focus.

   c) Repeat part (a) at least two more times. The first time, move the focus closer to the directrix. The second time, move the focus further away. What happens to the shape of the parabola as the focus is moved?

   d) What happens if you locate the focus below the horizontal directrix?

   e) How could you get the parabola to open to the right? to the left?

Exercises 8.3

2. If not given, find the focus (F), the directrix (d), the vertex (V), and an equation of each of the parabolic graphs.

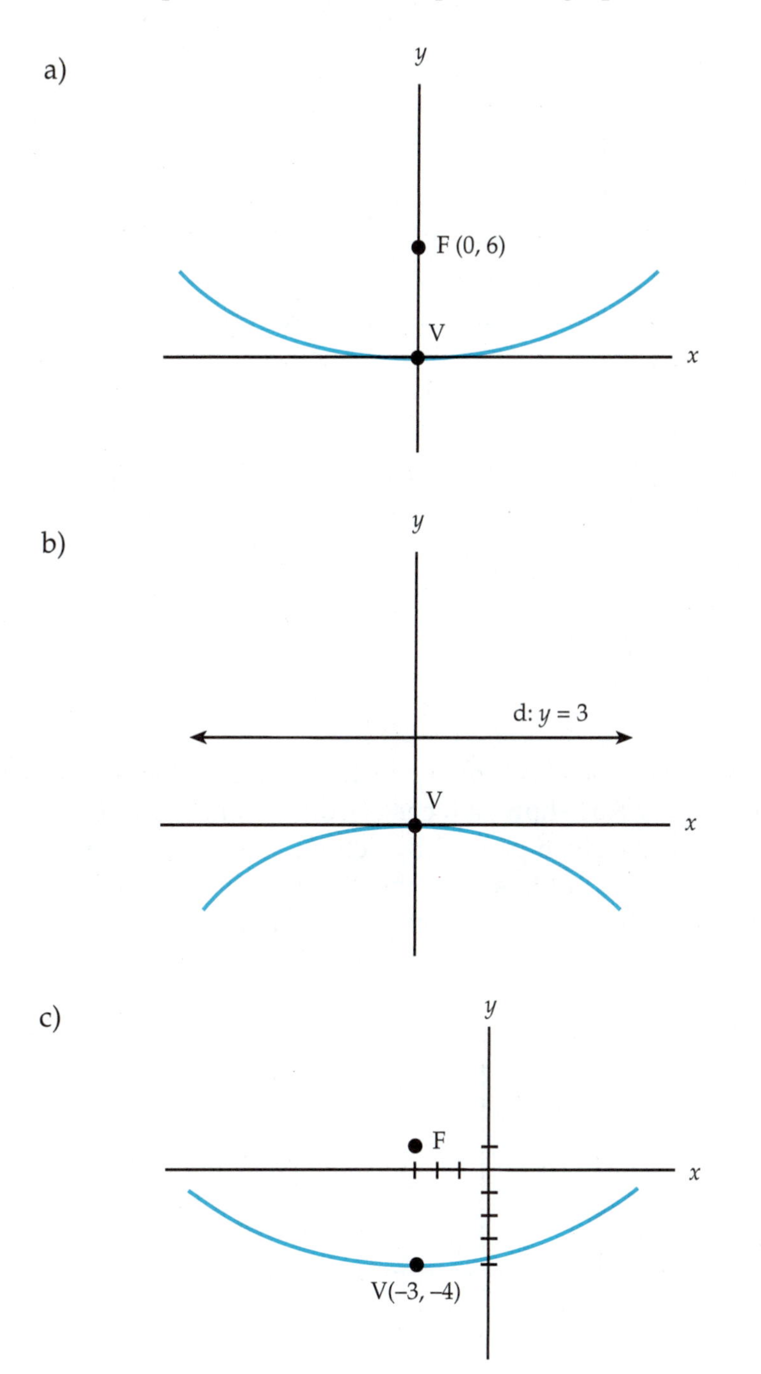

Exercises 8.3

d)

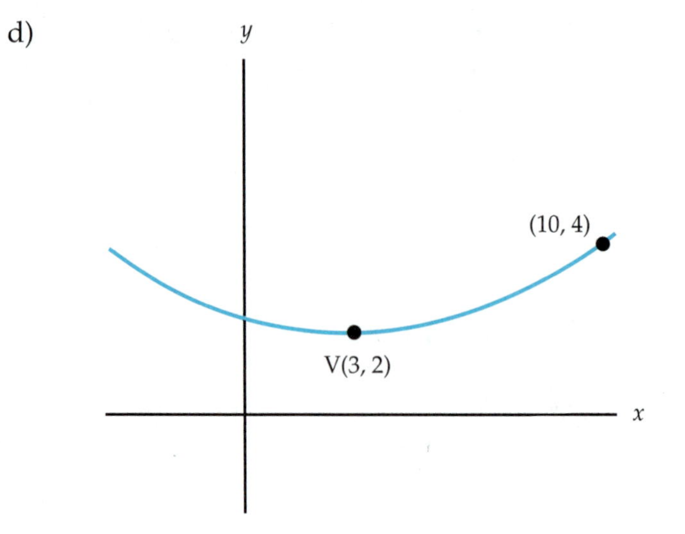

3. In parts (a)–(c), find the vertex, the focus, and the directrix for each parabola.

   a) $x^2 = 10y$

   b) $(x - 5)^2 = -12(y + 1)$

   c) $x^2 + 8x = 4y - 8$

4. Not all parabolas have axes of symmetry parallel to the y-axis. For example, **Figure 8.32** shows a locus of points equidistant from a given line d and a given point F. By definition, you know that this locus is a parabola, but this time, the directrix is vertical rather than horizontal and the axis of symmetry is horizontal.

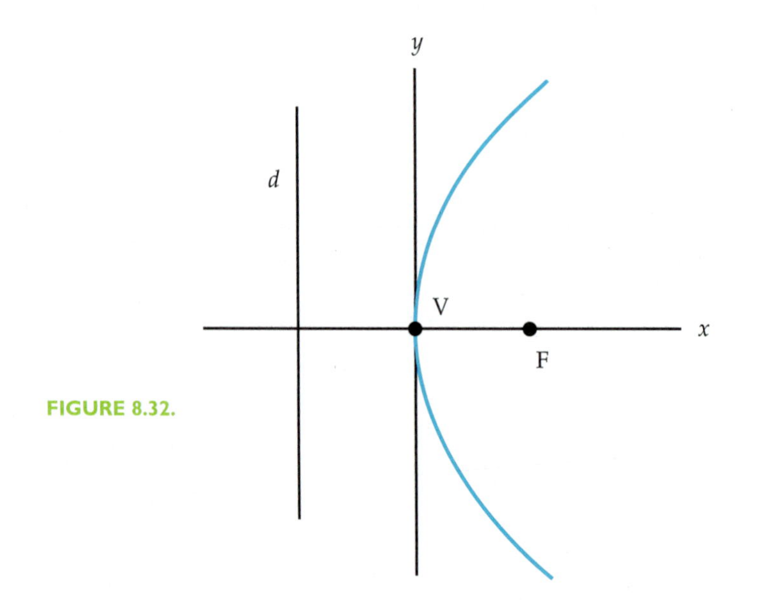

FIGURE 8.32.

Exercises 8.3

a) Is the parabola in Figure 8.32 a function? Explain why or why not.

b) Find equations for the directrix and the parabola if the focus is the point (4, 0) and the vertex is the origin.

c) If a parabola has vertex (0, 0), focus $(p, 0)$, and directrix $x = -p$, what is an equation of the parabola?

d) Given a parabola with vertex (h, k) and focus $(h + p, k)$, what is an equation of the parabola and an equation of the directrix?

e) Consider the parabola $(y - k)^2 = 4p(x - h)$. Which way does the parabola open if $p > 0$? If $p < 0$?

f) Graph $y^2 = 7x$. Give the focus, and an equation of the directrix.

5. If not given, find the focus (F), the directrix (d), the vertex (V), and an equation of each of the following graphs.

a)

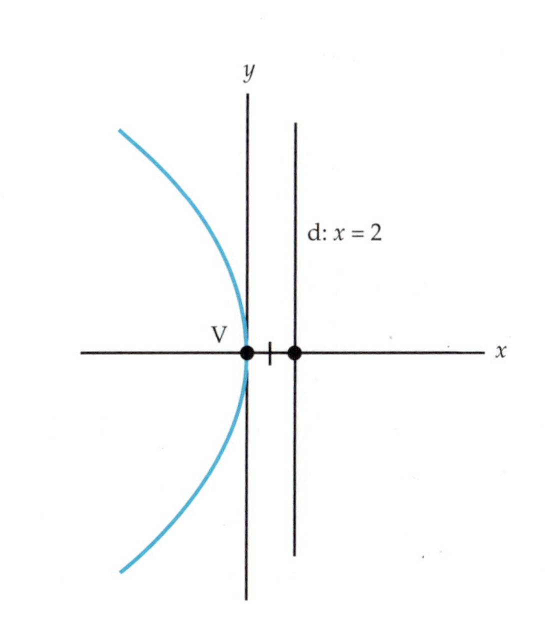

Exercises 8.3

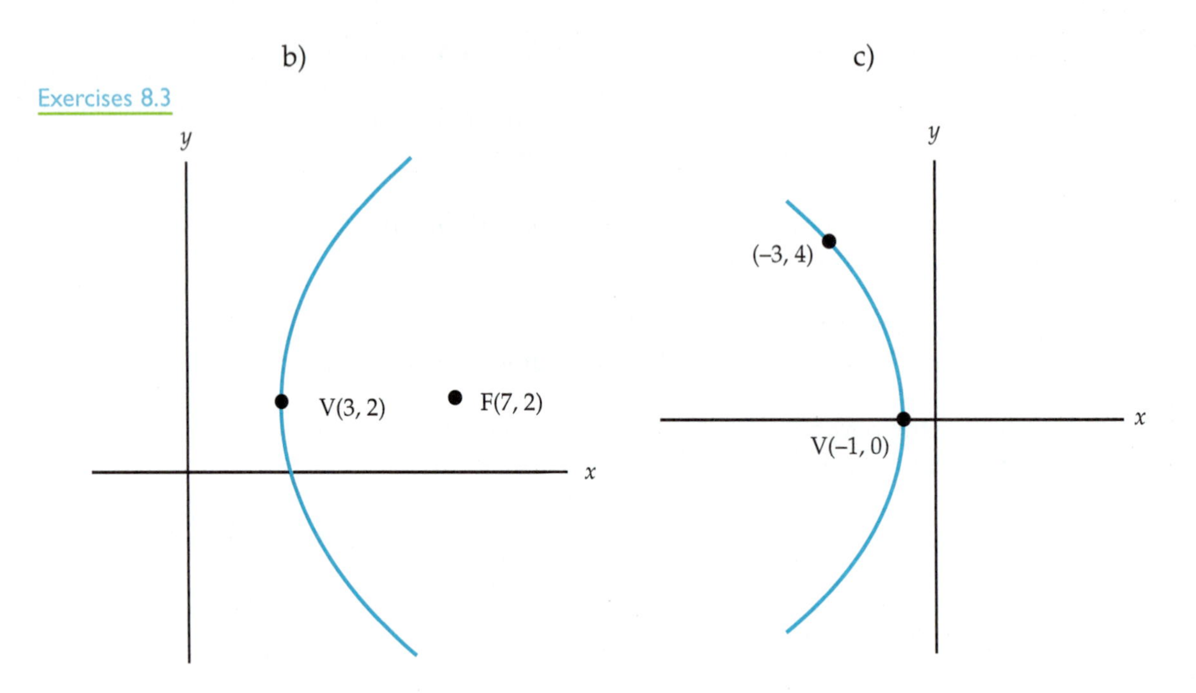

6. A metal reflecting shade used as part of a work light is shaped like a paraboloid of revolution. It is 10.5 inches across its opening and is 4 inches deep. To help choose a bulb size that yields optimal reflection, find the location of the focus.

7. Paper Folding a Parabola:

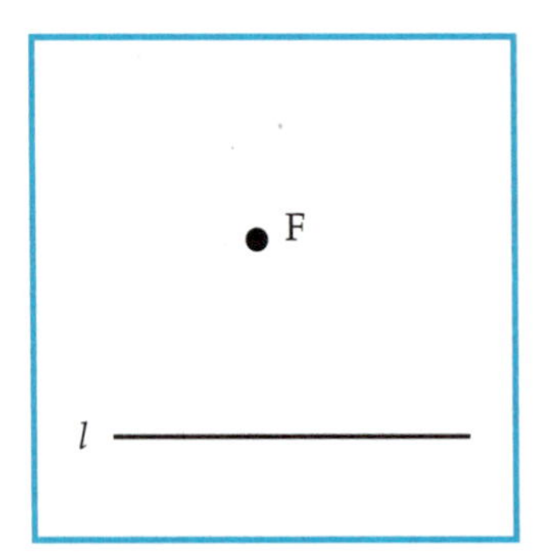

FIGURE 8.33.

a) Cut a piece of waxed paper approximately 6″ x 6″ or use a piece of precut patty paper, then follow the steps below:

Step 1: Draw any line l on your paper (this will be the directrix of your parabola), and draw any point F (focus) that is not on the line (see **Figure 8.33**).

Step 2: Fold a line perpendicular to line l. Mark the point of intersection of l and the perpendicular line point G.

Step 3: Fold the paper so that point F and point G coincide. Crease the line formed.

Step 4: Label the point of intersection formed by the crease (step 3) and the perpendicular line (step 2), P.

b) How do you know that point P is a point on the parabola?

c) Repeat the folding activities in steps 2–4 of part (a) fifteen to twenty more times. All of your points marked P will be on the parabola with focus F and directrix *l*. Note the creases formed by folding F and G onto each other. These creases provide an outline or **envelope** of the parabola. Also notice that each of these creases is tangent to the parabola.

Exercises 8.3

In **Figure 8.34,** line *t* is tangent to the parabola. With your knowledge of the reflection properties of parabolas, what observations can you make about the angle α formed by the ray parallel to the axis of symmetry that intersects the parabola at P and the tangent (*t*) to the parabola at P, and the angle β formed by the tangent (*t*) and the line segment PF?

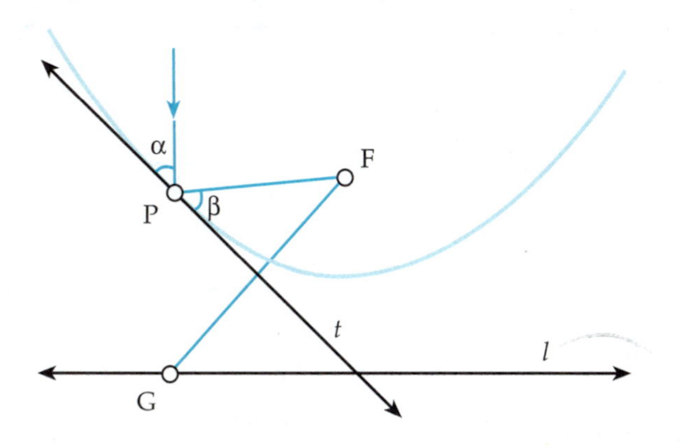

FIGURE 8.34.

8. Match the graphs in **Figures 8.35–8.42** to the following equations.

a) $x^2 = 8y$

b) $y^2 = 8x$

c) $x^2 = -8y$

d) $y^2 = -8x$

e) $(x - 3)^2 = 8(y + 1)$

f) $(x - 3)^2 = -8(y - 1)$

g) $(x + 3)^2 = -8(y - 1)$

h) $(x + 3)^2 = 8(y + 1)$

Exercises 8.3

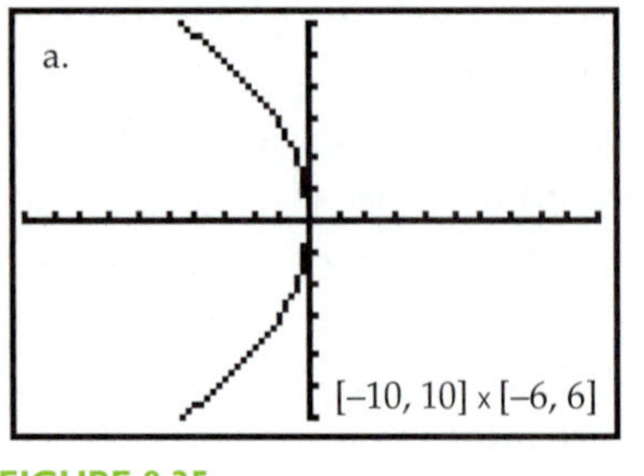

FIGURE 8.35.
For Exercise 8.

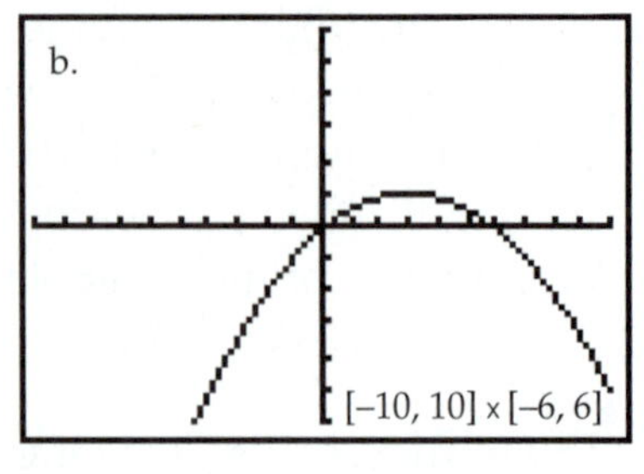

FIGURE 8.36.
For Exercise 8.

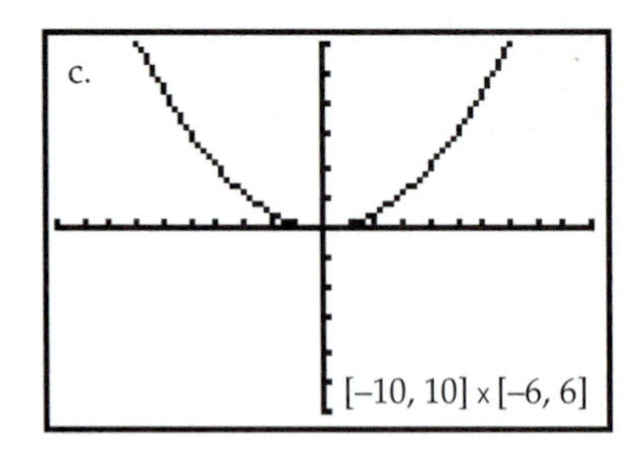

FIGURE 8.37.
For Exercise 8.

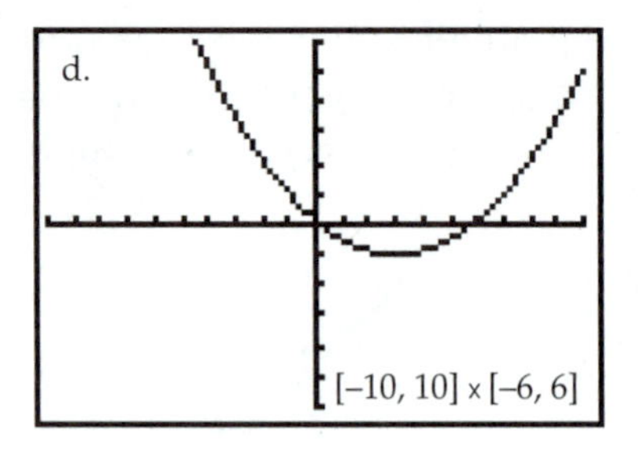

FIGURE 8.38.
For Exercise 8.

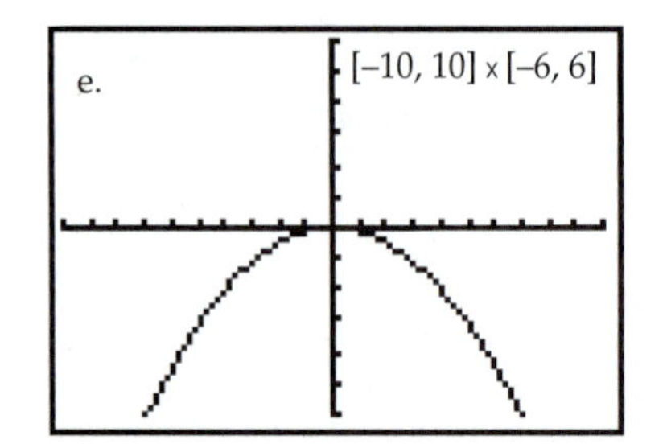

FIGURE 8.39.
For Exercise 8.

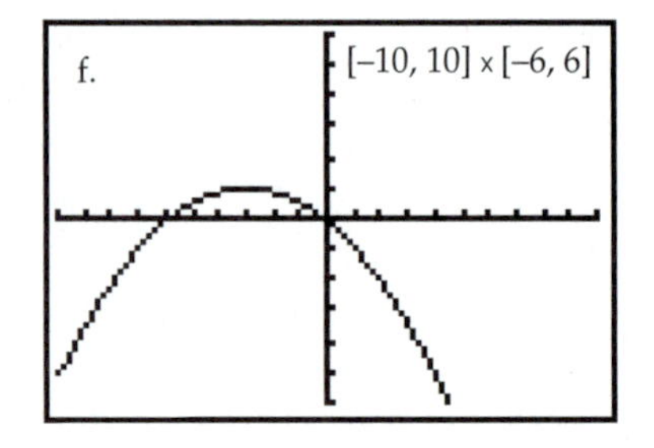

FIGURE 8.40.
For Exercise 8.

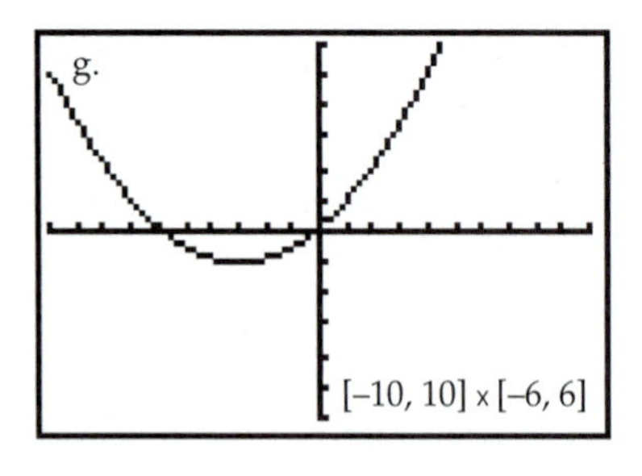

FIGURE 8.41.
For Exercise 8.

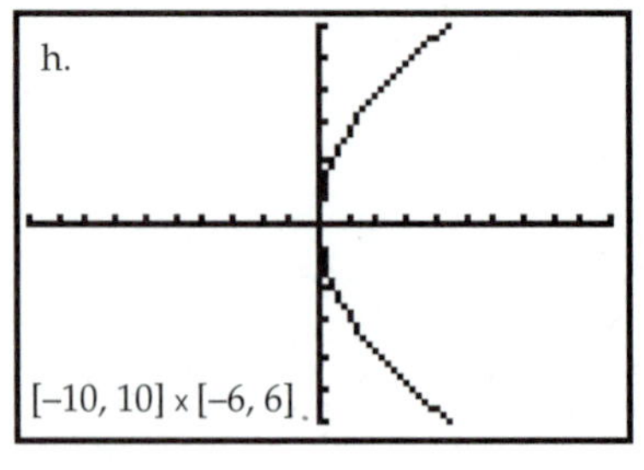

FIGURE 8.42.
For Exercise 8.

Exercises 8.3

9. Friends of yours are having trouble using the function grapher of their calculator to graph $y^2 = 16x$. Explain to them what to do and why it has to be done that way.

10. a) Sketch the parabola with focus (2, 4) and directrix $y = -6$.

    b) Identify a point, other than the vertex of the parabola, that appears to be on your sketch of the parabola.

    c) Find the equation of the parabola and verify whether or not your point from part (b) is on the parabola.

    d) If your guess from part (b) was not on the parabola, use your equation for the parabola to find a point that is on it. Use the same x-coordinate as your guess.

    e) When given an equation for a parabola and asked to sketch the graph, what can you do to get a more accurate graph?

11. For each parabola in parts (a)–(f), find the vertex, focus, and directrix.

    a) $x^2 = 24y$

    b) $(x - 4)^2 = -12y$

    c) $(y + 1)^2 = 4(x + 2)$

    d) $x^2 - 6x = 2y + 1$

    e) $y^2 + 4y - x + 6 = 0$

    f) $Ax^2 + Ey = 0, A \neq 0, E \neq 0$

12. While rummaging through the basement, Roberto found an old car's headlight. After he examining it closely, he suspected that the light source might be in the wrong place. According to what he knew about parabolas, the bulb seemed much too close to the glass that covered the headlight. Out of curiosity, he measured the diameter and the depth of the headlight. Use Roberto's measurements as shown in **Figure 8.43** to calculate where the light source should be placed. From your calculations, do you think Roberto was right in questioning the location of the bulb?

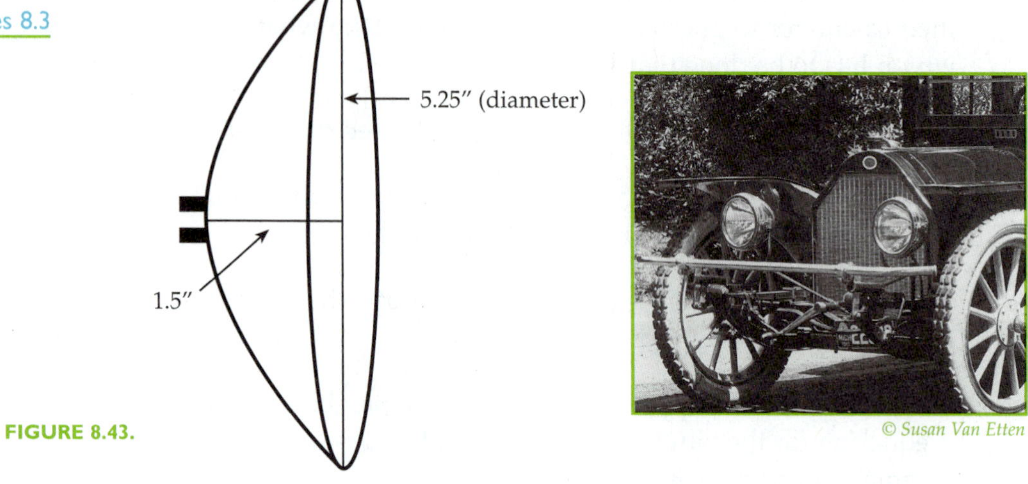

FIGURE 8.43.

13. The pattern created by the interference of parallel waves and concentric circular waves forms a group of parabolas (see **Figure 8.44**). Assume the radii of the circular waves increase by one unit as the circles get larger and the parallel lines are one unit apart. Explain why the darkened points are points on parabolas. (Hint: Let the focus of each parabola be located at the center of the circles.)

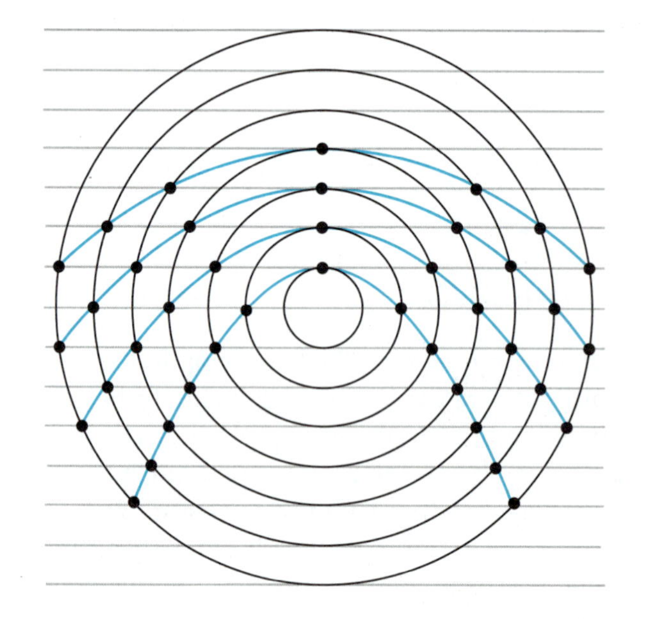

FIGURE 8.44.

Exercises 8.3

14. While browsing through a store, Naomi noticed a Digital Satellite System. She was intrigued by the dish's appearance because the focus was located outside the dish itself. She asked the salesperson to measure it for her. The following measurements were taken: diameter: 20.5 inches; depth: 2.25 inches. According to the salesperson's measurements, where should the receiver be located? For a paraboloid of revolution with these dimensions, does it make sense that the receiver was located outside the dish?

15. The main cable of the Brooklyn Bridge, one of the world's most famous bridges, is suspended between two towers. These towers are approximately 1596 feet apart and rise approximately 142 feet above the roadway as shown in **Figure 8.45**. Assuming the main cable of this bridge is in the shape of a parabola and it touches the roadway midway between the towers, find the height of the cable above the roadway at a point 100 feet from either of the towers.

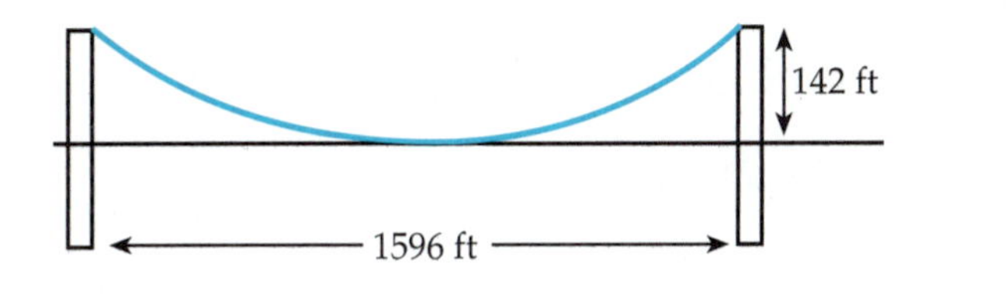

FIGURE 8.45.

16. **Investigation**. The standard form for a parabola contains three control numbers. The values of h and k give the location of the vertex of the parabola, and the value of a controls the "size" of the parabola. Large a's correspond to steep (narrow) graphs, and small a's make for shallow (wide) parabolas. In fact, since $p = 1/(4a)$, knowing a and h and k allows you to draw three points on the parabola very quickly—namely the vertex and the two points "directly beside" the focus. Still, especially if a is even moderately large, these three points don't give a very clear picture of the overall shape of the graph. However, standard form for a parabola looks exactly like the point-slope form of a line, except for the square, and knowing the slope of a line you can add as many points as you want to its graph with very little effort.

Exercises 8.3

Complete **Table 8.2** for each of the equations (a)–(d). For the equation (d), use x-values of h, $h + 1$, $h + 2$, $h + 3$, etc. Note that a few y and Δy values for (a) are shown as examples.

x	y	Δy
4	3	1
5	4	3
6	5	
7		
8		
9		

TABLE 8.2.

a) $y = 1(x - 4)^2 + 3$

b) $y = 2(x - 4)^2 + 3$

c) $y = -2(x - 4)^2 + 3$

d) $y = a(x - h)^2 + k$

Explain how a can be used almost like slope to get a sequence of points on the parabola extending out from the vertex.

LESSON 8.4

Modeling with Ellipses

Thus far in your exploration of conic sections, you've studied circles and parabolas. In this lesson, you will be introduced to another conic, the ellipse. In the words of mathematician Dr. Lee Whitt, "I am infatuated with the ellipse! It is my kind of curve—not too straight and not too twisted. It has style and personality. Indeed it has an identity and a presence you will not believe." *The UMAP Journal* 2 (Summer 1983): 157.

Activity 8.4

For this activity you will need a piece of string, two thumbtacks or pins, a piece of cardboard, and a piece of paper.

1. Place a piece of paper over the cardboard and use a marker to draw two points, F_1 and F_2, approximately 6 cm apart. Stick thumbtacks into the cardboard at the indicated points and attach a piece of string approximately 9 cm long to the tacks. Using the sharpened end of a pencil, pull the string taut. Keep the string taut and construct a locus of points by moving the pencil around F_1 and F_2. (see **Figure 8.46**) Describe your final drawing.

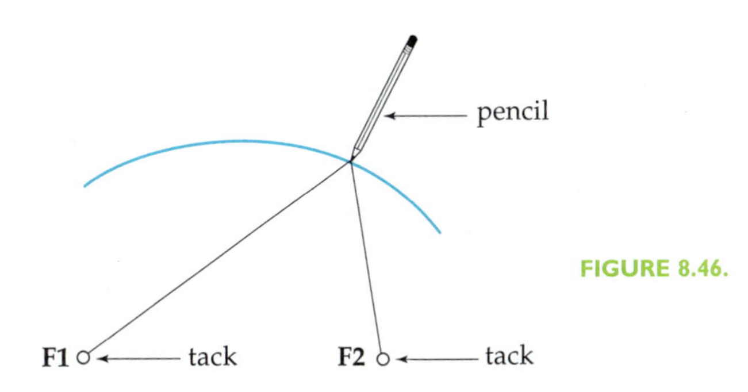

FIGURE 8.46.

2. To help determine a mathematical description of the locus, pick at least four points on the figure and measure their distances from F_1 and F_2. Describe what patterns you discover.

3. How do you know that the sum of the distances from F_1 and F_2 to any point P on your locus is a constant?

4. Repeat the construction that you did in part (a), keeping the length of the string the same but moving F_1 and F_2 closer together. What do you notice about the shape of the locus?

5. Repeat the construction. Again, keep the length of the string the same, but move the points further away from each other. What do you notice about the shape?

THE ELLIPSE

FIGURE 8.47.
Elliptical wing edges of the British Spitfire.

FIGURE 8.48.
Elliptical gears.

In Activity 8.4, you discovered a locus of points that had the appearance of a flattened circle. This locus of points in a plane such that the sum of the distances from each point on the locus to two fixed points is a constant is known as an **ellipse**, and the two fixed points are called **foci** (singular for foci is focus).

Ellipses have found many applications in the fields of science and engineering. It is possible to find bridges that are constructed with semi-elliptical arches. Some airplane wings, such as the British Spitfire in World War II (see **Figure 8.47**), are elliptical in design, and some gears for machinery (see **Figure 8.48**) are shaped like ellipses. Ellipses are even found in everyday situations. For example, if a glass or cup of liquid is tilted, the surface of the water forms an ellipse.

One very important example of an ellipse is found in the orbits of planets in our solar system. The German astronomer Johannes Kepler spent much of his life analyzing planetary data, much of which were collected by the Danish astronomer Tycho Brahe. Kepler was very disappointed when he discovered that the orbits were elliptical because he and others of his time believed that God created only circular orbits. In 1609 Kepler's first law was published; it stated that the planets follow elliptical paths with the Sun at one focus.

Despite Kepler's analysis of the motion of the planets, he never recognized the causes for the elliptical orbits. It wasn't until Sir Issac Newton applied the inverse square law of gravitation in the mid-1600s that an explanation of planetary motion was given. Through Newton's work, it is known that the orbits of the planets and comets around the Sun, of satellites around the Earth, and of moons around the planets are all ellipses.

FINDING THE EQUATION

Figure 8.49 shows an ellipse with foci F_1 and F_2. The line containing the foci intersects the ellipse at points A and B, which are called vertices. The line segment AB joining the **vertices** and containing the foci is called the **major axis.** The midpoint O of the segment joining the two foci is called the **center** of the ellipse. The line through the center, perpendicular to the major axis intersects the ellipse at points C and D. The line segment CD joining these two points of intersection is called the **minor axis**.

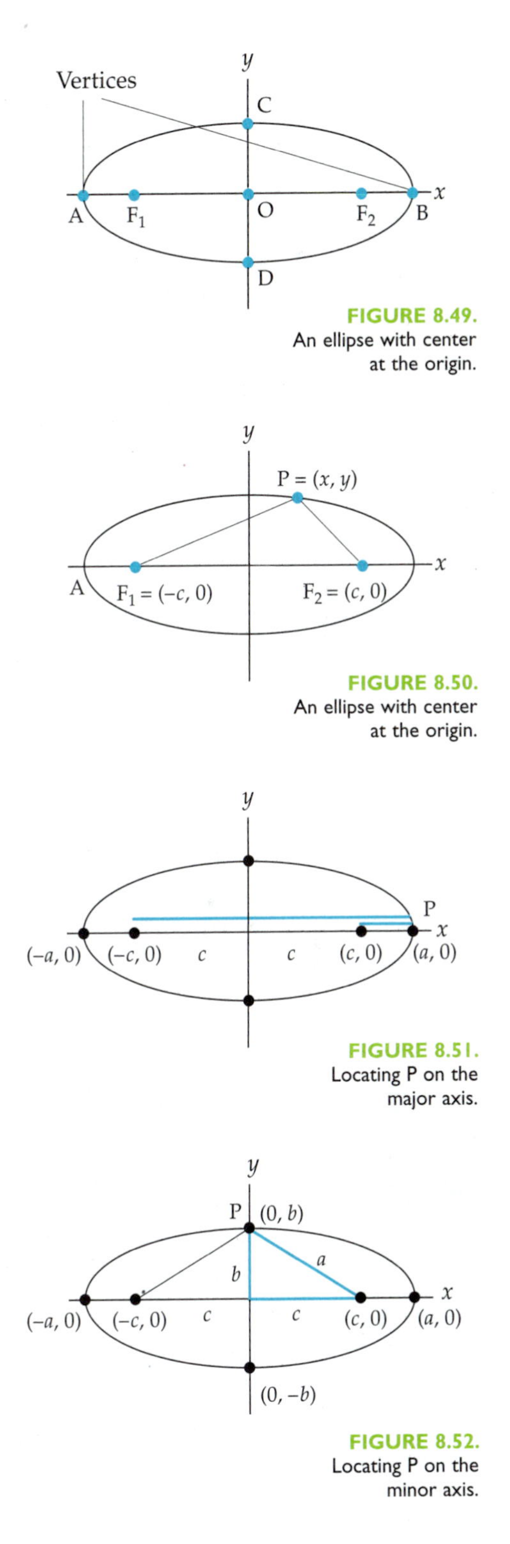

FIGURE 8.49. An ellipse with center at the origin.

FIGURE 8.50. An ellipse with center at the origin.

FIGURE 8.51. Locating P on the major axis.

FIGURE 8.52. Locating P on the minor axis.

To find an equation of an ellipse, it is easiest to consider a locus with center at the origin and major axis on the x-axis. By definition, you know that $PF_1 + PF_2$ = some constant. To simplify the algebra, let the constant equal $2a$. Thus

(1) $PF_1 + PF_2 = 2a$.

Again for convenience, locate the foci at $(0, c)$ and $(0, -c)$, where $c > 0$ is the distance from the center to a focus. **Figure 8.50** illustrates the situation thus far.

Now consider two special cases. First, choose the point P so that it is at one endpoint of the major axis. **Figure 8.51** shows it on the positive x-axis, with the two distances PF_1 and PF_2 highlighted. Since the distance from the right focus to P is equal to the distance from the left focus to the left endpoint of the major axis, the major axis must be exactly $2a$ units long.

As a second special case, choose P at one endpoint of the minor axis. **Figure 8.52** shows it on the positive y-axis. Note that in this position, the distance PF_2 forms the hypotenuse of a right triangle, with one leg being c. For convenience, define the length of the other leg (half the length of the minor axis) to be b, as illustrated in Figure 8.52. Then $a^2 - c^2 = b^2$.

Now return to the general case shown in Figure 8.50. The goal is to change equation (1) into a useful form.

Remember, the coordinates of P are (x, y). Thus,

(2) $\sqrt{(x+c)^2+y^2}+\sqrt{(x-c)^2+y^2}=2a.$

While this is a valid equation for the general ellipse in Figure 8.50, an equation without radicals is easier to work with. So, rewriting,

(3) $\sqrt{(x+c)^2+y^2}=2a-\sqrt{(x-c)^2+y^2}$

(4) $(x+c)^2+y^2=4a^2-4a\sqrt{(x-c)^2+y^2}+(x-c)^2+y^2$

(5) $x^2+2cx+c^2+y^2=4a^2-4a\sqrt{(x-c)^2+y^2}+x^2-2cx+c^2+y^2$

(6) $4cx-4a^2=-4a\sqrt{(x-c)^2+y^2}$

(7) $cx-a^2=-a\sqrt{(x-c)^2+y^2}$

(8) $c^2x^2-2a^2cx+a^4=a^2[(x-c)^2+y^2]$

To simplify further, expand the parentheses and collect all the terms with variables (x and y) on the left-hand side of the equation.

(9) $c^2x^2-2a^2cx+a^4=a^2[x^2-2cx+c^2+y^2]$

(10) $c^2x^2-2a^2cx+a^4=a^2x^2-2a^2cx+a^2c^2+a^2y^2$

(11) $c^2x^2-a^2x^2-a^2y^2=a^2c^2-a^4$

Isolating the variables,

(12) $(c^2-a^2)x^2-a^2y^2=a^2c^2-a^4$

(13) $(a^2-c^2)x^2+a^2y^2=a^2(a^2-c^2)$

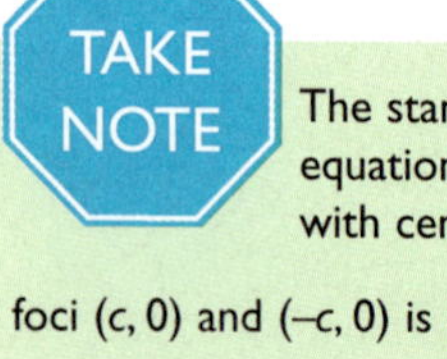

The standard form equation of an ellipse with center (0, 0) and foci $(c, 0)$ and $(-c, 0)$ is $\frac{x^2}{a^2}+\frac{y^2}{b^2}=1$, where $a > b > 0$ and $b^2 = a^2 - c^2$.

But $a^2 - c^2 = b^2$. Substituting this into equation (13) gives $b^2x^2 + a^2y^2 = a^2b^2$. Dividing to get 1 on the right-hand side gives the standard form for an equation of an ellipse.

By substitution, you can verify that the points $(a, 0)$, $(-a, 0)$, $(0, b)$ and $(0, -b)$ are the four intercepts of the ellipse, as noted in the special cases discussed above. (Knowing this will help you graph an ellipse quickly when it is in standard form.)

EXAMPLE 6

Find the foci and graph the ellipse whose equation is $\frac{x^2}{25} + \frac{y^2}{16} = 1$.

SOLUTION:

You know the vertices are (±5, 0) and the endpoints of the minor axis are (0, ±4). So a quick sketch produces **Figure 8.53**.

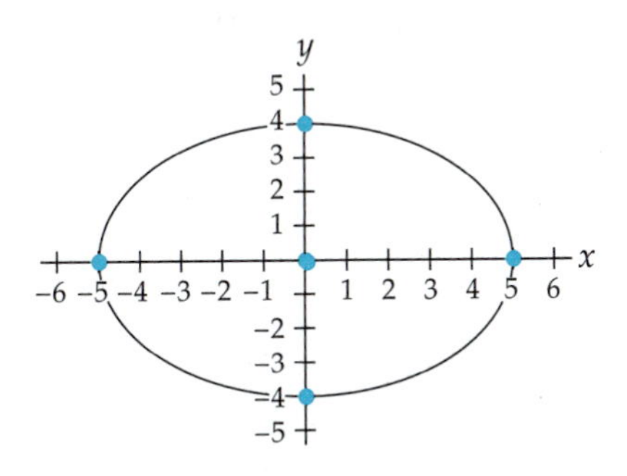

FIGURE 8.53.

From the relationship $b^2 = a^2 - c^2$, the foci of the above ellipse can be identified. Since $b = 4$, $a = 5$, then $c^2 = 5^2 - 4^2$ and $c = 3$. The foci must then be the points (±3, 0).

MAJOR AXIS ALONG THE y-AXIS

If an ellipse has the y-axis as its major axis, the same procedure that was used to find the equation of the ellipse whose major axis is along the x-axis can be used again (see **Figure 8.54**).

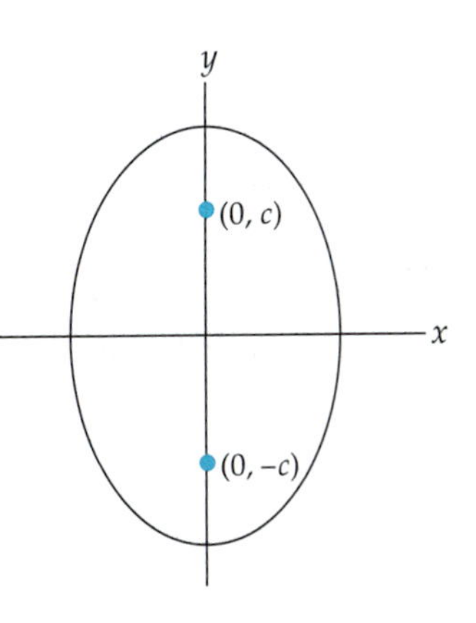

FIGURE 8.54.
Ellipse with the y-axis as its major axis.

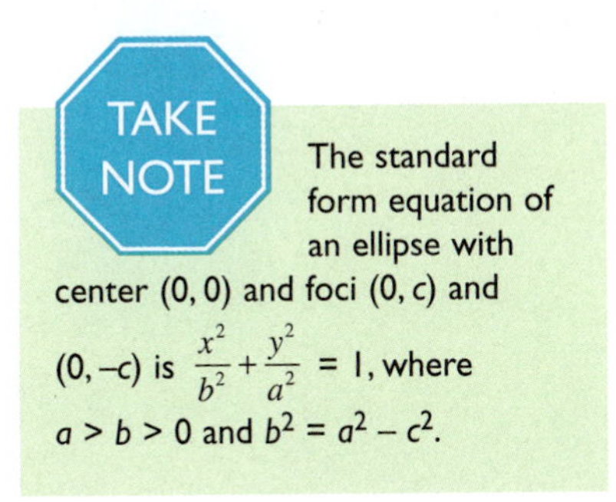

TAKE NOTE

The standard form equation of an ellipse with center (0, 0) and foci (0, c) and (0, −c) is $\frac{x^2}{b^2} + \frac{y^2}{a^2} = 1$, where $a > b > 0$ and $b^2 = a^2 - c^2$.

EXAMPLE 7

Find the equation of the ellipse in Figure 8.54 if $c = 2$ and the length of the major axis is 6.

SOLUTION:

If the length of the major axis is 6, then the y-intercepts of the ellipse are $(0, \pm 3)$ and $a = 3$. Since $b^2 = a^2 - c^2$ and $c = 2$, then $b = \sqrt{5}$. The equation of the ellipse is $\frac{x^2}{5} + \frac{y^2}{9} = 1$.

FIGURE 8.55. Ellipse shifted h units horizontally, k units vertically.

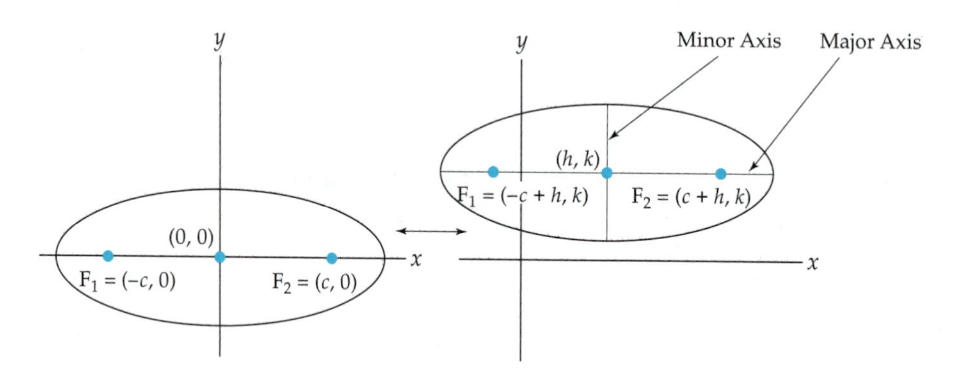

Figure 8.55 shows an ellipse with major axis along the x-axis and center $(0, 0)$ that has been shifted h units horizontally and k units vertically. The result is an ellipse with center (h, k), major axis parallel to the x-axis and minor axis parallel to the y-axis. Its equation is $\frac{(x-h)^2}{a^2} + \frac{(y-k)^2}{b^2} = 1$ where $a > b > 0$ and $b^2 = a^2 - c^2$.

If the ellipse has its major axis along the y-axis prior to the shift of h units horizontally and k units vertically, the result is an ellipse with an equation of $\frac{(x-h)^2}{b^2} + \frac{(y-k)^2}{a^2} = 1$ where $a > b > 0$ and $b^2 = a^2 - c^2$.

EXAMPLE 8

Find the equation of an ellipse with center at (1, 3), a focus at (4, 3), and one endpoint of the minor axis at (1, 6).

SOLUTION:

First make a sketch of the given information as in **Figure 8.56**.

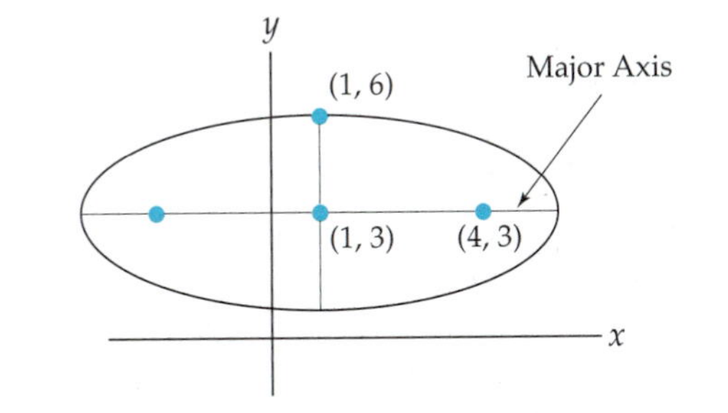

FIGURE 8.56.

From that information, you know that $c = 3$ and $b = 3$. You also know that $a^2 = b^2 + c^2$, so $a = \sqrt{18}$. The equation of the ellipse is $\frac{(x-1)^2}{18} + \frac{(y-3)^2}{9} = 1$.

REFLECTION PROPERTY

Ellipses, like parabolas, have an interesting reflective property. According to this property, if a light source is placed at one focus, it will reflect off the ellipse and concentrate at the other focus (see **Figure 8.57**).

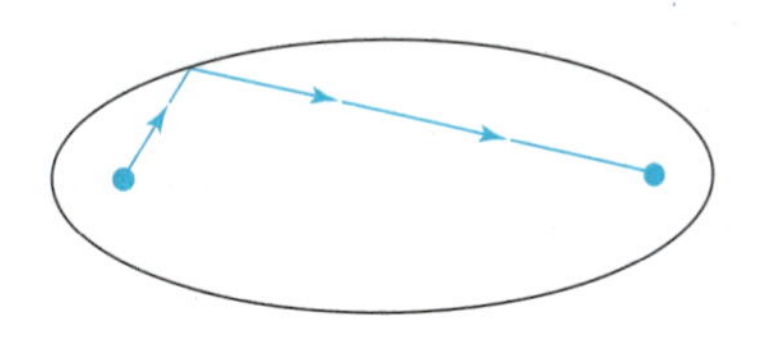

FIGURE 8.57.
The focal property of the ellipse.

The Statuary Hall of the Capitol in Washington, D.C., the entrance to the Oyster Bar on the lower level of New York's Grand Central Station, St. Paul's Cathedral in London, the Mormon Tabernacle in Salt Lake City, the Paris subway, the Tower of London, and the Taj Mahal all demonstrate this reflection property in what are known as "Whispering Galleries." These galleries or rooms have elliptical ceilings so that if you speak softly at one focus of the room, the sound is reflected off the ceiling to the other focus where it can be easily heard. Even in rooms where sounds from other locations in the room are loud, it doesn't interfere with the sounds reflected between the foci.

Exercises 8.4

1. Consider the ellipse $\frac{x^2}{25} + \frac{y^2}{9} = 1$.

   a) Find the center.

   b) Find the foci.

   c) Find the endpoints of the major axis.

   d) Find the endpoints of the minor axis.

   e) Find the lengths of the major and minor axes.

   f) Sketch the graph of the ellipse.

2. Consider the ellipse $\frac{x^2}{a^2} + \frac{y^2}{b^2} = 1$, where $a > b$.

   a) Find the center.

   b) Which axis is the major axis?

   c) What is the length of the major axis?

   d) What is the length of the minor axis?

   e) What are the endpoints of the major axis?

3. Consider the graph in **Figure 8.58**.

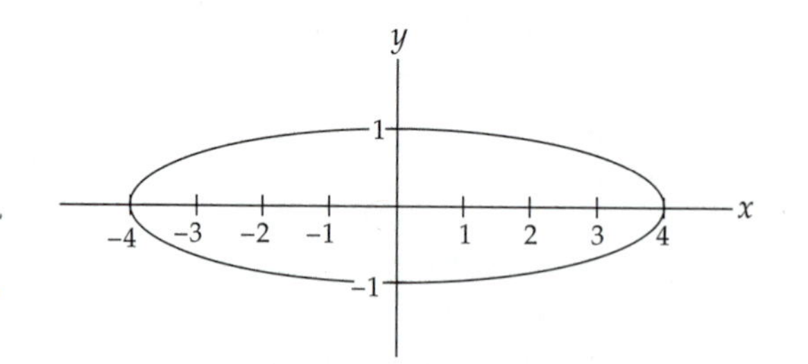

FIGURE 8.58.

   a) Find an equation of the graph.

   b) Give the coordinates of the foci.

Exercises 8.4

4. For this exercise, you will need a compass and straight edge, a geometric drawing utility such as Geometer's Sketchpad or Cabri, or a Plexiglas mirror such as a Mira.

   a) Use your method of choice and follow the steps below to construct points on an ellipse.

      Step 1: Draw a circle and label its center O. Construct a point G on the circle and a point F inside the circle, preferably such that O, G, and F are not collinear.

      Step 2: Construct segment GF and its perpendicular bisector.

      Step 3: Construct segment OG and label its intersection with the perpendicular bisector from step 2, point P.

      Step 4: (If you are using a drawing utility) Trace the path of P as you move point G on the circle.

      Step 4: (If you are not using a drawing utility) Choosing different locations for G on the circle but keeping the location of F the same, repeat steps 2 and 3 at least eight more times. Label your intersection points P_1–P_8.

   b) **Figure 8.59** was constructed using steps 1–3 from part (a) of this exercise. Use this figure to explain why the locus of point P is an ellipse with foci O and F. (Note: Segment PF has been drawn to help with your argument.)

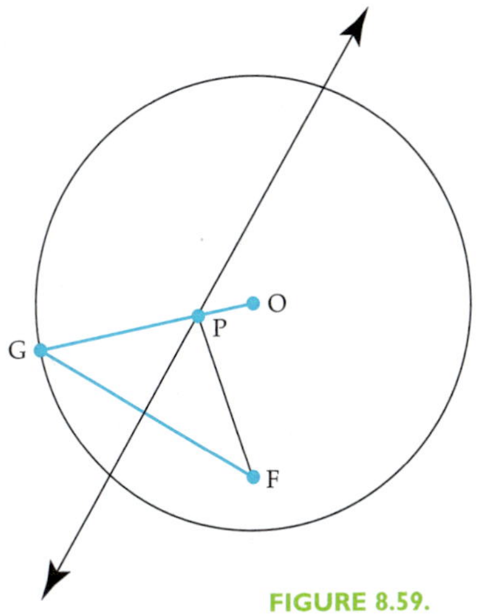

FIGURE 8.59.

5. A elliptical ceiling of a whispering gallery is 10 meters long. If the foci are located 2 meters from the center of the ellipse and 2 meters above the ground, how high is the ceiling at the center?

6. Consider the ellipse $\frac{x^2}{9} + \frac{y^2}{36} = 1$.

   a) Find the center.

   b) Find the foci.

   c) Find the endpoints of the major axis.

   d) Find the endpoints of the minor axis.

Exercises 8.4

e) Find the lengths of the major and minor axes.

f) Sketch the graph of the ellipse.

7. Does the equation $\frac{x^2}{4} + \frac{y^2}{16} = 1$ represent a function? Explain why or why not.

8. Explain how to graph the ellipse in Exercise 7 if your calculator does not have a feature for graphing conics such as circles and ellipses.

9. Consider the ellipse $\frac{x^2}{36} + \frac{y^2}{12} = 1$.

   a) Find the x- and y-intercepts.

   b) Find three other points on the ellipse.

   c) Is the point $\left(2\sqrt{6}, 2\right)$ on the ellipse? Show why or why not.

10. Construct an ellipse by paper folding:

    Step 1: Cut out a circle with radius about 5″ and label the center O.

    Step 2: Pick a point inside the circle and label it F.

    Step 3: Select a point on the circle and label it G.

    Step 4: Fold point G onto point F and crease the paper (see **Figure 8.60**).

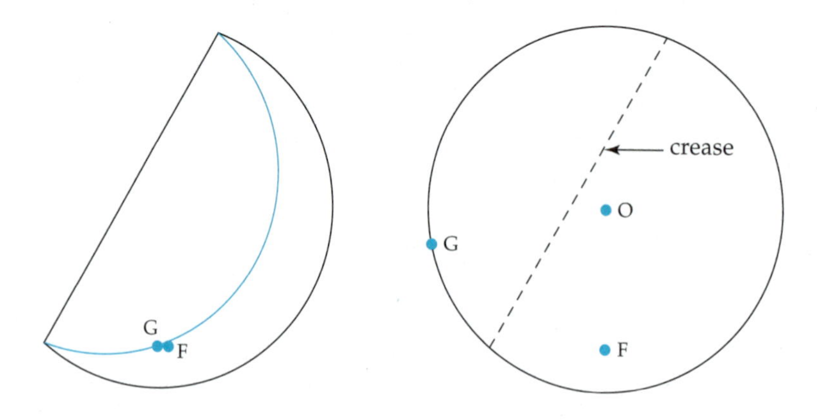

FIGURE 8.60.

Step 5: Use a straight edge to mark the point where segment OG intersects your crease. Label the point P_1 (see **Figure 8.61**).

Exercises 8.4

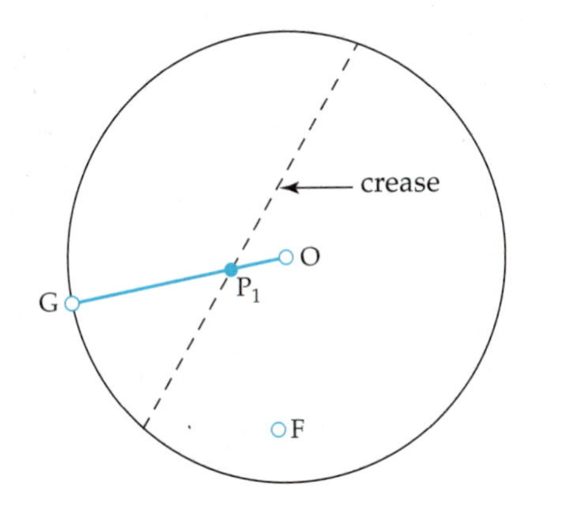

FIGURE 8.61.

Step 6: Repeat steps 3–5, choosing at least seven additional locations around the circle for G. Label the points where the crease intersects the segment OG, P_2, P_3, ...

Step 7: Connect your points with a smooth curve.

a) Explain why you think the curve you constructed is an ellipse.

b) Compare your ellipse to others in the classroom. Are all of your curves the same shape and size? Are all of your curves ellipses? What causes the different shapes?

c) Locate the crease on your paper that produced point P_1 and label the endpoints A and B. With a marker draw segments OG and P_1F (see **Figure 8.62**).

Describe the relationship between the crease AB and your ellipse.

d) With the help of a protractor, explain why a ray of light passing through the focus at F would be reflected through the other focus at O.

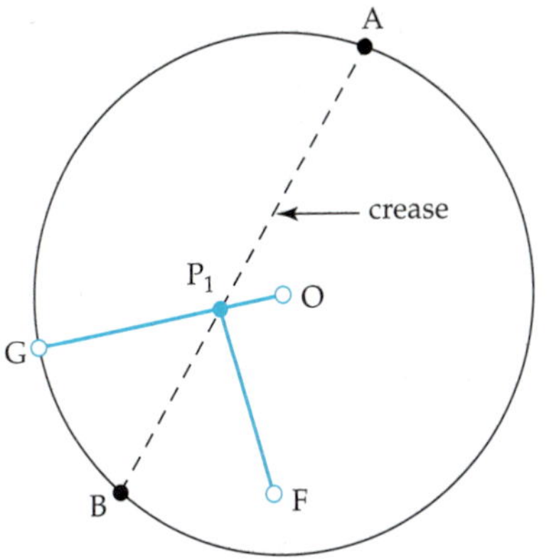

FIGURE 8.62.

11. Describe the curve whose equation is $\frac{x^2}{a^2} + \frac{y^2}{b^2} = 1$ where $a, b \neq 0$ and $a = b$.

Exercises 8.4

12. The orbit of the Earth around the Sun is an ellipse with the Sun located at one focus. The distance from the center of the ellipse to the Sun is 1.5 million miles and the length of the major axis is 186 million miles.

    a) Find the shortest distance from the Earth to the Sun.

    b) Write an equation for the Earth's elliptical orbit around the Sun.

13. a) Complete **Table 8.3.**

TABLE 8.3.

Equations of Ellipses	
Standard Form	**Expanded Form**
$\frac{(x-1)^2}{4}+\frac{y^2}{9}=1$	$9x^2-18x+4y^2-27=0$
$\frac{x^2}{2}+\frac{y^2}{25}=1$	
$\frac{x^2}{6}+\frac{(y+3)^2}{2}=1$	
$\frac{(x+2)^2}{3}+\frac{(y-1)^2}{7}=1$	
$\frac{x^2}{a^2}+\frac{y^2}{b^2}=1$	
	$3x^2+2y^2=24$

    b) From the table in part (a), what generalizations can you make about the equation of an ellipse. How can you tell just by looking at an equation if it is an ellipse?

14. Identify each equation as either an ellipse or not an ellipse. If the equation is an ellipse, find the center, foci, and endpoints of the major and minor axes. Then sketch the graph.

    a) $\frac{(x+2)^2}{4}+\frac{(y-3)^2}{9}=1$

    b) $x^2+10x+y^2=50$

    c) $(x-5)^2+4(y+1)^2=100$

    d) $4x^2+8x+y^2-4y+4=0$

15. Find an equation of an ellipse with the following characteristics. Graph the curve.

    a) Length of the major axis is 12; foci $(-1, 2)$ and $(7, 2)$.

b) Endpoints of the major axis (–1, 7) and (–1, –11); length of the minor axis is 6.

c) Endpoints of the major axis are (2, 2) and (10, 2); endpoints of the minor axis are (6, 5) and (6, –1).

16. Conic paper is designed with two sets of concentric circles that are uniformly spaced. On a sheet of conic paper, mark the centers of the two circles and let them represent the foci of your ellipse. Then locate and mark points whose sum of the distances from the two foci is 12. Use those points to sketch the ellipse.

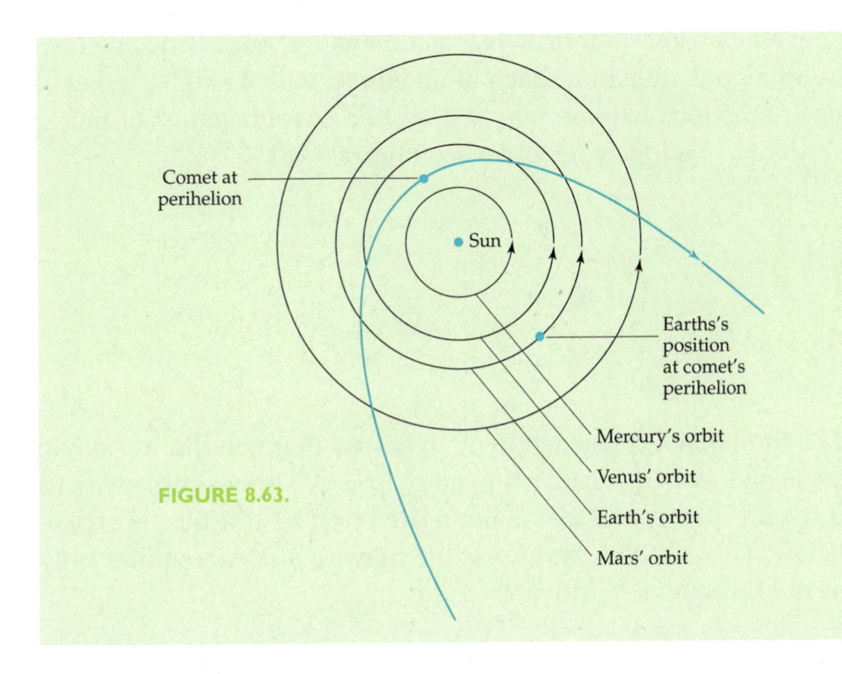

FIGURE 8.63.

17. The orbit of a comet is always a conic section. It can either be an ellipse, a parabola, or a hyperbola. Only if its orbit is an ellipse will the comet return periodically to the vicinity of the Sun. In the case of Halley's Comet, the orbit is elliptical. Refer to **Figure 8.64** and answer the following questions:

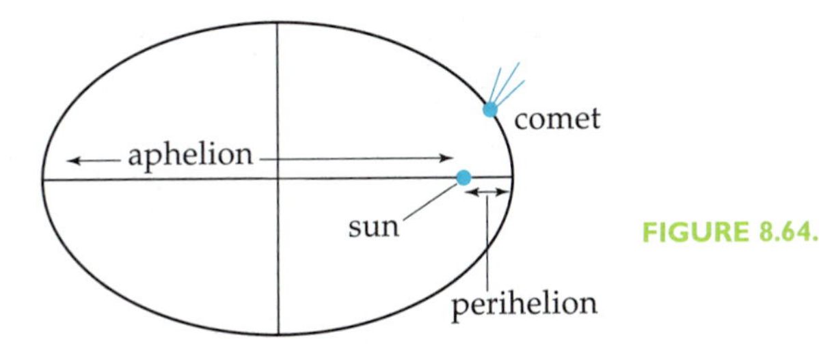

FIGURE 8.64.

Exercises 8.4

HALLEY'S COMET

The most famous of all comets, Halley's Comet, was last visible in 1986. The comet, a "50 billion ton flying mountain of ice," comes close to the earth approximately every 76 years. The sightings in 1986 were some of the least spectacular in history due to the relative positions of the Earth, the Sun, and the comet, as shown in **Figure 8.63.** At perihelion (the comet's closest distance to the Sun), the comet and Earth were on opposite sides of the Sun, obscuring the view of the comet when it was putting on its best show.

You probably did not see the comet in 1986 but never fear, it will be at aphelion (the comet's furthest distance from the Sun) in 2024 and will visit close to Earth again in 2062.

Exercises 8.4

a) If the aphelion, the comet's furthest distance from the Sun, is 35.3029 AU and the perihelion is 0.587 AU, how close to the Sun does the comet come in miles? (AU refers to astronomical unit, and 1 AU $\approx$ 92.96 million miles.)

b) How far is the center of the comet's orbit from the Sun in AUs?

c) What is the length of the minor axis of the orbit?

d) How far from the Sun is the comet (in AUs) when it is at one of the endpoints of its minor axis?

18. One really "smart" fourth year mathematics student decided to build a pool table in a shape of an ellipse with a single pocket at one of the foci. Explain why it might be advantageous for the student to build such a table (see **Figure 8.65**).

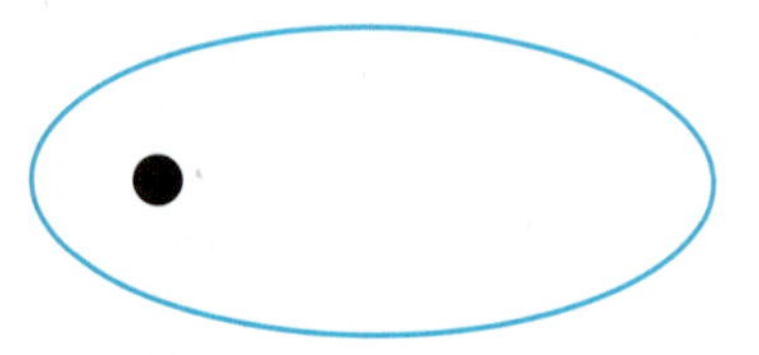

FIGURE 8.65.
Pool table.

19. The thumbtack construction of an ellipse that you did in Activity 8.4 is one way of constructing an ellipse. Another simple way to draw an ellipse, one that is often preferred by drafters, is explained below. This method, known as the *trammel method*, requires only a strip of plastic or cardboard.

Use the directions below to construct a trammel, and then use it to draw an ellipse.

Step 1: Draw two lines that are perpendicular to each other. Let one of the perpendicular lines be the line that contains the major axis of the ellipse and the other be the line that contains the minor axis. The center of the ellipse will be located at the intersection of the two lines.

Step 2: On a piece of plastic or cardboard (see **Figure 8.66**), locate three points M, N and P such that MP = a (half the length of your major axis) and NP = b (half the length of the minor axis).

Exercises 8.4

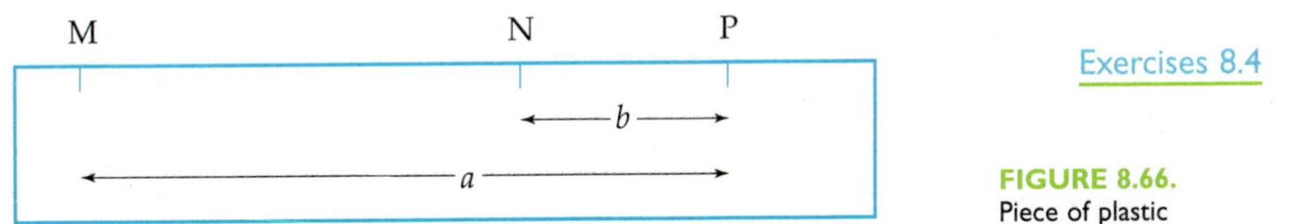

FIGURE 8.66. Piece of plastic or cardboard.

Step 3: Point P will be on the ellipse as long as M is on the minor axis and N is on the major axis as shown in **Figure 8.67**. Move M up and down the minor axis (vertical in this case) while N moves along the major axis (horizontal). As you move the trammel along the axes, mark as many Ps as needed to get an accurate graph.

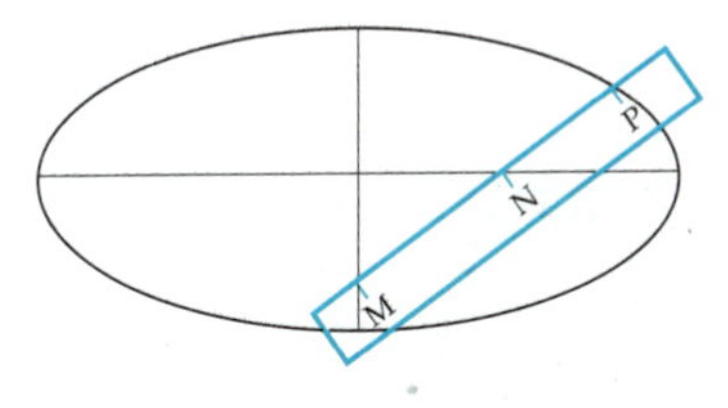

FIGURE 8.67. Constructing the ellipse.

20. Suppose two ellipses intersect each other.

a) Is it possible for them to intersect in exactly one point? If so, make a sketch showing the ellipses and the point of intersection.

b) Is it possible for them to intersect in exactly two points? Again, if so, make a sketch.

c) Are any other number of intersections possible? If so, make sketches.

21. Suppose the orbits of two comets moving around a sun are modeled by the following equations: $\frac{x^2}{100} + \frac{y^2}{64} = 1$ and $\frac{(x-6)^2}{48} + \frac{(y+4)^2}{64} = 1$.

a) Graph the orbits of the comets.

b) Use a graphing calculator to find an approximation of the points of intersection of the two ellipses.

c) Find the location of the sun. (Assume the sun is at one focus of each ellipse.)

22. **Investigation:** Use **Handout 8.4**, a CBL, a motion detector, and a graphing calculator to examine the motion of a swinging pendulum by plotting velocity vs. position of the pendulum.

Modeling with Hyperbolas

The **hyperbola** is the only conic section yet to be studied. A hyperbola is defined as the locus of points in a plane such that the difference of the distances from each point on the locus to two fixed points is a constant. As with the ellipse, the two fixed points are called **foci**. (Recall that distance is always a non-negative number.)

Activity 8.5

1. As you worked your way through previous lessons in this chapter, you used several methods to construct certain curves. Choose a method (e.g. ruler, waxed paper, geometric drawing utility, conic paper) and from its definition, construct a hyperbola with the foci as shown in **Figure 8.68.**

FIGURE 8.68. ←approximately 2"→

2. Compare and contrast the curve you just created to that of an ellipse.

3. Find the midpoint of the segment joining the foci. This point is called the **center** of the hyperbola. Every hyperbola has two asymptotes that pass through the center of the hyperbola. Use a ruler to help you sketch where you think the asymptotes are located. (Hint: Recall from Chapter 3 that an asymptote is a line that the graph of a curve approaches as x gets infinitely large in either the positive or negative direction, except for vertical asymptotes.)

THE HYPERBOLA

As mentioned above, a hyperbola is a locus of points in a plane such that the difference of the distances from each point on the locus to two fixed points is a constant. And as you discovered in Activity 8.5, a hyperbola consists of two separate curves, called

branches (see **Figure 8.69**). The line that passes through the foci is called the **transverse axis,** and the midpoint of the segment joining the foci is the **center** of the hyperbola. The line through the center, perpendicular to the transverse axis is called the **conjugate axis.**

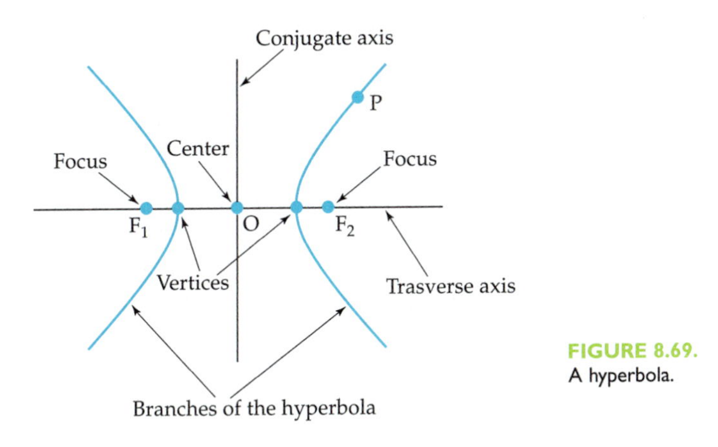

FIGURE 8.69.
A hyperbola.

In Figure 8.69, the point P is on the hyperbola. This point, as well as every other point on either branch of the curve, is some constant k units closer to one of the foci than to the other. That is, $|PF_1 - PF_2| = k$.

As with the other conics—the circle, the parabola, and the ellipse—applications of the hyperbola abound. For example, the orbit of a planet or comet is an ellipse if the planet or comet revolves about the Sun. If it doesn't, the path is a hyperbola.

When you get an opportunity, examine the patterns in a pool or pond when you drop two pebbles into the water. As shown in **Figure 8.70**, it is possible to see both ellipses and hyperbolas in the interference patterns formed by the waves.

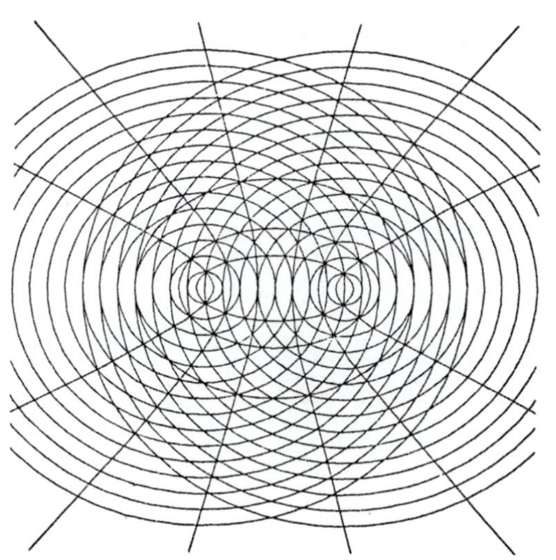

FIGURE 8.70.
Patterns formed by two wave sets.

A NOTABLE QUOTE

"The hyperbola is frequently in the limelight as the graph of equations in physics, chemistry, biology, economics and engineering. It has a starring role in Einstein's theory of special relativity where an observer in an inertial reference frame sees a particle in a parallel force field follow a hyperbolic path in space-time." Dr. Lee Whitt, mathematician, *The UMAP Journal* 5 (Spring 1984): 9.

Look carefully at the shadows formed on the wall by a lamp (see **Figure 8.71**). The shadows are most likely hyperbolic in shape.

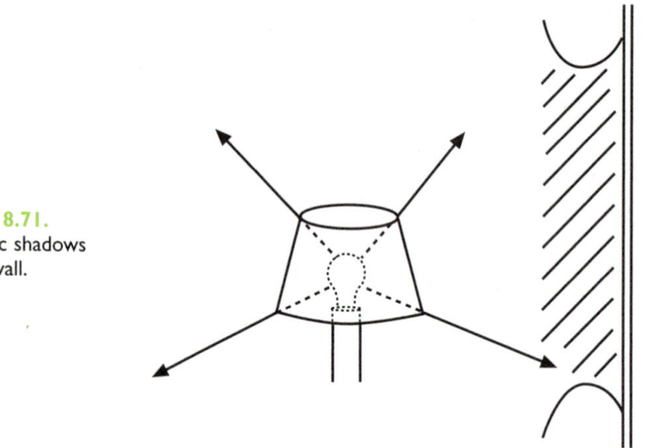

FIGURE 8.71.
Hyperbolic shadows on a flat wall.

HYPERBOLA CLUES

As strange as it may seem, hyperbolas are the signal to police or other investigators that metallic objects lie below the surface of the earth. Ground Penetrating Radar (GPR) systems are used to map areas where people are looking for information underground.

Figure 8.72 shows data collected in conjunction with a police investigation in Canada. As the police officers scan the area, the data are displayed in real-time on a computer screen. From the data, investigators know exactly where to dig for evidence.

The data were collected in the front yard of a home. Signs of digging are shown on the left side. The hyperbola on the right indicates a metallic object.

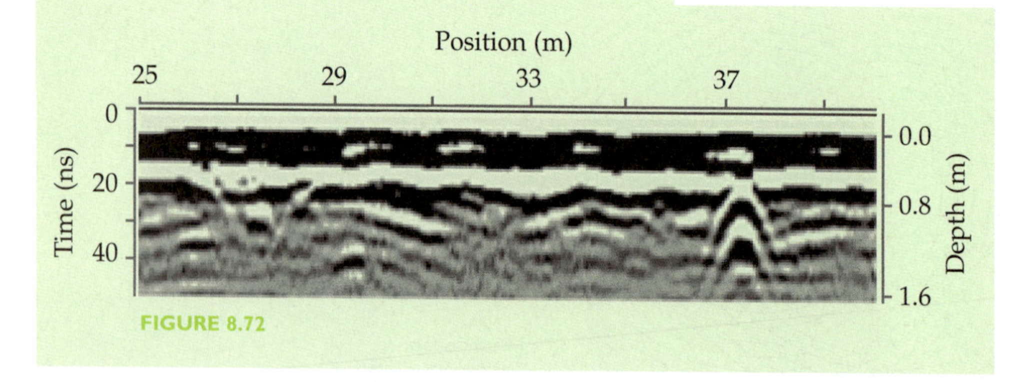

FIGURE 8.72

One of the most significant applications of the hyperbola is found in navigation. The LOng RAnge Navigation system, known as LORAN, is based on the time difference between the reception of signals sent simultaneously from two stations. Even though LORAN is taking a back seat to GPSs (global positioning systems), it is still in extensive use in aircraft, boats, and ships. A more detailed mathematical discussion of this system appears later in Exercises 11 and 12.

THE EQUATION FOR THE HYPERBOLA

An equation of a hyperbola can be found by using its definition and the distance formula, as in the case of the ellipse. If the foci are located on one of the coordinate axes, say the x-axis, and the center is at the origin as shown in **Figure 8.73**, finding the equation is not difficult. The details of the derivation of the equation are left for Exercise 8.

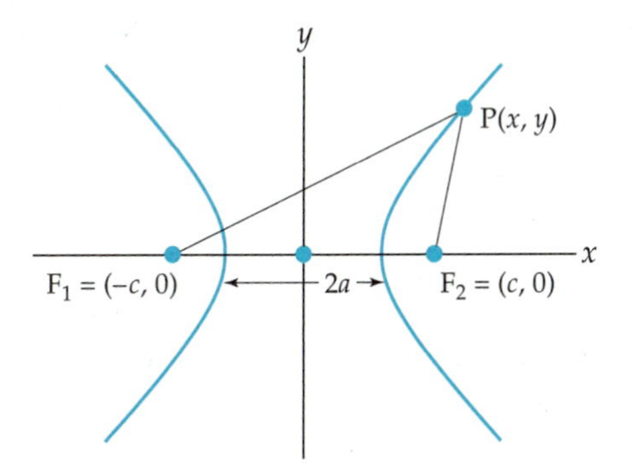

FIGURE 8.73.
A hyperbola with foci at $(-c, 0)$ and $(c, 0)$.

Note that the x- and y-intercepts can be found, if there are any, by letting $y = 0$ and $x = 0$ respectively. In Figure 8.73, the x-intercepts can be found by solving $\frac{x^2}{a^2} = 1$. Hence, the intercepts, also known as **vertices**, are $(-a, 0)$ and $(a, 0)$. The y-intercepts can be found by solving $\frac{y^2}{b^2} = -1$. Since this equation has no real solutions, you can conclude that there are no y-intercepts for the hyperbola in the figure. In Activity 8.5, you noticed that the hyperbola lies entirely within two sectors of the plane bounded by two straight lines called asymptotes. **Figure 8.74** suggests that the asymptotes are the diagonals of a rectangle (called the **central rectangle**) formed by the lines $x = -a$, $x = a$, $y = b$, and $y = -b$, and the equations of the asymptotes are $y = \frac{b}{a}x$ and $y = -\frac{b}{a}x$. Although the asymptotes are not part of the hyperbola, just as the foci and center are not part of the curve, they do help guide you when you sketch the graph. When you are asked to sketch a hyperbola, it is always a good idea to draw the central rectangle and the asymptotes.

TAKE NOTE

The standard form equation of a hyperbola with center $(0, 0)$ and foci $(c, 0)$ and $(-c, 0)$ $(-c, 0)$ is $\frac{x^2}{a^2} - \frac{y^2}{b^2} = 1$, where $a < c$ and $b^2 = c^2 - a^2$.

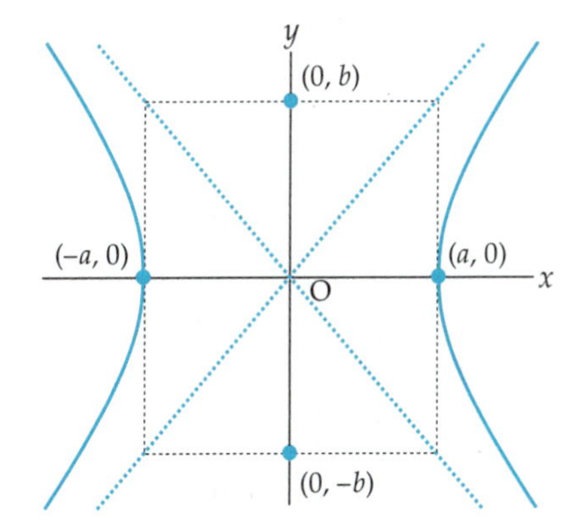

FIGURE 8.74.
Asymptotes of the hyperbola.

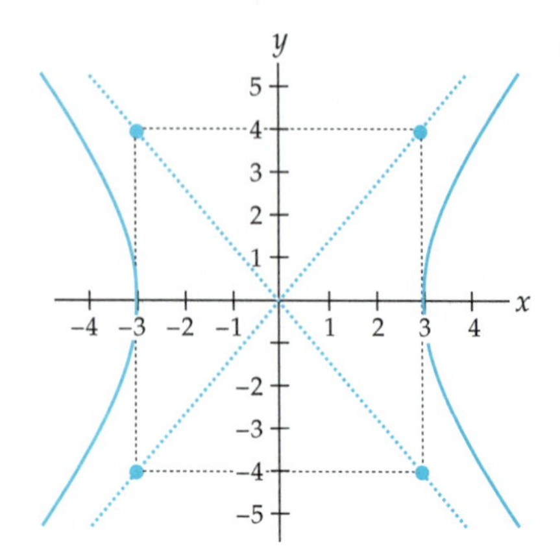

FIGURE 8.75.

EXAMPLE 9

Find an equation for the hyperbola with foci (–5, 0) and (5, 0) and vertex at (3, 0). Sketch the graph and give the equations of the asymptotes.

SOLUTION:

You know that $c = 5$, $a = 3$ and that $b^2 = c^2 - a^2$. So $b = 4$. Therefore, $\frac{x^2}{9} - \frac{y^2}{16} = 1$. The asymptotes are $y = \pm\frac{b}{a}x$ or $y = \pm\frac{4}{3}x$. The graph is shown in **Figure 8.75**.

As in the study of the ellipse, the foci can be located on the y-axis rather than the x-axis (see **Figure 8.76**).

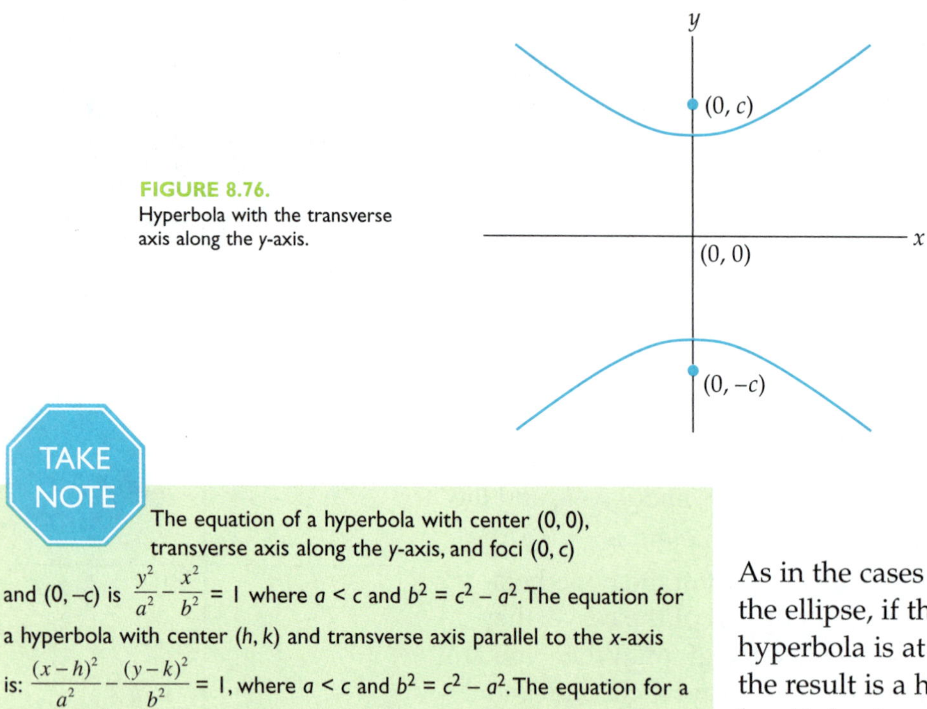

FIGURE 8.76.
Hyperbola with the transverse axis along the y-axis.

TAKE NOTE

The equation of a hyperbola with center (0, 0), transverse axis along the y-axis, and foci $(0, c)$ and $(0, -c)$ is $\frac{y^2}{a^2} - \frac{x^2}{b^2} = 1$ where $a < c$ and $b^2 = c^2 - a^2$. The equation for a hyperbola with center (h, k) and transverse axis parallel to the x-axis is: $\frac{(x-h)^2}{a^2} - \frac{(y-k)^2}{b^2} = 1$, where $a < c$ and $b^2 = c^2 - a^2$. The equation for a hyperbola with center (h, k) and transverse axis parallel to the y-axis is: $\frac{(y-k)^2}{a^2} - \frac{(x-h)^2}{b^2} = 1$ where $a < c$ and $b^2 = c^2 - a^2$.

As in the cases of the parabola and the ellipse, if the center of the hyperbola is at some point (h, k), the result is a hyperbola shifted h units horizontally and k units vertically.

EXAMPLE 10

Find an equation for the hyperbola with the center at (1, 2), one focus at (4, 2) and one vertex at (3, 2). Find the equations of the asymptotes. Sketch the graph.

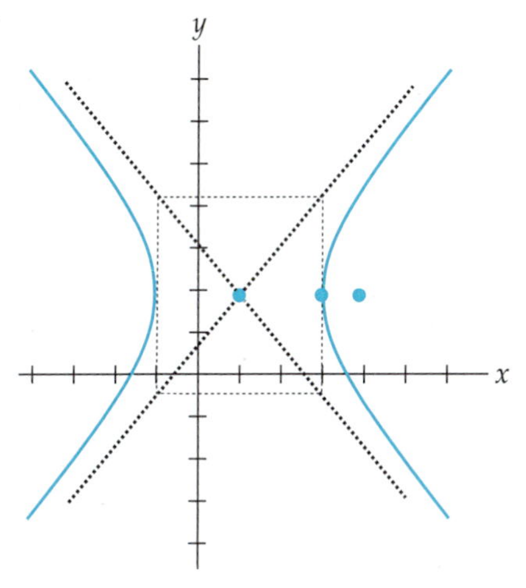

FIGURE 8.77.

SOLUTION:

Begin by making a sketch of the given information.

From your sketch, you know that $c = 3$, $a = 2$, and that the transverse axis is parallel to the x-axis. Since $b^2 = c^2 - a^2$, you know that $b^2 = 5$. The equation is $\frac{(x-1)^2}{4} - \frac{(y-2)^2}{5} = 1$.

Activity 8.6

In Lessons 8.3 and 8.4 you explored the reflection properties of the parabola and the ellipse. In this activity, you will investigate the reflection properties of the hyperbola.

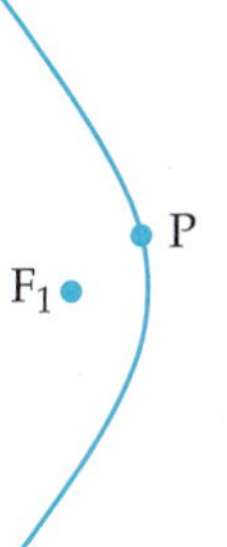

FIGURE 8.78. Hyperbola.

As shown in the **Figure 8.78**, the hyperbola divides the plane into three regions, one containing the focus F_1, one containing the focus F_2, and one containing no foci.

1. **Handout 8.5** contains a sketch of a hyperbola. On that sketch, let P be a point on the branch of the hyperbola that isolates F_1. Draw the tangent to the hyperbola at point P.

2. Suppose a light source is located at F_1. Draw the segment F_1P. Use your prior knowledge of how light reflects off a line, and with the help of a protractor, draw the line PQ to represent a ray of light reflected off the hyperbola at P.

3. Repeat Steps 1 and 2 for at least two other points P on the hyperbola. What do you notice about the lines PQ?

4. Again, draw a hyperbola with two branches and foci at F_1 and F_2. Let R be a point in the region that contains no foci. (This time, imagine that the reflective part of the hyperbola is on its convex side.) Draw segment RF_1, and label the point where the segment intersects the hyperbola P. Then draw the tangent to the hyperbola at point P.

   Suppose a light source is located at point R and hits the hyperbolic mirror at P. Draw a line to show how the light is reflected. Repeat this procedure for at least three different points R. What do you notice about the lines that are reflected off the hyperbola?

5. Write a short summary of what you found out about the reflection properties of the hyperbola.

Exercises 8.5

1. Sketch the graph of the equation. Show the asymptotes in your sketch. Write the equations of the asymptotes.

   a) $\frac{x^2}{25} - \frac{y^2}{144} = 1$

   b) $\frac{y^2}{25} - \frac{x^2}{144} = 1$

   c) $4x^2 - 9y^2 = 36$

2. Find an equation of the hyperbola with the following characteristics:

   a) Center at the origin, vertex (–2, 0), and focus (4, 0).

   b) Center (0, 0), vertex (0, 6), and the equation of one of the asymptotes $y = \frac{3}{4}x$.

3. Discuss the symmetry of a hyperbola with its center at the origin.

4. For this exercise, you will need a compass and straight edge, a geometric drawing utility such as Geometer's Sketchpad or Cabri, or a Plexiglas mirror such as a Mira.

   a) Use your method of choice and follow the steps below to construct points on a hyperbola.

      Step 1: Draw a circle and label its center O. Construct a point G on the circle and a point F outside the circle. (Hint: It is preferable that these points not be collinear.)

      Step 2: Construct segment GF and its perpendicular bisector.

      Step 3: Construct the **line** OG. Find the intersection of the line OG and the perpendicular bisector from step 2. Label this point P.

      Step 4: (If you are using a drawing utility) Trace the path of P as you move point G on the circle.

Exercises 8.5

Step 4: (If you are not using a drawing utility) Keep F fixed and choose different locations for G on the circle. Repeat steps 2 and 3 at least eight more times. Label your intersection points P_1–P_8. Make sure that you choose enough points in different locations so that you get points on both branches of the hyperbola.

b) **Figure 8.79** was constructed using steps 1–3 from part (a) of this exercise. Use the figure to explain why the locus of point P is a hyperbola with foci O and F. (Note: Segment PF has been drawn to help you with your argument.)

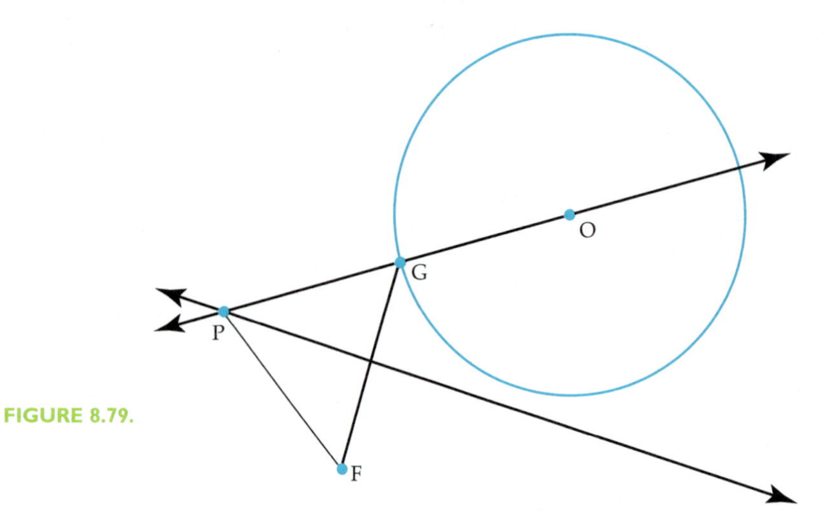

FIGURE 8.79.

5. A hyperbola where $a = b$ is called an **equilateral** hyperbola.

   a) Find an equation for the equilateral hyperbola with foci (–6, 0) and (6, 0).

   b) What can you say about the angles formed by the intersection of the asymptotes?

6. Sketch the graph of each equation. Show the asymptotes in your sketch. Write the equations of the asymptotes.

   a) $\frac{(x+4)^2}{9} - \frac{(y-1)^2}{25} = 1$

   b) $(y + 4)^2 - 16(x + 2)^2 = 64$

   c) $9x^2 - 4y^2 - 54x - 8y + 41 = 0$

Exercises 8.5

7. Find an equation of the hyperbola with the following characteristics:

   a) Center (1, –3), focus (1, 1), vertex (1, –5).

   b) Foci (–6, 2) and (2, 2) and contains the point (1, 2).

8. If (x, y) is any point on a hyperbola with center at the origin and foci $(-c, 0)$ and $(c, 0)$, and if the constant difference of the distances between the point (x, y) and the foci is $2a$, verify that an equation for the given hyperbola is $\frac{x^2}{a^2} - \frac{y^2}{b^2} = 1$.

9. Constructing a hyperbola by paper folding:

   Step 1: If waxed paper or patty paper is available, copy the circle given to you onto it, otherwise, use the circle on the paper in **Handout 8.2**. Label its center O.

   Step 2: Choose a point outside the circle and label it F.

   Step 3: Select a point on the circle and label it G.

   Step 4: Fold point F onto point G and crease the paper as shown in **Figure 8.80**.

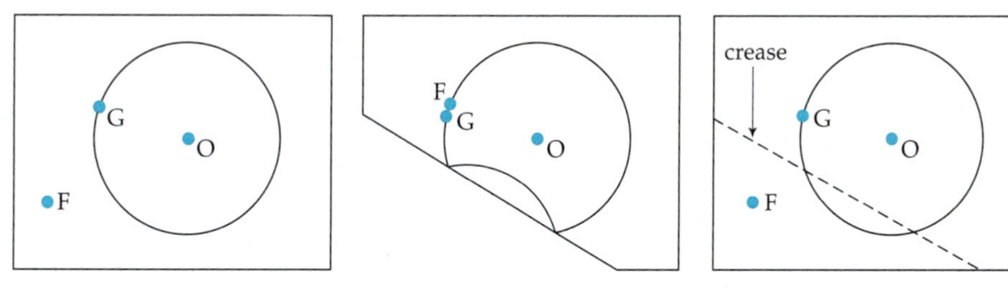

FIGURE 8.80.

   Step 5: Use a straight edge to mark the point where segment OG intersects your crease. Label the point P_1 (see **Figure 8.81**).

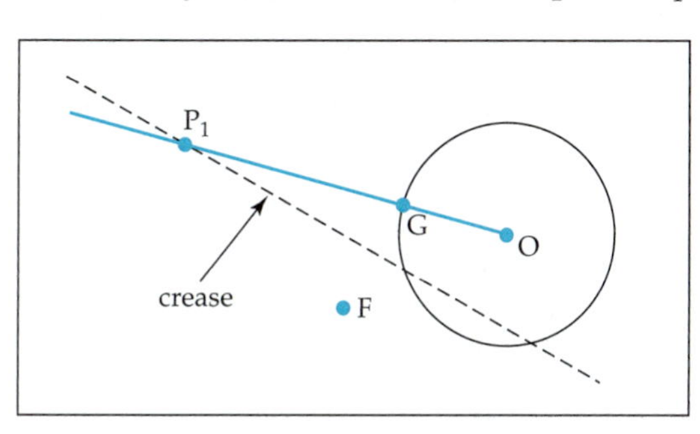

FIGURE 8.81.

Exercises 8.5

Step 6: Repeat steps 3–5 choosing seven or more additional locations around the circle for G. Label the points where the crease intersects the segment OG, P_2, P_3, ...

Step 7: Connect your points with smooth curves to show the two branches of the hyperbola.

a) Explain why the curve you constructed is a hyperbola.

b) Compare the shape of your hyperbola to others in the classroom. Are all of the curves congruent? Are all of your curves hyperbolas? What causes the different shapes?

c) Locate the crease on your paper that produced point P_1 and label the endpoints A and B. With your marker draw the segment P_1F and the line OP_1 that contains G (see **Figure 8.82**).

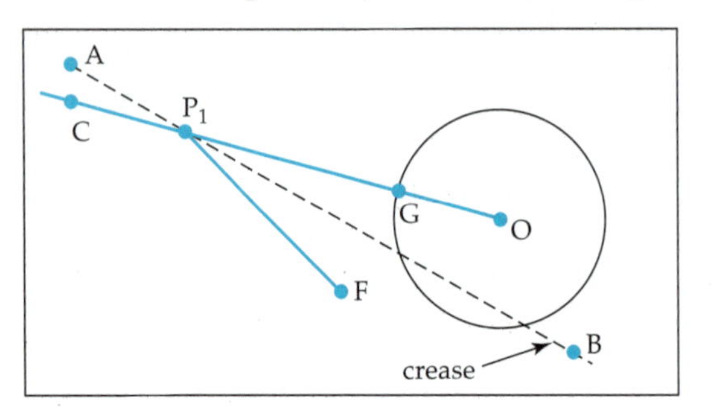

FIGURE 8.82.

Describe the relationship between the crease AB and your hyperbola.

d) With the help of a protractor, explain why a ray of light FP_1 would be reflected along a line P_1C, a line that contains the other focus O.

10. For each of the following equations, identify the transverse axis and find the center, the vertices, the foci, and the equations of the asymptotes.

a) $\frac{(y+2)^2}{18} - \frac{(x-5)^2}{9} = 1$

b) $3(x-4)^2 - 18(y-1)^2 = 27$

c) $x^2 - y^2 + 4x - 2y - 22 = 0.$

Exercises 8.5

11. a) Sketch the graph of $\frac{x^2}{4} - \frac{y^2}{9} = 1$ and its asymptotes.

    b) Write equations for the asymptotes.

    c) Write equations of three other hyperbolas that have the same asymptotes as the hyperbola in part (a).

    d) Write equations of three hyperbolas with asymptotes $y = \pm\frac{b}{a}x$.

    e) Write a general equation for all hyperbolas with asymptotes $y = \pm\frac{b}{a}x$.

    f) Write equations of three hyperbolas with asymptotes $3y + 5x = 19$ and $3y - 5x = -1$.

12. In Lesson 1.2, you explored graphs qualitatively. Consider the following context in a similar qualitative manner.

    Suppose two people A and B are outside when they hear an explosion, and person A hears it approximately 1 second after person B.

    a) What do you know about the locations of A and B relative to the explosion?

    b) Since the speed of sound travels at a constant rate, the location of the explosion can be modeled by using a hyperbola with foci at A and B as shown **Figure 8.83**.

    Where could the explosion X be located? Explain why a hyperbola is an appropriate model for this situation.

    c) Suppose that a third person C hears the explosion one second after A does and that C's location relative to A and B is as shown in **Figure 6.84**. Make a qualitative sketch that shows the possible location(s) for the explosion.

    d) How many locations for X did you find? If you found exactly one location for X, is it possible for A, B, and C to be located in such a way that you could not have found a unique location? If so, make a sketch to explain.

A B

FIGURE 8.83.

Corbis

Exercises 8.5

13. The mathematical basis of LORAN (LOng RAnge Navigation system), used throughout the world to locate geographical positions of ships and planes, is very similar to your explorations in Exercise 12. For a ship to locate its position, its navigator must calculate the time difference between the arrival of two radio signals sent from two different LORAN transmitters. This determines a hyperbolic curve on which the ship is situated. It then takes another time difference from a second set of transmitters whose hyperbolic curve intersects that of the first pair. Once the intersection is determined, a fix of the ship's position can be obtained. (Note: The speed of each radio signal is 186,000 miles per second.)

    Suppose three LORAN stations are positioned as shown in **Figure 8.85**.

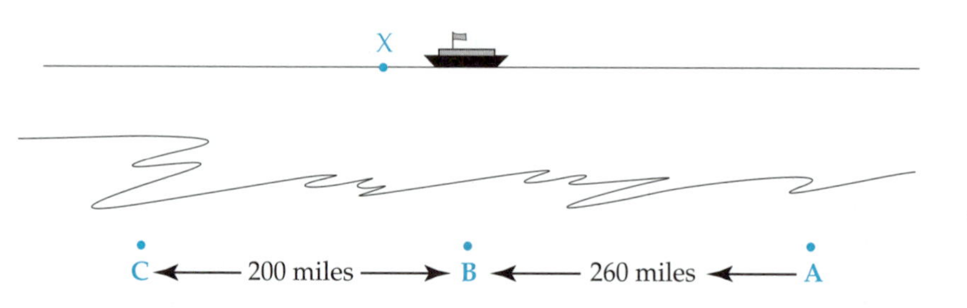

FIGURE 8.85.

    If the navigator of the ship X records a time difference of 0.000860 seconds between the signals from transmitters A and B, and a time difference of 0.000645 seconds between the signals from transmitters B and C, how far is the ship from the coastline? (Note: Assume the coast is a straight line through A, B, and C.)

14. Telescopes and large tracking antennas often use reflectors, called compound reflectors, that have more than one reflective surface. One advantage of this system is that telescopes with compound reflectors are smaller than ordinary telescopes with the same power. In antennas with compound reflectors, the collection devices are located closer to the support for the base, decreasing the torque exerted on the mounting hardware.

NASA

One type of these systems uses the Cassegrain principle, which consists of a primary parabolic reflector and a secondary hyperbolic reflector. For this system to work, both reflectors must share a focus.

Consider **Figure 8.86**.

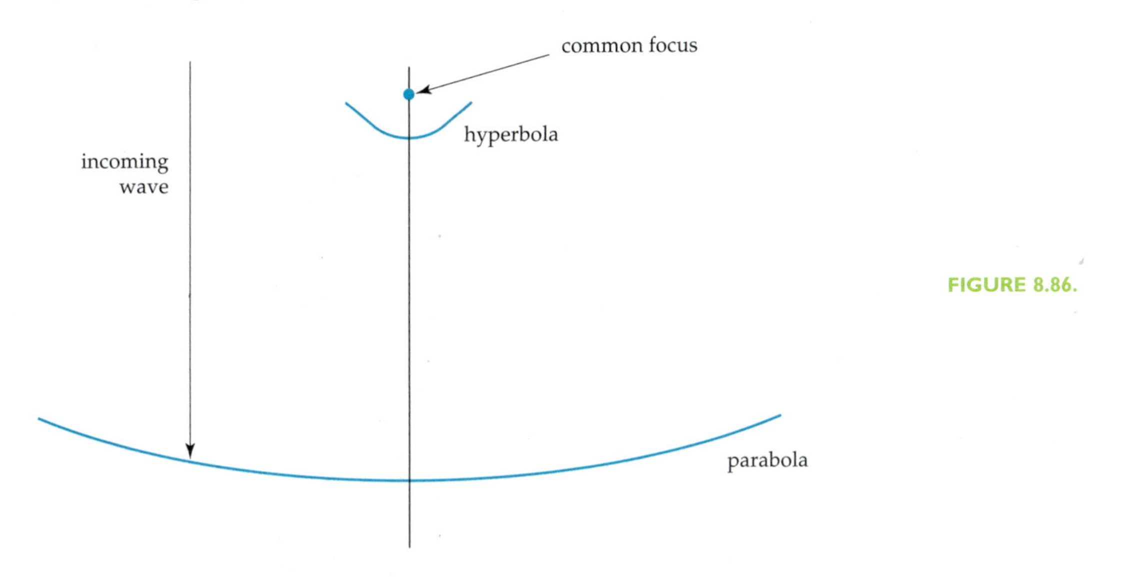

FIGURE 8.86.

Redraw the figure on your paper and use your knowledge of the reflective properties of the parabola and the hyperbola to show where the cone of the antenna (the receiver) should be located so that it can capture the signal. Explain why you placed the cone where you did.

15. Use a sheet of conic paper and let the two centers of the circles represent the foci of the desired hyperbola. Locate and mark points whose difference in distances from the two foci is 6 units (7 and 1, 8 and 2, 9 and 3, etc.). Then use these points to sketch the hyperbola.

16. According to the 1998 Internal Revenue tax code, personal moving expenses are tax-deductible if the distance between the taxpayer's new principal place of work and his/her former residence is at least 50 miles greater than the distance between his/her former principal place of work and his former residence.

Exercises 8.5

a) Consider **Figure 8.87**.

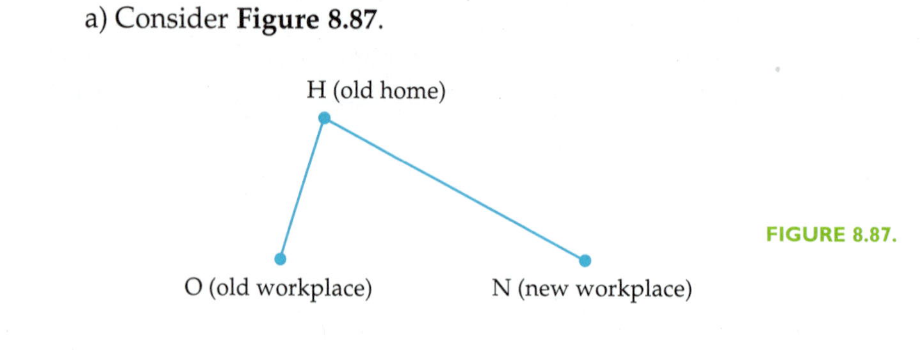

FIGURE 8.87.

Write an algebraic statement that must be true in order to deduct moving expenses on your income tax.

b) Assume that the former principal workplace is located at point A in **Figure 8.88** and that the new principal workplace is located at point B. Draw a picture that shows all possible locations for the former residence that would qualify for a tax deduction. Explain your reasoning.

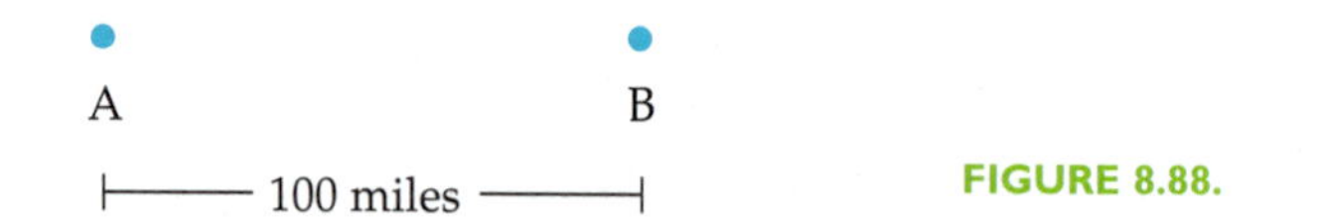

FIGURE 8.88.

c) Find an equation for the locus of points in a plane such that the difference of the distances from each point on the locus to the old and new workplaces is 50 miles. Assume that the distance between old workplace A and the new workplace B is 100 miles.

CHAPTER

8

ANALYTIC GEOMETRY

Chapter 8 Review

1. Write a summary of the important mathematical ideas found in Chapter 8.

2. Sketch the locus of points one unit distance from the circle whose center is O and whose radius is 3 (see **Figure 8.89**).

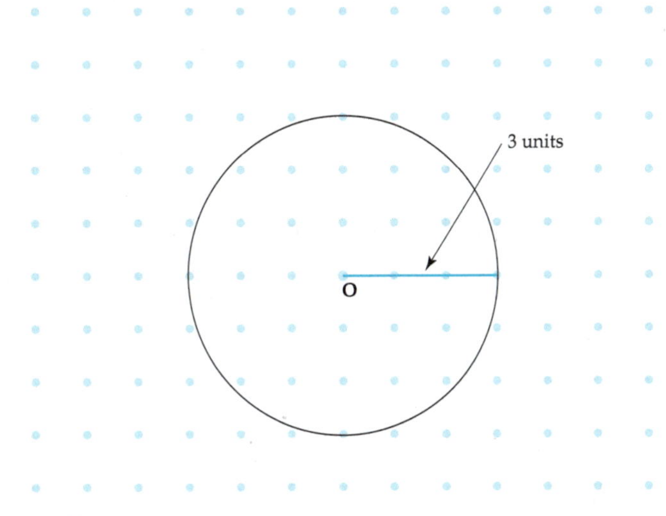

FIGURE 8.89.

3. Consider a circle with its center at the origin. Find an equation of the line that is tangent to the circle at the point (–5, 4).

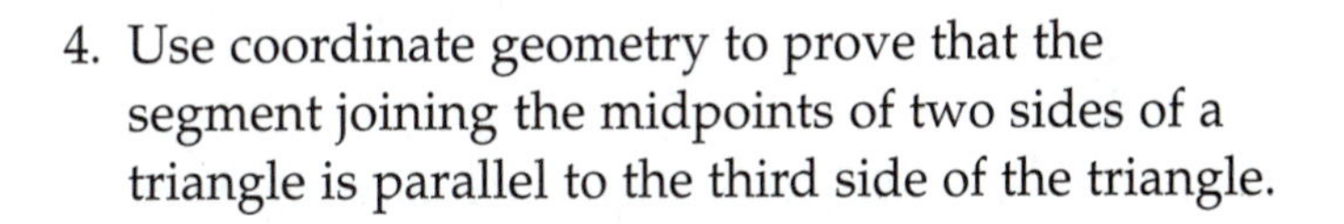

4. Use coordinate geometry to prove that the segment joining the midpoints of two sides of a triangle is parallel to the third side of the triangle.

5. From your work in Exercise 4, what can you say about the length of the segment of the line joining the midpoints of two sides and the length of the third side?

6. A 20-meter pole and a 30-meter pole are positioned 10 meters away from each other (see **Figure 8.90**). Wires attach the top of one pole to the bottom of the other.

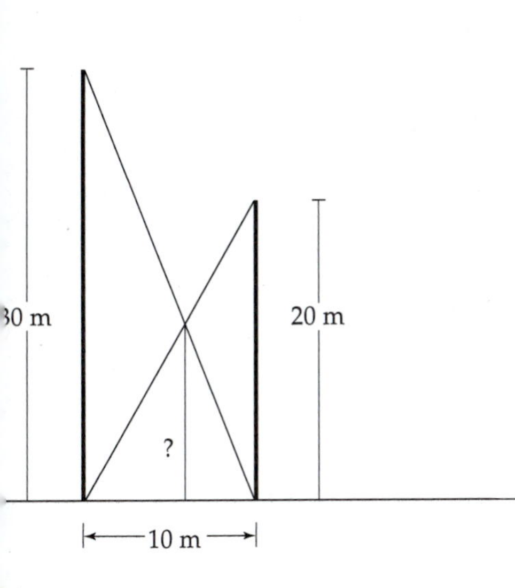

FIGURE 8.90.

a) How high off the ground do the two wires intersect?

b) Predict what will happen to the height of the intersection if the poles are moved apart to a distance of 20 meters.

c) Determine the height of the intersection if the poles are moved so the distance between them is 20 meters.

d) What do you think happens to the height of the intersection if the poles are moved c meters apart? Show that your conjecture is correct.

e) What can you say about the relationships among the heights of the poles, the distance between the poles, and the height of the intersection of the wires?

7. Consider the equation $Ax^2 + Cy^2 + Dx + Ey + F = 0$.

a) If the equation represents a parabola, what do you know about A and C?

b) If the equation represents an ellipse, what do you know about A and C?

c) If the equation represents a hyperbola, what do you know about A and C?

d) If the equation represents a circle, what do you know about A and C?

8. For each of the following equations, identify the conic it describes and graph it.

- If it is a circle, give its center and radius.
- If it is a parabola, give its vertex, focus, and directrix.
- If it is an ellipse, give its center, foci, vertices, and the endpoints of the major and minor axes.
- If it is a hyperbola, give its center, foci, vertices, and asymptotes.

a) $4x^2 + 9y^2 - 16x = 20$

b) $x^2 = 3y^2 - 9$

c) $2x^2 + y - 8x = 0$

d) $x^2 + y^2 - 6x + 10y - 30 = 0$

9. Write an equation for each of the following descriptions. Sketch the graph.

a) Hyperbola: center (0, 0), focus (3, 0), and vertex (2, 0)

b) Parabola: focus (4, 6) and directrix $x = -2$

c) Circle: center (–3, –5) and radius of length 9

d) Ellipse: focus (6, 3), an endpoint of the major axis (–4, 3), and center at (2, 3)

10. As **Figure 8.91** shows, a bridge is to be built across a river that is 40 meters wide, and the center of the bridge must rise 15 meters above the river.

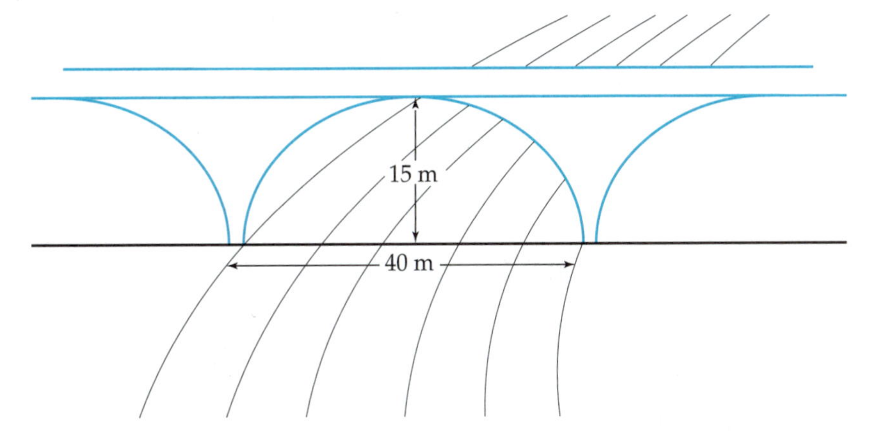

FIGURE 8.91.

Two plans are being considered: a parabolic arch bridge and a semi-elliptical arch bridge.

a) Determine an equation of the parabola that meets the given criteria.

b) Determine an equation of the ellipse that meets the given criteria.

Assume that the most common-sized boat that travels this river has a width of 16 meters.

c) How tall can the boat be if the bridge is the parabolic arch from part (a)?

d) How tall can the boat be if the bridge is the semi-elliptical arch from part (b)?

e) Which type of arch would you recommend for the bridge? Explain your reasoning.

11. In your study of conics, you examined the reflective properties of hyperboloids and paraboloids, three-dimensional objects formed by rotating the hyperbola and parabola about the axis of symmetry that contains the focus or foci. Likewise, the sphere can be thought of as a three-dimensional object formed by rotating a circle about a diameter.

a) Assume that a light source is located at the center of a sphere. Describe how the light will reflect if the inside of the sphere has a reflective surface (see **Figure 8.92**).

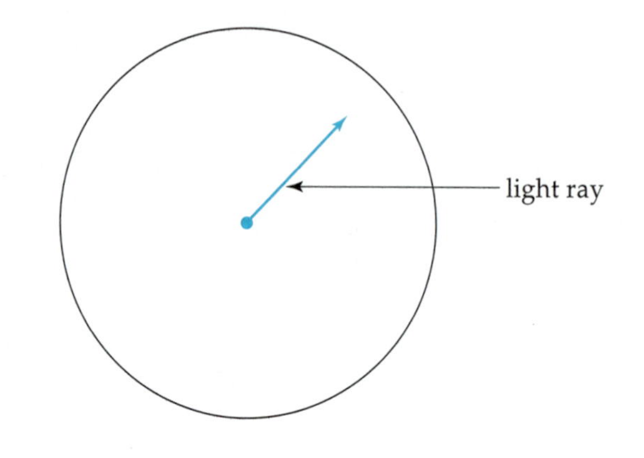

FIGURE 8.92.

b) What happens if the light source is located at any point other than the center?

12. The following problem is a favorite of Dr. Henry Pollak, renowned mathematical modeler. Suppose two people in the same town are on the phone and decide to meet for lunch. One says, "Look, you always walk twice as fast as I do. We should meet some place that is twice as far from you as it is from me!" Use coordinate geometry to describe where they should meet. (Hint: Put the locations of the houses at $(-1, 0)$ and $(1, 0)$, let the meeting point $P = (x, y)$. Use the distance formula to find the equation of all points twice as far from one house as from the other. Dr. Pollak thinks you'll be surprised.)

13. All four conics can be generated using a flashlight to create part of a cone and any flat surface such as a table or the wall to create a plane.

    a) Use a flashlight and a flat surface to produce images of a circle, an ellipse, a parabola, and one branch of a hyperbola.

    b) Write a paragraph explaining what you did and how you know that the curves were what they were supposed to be.

14. Recall the introduction to Lesson 8.3 on Conic Sections. In the reading, it mentioned that our study would "focus on the conics formed when the plane does not pass through the vertex." If the intersecting plane passes through the vertex, what is usually called **degenerate conics** are formed.

    a) What are the geometric possibilities formed by passing the plane through the cone in such a way that it contains the vertex?

    b) Expand the equation $(x-2)^2 + (y-9)^2 = 0$ and identify the conic. Rewrite the standard form equation to sketch the graph. What is the solution set?

    c) Complete the **Table 8.4**.

Standard Form	Expanded Form	Conic	Solution Set
$x^2 + y^2 = -4$			
$\frac{x^2}{4} + \frac{y^2}{9} = 0$			
$\frac{x^2}{9} + \frac{y^2}{4} = -1$			
$\frac{x^2}{4} - \frac{y^2}{9} = 0$			
$\frac{x^2}{9} - \frac{y^2}{4} = -1$			

TABLE 8.4.

Chapter 9

COUNTING AND THE BINOMIAL THEOREM

© Susan Van Etten

CHAPTER INTRODUCTION

Lotteries are a popular but controversial form of entertainment in the United States. An estimated 30 million Americans play lotteries regularly. Americans spend over $40 billion annually on lotteries.

Most lotteries are run by states or groups of states. A typical lottery returns 50-60% of its proceeds in prizes and pays 30-40% to government programs, with the rest used for administrative costs and commissions to ticket sellers. Some states, such as California, place a large portion of lottery revenues in the state's general fund. Florida and New Hampshire are examples of states that spend all the revenues on education.

In some lotteries, the probability of winning is small, and the jackpot is large; in others, the probability of winning is much higher, but the jackpot may be only a few hundred dollars. When a lottery with a large jackpot has no winner, the jackpot accumulates. In 1998, for example, a multi-state lottery jackpot had grown to $296 million before being won.

Many people are concerned about the impact of lotteries on gambling addiction. A study in Iowa found that about 5% of the state's population suffered from serious gambling addiction and another 23% had problems that were less severe. Minnesota is an example of a state that earmarks part of its lottery proceeds for treatment of gambling addiction. Another concern of lottery critics is that lotteries place a disproportionate share of the cost of government on people with low incomes. A study in Georgia, for example, found that families with incomes under $20,000 spent an average of about $250 annually on lottery tickets, but families with incomes over $40,000 spent an average of about $100.

Whether or not you approve of lotteries, it is undeniable that they pose challenging mathematical questions. For example, an agency that runs a lottery must determine the odds of winning, which requires the calculation of the number of different ways that a participant can enter. Counting methods and probability are two essential tools for the development of mathematical models that govern the operation of lotteries.

LESSON 9.1

Counting Basics

Many states run a daily lottery game in which the contestant selects three digits from 0–9 (digits may be repeated). For a small bet—usually $1—the contestant earns a chance to win by matching a randomly generated three-digit number exactly, or by matching the digits in a different order. In Iowa, for example, the prize is $600 for a perfect match, and $100 for a match in any order.

All lottery games raise counting questions in which order is an essential consideration. This lesson considers basic mathematical techniques for answering questions such as: In how many ways can a person enter a lottery when order matters? In how many ways can a person enter a lottery when it does not?

Activity 9.1

Your challenge in this activity is to investigate the number of different ways in which lottery entries can be made. One way to do this is by making a complete list of all possible entries. However, since the number of possible entries is often quite large, this method can be time-consuming and impractical. Therefore, it is best to start with some simple, hypothetical lotteries, make a full list, and look for a general principle that can be extended to larger lotteries.

With the goal of developing general principles in mind, investigate each of the following.

1. Consider a lottery in which an entry consists of one of the letters A, B, C, D, E followed by one of the numbers 0, 1, 2, 3. In how many ways can a contestant fill out a ticket? How is the number of ways of filling out the ticket related to the number of ways of choosing the letter and the number of ways of choosing the number?

2. Apply the principle you described in 1 to a lottery in which an entry consists of a digit from 0–9 and another digit from 0–9. In other words, apply the principle to a two-digit number. (Remember, repetition of digits is allowed.) In how many ways can you make your selection? Show how to extend your answer to determine the number of ways of entering the Iowa lottery described on the previous page. Some states, such as Connecticut, have a four-digit lottery. Show how to extend your previous answer to find the number of ways of entering a four-digit lottery.

3. Suppose the three-digit Iowa lottery did not let the contestant repeat a number. That is, you could enter 7-8-9, but not 7-8-8. How should the principle you developed in 1 be modified to determine the number of entries in this case?

4. In most states, you must indicate whether your entry is for a match in the correct order or a match in any order. Suppose you enter the digits 7-8-9 and choose a match in any order. List all the ways you could win. Also show how the principle you developed in 1 could be used to determine the number of items in your list.

5. Lotteries with large prizes usually require a selection of several numbers from many more than 1–10, and do not allow repetition. For example, you might be required to pick five numbers from 1–40. You could pick 4-12-23-31-37, but you could not pick 4-12-4-23-31. As a simpler example, consider a lottery that requires you to pick three distinct numbers from 1–5. Show how to determine the number of ways of entering without listing them all.

6. In lotteries of the type described in 5, the order does not matter. That is, picking 4-12-23-31-37 is the same as picking 12-4-23-31-37. Apply your analysis in 4 to your results in 5 to find the number of entries if the order does not matter.

7. Summarize the principles that you have developed in this activity for determining the number of ways of filling out a lottery ticket.

THE BASIC MULTIPLICATION PRINCIPLE

Many people are interested in probability. Lottery officials want to know the probability that someone will win. Meteorologists must determine the probability of good or bad weather. To set policy rates, insurance companies must estimate the probability of claims.

For insurance companies, finding the probability that a claim will be made against a policy requires an analysis of data about past claims. But for others such as lottery officials, calculation of probabilities is based on an exact counting of all ways in which a player could make an entry.

The simplest method of counting all the ways in which something can happen is to list them all. This method is satisfactory if the number of items in the list is small. Unfortunately, that is not the case in most modeling situations. Thus, it is often necessary to count without listing, or to put it another way, without counting.

The simplest counting technique (after listing all possibilities) is called the **basic multiplication principle**. It says that if events A and B can occur in a and b ways, respectively, then the sequence of events A, B can occur in $a \times b$ ways.

EXAMPLE 1

Consider a simple game in which the player rolls a die and flips a coin. Determine the number of outcomes in the game.

SOLUTION:

Since a die can fall in 6 ways and a coin can fall in 2 ways, there are $6 \times 2 = 12$ ways the two can fall. Of course, since the number is small, it is reasonable to verify this result by listing all twelve (see **Figure 9.1**).

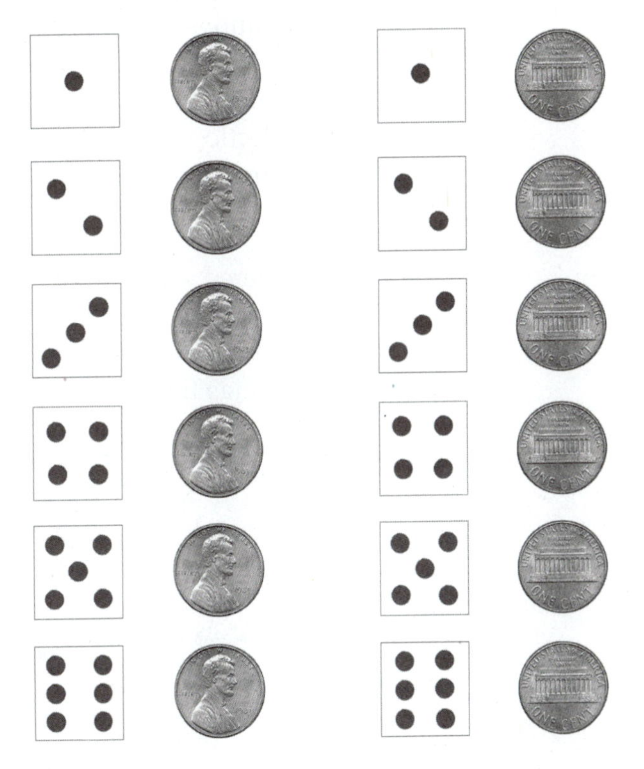

FIGURE 9.1. The twelve ways a die and coin can fall.

Although the multiplication principle is often stated for two events, it can be extended to three or more. For example, if three coins are tossed in succession, there are 2 x 2 x 2 = 8 different outcomes since the first coin can fall in 2 ways, the second coin can fall in 2 ways, and the third coin can fall in 2 ways. If a full listing is desired, it can be made by employing a systematic procedure such as a tree diagram (see **Figure 9.2**).

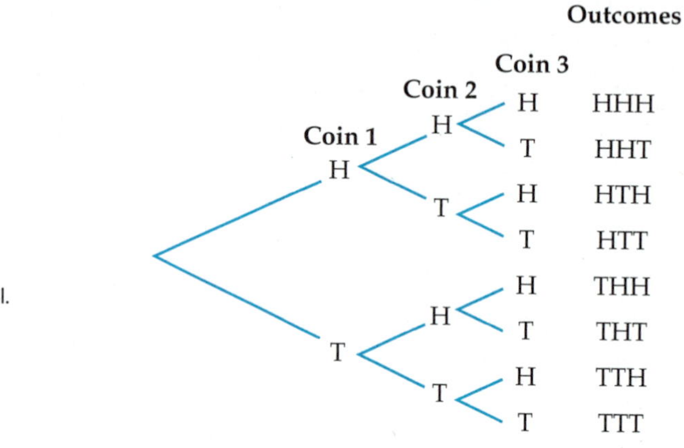

FIGURE 9.2. The eight ways three coins can fall.

Permutations

A **permutation** is an *arrangement* of several distinct objects. That is, there is a first, a second, a third, and so forth. Counting the number of permutations of a collection of objects is done by applying the basic multiplication principle. For example, in a lottery in which a person must select three different digits from 0–9, there are 10 choices for the first selection. Since the choices must be different, there are 9 choices for the second selection and 8 for the third. Therefore, the game can be entered in 10 x 9 x 8 = 720 ways.

The ordered selection of three different digits from ten available is sometimes described as a permutation of three things from a group of 10. $P(10, 3)$ and ${}_{10}P_3$ each represent the total number of permutations of three things from a group of 10. Thus, $P(10, 3) = {}_{10}P_3 = 720$.

If the three-digit lottery were a ten-digit lottery in which the player arranged all ten of the digits 0–9, there would be 10 x 9 x 8 x 7 x 6 x 5 x 4 x 3 x 2 x 1 = 10! = 3,628,800 ways of entering. (Recall that 10! is called the *factorial* of 10, or 10 factorial.)

Since $10 \times 9 \times 8 = \dfrac{10\times9\times8\times7\times6\times5\times4\times3\times2\times1}{7\times6\times5\times4\times3\times2\times1}$, ${}_{10}P_3 = \dfrac{10!}{7!}$.

In general, ${}_{m}P_{n} = \dfrac{m!}{(m-n)!}$.

EXAMPLE 2

Find the number of different ordered selections of three different letters from the 26 letters of the alphabet.

SOLUTION:

$$P(26, 3) = {}_{26}P_3 = \frac{26!}{(26-3)!} = \frac{26!}{23!} = 26 \times 25 \times 24 = 15{,}600.$$

The permutation formula gives a puzzling result in some cases. For example, consider the number of ways of arranging the letters A, B, C, D, E. By the multiplication principle, there are $5 \times 4 \times 3 \times 2 \times 1 = 5! = 120$ arrangements. However, the permutation formula gives ${}_5P_5 = \frac{5!}{(5-5)!} = \frac{5!}{0!}$; but 0! is meaningless. Therefore, 0! is defined to be 1.

Combinations

In some counting situations, order does not matter. For example, in some three-digit lottery games, the player wins by matching all three digits in any order. Thus, the player has to select three digits, but does not have to arrange them in a particular order.

Consider the question of counting the number of ways of choosing three different digits in a lottery game with unordered matches. For any set of winning digits, say 7-2-9, there are $3 \times 2 \times 1 = 6$ ways of arranging them. Therefore, the number of ways of entering when unordered matches are allowed is $\frac{1}{6}$ the number of arrangements of three different digits from 0–9: $\frac{720}{6} = 120$.

An unordered selection of several objects from a group of distinct objects is called a **combination**.

${}_{10}C_3$, $C(10, 3)$, $\binom{10}{3}$, and $\binom{10}{3,7}$ are all used to represent the total number of combinations of three things from a group of 10. The total number of combinations of three things from a group of 10 is $\frac{720}{6} = \frac{{}_{10}P_3}{3!} = 120$. In general ${}_mC_n = \frac{{}_mP_n}{n!} = \frac{m!}{(m-n)!n!}$.

EXAMPLE 3

Find the number of unordered selections of three different letters from the 26 letters of the alphabet.

SOLUTION:

$$C(26, 3) = {}_{26}C_3 = \binom{26}{3} = \frac{26!}{(26-3)!3!} = \frac{26!}{23!3!} = \frac{26 \times 25 \times 24}{3 \times 2 \times 1} = 2600.$$

When you examine a counting problem, here are a few tips.

- When repetition is allowed, try the basic multiplication principle.
- If you are uncertain whether order matters, try making a partial list of items that are considered different.
- Remember that counting permutations (arrangements) is a special case of the multiplication principle in which repetition is not allowed.
- Remember that a combination is an unordered selection.

Exercises 9.1

1. a) Discuss the counting technique you would use to determine the number of different committees of three students that can be selected from your class.

   b) Repeat part (a) if the positions on the committee are different—for example, a chair, a vice-chair, and a recorder.

   c) Determine the number of three-student committees in your class in both cases.

2. You may have studied election methods in a previous course (for example, see *Mathematics Modeling Our World*, Course 1, Unit 1, Pick a Winner).

   a) In some voting systems, voters are asked to rank the candidates. In how many ways can a voter rank five candidates? Which counting technique did you use to obtain your answer?

   b) In a system called approval voting, a voter does not rank the candidates. Instead, the voter marks all the candidates of which the voter approves. If there are five candidates in an election, in how many ways can a voter approve of two? Of three?

3. Computer and graphing calculator screens are composed of pixels. How many pixels are there on a graphing calculator screen that is 95 pixels wide and 63 pixels high? What counting technique did you use to obtain your answer?

4. You may have studied graphs (or networks) in a previous course (for example, see *Mathematics: Modeling Our World*, Course 2, Unit 3, Hidden Connections). Basically, a graph is a collection of vertices and edges that connect some of the vertices. A complete graph is one in which every pair of vertices is connected; **Figure 9.3** is a complete graph with four vertices.

   FIGURE 9.3. A complete graph with four vertices.

   a) How many edges are there in a complete graph with five vertices? In a complete graph with 10 vertices? In a complete graph with n vertices?

   b) When you draw an edge, are you selecting two vertices or arranging them? How does your answer justify the calculation you made to answer part (a)?

Exercises 9.1

© Susan Van Etten

5. A Scrabble® player is trying to make a word from seven different tiles (letters).

   a) How many different "words" are possible if all seven tiles are used?

   b) How many different "words" can be made with five of the tiles?

6. Pick 6 Lotto is one of the games in the New Jersey Lottery. A player picks six different numbers from 1–44 and wins the jackpot if the selected numbers match in any order.

   a) In how many ways can a player enter New Jersey Pick 6 Lotto?

   b) The form used to enter New Jersey Pick 6 Lotto has room for five entries. If the form is 0.003 inches thick, how high would a stack of forms be that made every possible entry? Show your calculation. Compare your result to the height or length of an object familiar to you.

   c) If you undertook the ambitious project of submitting every possible entry in New Jersey Pick 6 Lotto and used a systematic procedure that allowed you to complete one entry every 5 seconds, how long would this task take? (Assume a number of working hours per day that you think is reasonable.) Each entry in this lottery costs $1. How large would the jackpot need to be to make this task worthwhile?

   d) Michigan Lotto is similar to New Jersey Pick 6 Lotto in that the player picks six different numbers and wins if they match in any order. However, in Michigan Lotto, the six numbers are selected from 1–49. Like New Jersey Pick 6 Lotto, Michigan Lotto allows up to five entries per form at a price of $1 each. Compare the results of your analysis of New Jersey Pick 6 Lotto in (a)–(c) to Michigan Lotto.

7. Traveling salesperson problems form an important collection of mathematical problems. They are so named because one such problem asks which is the cheapest way for a traveling salesperson to visit customers in several cities and return home. For example, **Figure 9.4** represents a 4-city problem. The numbers written along the edges are the costs of flights in dollars. Usually one of the cities—for example, Phoenix—is the home of the salesperson, and the trip must start and end there.

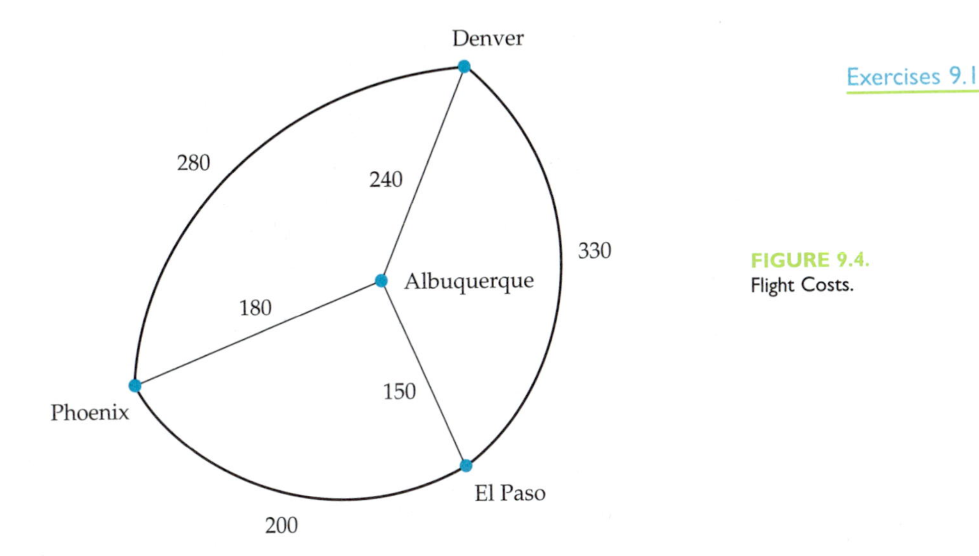

Exercises 9.1

FIGURE 9.4.
Flight Costs.

Mathematicians have been unable to find an efficient method for solving such problems. One inefficient way to find the cheapest trip is to list all possibilities and pick the cheapest. If the number of cities is large, a computer could be used to do the listing and find the lowest total cost.

a) Make a complete list of all possible trips in this example that start and end in Phoenix. Use your list to find the cheapest trip.

b) Adapt one of the counting methods you learned in this lesson to the problem of determining the number of different total costs in a problem of this type with five cities. Explain your procedure.

c) How many different costs are possible in a 15-city traveling salesperson problem? How long would it take a computer to list them all if the computer can list 1000 trips and the related total costs in one second? If the computer can do one million trips per second?

8. Combination locks are popular for school lockers. Often these locks consist of a rotary dial with marks numbered 0–40. The user must dial a three-number combination to open the lock.

Corbis

a) If a number is not repeated, how many "combinations" are possible?

b) How many "combinations" are possible if a number can be repeated?

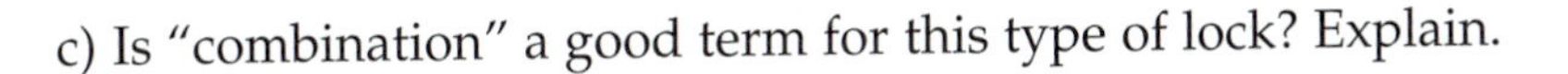
c) Is "combination" a good term for this type of lock? Explain.

Exercises 9.1

9. A crowd is waiting to purchase tickets for a concert. When only 10 tickets remain, the sales desk announces that they will be awarded at random to those who are still waiting. If 40 people are waiting, in how many ways can the tickets be awarded?

10. The zip codes used by the U. S. Postal Service originally consisted of five digits from 0–9.

    a) If any five-digit zip code except 00000 is legal, how many five-digit zip codes are possible?

    b) Zip codes today consist of a five-digit zip code plus an additional four or six digits and are often referred to as zip + four or zip + six. How many of each are possible?

11. A binary code is one in which each character can be one of two. For example, each bar in the series of bars used by the U. S. Postal Service to make zip codes machine readable is either long or short.

    a) Suppose a binary coding system uses a series of five lights, each of which may be on or off. A sample code is shown in **Figure 9.5.**

FIGURE 9.5. A binary code.

    How many different codes are possible? Why are these codes arrangements?

    b) How many different two-light codes such as the one in Figure 9.5 are possible? Why is order unimportant in counting these?

    c) How many one-light, three-light, and four-light codes are there?

    d) What is the value of ${}_5C_0$, and what is its meaning in this context? Answer the same questions for ${}_5C_5$.

    e) What does this exercise tell you about the sum of all possible unordered selections of objects from a group of 5? From a group of n?

Exercises 9.1

12. Portions of many cities resemble a rectangular grid. **Figure 9.6** represents a portion of such a city. The points on the grid are intersections; to avoid clutter, streets are not drawn. A person wants to walk from A to C by following streets (grid lines), but not walking any more blocks than necessary. One possible path is shown: walk two blocks East and three blocks South.

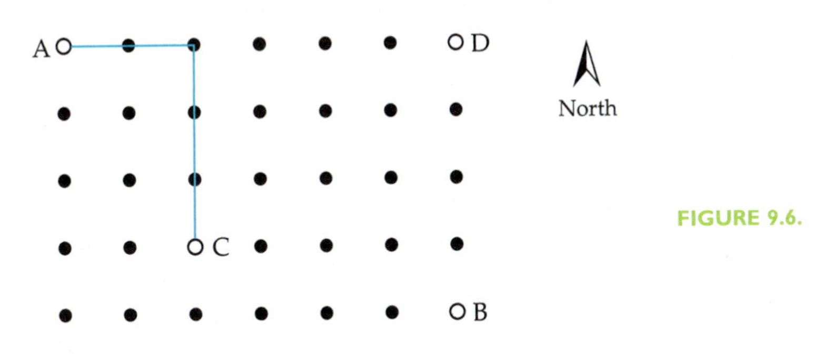

FIGURE 9.6.

a) Find all possible paths from A to C. Since it would be difficult to distinguish the paths if you drew them all on one grid, you may find a shorthand helpful: the path shown can be represented by *EESSS*, where *E* is a one-block walk East and *S* is a one-block walk South.

b) Explain how to use counting techniques to find the total number of paths in (a) without making a list.

c) Without making a list, find the total number of paths from C to B and the total number of paths from A to B in Figure 9.6.

d) Someone incorrectly counts the total number of paths from A to B as $2^{10} = 1024$. What kind of paths did this person count?

e) How many paths are there from A to D in Figure 9.6? In the shorthand of part (a), how many *S*s and *E*s are there in the representation of such a path?

13. Each year in the United States, medical school graduates are assigned to hospitals for their residencies. Each graduate ranks the available hospitals, and each hospital ranks the graduates. A computer algorithm is used to create the matches.

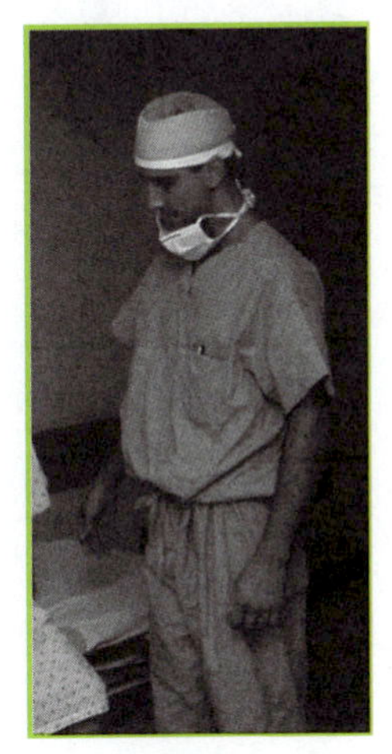

Exercises 9.1

a) In how many ways can five graduates be assigned to five hospitals if each hospital has exactly one residency available?

b) If you are one of the residents, in how many of these matchings do you receive your first choice of hospitals? Explain.

c) If there are only five residency openings (at five different hospitals), but 7 graduates, in how many ways can the assignments be made?

14. Although most people enter a lottery by selecting numbers, some people enter by eliminating numbers.

a) Consider a mini-lottery in which the participant selects four different numbers from 1–12. Show how to count the number of ways of entering.

b) Show how to count the number of ways of entering by selecting eight numbers to eliminate.

c) In general, what can you say about ${}_mC_n$ and ${}_mC_{m-n}$?

15. Assume that n is a positive integer.

a) Interpret $\binom{n}{0}$, $\binom{n}{1}$, and $\binom{n}{n}$ in the context of committee selection (Exercise 1).

b) Interpret $\binom{n}{0}$, $\binom{n}{1}$, and $\binom{n}{n}$ in the context of paths (Exercise 12).

c) How are $\binom{n}{0}$ and $\binom{n+1}{0}$ related?

d) How are $\binom{n}{1}$ and $\binom{n+1}{1}$ related?

e) How are $\binom{n}{n}$ and $\binom{n+1}{n+1}$ related?

Compound Events

Situations sometimes occur in which events cannot be counted by a single application of one of the techniques you used in Lesson 9.1. This lesson extends the counting techniques of the previous lesson to handle more complicated situations.

Activity 9.2

Consider the selection of a committee from the members of your class under the condition that two of the committee members be girls and two of them be boys. Since the number of such committees is likely to be fairly large, in this activity you will consider a simplified version.

A committee of two girls and two boys is to be selected from eight students, five of whom are boys and three of whom are girls: Juan, Sol, Karl, Roland, Ben, Elena, Gena, Miriam.

Your goal in this activity is to determine the number of different committees possible. The number of students has been kept small so that making a full list of all possibilities is a reasonable approach; you can divide up the task if you are working in a group.

Here are several questions to answer as you work on this problem:

1. In how many ways can a committee of two boys be selected from the boys in this group?

2. In how many ways can a committee of two girls be selected from the girls in this group?

3. How is the number of committees of two boys and two girls related to the number of two-boy committees and the number of two-girl committees you found in 1 and 2?

4. Consider a different type of committee. Suppose that a two-person committee composed of either two boys or two girls is being selected. In how many ways can this be done, and how is the number related to your answers in 1 and 2?

5. Summarize the uses you made of addition and multiplication as you answered the previous questions. In what kinds of situations is each operation used?

6. Use the shortcuts you described in the previous question to determine the number of committees of three boys and one girl from this same group. Also determine the number of ways of choosing a committee of three boys or a one-girl committee.

THE MULTIPLICATION PRINCIPLE REVISITED

Counting the number of ways an event can occur is sometimes done by counting two or more component events, then performing one or more additional calculations. For example, some codes such as automobile vehicle identification numbers (VINs) are composed of both letters and numbers. Since VINs are quite long, consider a simpler code that consists of any two different letters followed by two different digits: AG72 is one such code.

The number of ways of selecting the letters of this code is ${}_{26}P_2 = 650$; the number of ways of selecting the digits is ${}_{10}P_2 = 90$. Imagine the process of making a complete list of all possible codes. For a given pair of letters, say AB, there are 90 different codes: AB01, AB02, and so forth (but not AB11, AB22, ...). Since there are 650 ways of selecting the two letters, the total number of codes is 650 x 90 = 58,500.

The total number of codes can be counted without making a list or even thinking about the listing process. A code consists of an ordered selection of two letters followed by an ordered selection of two numbers. Therefore, the basic multiplication principle can be applied to the two component counts: 650 x 90 = 58,500. Note that a verbal description of the relationship between the code and its components says that the code consists of two letters *and* two digits. The word *and* in the verbal description of the way in which component events produce a compound event is an indication that component counts should be multiplied.

This particular problem can be solved also by applying the basic multiplication principle to the sequence of individual characters in the code. Since letters and digits may not be repeated, there are 26 choices for the first character, 25 for the second, 10 for the third, and 9 for the fourth. Therefore, the multiplication principle says that the total number of codes is 26 x 25 x 10 x 9 = 58,500.

THE ADDITION PRINCIPLE

Now consider a change in the coding scheme used previously. Instead of a four-character code composed of two different letters followed by two different numbers, this code is composed of two characters that may be either two different letters or two different numbers. Samples are AG, 72, XT, and 05.

In a list of all possibilities, there would be ${}_{26}P_2 = 650$ two-letter codes and ${}_{10}P_2 = 90$ two-digit codes. Thus the total number of codes in the list is 650 + 90 = 740.

The **addition principle** says that when a compound event is composed of one or the other of two components, the number of ways the event can occur is the sum of the number of ways the two components can occur, provided the component events have nothing in common. Events that have nothing in common are called **mutually exclusive**. (What to do when component events are not mutually exclusive is considered in this lesson's exercises.)

EXAMPLE 4

Lottery A requires a contestant to pick three digits from 0–9. Repetitions are not allowed and only matches in the proper order win. Lottery B is similar, but participants pick four digits that may be repeated. In how many ways can a participant enter both lotteries by making exactly one entry in each lottery?

SOLUTION:

Lottery A can be entered in ${}_{9}P_3 = 720$ ways; Lottery B can be entered in 10 x 10 x 10 x 10 = 10^4 = 10,000 ways. There are 720 x 10,000 = 7,200,000 ways to make single entries in both.

EXAMPLE 5

Determine the number of different entries if the participant makes exactly one entry in either Lottery A or Lottery B in Example 4.

SOLUTION:

There are 720 + 10,000 = 10,720 different entries possible.

PEOPLE AND MATH

Have you ever stopped to think about all of the situations in life that depend upon scheduling? The timing of your classes and the synchronization of traffic lights, for example, affect you every day. The commercial sector includes the challenges of scheduling personnel and appointments, and determining how many possible routes and schedules there can be for shipping via plane, train, or boat. If you are involved in solving these or other similar problems in your future career, you will be working with the mathematics of **combinatorics**. Related to the study of counting methods, combinatorics is concerned with solving discretely structured problems for which the number of possibilities is finite, but possibly quite large. Modern computer science, requiring discrete formulations of problems, is just one area where this math is applied. Combinatorics is also closely intertwined with optimization, and logistics whenever you are trying to determine not only "how many" possiblities there are, but also which are "the best," or the most efficient and least costly.

Exercises 9.2

1. Some states have license plates composed of three letters followed by three digits.

   a) How many different plates are possible if letters and digits may not be repeated?

   b) How many different plates are possible if letters and digits may be repeated?

   c) There are 26 letters and 10 digits, a total of 36 characters. What kinds of plates are counted by ${}_{36}P_6$?

   d) How many plates of four different letters followed by two different digits are possible?

   e) If plates may be three different letters followed by three different digits or four different letters followed by two different digits, how many plates are possible?

© Susan Van Etten

2. Powerball is a multi-state lottery in which players select five different numbers from 1–49 and a single number from 1–42. The jackpot is shared by those who match all five in any order and also match the special sixth number (the powerball). In how many ways can a player enter Powerball?

3. Find the number of different committees of each type that are possible in your class.

   a) Four-person committees.

   b) Four-person committees with two boys and two girls.

   c) Two-person committees that are either all boys or all girls.

4. The number of different seating arrangements in a classroom can be quite large.

   a) Find the number of different seating arrangements in a classroom with 25 seats and 25 students.

   b) Find the number of different seating arrangements in a classroom with 25 seats and 20 students.

Exercises 9.2

c) To get an idea of the size of these answers, compare the number of seating arrangements you found in part (a) to the height of a stack of paper. The number of sheets in the stack is the answer you found in part (a), and the thickness of each sheet is 0.003 inches. (The distance from the Earth to the Sun is about 93,000,000 miles.)

5. In many lottery games, there are prizes other than the jackpot. In Pennsylvania's Super 6 Lotto, for example, there are three additional prize categories.

   a) A player in Super 6 Lotto chooses six numbers from 1–69. In how many ways can a player enter?

   b) The Super 6 Lotto awards the jackpot to those who match all six winning numbers, but a portion of the proceeds is shared by those who match five of the six. In how many ways can a player earn a share of this prize? (Hint: The player must match five of the winning numbers and one of the non-winning numbers.)

   c) Another portion of the proceeds is shared by those who match four of the six and yet another portion by those who match three of the six. In how many ways can a player win a prize of any type in Super 6 Lotto?

   d) Of the money allocated to prizes in Super 6 Lotto, 76% goes to those who match all six, 8% goes to those who match five of the six, 7.5% to those who match four of the six, and 8.5% to those who match three of the six. Why do you think more money is allocated to three-match prizes than to four- or five-match prizes?

6. The coach of a football team has eight candidates for four starting positions in the defensive line, six candidates for three starting linebacker positions, and seven candidates for four defensive back positions. No one is a candidate for more than one of the three types of positions.

   a) Write a one-sentence description that breaks down the eleven-person defensive unit into the three types of positions. Should the word *and* or the word *or* be used in a proper description?

   b) How many different starting teams can the coach field if the positions within the line, linebacker, and backfield categories are not differentiated?

c) How many different starting teams can the coach field if the positions within the line, linebacker, and backfield categories are considered different?

Exercises 9.2

7. The addition principle as stated in this lesson assumes that events are mutually exclusive. To see how the principle is applied when two events have something in common, consider a school at which 28 students play basketball, 25 play volleyball, and 7 play both sports.

a) On a copy of **Figure 9.7**, write the correct number of students in each of the three regions.

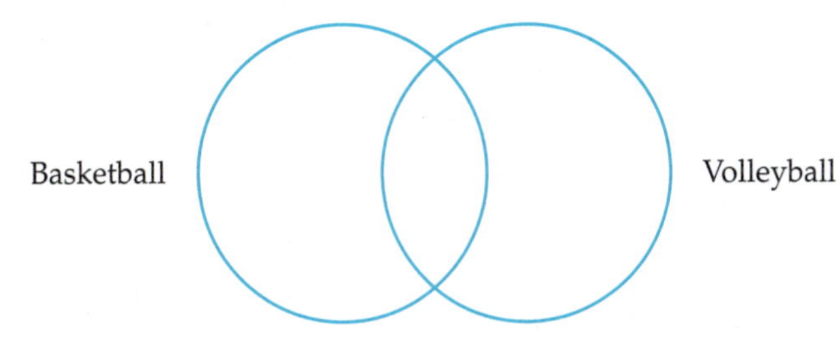

FIGURE 9.7.

b) What is the total number of students who participate in basketball and volleyball?

c) Can the total number of students be found by applying the addition principle as stated in this lesson? That is, do you just add the number of people who play each sport? If your answer is no, explain how the addition principle should be modified when two events have something in common.

8. An ice cream store offers 27 flavors, from which a customer can choose a one-scoop, two-scoop, or three-scoop cone. The scoops in a multi-scoop cone can be of the same flavor or of different flavors.

a) How many different cones are possible if all scoops are the same flavor? Justify your answer.

b) How many different multi-scoop cones are possible if all scoops are different? Justify your answer.

c) How many different three-scoop cones are possible with two scoops of one flavor and one scoop of a different flavor? How many different two-scoop cones are possible with one scoop of one flavor and one scoop of a different flavor? Justify your answers.

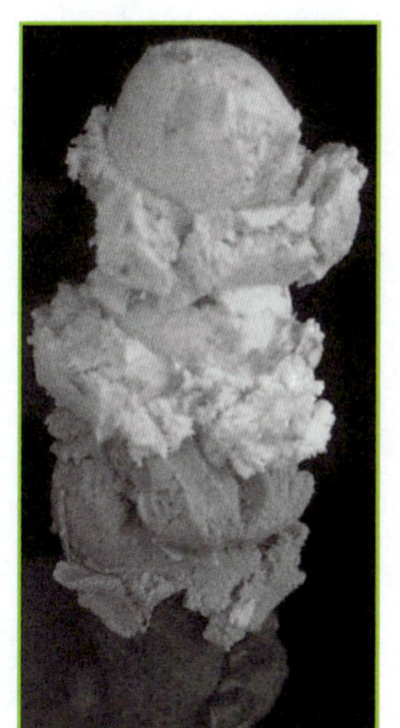

Exercises 9.2

d) This store wants to advertise the number of different cones a customer can create. What number should be advertised?

9. UPC bar codes are familiar to everyone, since they are on nearly every product that can be purchased in stores. The bars represent two groups of five digits. The first group identifies the product's manufacturer; the second group represents the product.

   a) How many manufacturers can be encoded?

   b) How many products can be encoded by each manufacturer?

   c) Explain why ${}_{10}P_5$ is not the correct number of manufacturers that can be encoded.

10. A teacher is preparing a ten-question true/false quiz and plans to prepare a different form of the quiz for each student by ordering the questions differently on each form.

    a) In how many ways can the teacher order the questions?

    b) The teacher has decided that five of the answers will be true and five will be false. In how many ways can the teacher arrange the questions by answer type?

    c) Explain why the answer to part (b) is not 2^{10}.

11. Dominoes are used for a variety of games and come in different-sized sets. The most common is the double-six set. In the double-six set, there is a domino for every possible pairing of 0–6 spots. For example, in a double-six set, you find a domino with six spots on one half and three on the other, another domino with five spots on one half and five spots on the other, and so forth.

    © Susan Van Etten

    a) The number of dominoes in a double-six set can be counted using the techniques you have learned in this chapter. Would you count the number of dominoes in a double-six set in a single calculation, or break the calculation into parts? Explain.

Exercises 9.2

b) How many dominoes are there in a double-six set?

c) Explain why 7^2 does not count the number of dominoes in a double-six set.

d) How many dominoes are there in a double-twelve set? (A double-twelve set has every possible number of spots from 0–12.)

12. Because medical tests are often expensive, sample pooling is a common practice when screening large groups of people for such things as steroid use. The logic is simple: if ten individual samples are tested together and the result is negative, all ten individual tests can be considered negative.

    a) How many different outcomes are there among ten individual samples? (For example, all ten people negative, first person positive, and the other nine negative.)

    b) In how many of the outcomes among ten people do three people test positive and the others negative?

13. Methods for counting arrangements of several objects must be modified if some of the objects are identical. For example, there are $7! = 5040$ distinct arrangements of the letters of *VERMONT*, but fewer arrangements of the letters of *ALABAMA* since four of the letters are identical.

    a) How many arrangements of the letters of *ALABAMA* are there? Explain?

    b) Extend the modification you described in part (a) to count the number of distinct arrangements of the letters of *MISSISSIPPI*.

    c) Five people have formed a basketball team and must decide how to assign positions. There must be two guards, two forwards, and one center. In how many ways can positions be assigned? Explain.

14. What is the value of $\binom{k}{0}+\binom{k}{1}$? Explain.

LESSON 9.3

The Binomial Theorem

Although you have studied only a few counting techniques, they form a powerful tool kit that can be applied to a wide range of problems. Therefore, it is not surprising that people occasionally feel confused about how to apply these techniques. In this lesson you will consider some simple counting situations in which there are only two possible outcomes in order to gain a deeper understanding of important counting concepts.

Although situations involving coin flipping or true/false tests may seem trivial, they involve some of the most important counting concepts. Think, for a minute, about the digital technology that is so common today. This technology is based on electronic coding of information in strings of 0s and 1s. A string of 0s and 1s is really no different from a string of Hs and Ts representing the flips of a coin, or a string of Ts and Fs representing the answers to a true/false quiz.

People who design digital technology must solve counting problems. Often the problems concern capacity—the number of items that can be coded with strings of a specific number of 0s and 1s, for example.

Activity 9.3

Although most digital technology uses millions of 0s and 1s to code information, the numbers in this activity are much smaller in order to keep things manageable.

1. Consider a string of four 0s and 1s. Explain how counting techniques can be used to determine the number of different codes that are possible.

2. Use counting techniques to break down the count in the previous question into counts by type: codes with four 0s, codes with three 0s, codes with two 0s, codes with one 0, codes with no 0s. Justify your choice of method.

3. You didn't need a complete list to answer the previous questions, but you could have made one. Show how a tree diagram can be used to find all possible codes of four 0s and 1s.

4. The tree diagram you just made lists all the possibilities you counted in the first question. The tree diagram can be modified slightly to reflect the counts you made in 2. **Figure 9.8** is a start that shows codes composed of two 0s and/or 1s.

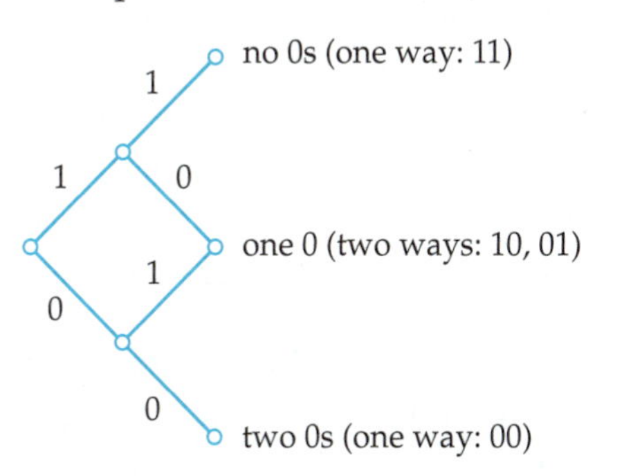

FIGURE 9.8.

This type of tree diagram is not as cluttered as the type you made in 3, since it does not require a path for every outcome. Extend this tree diagram to codes of four 0s and 1s to show that it gives your answers in 2. (Hint: Keep running totals rather than waiting until the diagram is drawn to count all the paths.)

5. The tree diagram in your answer to 4 can be adapted to raise an algebraic binomial such as $(x + y)$ to a power. To do so, begin by replacing the 1s and 0s with xs and ys. Then multiply along the branches to find and count the terms. For example, $(x + y)^1$ can be represented by the simple tree diagram in **Figure 9.9**.

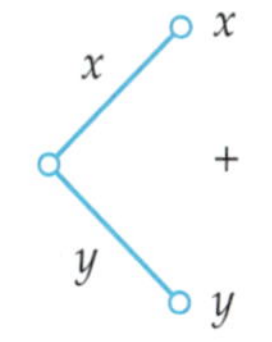

FIGURE 9.9.

Since $(x + y)^2 = (x + y)(x + y)$, and since $(x + y)(x + y) = x(x + y) + y(x + y)$, $(x + y)^2$ can be found by extending the previous tree diagram as shown in **Figure 9.10**.

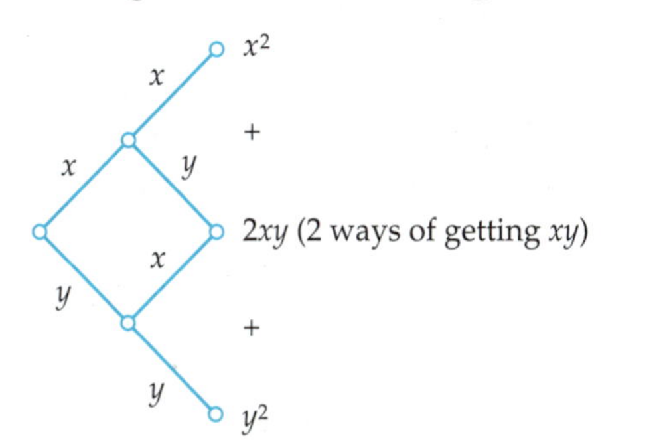

FIGURE 9.10.

Modify the diagram you made in 4 to find $(x + y)^4$.

6. Describe any patterns you noticed in constructing the previous tree diagram. Discuss how the patterns could be used to raise $(x + y)$ to the fifth power.

THE BINOMIAL THEOREM

In mathematics, the term *binomial* has a special meaning. It refers to an algebraic expression with two terms, such as $x + y$ or $2a - 4$. The abstract and specialized nature of mathematical concepts often masks their utility in the real world. But *bi* is a common prefix in English: binoculars, binary, bicentennial, bilateral, bilingual, biplane, and bifocal are all related to binomial in that they refer to two of something.

The multitude of real-world events in which there are only two possibilities explains why the prefix *bi* is common: true/false tests, polls in which questions have yes/no answers, and so many electrical devices that can be either on or off.

Some of the most dramatic changes of the last few decades are due to computers and related technologies. Computer chips seem to be everywhere: in cars, graphing calculators, and compact disc players, to name a few. The digital revolution is a binary revolution. Deep inside every computer chip, information is coded by millions of miniature switches that are either on or off.

Digital coding gives rise to many counting questions. Often these questions are about capacity: How much information can be stored on a compact disc? How many programs can a graphing calculator hold?

Surprising as it may seem, the basis for the answers to these and many similar questions lies in the way mathematicians raise binomials such as $x + y$ to powers.

As an example, consider $(x + y)^3$. The expanded form of $(x + y)^3$ has four terms: x^3, x^2y, xy^2, and y^3. The tricky question is: How many times do each of these terms occur when the product $(x + y)(x + y)(x + y)$ is calculated?

When counting requires a full list, a tree diagram helps produce the list in a systematic way. The diagram can be constructed so that like terms arrive at the same vertex (see **Figure 9.11**).

Drawing the tree diagram in this way avoids clutter. Since the primary concern is with the number of terms, there is no need to have the diagram terminate in $2^3 = 8$ branches with, for example, a separate branch for each of the three x^2y-terms.

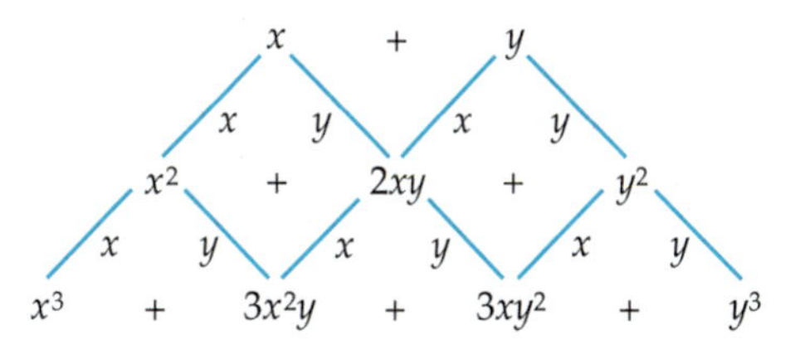

FIGURE 9.11. A tree diagram for $(x + y)^3$.

DISCUSSION/REFLECTION

The tree diagram in Figure 9.11 shows that determining the coefficient of each term in the expansion of $(x + y)^3$ is equivalent to counting the number of paths in the tree diagram. Can you find the paths that lead to each term in the expansion?

The problem of counting the terms in a binomial expansion is identical to the problem of counting the arrangements of true and false answers on a true/false quiz, to the problem of counting the ways coins can fall, and to the problem of counting the number of different binary codes that are possible.

Consider the problem of counting the number of ways two trues and one false can occur on a three-question true/false quiz. There are three: TTF, TFT, and FTT. Similarly, there are three ways the term x^2y can occur in the expansion of $(x + y)^3$: xxy, xyx, and yxx.

But counting the number of terms with two xs and one y can be done with combinations: there are ${}_3C_2 = 3$ ways of selecting the positions for the xs (or, if you prefer, ${}_3C_1 = 3$ ways of selecting the position for the y).

The connection between the expansion of $(x + y)^n$ and counting techniques is called the binomial theorem.

The Binomial Theorem

$$(x + y)^n = {}_nC_0x^n + {}_nC_1x^{n-1}y + {}_nC_2x^{n-2}y^2 + \ldots + {}_nC_{n-1}xy^{n-1} + {}_nC_ny^n.$$

EXAMPLE 6

Expand $(a - 2b)^4$.

SOLUTION:

$(a - 2b)^4 =$
${}_4C_0a^4(-2b)^0 + {}_4C_1a^3(-2b)^1 + {}_4C_2a^2(-2b)^2 + {}_4C_3a^1(-2b)^3 + {}_4C_4a^0(-2b)^4 =$
$a^4 - 8a^3b + 24a^2b^2 - 32ab^3 + 16b^4.$

With the aid of the binomial theorem, it is possible to determine any term of a binomial expansion without calculating the entire expansion. Note that in any term, the sum of the powers of x and y is n.

EXAMPLE 7

Find the term in the expansion of $(x + y)^{10}$ that contains x^4.

SOLUTION:

Since the sum of the powers of x and y in any term must be 10, the term that contains x^4 also contains y^6. The number of x^4y^6-terms in the expansion is ${}_{10}C_6 = 210 = {}_{10}C_4$. Thus, the desired term is $210x^4y^6$.

Exercises 9.3

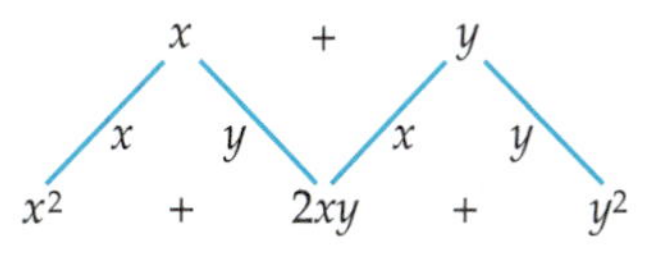

FIGURE 9.12.

1. The tree diagram in **Figure 9.12** shows the expansion of $(x + y)^2$. Use the distributive property to find the product $(x + y)(x + y)$ and explain how the process is represented in the diagram.

2. Revisit Exercise 12 of Lesson 9.1. How is the path counting in Exercise 12 related to the binomial counting in this lesson (Lesson 9.3)?

FIGURE 9.13.

3. The tree diagram used in **Figure 9.13** is a variation on Pascal's Triangle, which is named for the French mathematician and philosopher Blaise Pascal (1623–1662). Although it bears Pascal's name, the triangle is known to have appeared in Chinese writings as early as 1303. Some historians believe that the Persian poet Omar Khayyam (1048–1122) may have had partial knowledge of the triangle.

   a) Explain how one row of the triangle is obtained from the previous.

   b) What is the next row of the triangle in Figure 9.13?

   c) Find the fifth number (counting from the left) in the tenth row of the triangle without extending the triangle to ten rows. Explain how you found it.

Omar Khayyam

4. Expand $(x + y)^6$.

5. a) How are ${}_nC_k$ and ${}_nC_{n-k}$ related?

   b) If ${}_nC_k$ occurs in a term of the expansion of $(x + y)^n$, what are the powers of x and y in that term? Explain.

6. a) Explore the relationship between the coefficients of two consecutive terms of a binomial expansion. For example, what is the ratio of ${}_{10}C_4$ to ${}_{10}C_3$?

   b) In general, what is the ratio of ${}_nC_{k+1}$ to ${}_nC_k$? Explain.

   c) Interpret the results of the previous two parts of this exercise. That is, if ${}_{15}C_5 = 3003$, how can you obtain ${}_{15}C_6$ recursively?

Exercises 9.3

7. Exercise 6 discussed a recursive relationship that exists within a row of Pascal's triangle. But recursive relationships also exist between rows; in fact, it is between-the-rows recursion that you used to generate rows of the triangle.

   a) Explain how the third number in the fourth row of Pascal's triangle is obtained from numbers in the previous row. Do not use specific numbers; describe them by row and position within the row.

   b) In terms of n and k, explain how the kth number in the nth row of Pascal's triangle is obtained from numbers in the preceding row.

   c) Since the numbers in Pascal's triangle are combinations, your previous answer can be written using combination symbols. Use combination symbols to describe how to obtain ${}_nC_k$ from combinations that occur in the previous row of Pascal's triangle.

   d) Apply factorial formulas to the expression you wrote for ${}_nC_k$ in part (c). Use symbol manipulation to verify that this expression equals ${}_nC_k$.

8. Use the binomial theorem to expand $(2x - 5)^3$. Calculate all powers of constants. Show your work.

9. The binomial theorem can be used to raise complex numbers to powers.

   a) Use the binomial theorem to calculate $(1 + i)^4$. (Recall that $i^2 = -1$.)

   b) Verify your previous answer by using polar form to calculate $(1 + i)^4$.

10. When drawn with branches, Pascal's triangle becomes a tree diagram (see **Figure 9.14**).

    The number at a vertex is the number of different paths that lead to that vertex.

    a) Find all the paths that lead to 6 in this tree diagram.

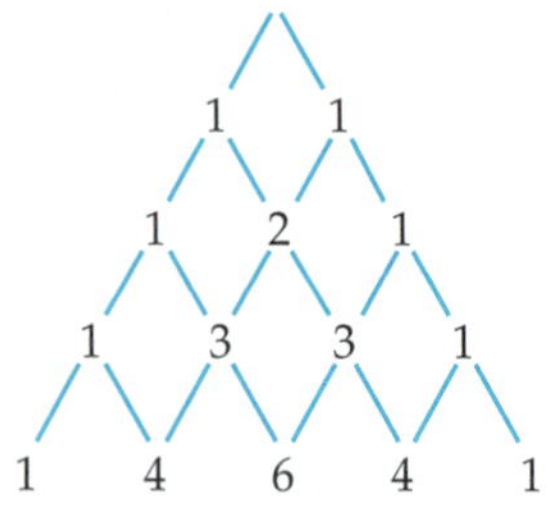

1 1
1 2 1
1 3 3 1
1 4 6 4 1

FIGURE 9.14.

b) Since the numbers in the triangle represent the number of paths, they can be thought of as frequencies. If, for example, you started at the top of the triangle and flipped a coin at each vertex to decide which of the two branches to take, then, over many repeats, you would reach the bottom vertices in the ratio of 1:4:6:4:1. Find the tenth row of the triangle and the percentage of trials that would end at each bottom vertex.

Exercises 9.3

c) Plot the percentages vs. the position in the tenth row. That is, plot the percentage associated with the first number of the row vs. 1, the percentage associated with the second number of the row vs. 2, and so forth. Describe the graph.

d) The "first" number in a row of Pascal's triangle is often considered the "zeroth" number. How would this numbering scheme change your graph?

11. In **Figure 9.15**, slanted rectangles have been used to group the numbers in Pascal's triangle.

a) Starting at the top and working downward, form a sequence made up of the sums of the numbers in the rectangles.

b) Describe the sequence recursively. That is, how is S_n obtained from its predecessors?

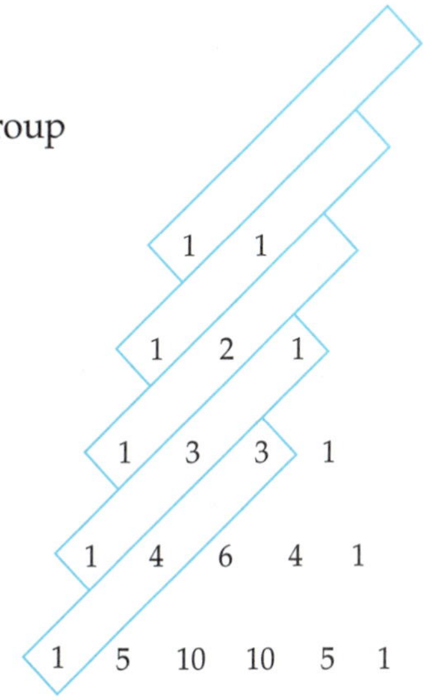

1 1
1 2 1
1 3 3 1
1 4 6 4 1
1 5 10 10 5 1

FIGURE 9.15.

Chapter 9 Review

1. Summarize the important mathematical ideas of this chapter.

2. No doubt you mentioned combinations in your summary in Item 1. Combinations have appeared in a variety of ways in this chapter. List several.

3. A standard deck of cards contains 13 different cards in each of four suits, for a total of 52 cards. Many card games use five-card hands. How many different five-card hands can be dealt from a standard deck?

4. Find the term of $(2a - b)^7$ that contains a^4.

5. a) In how many ways can you and two of your friends each pick a number from 1–10?

   b) In how many ways can you and two of your friends each pick a different number from 1–10?

6. The telephone area codes first used in 1947 permitted any first digit except 0 and 1. The second digit could be only 0 or 1, and the third could be any digit except 0.

   a) How many area codes were possible under the 1947 rules?

   b) In 1995 the rules were changed because the old rules did not allow enough area codes. The new rules allow any second digit. How many area codes are possible now?

   c) Assuming that each area code can have any exchange (the first three digits of a phone number), and each exchange can have any four-digit phone number, how many phones can each area code support?

7. The Oregon Megabucks lottery requires a participant to choose six different numbers from 1–44. The New Mexico Roadrunner Cash lottery requires a participant to choose five different numbers from 1–31.

   a) Which lottery would you expect to have the larger jackpot? Explain.

   b) In Oregon Megabucks, participants cannot make a single entry; they must make two. How does this requirement change your answer to part (a)?

c) Oregon Megabucks awards smaller prizes when an entry matches four or five of the six winning numbers. In how many ways can a single entry win in Oregon?

8. How many different sums of money can be made from a $1 bill, a $5 bill, a $10 bill, and a $20 bill?

9. A fast-food chain once advertised that there were "256 ways to have your single." They had 8 available ingredients that a customer could add to hamburgers. Do you agree with the claim? Explain.

10. In how many ways can the manager of a baseball team set the batting order of the nine starting players if the pitcher must bat last?

11. Expand $(x - 3)^4$.

12. A rectangular planter is divided into four sections, each of which can hold a single plant. (Assume the planter's front is distinguishable from its back.)

    a) In how many ways can you plant two tulips and two geraniums if both tulips are the same color and both geraniums are the same color?

    b) In how many ways can you plant two tulips and two geraniums if each tulip is a different color and each geranium is a different color?

13. a) How many handshakes are possible among the students in your classroom if each student shakes hands with every other student?

    b) How many handshakes are possible if each student shakes hands only with students of the same sex?

    c) How many handshakes are possible between students of opposite sexes?

14. Two prizes are being awarded to two different people in a group of eight.

    a) In how many ways can the prizes be awarded if the prizes are different?

    b) In how many ways can the prizes be awarded if the prizes are the same?

15. Discuss at least two different ways of determining the coefficients of a binomial expansion such as $(x + y)^5$.

Chapter 10

MODELING CHANGE WITH DISCRETE DYNAMICAL SYSTEMS

CHAPTER INTRODUCTION

Mathematical models are often created to help people predict the value of a variable at some time in the future. The variable might be population, a real estate value, the number of people with a communicable disease, or the location of a moving object such as an atom. This chapter is concerned with modeling change using the form:

$$\textit{future value} = \textit{present value} + \textit{change}.$$

A future value is often estimated from what is known (the present value) by adding change that has been carefully observed and modeled. In many cases, the basis for modeling change is the related form:

$$\textit{change} = \textit{current value} - \textit{previous value}.$$

Collecting and plotting data over intervals can uncover patterns that help build models that capture the trend of the data. As you will soon see, this form leads to a **difference equation** that describes change.

Lesson 10.1 examines behavior that can be modeled *exactly* by difference equations. More typically, the modeler approximates behavior, so Lesson 10.2 discusses *approximating* change that has been carefully observed and plotted. A major goal of model construction is predicting the *long-term behavior* of the variable under study. Therefore, Lesson 10.3 is concerned with building and using numerical solutions to difference equations to help predict future behavior. Also it is concerned with assessing the sensitivity of predictions to changes in the surrounding conditions. In Lesson 10.4, you will model a variety of interactive systems, including economic, political, and ecological systems involving predator and prey.

LESSON 10.1

Modeling Change with Difference Equations

This lesson's primary concern is behavior that can be modeled exactly. For example, mathematical models that predict the balance of a loan at some point in the future give exact results. Outside the financial world, however, approximate models of behavior are more common. The opening activity involves approximation and serves as an introduction to the entire chapter.

Activity 10.1

In the Battle of Trafalgar in 1805, a combined French and Spanish naval force under Napoleon fought a British naval force under Admiral Nelson. Initially, the French-Spanish force had 33 ships, and the British had 27 ships.

Modeling assumptions for this activity: A naval battle takes place in stages, or encounters. During an encounter, each side suffers a loss equal to 10% of the number of ships in the opposing force. Fractional values are meaningful and indicate that one or more ships are not at full capacity.

1. How many stages are required to conclude the naval battle between the French-Spanish and British forces? Who wins? How many ships does each side have at the end of the conflict? (Suggestion: Use a computer or calculator to prepare a table showing the number of ships in each force at each stage.)

2. Napoleon's force of 33 ships was arranged essentially along a line separated into three groups as shown in **Figure 10.1**.

Force B = 17 Force A = 3 Force C = 13

FIGURE 10.1.
Configuration of Napoleon's fleet.

© Susan Van Etten

Lord Nelson's strategy was to engage Force A with 13 ships (holding 14 in reserve). He then planned to combine those ships that survived the skirmish against Force A with the 14 ships in reserve to engage Force B. Finally, after the battle with Force B, he planned to use all remaining ships to engage Force C.

How many stages are required to conclude each of the three naval battles that Lord Nelson planned between the French-Spanish and British forces? Who wins? How many ships does each side have at the end of the conflict?

3. Experiment with other strategies. Can you improve on Lord Nelson's?

DIFFERENCE EQUATIONS

The modeling you did in Activity 10.1 approximates the behavior of the naval forces involved in the Battle of Trafalgar. Approximation is inevitable in many situations, because, for example, no model can completely predict the results of a naval battle. However, the inevitable error associated with approximation can obscure the essential ideas, so first consider some situations that can be modeled exactly.

Suppose you invest $1000 in a savings certificate that pays interest at 6% per year, compounded monthly. That is, every month $\frac{1}{2}$% of the current value of the certificate is added to the account. Since the balance in any month is the balance of the previous month plus 0.5% of that balance, the following sequence of numbers represents the monthly values, correct to the penny.

$A = \{1000, 1005, 1010.03, 1015.08, \ldots\}$

In general, a **sequence** is an ordered list of values. Individual members of a sequence are called **terms**. The terms are usually indexed beginning with 0 or 1, depending on the context. A term's index is written as a subscript.

For convenience, each term a_n of the sequence A can be indexed with the number n of months elapsed. It is common practice to show the month number and the associated balance in a table (see **Table 10.1**).

TABLE 10.1. Monthly balances in bank account.

n	0	1	2	3	...
a_n	1000	1005	1010.03	1015.08	...

A **difference equation** is a symbolic representation of the relationship between terms in a sequence. If the change that takes place is known, the form *future value = present value + change* can be used to write a difference equation.

In the savings account example, *future value = present value + change* becomes $a_{n+1} = a_n + change$, where the change taking place each month is the addition of $\frac{1}{2}$% interest. So the difference equation is $a_{n+1} = a_n + 0.005a_n$.

A difference equation is a convenient formula for computing future values. For example, suppose the savings account begins with an investment of \$2000. Then, the **initial value** $a_0 = 2000$ coupled with the difference equation $a_{n+1} = a_n + 0.005a_n$ yields the model

$$a_{n+1} = a_n + 0.005a_n$$
$$a_0 = 2000. \qquad (1)$$

The difference equation makes it easy to calculate $a_1 = a_0 + 0.005a_0 = 2000 + 0.005 \times 2000 = 2010$ and successive terms in the sequence with either a spreadsheet or the Seq mode of a calculator (see **Table 10.2**).

n	a_n
0	\$2000.00
1	\$2010.00
2	\$2020.05
3	\$2030.15
4	\$2040.30
5	\$2050.50
6	\$2060.76
7	\$2071.06
8	\$2081.41
9	\$2091.82
10	\$2102.28
11	\$2112.79
12	\$2123.36
...	...

TABLE 10.2. Balances calculated from difference equation.

The values defined by a difference equation are called a sequence or a **dynamical system.** Many problems in finance follow rules that can be precisely modeled using dynamical systems.

Example 1

Your grandparents cash in an IRA presently worth \$150,000. They place the money in a bank account paying monthly interest at the rate of $\frac{1}{2}$% of the current value of the account. They wish to withdraw \$1500 each month to supplement their income. How long will their IRA money last?

Corbis

Solution

The initial value in the account is $a_0 = 150{,}000$. The change taking place each month is that your grandparents receive interest worth $\frac{1}{2}$% of the current value and withdraw \$1500. Thus $a_1 = a_0 + (0.005a_0 - 1500)$ and, more generally,

$$a_{n+1} = a_n + (0.005a_n - 1500)$$
$$a_0 = 150{,}000. \qquad (2)$$

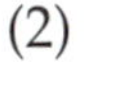

The value of the bank account each month can be calculated with a calculator or computer (see **Table 10.3**).

Scrolling the table shows that the account becomes negative in the 139th month (see **Table 10.4**). (Note that spreadsheets often indicate a negative value by enclosing it in parentheses.)

n	a_n
0	$150,000.00
1	$149,250.00
2	$148,496.25
3	$147,738.73
4	$146,977.42
5	$146,212.31
6	$145,443.37
7	$144,670.59
8	$143,893.94
9	$143,113.41
10	$142,328.98
11	$141,540.63
12	$140,748.33
...	...

TABLE 10.3.
Balance of IRA funds.

125	$20,199.18
126	$18,800.18
127	$17,394.18
128	$15,981.15
129	$14,561.06
130	$13,133.86
131	$11,699.53
132	$10,258.03
133	$8809.32
134	$7353.37
135	$5890.13
136	$4419.58
137	$2941.68
138	$1456.39
139	$ (36.33)

TABLE 10.4.
Final months of the IRA funds.

After 138 months, your grandparents will have $1456.39 remaining.

PROBLEM SOLVING WITH ITERATION

The calculation of a term of a sequence from previous terms is called an **iteration**. Although repeated iteration is a tedious problem-solving technique when done by hand, it becomes simple enough to be useful when done on a computer or calculator. For example, questions that arise in financial applications can be difficult to answer directly from a difference equation. Computer or calculator iteration makes a guess-and-check approach reasonable.

EXAMPLE 2

© Susan Van Etten

Your sister and her husband are contemplating whether to buy a home or rent. They prefer to buy if they can find a nice home, but are not sure of their price range. Their budget allows them to spend about $1200 for either rent or a mortgage (home loan) payment. They will commit current savings to covering the down payment, but are very concerned about the monthly payments. They can get a mortgage at 8% per year, (so interest is charged at the rate of $\frac{2}{3}$% per month). The length of the mortgage is 20 years, or 240 months. How expensive a home can your sister and her husband buy?

SOLUTION

The change taking place each month is that the balance of the loan is increased by $\frac{2}{3}$% of the current value and reduced by the monthly payment of $1200. Therefore, the difference equation is
$a_{n+1} = a_n + 0.006\overline{6}a_n - 1200.$

But in this situation a_0, which represents the price of the home after the down payment, is unknown. What loan will cause $a_{240} = 0$? Consider a $120,000 loan (after the down payment.) The model

$a_{n+1} = a_n + 0.006\overline{6}a_n - 1200.$
$a_0 = 120{,}000$

can be iterated on a calculator or computer to see if $a_{240} = 0$ (see **Table 10.5**).

If you continue to iterate, depending on the accuracy with which you approximate $\frac{2}{3}$% monthly interest, you reach a value of the loan after 20 years (240 months) of approximately $–115,608.19 (see **Table 10.6**). The negative sign indicates that the loan had been paid off sometime earlier, but the model is not valid if a_n becomes negative.

In fact, the loan would be paid off after 165 monthly payments of $1200 (see **Table 10.7**), plus a final payment of approximately $406.82 (plus $\frac{2}{3}$%, or $409.53).

Experimenting with different values of a_0 shows, for example, that an initial value of $143,500 results in a debt of $171.70 after 240 months (see **Table 10.8**).

Therefore, your sister and her husband can afford a mortgage of approximately $143,000.

n	a_n
0	$120,000.00
1	$119,600.00
2	$119,197.33
3	$118,791.98
4	$118,383.93
5	$117,973.15
6	$117,559.64
7	$117,143.37
8	$116,724.33
9	$116,302.49
10	$115,877.84
11	$115,450.36
12	$115,020.03
...	...

TABLE 10.5.
Iterating a loan balance.

n	a_n
226	$ (89,349.91)
227	$ (91,145.57)
228	$ (92,953.21)
229	$ (94,772.90)
230	$ (96,604.72)
231	$ (98,448.75)
232	$ (100,305.07)
233	$ (102,173.77)
234	$ (104,054.93)
235	$ (105,948.63)
236	$ (107,854.96)
237	$ (109,773.99)
238	$ (111,705.81)
239	$ (113,650.52)
240	$ (115,608.19)

TABLE 10.6.
Final months of $120,000 loan iteration.

161	$ 5117.20
162	$ 3951.32
163	$ 2777.66
164	$ 1596.18
165	**$ 406.82**
166	$ (790.47)
167	$ (1995.74)
168	$ (3209.05)
169	$ (4430.44)
170	$ (5659.98)
171	$ (6897.71)

TABLE 10.7.
Finding the payoff point.

228	$13,953.48
229	$12,846.50
230	$11,732.14
231	$10,610.36
232	$ 9481.09
233	$ 8344.30
234	$ 7199.93
235	$ 6047.93
236	$ 4888.25
237	$ 3720.84
238	$ 2545.64
239	$ 1362.61
240	$ 171.70

TABLE 10.8.
Checking 20-year balance for $143,500 loan.

Exercises 10.1

1. By substituting $n = 0, 1, 2, 3, 4$ write out the first four equations represented by each difference equation, then evaluate them to list the first five terms of the sequence they define.

   a) $a_{n+1} = 5a_n,\ a_0 = 2$

   b) $a_{n+1} = 3a_n + 6,\ a_0 = 0$

   c) $a_{n+1} = a_n(a_n + 2),\ a_0 = 2$

   d) $a_{n+1} = 2a_n^{\ 2},\ a_0 = 1$

2. In Example 1, suppose your grandparents want the IRA to last for 20 years. Experiment with different monthly withdrawals to find one that depletes the account in 20 years.

3. In Example 2, your sister and her husband have re-evaluated their budget and decided they can afford a mortgage payment of \$1300 per month. How expensive a home can they afford (after down payment)?

4. You currently have \$2000 in a savings account that pays 4% interest per year compounded monthly. At the end of each month you add \$100 to the account. How much will the account be worth after one year?

5. You presently owe \$500 on a credit card that charges 1.5% interest each month. You pay \$25 each month and make no new charges. Will you pay off the credit card? If so, approximately when?

6. You owe \$1700 on a credit card that charges 1.5% interest each month. You pay \$25 each month and make no new charges. Will you pay off the credit card? If so, approximately when?

7. Your parents are considering a 30-year mortgage that charges 6% per year compounded monthly. They wish to borrow \$150,000. Develop a model in terms of a monthly payment p that allows the mortgage to be paid off after 360 payments. Approximately what is p? (Hint: What is a_0? a_{360}?)

Exercises 10.1

8. Your grandparents have an annuity. The value of the annuity increases each month as $\frac{1}{2}\%$ of the previous month's balance is earned as interest. Your grandparents withdraw \$1000 each month for living expenses. They currently have \$50,000 in the annuity. How does the monthly withdrawal compare to the interest? Will the annuity run out of money? If so, approximately when?

PEOPLE AND MATH

An important career that involves modeling future change is that of an **ecologist**. Think of the myriad of interactive systems that function in a specific environment such as a rain forest. As conditions including pollution or overpopulation occur in that or any other environment, there are resulting changes in soil, air, water, and plant and animal populations. Over time, careful observation and inventory of existing species by ecologists provide important data that can be analyzed to help model, or predict, future changes. In some cases, the predictions of ecologists help to prompt changes that can save species on the brink of extinction.

LESSON 10.2

Approximating Change with Difference Equations

Lesson 10.1 used the form *future value = present value + change* to model financial situations exactly. More typically, mathematically describing change is not a precise procedure. When change is not exact, mathematicians plot the change, observe a pattern, and then approximate the pattern in mathematical terms. Once the change has been approximated, an analysis of the error can help judge the propriety of the model used for the approximation.

Modeling change is the art of selecting a function to approximate the change. To isolate the change for study, the form *change = current value − previous value* is useful, as you will see in this lesson.

Activity 10.2

The illustration below shows a golf ball bouncing on a hard surface.

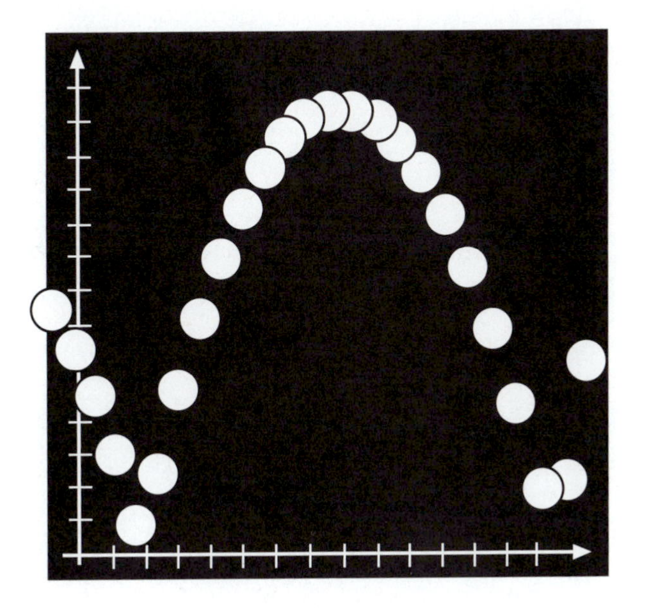

1. Perform an experiment to measure the height of a bouncing ball over time. For example, you might use a calculator-based laboratory with a motion detector. Build a table showing height measurements for a single bounce of the ball.

2. Create a table showing the change that occurs between successive entries in your table of data in Item 1. Number the entries in your table from 0. Graph the data.

3. Describe the change verbally. Can you create a model for the relationship between the change and the observation number?

FIRST DIFFERENCES

To see how change can be approximated, reconsider the sequence of savings account balances A = {1000, 1005, 1010.03, 1015.08,…} from Lesson 10.1. The sequence is also shown in **Table 10.9**.

TABLE 10.9.
Savings account data repeated from Table 10.1.

n	0	1	2	3	...
a_n	1000	1005	1010.03	1015.08	...

The change can be described by calculating the **first differences** of $A = \{a_0, a_1, a_2, a_3, \ldots\}$:

$$\Delta a_0 = a_1 - a_0$$

$$\Delta a_1 = a_2 - a_1$$

$$\Delta a_2 = a_3 - a_2$$

...

where the symbol Δa_0 is read as "delta *a* sub zero". In general, the *n*th **first difference** is $(\Delta a_n = a_{n+1} - a_n)$.

For the savings account data in Table 10.9,

$$\Delta a_0 = a_1 - a_0 = 1005 - 1000 = 5$$

$$\Delta a_1 = a_2 - a_1 = 1010.03 - 1005 = 5.03$$

$$\Delta a_2 = a_3 - a_2 = 1015.08 - 1010.03 = 5.05$$

...

Note that in this case the first differences are increasing.

Figure 10.2 shows a geometric interpretation of the first difference as a rise or fall; that is, the change in the height of the graph of the sequence during one time period.

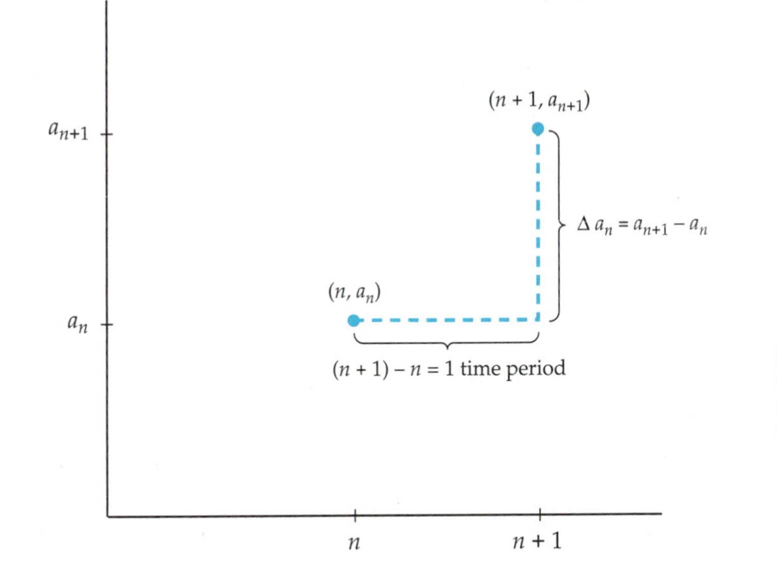

FIGURE 10.2.
The first difference of a sequence is the rise in the graph.

CONSTANT CHANGE

The simplest change is constant over time.

Mechanical springs are used in a variety of applications, including automobile suspension systems. The reaction of the spring to various loads must be modeled in order to design a vehicle such as a tank, dump truck, utility vehicle, or luxury car that responds to road conditions in a desired manner. **Table 10.10** shows the results of an experiment to measure the elongation (see **Figure 10.3**) of a spring. The elongation is shown as a function of the number n of units of mass placed on the spring.

Number n of Units of Mass	Elongation in inches a_n
0	0
1	0.875
2	1.721
3	2.641
4	3.531
5	4.391
6	5.241
7	6.120
8	6.992
9	7.869
10	8.741

TABLE 10.10.
Elongation of a spring for various masses.

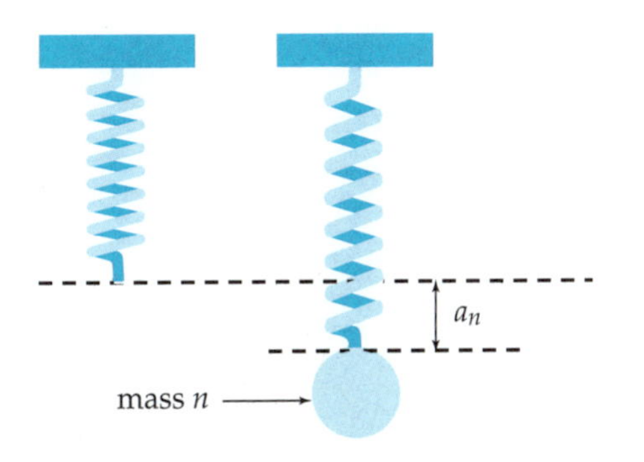

FIGURE 10.3.
The elongation is measured from the equilibrium position of the spring without attached mass.

n	a_n	Δa_n
0	0.000	0.875
1	0.875	0.846
2	1.721	0.920
3	2.641	0.890
4	3.531	0.860
5	4.391	0.850
6	5.241	0.879
7	6.120	0.872
8	6.992	0.877
9	7.869	0.872
10	8.741	

TABLE 10.11.

Table 10.11 shows the change in the elongation Δa_n.

The change Δa_n does not appear constant. However, since there is usually some variation in a physical experiment, it seems reasonable to consider a model in which the change in elongation with each unit of mass added is *approximately* constant.

EXAMPLE 3

Based on an assumption of constant change for the elongation data in Table 10.11, find a difference equation model. Examine the appropriateness of the model.

SOLUTION:

n	Observation	Prediction	Error
0	0.000	0.000	0.000
1	0.875	0.874	0.001
2	1.721	1.748	−0.027
3	2.641	2.622	0.019
4	3.531	3.496	0.035
5	4.391	4.370	0.021
6	5.241	5.244	−0.003
7	6.120	6.118	0.002
8	6.992	6.992	0.000
9	7.869	7.866	0.003
10	8.741	8.740	0.001

TABLE 10.12.

The average change is approximately 0.874, which gives the model $a_{n+1} = a_n + \Delta a_n$, or

$a_{n+1} = a_n + 0.874$

$a_0 = 0.$

The model predicts the first term of the sequence to be $a_1 = a_0 + 0.874 = 0.874$. Other predictions and the errors (residuals) are shown in **Table 10.12.**

The errors are relatively small except when $n = 2$, 3, 4, 5, which could be due to an anomaly such as measurement error. The situation needs further investigation before the model can be accepted or abandoned.

LINEAR CHANGE

Change sometimes exhibits a linear pattern.

How far does an automobile travel once the brakes have been applied? Consider the data in **Table 10.13**, where n is the speed of an automobile in increments of 5 mph ($n = 2$ represents 10 mph), and a_n is the distance in feet required to stop the automobile once the brakes have been applied.

Table 10.14 shows the change in distance Δa_n.

TABLE 10.13.

n	a_n
1	3
2	6
3	11
4	21
5	32
6	47
7	65
8	87
9	112
10	140
11	171
12	204
13	241
14	282
15	325
16	376

TABLE 10.14.

n	a_n	Δa_n
1	3	3
2	6	5
3	11	10
4	21	11
5	32	15
6	47	18
7	65	22
8	87	25
9	112	28
10	140	31
11	171	33
12	204	37
13	241	41
14	282	43
15	325	45
16	370	

Figure 10.4 shows a plot of Δa_n versus n. The graph is approximately linear.

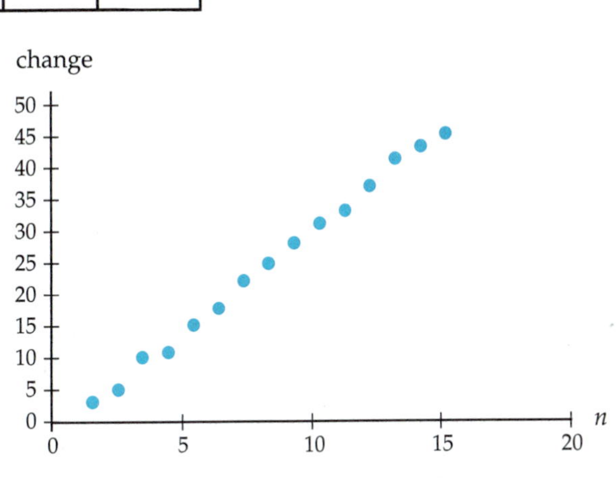

FIGURE 10.4. Graph of first differences for braking data.

EXAMPLE 4

n	Observation	Prediction	Error
1	3	3	0
2	6	6	0
3	11	12	–1
4	21	21	0
5	32	33	–1
6	47	48	–1
7	65	66	–1
8	87	87	0
9	112	111	1
10	140	138	2
11	171	168	3
12	204	201	3
13	241	237	4
14	282	276	6
15	325	318	7
16	370	363	7

TABLE 10.15.

Under the assumption of linear change, find a difference equation model for the stopping distance data in Table 8.14. Examine the appropriateness of the model.

SOLUTION:

The graph approximates a line through the origin with slope 3. Since a line through the origin with slope 3 has the equation $y = 3x$, the change can be modeled as $\Delta a_n = 3n$. Therefore, the stopping distance can be modeled as $a_{n+1} = a_n + \Delta a_n = a_n + 3n$, which gives the difference equation model

$$a_{n+1} = a_n + 3n$$

$$a_1 = 3.$$

The predictions and errors for the model are shown in **Table 10.15**.

The error increases as n increases, which can also be seen in a residual plot (**Figure 10.5**). In this case, the lack of randomness in the residuals raises questions about the assumption of linear change or the estimated slopes used in the model.

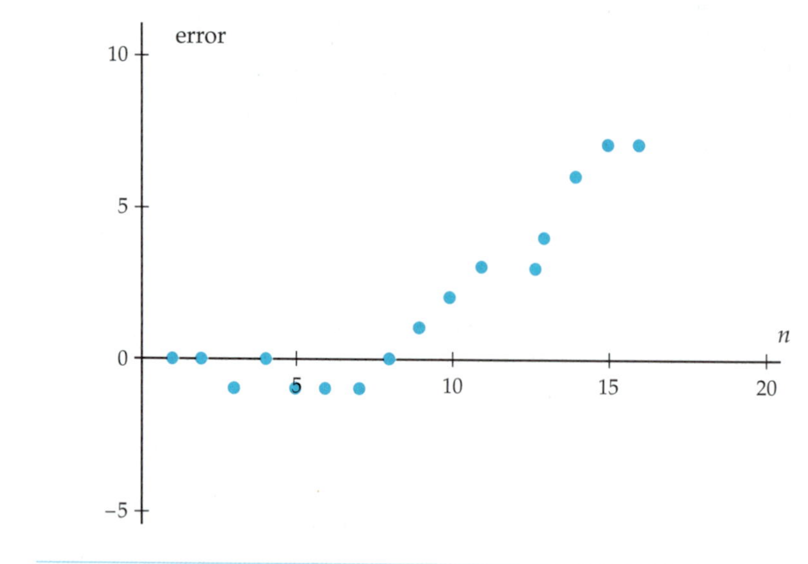

FIGURE 10.5. Residual plot for Table 10.15.

PROPORTIONAL CHANGE

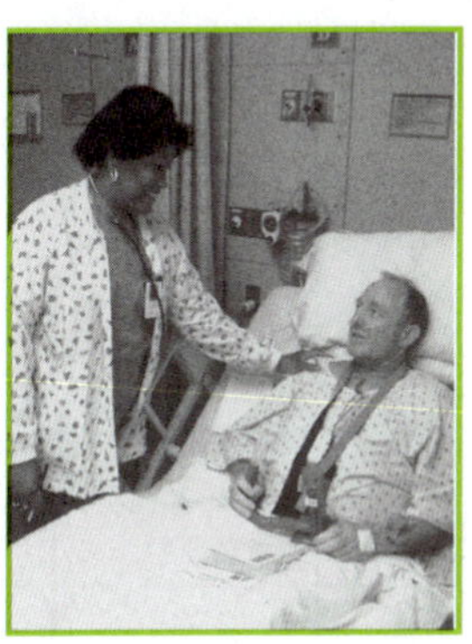
© Susan Van Etten

A third type of change that is commonly encountered is proportional change.

Digoxin is used in the treatment of heart disease. Doctors must prescribe an amount of medicine that keeps the concentration of digoxin in the bloodstream above an **effective level** without exceeding a **safe level** (there is variation among patients). To begin building a model, consider the rate of decay of digoxin in the bloodstream. Suppose an initial dosage of 0.5 mg is in the bloodstream. In **Table 10.16**, n represents the number of days after taking the initial dosage, and a_n represents the amount of digoxin remaining in the bloodstream for a particular patient.

n	0	1	2	3	4	5	6	7	8
a_n	0.500	0.345	0.238	0.164	0.113	0.078	0.054	0.037	0.026

TABLE 10.16. Digoxin remaining in bloodstream.

Table 10.17 shows the change each day.

n	a_n	Δa_n
0	0.500	−0.155
1	0.345	−0.107
2	0.238	−0.074
3	0.164	−0.051
4	0.113	−0.035
5	0.078	−0.024
6	0.054	−0.017
7	0.037	−0.011
8	0.026	

TABLE 10.17. Change in digoxin.

Figure 10.6 shows that the amount of change during a time interval is approximately proportional to the amount of digoxin present at the beginning of the time interval. The plot of Δa_n versus a_n is approximately linear.

n	Δa_n	$\frac{\Delta a_n}{a_n}$
0.5	−0.155	−0.310
0.345	−0.107	−0.310
0.238	−0.074	−0.311
0.164	−0.051	−0.311
0.113	−0.035	−0.310
0.078	−0.024	−0.308
0.054	−0.017	−0.315
0.037	−0.011	−0.297

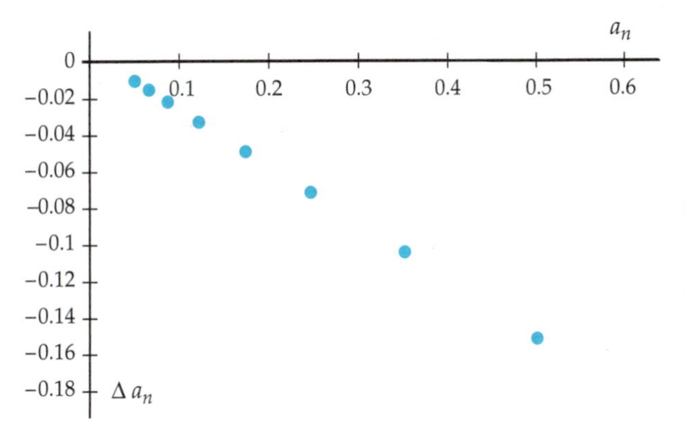

FIGURE 10.6. Change in digoxin is approximately proportional to amount remaining.

EXAMPLE 5

Under the assumption of proportional change, find a difference equation model for the digoxin decay data in Table 10.17. Examine the appropriateness of the model.

SOLUTION:

The graph in Figure 10.6 approximates a line through the origin with slope $k \approx -0.31$. Since the graph shows Δa_n as a function of a_n, $\Delta a_n = -0.31a_n$. Therefore, $a_{n+1} = a_n - 0.31a_n$.

A difference equation model for the decay of digoxin in the bloodstream given an initial dosage of 0.5 mg is:

$a_{n+1} = a_n - 0.31a_n = 0.69a_n$

$a_0 = 0.5.$

n	Observation	Prediction
0	0.500	0.500
1	0.345	0.345
2	0.238	0.238
3	0.164	0.164
4	0.113	0.113
5	0.078	0.078
6	0.054	0.054
7	0.037	0.037
8	0.026	0.026

TABLE 10.18.
Comparing digoxin model to data.

Table 10.18 shows that the predictions match the observations. The lack of error indicates that the assumption of proportional change is indeed correct.

The examples in this lesson demonstrate how to show that change is approximately constant, linear, or proportional to the current amount. Additionally, they demonstrate how to build a model based on the assumption of one of these three types of change. In practice, one typically experiments with a data set before finding a function that provides an approximation of the change being observed that is good enough for the purposes at hand. The more mathematics you know, the better you can approximate change in the world around you!

Exercises 10.2

1. Consider the hypothetical data in **Table 10.19**.

TABLE 10.19.

n	0	1	2	3	4	5	6	7
a_n	15	15.8	16.6	17.4	18.2	19	19.8	20.6

a) Show that the change $\Delta a_n = a_{n+1} - a_n$ is constant.

b) Write a difference equation for these data.

c) Show that these data can be calculated directly by the linear formula $a_{n+1} = 0.8a_n + 50$.

d) Of what type function is constant change a characteristic? Explain.

2. Consider the hypothetical data in **Table 10.20**.

TABLE 10.20.

n	0	1	2	3	4	5	6	7
a_n	6	13	24	39	58	81	108	139

a) Show that the change $\Delta a_n = a_{n+1} - a_n$ is linear.

b) Find a difference equation model for these data.

c) Show that the nth term can be calculated directly by the quadratic function $a_n = 2n^2 + 5n + 6$.

d) Of what type function is linear change a characteristic? Explain.

3. Consider the hypothetical data in **Table 10.21**.

TABLE 10.21.

n	0	1	2	3	4	5	6	7
a_n	5	15	45	135	405	1215	3645	10,935

a) Show that the change $\Delta a_n = a_{n+1} - a_n$ is proportional to the amount present a_n.

b) Find a difference equation model for these data.

c) Show that the nth term can be calculated directly by the exponential function $a_n = 5(3)^n$.

Exercises 10.2

d) Of what type function is proportional change a characteristic? Explain.

4. How far does an automobile travel after a driver perceives the need to stop, but before the brakes are applied (driver reaction distance)? Consider the data in **Table 10.22**, where n is the speed of an automobile in increments of 5 mph ($n = 2$ represents 10 mph), and a_n is the distance in feet the automobile travels before the brakes are applied.

TABLE 10.22.

n	1	2	3	4	5	6	7	8	9	10	11	12	13	14	15	16
a_n	6	11	16	22	28	33	39	44	50	55	61	66	72	77	83	88

a) Build a difference equation model.

b) Compare the observations with your model's predictions.

5. See the photograph of a golf ball bouncing on a hard surface that appeared at the beginning of Activity 10.2. A bounce on a hard surface occurred between the fifth and sixth images. The strobe light flashed every 0.03 seconds. Beginning with the sixth image, the vertical height of the golf ball is recorded in **Table 10.23** versus the time measured in increments of 0.03 seconds.

a) Build a difference equation that models the height of the golf ball.

b) Compare the observations with your model's predictions.

TABLE 10.23.

Time in 0.03 sec	0	1	2	3	4	5	6	7	8
Height	1.7	4.3	6.4	8.2	9.7	10.9	11.7	12.3	12.6

Time in 0.03 sec	9	10	11	12	13	14	15	16	17
Height	12.5	12.2	11.6	10.7	9.5	7.9	5.9	3.6	0.9

Exercises 10.2

6. A radioactive dye is injected into a patient's veins to facilitate an x-ray procedure. Monitoring the radioactivity over the course of several minutes yielded the data in **Table 10.24.**

   a) Build a mathematical model using difference equations.

   b) Compare observations with your model's predictions. Use your model to predict when the radioactivity will be below 500 counts per minute.

7. A drug administered to laboratory animals decreases in concentration (in parts per million) in the animal tissue. Concentrations are shown in **Table 10.25**.

   a) Determine whether a proportional change model would be suitable for these data.

   b) Since a model for these data will be used to establish a dose to maintain a level of 500–1000 ppm, the model should give higher weight to the first 5–6 data pairs. Build a model that does so.

   c) Compare the observations with your model's predictions.

8. Use the model developed in Exercise 7 to prescribe an initial dosage and a maintenance dosage that keeps the concentration above the effective level of 500 ppm, but below a safe level of 1000 ppm. (Hint: Experiment with different values until you have results you think are satisfactory.)

9. You have seen that proportional change is described in general by $a_{n+1} - a_n = ka_n$, where k is a constant describing the growth in the sequence's change. You have also seen that proportional change occurs when sequences exhibit exponential growth. In general, how is the base of the exponential growth related to k? Explain.

Time (min)	Radioactivity (cpm)
0	10,023
1	8174
2	6693
3	5500
4	4489
5	3683
6	3061
7	2497
8	2045
9	1645
10	1326

TABLE 10.24.

Time (days)	Concentration (ppm)
0	853
1	587
2	390
3	274
4	189
5	130
6	97
7	67
8	50
9	40
10	31

TABLE 10.25.

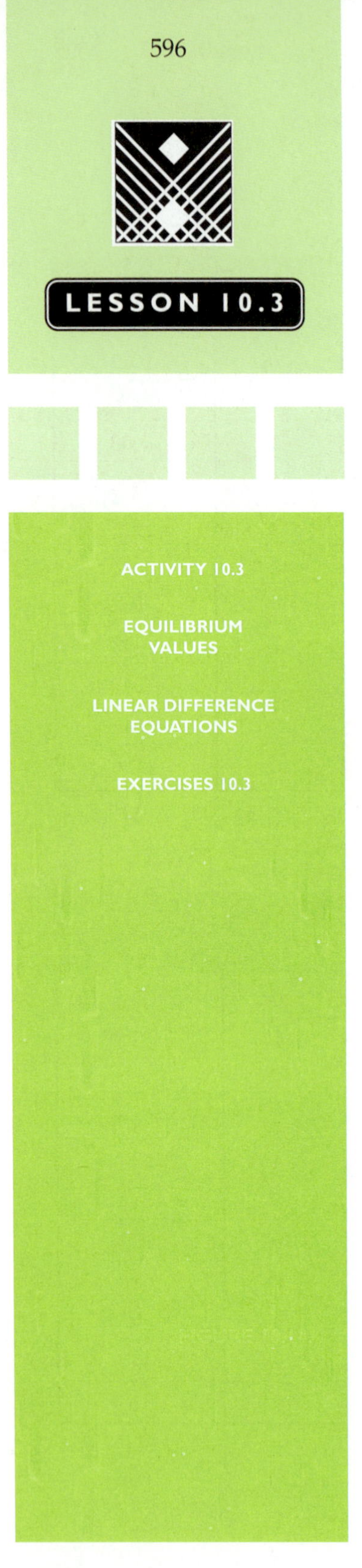

Numerical Solutions

Among the models you studied in Lesson 10.2 are those in which the change during a period is proportional to the amount present. In these models, change has the form $\Delta a_n = a_{n+1} - a_n = ka_n$, or more generally, $a_{n+1} = ra_n$ where r is a positive or negative constant.

In this lesson you will build numerical solutions to proportional change difference equations and other types of difference equations by starting with an initial value and iterating a sufficient number of subsequent values to determine the patterns involved. In some cases, the behavior predicted by the difference equations can be characterized by the mathematical structure of the system. In other cases, wild variations in the behavior are caused by only small changes in the initial values of the difference equation or in the proportionality constant.

Activity 10.3

Difference equations often approximate real-world behavior that people are trying to understand. Often the goal is to predict future behavior. Consequently, the long-term behavior of the system is of great interest. For example, does the sequence under study:

Grow without bound? (See **Figure 10.7a**.)

Become negative without bound? (See **Figure 10.7b**.)

Approach a limiting value? (See **Figure 10.7c**.)

Oscillate with increasing amplitude? (See **Figure 10.7d**.)

Oscillate while approaching a limit? (See **Figure 10.7e**.)

Demonstrate periodicity? (See **Figure 10.7f**.)

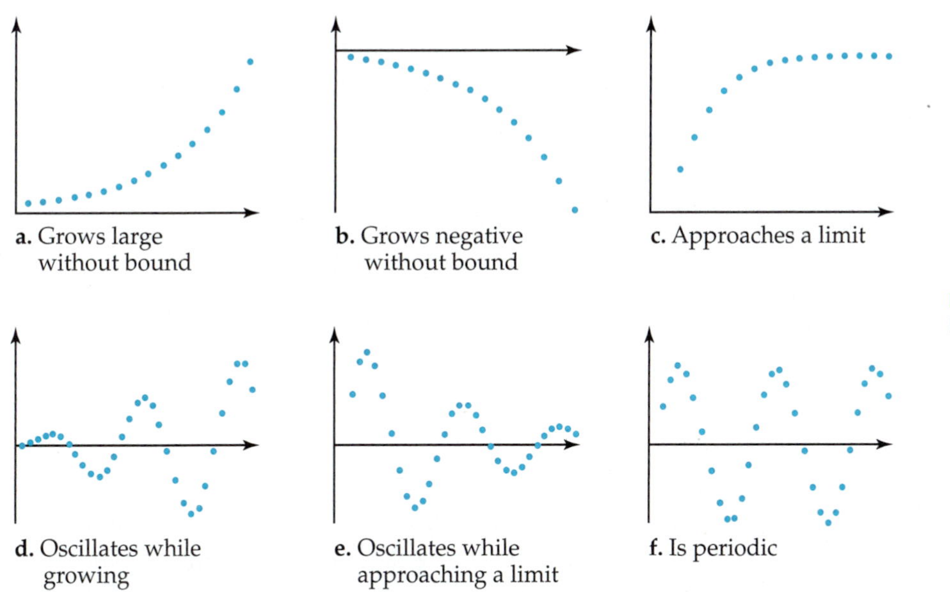

FIGURE 10.7.
Long-term behavior.

Your goal in this activity is to gain insight into the nature of the long-term behavior of difference equations of the form $a_{n+1} = ra_n$ for all possible values of r, and any initial value a_0. Are there critical values of r for which the nature of the behavior (as described in Figure 10.7) changes?

Experiment with different values of r and describe the long-term behavior you observe. Include at least one example for each of the following cases.

1. $r = 0$
2. $r = 1$
3. $r < 0$
4. $|r| < 1$
5. $|r| > 1$

EQUILIBRIUM VALUES

A value for which a dynamical system remains constant, once reached, is called an **equilibrium value**. In other words, for an equilibrium value a to occur, $a_{n+1} = a_n = a$. Equilibrium values are useful in analyzing long-term behavior.

Consider difference equations of the form $a_{n+1} = ra_n$.
Substitute $a_{n+1} = a_n = a$ to find the equilibrium value(s) a.
$a_{n+1} = ra_n$ becomes

$a = ra$, so

$a(1 - r) = 0$, or

$a = 0$ if $r \neq 1$.

If the sequence begins with the equilibrium value $a = 0$, it remains constant. Note that if $r = 1$, the sequence is constant for any initial value, including the equilibrium value 0.

LINEAR DIFFERENCE EQUATIONS

A linear difference equation is one having the form $a_{n+1} = ra_n + b$, where r and b are positive or negative constants. They are called linear because of the form's similarity to that of a linear function $f(x) = mx + b$. (Note that a difference equation of the form $a_{n+1} = ra_n$ is a special case with $b = 0$.)

To find the equilibrium value(s) a, substitute $a_{n+1} = a_n = a$:

$a_{n+1} = ra_n + b$

$a = ra + b$

$a = \frac{b}{1-r}$, if $r \neq 1$.

Equilibrium values are useful in answering questions in situations that can be modeled with difference equations. In some situations, an equilibrium value is **stable**. That is, the sequence approaches the equilibrium value even though the sequence does not start there. In other cases, the equilibrium value is **unstable**. That is, the sequence does not approach the equilibrium value unless it starts there exactly. The following examples demonstrate stable and unstable equilibrium values.

EXAMPLE 6

Your grandparents wish to cash in IRAs to place in a savings account so they can make monthly withdrawals of \$1500 to supplement their income. They plan to place the money in a bank account paying monthly interest at the rate of $\frac{1}{2}$% of the current value of the account. What value of IRA should your grandparents cash in?

SOLUTION:

The change taking place is that each month your grandparents receive interest worth $\frac{1}{2}$% of the current value and withdraw \$1500. Thus
$a_1 = a_0 + 0.005a_0 - 1500$ and, more generally,

$a_{n+1} = a_n + 0.005a_n - 1500$

$a_0 = ?$

For the difference equation $a_{n+1} = ra_n + b$, the equilibrium value a is $a = \frac{b}{1-r}$. Substituting gives

$$a = \frac{b}{1-r} = \frac{-1500}{1-1.005} = 300{,}000$$

Consider three different starting values:

Case A: $a_0 = 301{,}000.00$

Case B: $a_0 = 300{,}000.00$

Case C: $a_0 = 299{,}000.00$.

Iterate the difference equation $a_{n+1} = a_n + 0.005a_n - 1500$ for each of the three starting values and graph the results (see **Figure 10.8**).

Note that for Case B, the value of the bank account remained at the equilibrium value as expected. For an initial value slightly above or below the equilibrium value, as in Cases A and C, the value of the account moved away from the equilibrium value. The equilibrium value $a = 300{,}000$ is unstable.

If your grandparents can afford \$300,000, the account will have the same value forever. Above \$300,000, the account will actually increase even though your grandparents withdraw \$1500 each month. Less than \$300,000 will cause the account to decrease to zero, in which case they will need to consider how long they want their money to last.

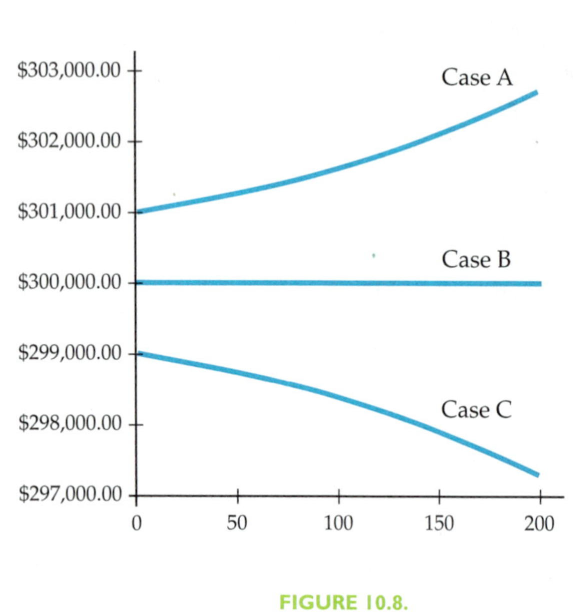

n	a_n **Case A**	a_n **Case B**	a_n **Case C**
0	$301,000.00	$300,000.00	$299,000.00
1	$301,005.00	$300,000.00	$298,995.00
2	$301,010.03	$300,000.00	$298,989.98
3	$301,015.08	$300,000.00	$298,984.92
4	$301,020.15	$300,000.00	$298,979.85
5	$301,025.25	$300,000.00	$298,974.75
6	$301,030.38	$300,000.00	$298,969.62
7	$301,035.53	$300,000.00	$298,964.47
8	$301,040.71	$300,000.00	$298,959.29
9	$301,045.91	$300,000.00	$298,954.09
10	$301,051.14	$300,000.00	$298,948.86
11	$301,056.40	$300,000.00	$298,943.60
12	$301,061.68	$300,000.00	$298,938.32
13	$301,066.99	$300,000.00	$298,933.01
14	$301,072.32	$300,000.00	$298,927.68
15	$301,077.68	$300,000.00	$298,922.32
16	$301,083.07	$300,000.00	$298,916.93

FIGURE 10.8.
A partial table and a graph.

EXAMPLE 7

Recall from Lesson 10.2 that digoxin is used in the treatment of heart disease. Doctors must prescribe an amount of medicine that keeps the concentration of digoxin in the bloodstream above an effective level without exceeding a safe level. The initial dosage was 0.5 mg in the bloodstream, and the amount of digoxin in the bloodstream over time is described by the data in **Table 10.26**, where n represents the number of days after taking the initial dosage, and a_n represents the amount of digoxin remaining in the bloodstream.

TABLE 10.26.
Digoxin levels.

n	0	1	2	3	4	5	6	7	8
a_n	0.5000	0.345	0.238	0.164	0.113	0.078	0.054	0.037	0.026

The difference equation model from Lesson 10.2 is

$a_{n+1} = 0.69a_n$

$a_0 = 0.5.$

What maintenance dosage—an amount to take each day to maintain an appropriate level of digoxin in the bloodstream—should the doctor prescribe if the level should not exceed 3.5 mg, but should not fall below 1 mg?

SOLUTION:

The model becomes

$a_{n+1} = 0.69a_n + b$

$a_0 = ?$

where b is the daily maintenance dose and a_0 is the initial dosage that the doctor prescribes. Begin by considering the equilibrium value. For this model, the equilibrium value is $a = \frac{b}{1-0.69} = \frac{b}{0.31}$.

Part I.

Many solutions to the maintenance dosage question exist. Choose an equilibrium value between 1 mg and 3.5 mg, say $a = 3$ mg, and determine b:

$b = 0.31a = 0.31(3) = 0.93.$

For the difference equation $a_{n+1} = 0.69a_n + 0.93$, the equilibrium value is 3. Select several starting dosages and determine the corresponding long-term behavior.

Case A : $a_0 = 3.5$

Case B: $a_0 = 3$

Case C: $a_0 = 2.5.$

The data that result from iterating the difference equation $a_{n+1} = 0.69a_n + 0.93$ for each of the three starting values and a graph of those data are shown in **Figure 10.9**.

n	a_n Case A	a_n Case B	a_n Case C
0	3.50	3.00	2.50
1	3.35	3.00	2.66
2	3.24	3.00	2.76
3	3.16	3.00	2.84
4	3.11	3.00	2.89
5	3.08	3.00	2.92
6	3.05	3.00	2.95
7	3.04	3.00	2.96
8	3.03	3.00	2.97
9	3.02	3.00	2.98
10	3.01	3.00	2.99
11	3.01	3.00	2.99
12	3.01	3.00	2.99
13	3.00	3.00	3.00
14	3.00	3.00	3.00
15	3.00	3.00	3.00
16	3.00	3.00	3.00

FIGURE 10.9.
Maintaining 3.0 mg.

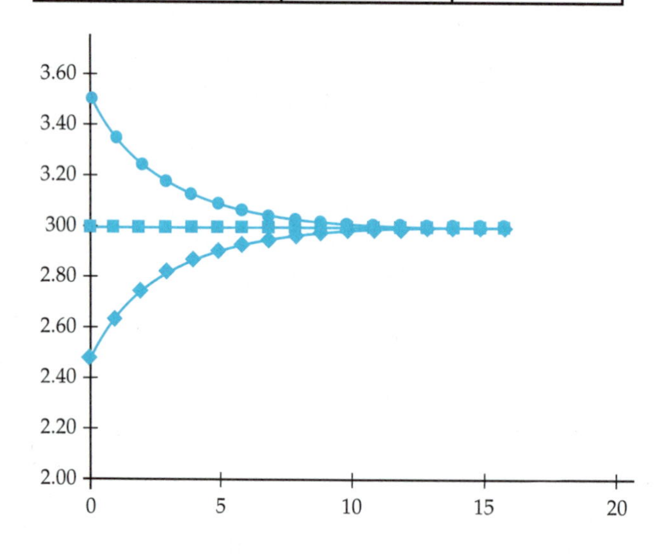

It is important to understand that the values in the table show the concentration just *after* a dose is administered. For example, for Case B the initial concentration level of 3.00 mg decays to 2.07 mg just before a dose is administered. The maintenance dosage of 0.93 equals that loss and restores the level to its maximum value of 3.00 mg.

For Case B, the level of digoxin remains at the equilibrium value as expected. For initial values slightly above or below the equilibrium value, as in Cases A and C, the digoxin level approaches the equilibrium value as time increases. The equilibrium value $a = 3$ is stable.

Part II.

If digoxin is prescribed only in integer amounts, you might use an initial dose of 2 mg and a maintenance dose of 1 mg. The resulting model is:

$a_{n+1} = 0.69a_n + 1$

$a_0 = 2,$

which has an equilibrium value at $\frac{1}{0.31} \approx 3.23$. Results of iterating this model are shown in **Table 10.27**.

Since these data represent the concentration just after a dose is taken, and since the maintenance dose is 1 mg, the before-dose data are 1 mg lower. The smallest before-dose concentration is approximately 1.38 mg, which occurs before the first maintenance dose. Thus, the concentration ranges from about 1.38 mg to about 3.23 mg, which is within the desired 1–3 mg range.

n	a_n
0	2.000
1	2.380
2	2.642
3	2.823
4	2.948
5	3.034
6	3.094
7	3.135
8	3.163
9	3.182
10	3.196
11	3.205
12	3.212
13	3.216
14	3.219
15	3.221
16	3.223
17	3.224
18	3.224
19	3.225
20	3.225

TABLE 10.27. Maintenance dose of 1 mg.

An important property of difference equations is that numerical solutions can be constructed for any difference equation given an initial value. In this lesson you have seen that long-term behavior can be sensitive to the starting value and to the values of the parameter r in difference equations of the form $a_{n+1} = ra_n + b$.

Exercises 10.3

1. For the following difference equation models, find an equilibrium value if one exists. If there is an equilibrium value, is it stable?

   a) $a_{n+1} = 1.1a_n$

   b) $a_{n+1} = 0.9a_n$

   c) $a_{n+1} = -0.9a_n$

   d) $a_{n+1} = a_n$

   e) $a_{n+1} = -1.2a_n + 50$

   f) $a_{n+1} = 0.8a_n + 100$

   g) $a_{n+1} = a_n + 50$

2. In general, describe equilibrium values if $r = 1$ in the difference equation $a_{n+1} = ra_n + b$.

3. Build numerical solutions for each of the following dynamical systems. Include both a table and a graph. Describe the long-term behavior.

   a) $a_{n+1} = -1.2a_n + 50$, $a_0 = 1000$.

   b) $a_{n+1} = 0.8a_n - 100$, $a_0 = 500$.

   c) $a_{n+1} = 0.8a_n - 100$, $a_0 = -500$.

   d) $a_{n+1} = -0.8a_n + 100$, $a_0 = 1000$.

4. You currently have \$5000 in a savings account that pays 0.5% interest each month. You add \$200 each month.

   a) Build a difference equation model to compute the value of the account.

   b) Build a numerical solution to determine when the account will reach \$20,000.

Exercises 10.3

5. You owe $500 on a credit card that charges 1.5% interest each month. You budget $50 each month, and plan to make no new charges until the card is paid off.

   a) Build a difference equation that models the credit card debt.

   b) Build a numerical solution to determine when the card will be paid off.

   c) Find the equilibrium value for your difference equation model. Interpret the equilibrium value in terms of a credit card debt.

6. Your grandparents have an annuity. The value of the annuity increases each month as 1% interest on the previous balance is deposited. Your grandparents withdraw $1100 each month for living expenses. Presently they have $50,000 in the annuity.

   a) Build a difference equation that models the value of the annuity.

   b) Build a numerical solution to determine when the annuity will be depleted.

   c) Find the equilibrium value for your difference equation model. Interpret the equilibrium value in terms of the value of the annuity and the monthly withdrawal.

7. In advanced courses, you can derive a formula for the nth term of the sequence that represents the **solution** to difference equations of the form $a_{n+1} = ra_n + b$ as

   $$a_n = cr^n + \frac{b}{1-r},$$

   where c is a constant that can be determined if an initial value is given. (Note that the first term is an exponential and the second term is the equilibrium value $a = \frac{b}{1-r}$.)

   a) For the special base $b = 0$, discuss both the difference equation and the solution.

   b) For the difference equation $a_{n+1} = 0.5a_n + 200$ with $a_0 = 100$, find the solution.

   c) Verify that the solution and the original difference equation in part (b) give the same results for several values of n.

Exercises 10.3

d) In Activity 10.3, you classified the nature of the solution to $a_{n+1} = ra_n$ based on the value of r. Now consider the solution form $a_n = cr^n + \frac{b}{1-r}$ to the difference equation $a_{n+1} = ra_n + b$. Based on the value of r, determine whether the equilibrium value $a = \frac{b}{1-r}$ will be approached. Classify the long-term behavior based on critical values of r.

e) Based on the solution form $a_n = cr^n + \frac{b}{1-r}$, describe the role b plays in equilibrium values and their stability.

8. The difference equation $a_{n+1} = r\,(1 - a_n)a_n = r\,(a_n - a_n^2)$ is considered a nonlinear difference equation because each term is squared to produce the next term. As an example, consider $a_{n+1} = r\,(1 - a_n)a_n$ when $a_0 = 0.2$.

   Iterate this sequence for each of the following values of r and describe the long-term behavior in each case. Interpret your results as though the difference equation modeled population.

   a) $r = 2$

   b) $r = 3.25$

   c) $r = 3.5$

   d) $r = 3.75$

   The four cases indicate a sensitivity to the value of the parameter r. Small changes in r can cause remarkably different solutions. The model quickly becomes useless for predictive purposes. The model is also sensitive to the initial conditions.

LESSON 10.4 Systems of Difference Equations

Many situations that can be modeled with difference equations have more than one dependent variable. For example, predator and prey populations are variables in an ecological system, the number of voters for each of several parties are variables in a political system, and price and supply are variables in an economic system. This lesson examines several systems with two variables. Equilibrium values will be of particular importance in obtaining predictions from these difference equation models.

Activity 10.4

You have been hired as a consultant to a car rental company. The company wishes to make a bid on a lucrative long-term contract being offered by a large travel agency. The travel agency is offering travel packages to the Orlando and Tampa areas of Florida. The package includes air travel, hotels, amusement park tickets, and a car for a week.

The car rental agency wants to submit a competitive bid. They would like to supply about 700 cars on a long-term basis. Their concern is estimating the number of cars they would have to transport between Orlando and Tampa to meet demand in both cities. Your job is to provide these forecasts.

You have requested and analyzed the historical records of the travel company. You determine that of the cars rented in Orlando, an average of 60% are returned to Orlando, while an average of 40% are returned to Tampa. Of those rented in Tampa, an average of 70% are returned to Tampa, and an average of 30% are returned to Orlando (see **Figure 10.10**).

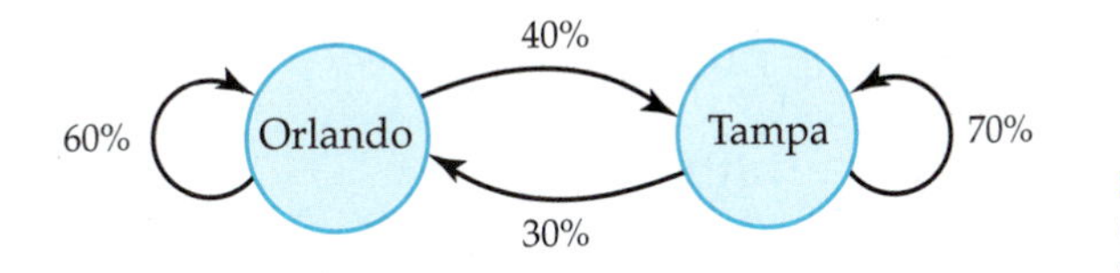

FIGURE 10.10. Car rental returns in Orlando and Tampa.

1. Let n represent the number of weeks, O_n the number of cars in Orlando at the end of week n, and T_n the number of cars in Tampa at the end of week n.

   Write a difference equation that describes the number of cars in Orlando at the end of week $n + 1$. Do the same for the number of cars in Tampa at the end of week $n + 1$.

2. Suppose the company plans to supply 700 cars for the contract. For each of the following initial conditions, iterate the model you built in Item 1 and describe the results.

   a) Orlando: 300 cars, Tampa: 400 cars.

   b) Orlando: 350 cars, Tampa: 350 cars.

   c) Orlando: 700 cars, Tampa: 0 cars.

   d) Orlando: 500 cars, Tampa: 200 cars.

   e) Orlando: 200 cars; Tampa: 500 cars.

   f) Orlando: 0 cars, Tampa: 700 cars.

© *Susan Van Etten*

3. Describe the long-term behavior of this system. Do you consider this system stable?

4. What advice would you give to the car rental company?

EQUILIBRIUM VALUES IN SYSTEMS OF DIFFERENCE EQUATIONS

A **system of difference equations** is a set of two or more difference equations in two or more variables. For example, the following equations, which describe rental car movement in Activity 10.4, form a system of difference equations.

$$O_{n+1} = 0.6O_n + 0.3T_n$$

$$T_{n+1} = 0.4O_n + 0.7T_n.$$

Equilibrium values are values for both O_n and T_n for which no change takes place in either O_{n+1} or T_{n+1}. Call the equilibrium values, if they exist, O and T. Then $O = O_n = O_{n+1}$ and $T = T_n = T_{n+1}$ simultaneously. Substitution in the system gives

$$O = 0.6O + 0.3T$$

$$T = 0.4O + 0.7T.$$

Each equation in this pair is equivalent to $O = \frac{3}{4}T$. For example, suppose the car rental company plans to supply 350 cars for the contract and divides the cars so the number in Orlando is three-fourths of the number in Tampa. Then,

$$O + T = 350$$

$$\frac{3}{4}T + T = 350$$

$$\frac{7}{4}T = 350$$

$$T = 200.$$

The pair $T = 200$, $O = 350 - 200 = 150$ forms an equilibrium value for the system. If these values are used as initial values:

$$O_{n+1} = 0.6O_n + 0.3T_n$$

$$T_{n+1} = 0.4O_n + 0.7T_n$$

$$O_0 = 150$$

$$T_0 = 200$$

the system remains constant:

$$O_1 = 0.6O_0 + 0.3T_0 = 0.6(150) + 0.3(200) = 150$$

$$T_1 = 0.4O_0 + 0.7T_0 = 0.4(150) + 0.7(200) = 200.$$

When equilibrium values exist, the system's behavior for values near the equilibrium should be examined. For such values, does the system:

a) remain close

b) approach the equilibrium value

c) not remain close?

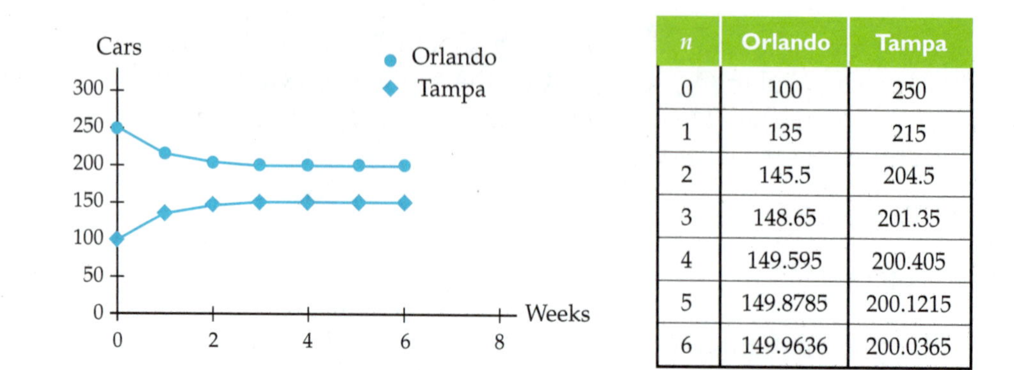

n	Orlando	Tampa
0	100	250
1	135	215
2	145.5	204.5
3	148.65	201.35
4	149.595	200.405
5	149.8785	200.1215
6	149.9636	200.0365

FIGURE 10.11. A stable equilibrium with 350 cars.

For example, if there are 100 cars in Orlando and 250 in Tampa initially, the system behaves as shown in **Figure 10.11**.

The equilibrium values are quickly approached; the equilibrium value of 100 cars in Orlando and 250 in Tampa appears stable.

Example 8

Consider a habitat that supports two competing species, hawks and owls. That is, the hawks and owls do not prey on one another, but compete for the same food sources. Let O_n represent the owl population after n years, and H_n represent the hawk population. Consider the growth of each species in the absence of constraints and competition. Assume that each demonstrates exponential growth:

$$O_{n+1} = aO_n$$

$$H_{n+1} = bH_n$$

where a and b are positive constants.

One popular model for the interaction between the two species assumes that the detrimental effect on the growth of each species in the next year is proportional to the product of O_n and H_n. Thus, consider the model:

$$\begin{aligned} O_{n+1} &= aO_n - cO_nH_n \\ H_{n+1} &= bH_n - dO_nH_n \end{aligned} \qquad (1)$$

where c and d are positive constants.

Assume ecologists have estimated the values for a, b, c, and d, resulting in the model:

$$O_{n+1} = 1.2O_n - 0.001O_nH_n$$

$$H_{n+1} = 1.3H_n - 0.002O_nH_n.$$

a) Show that equilibrium values exist at $(O, H) = (0, 0)$ and $(O, H) = (150, 200)$.

b) Determine whether the equilibrium value $(O, H) = (150, 200)$ is stable.

SOLUTION:

a) To find equilibrium values O and H, set $O = O_{n+1} = O_n$ and $H = H_{n+1} = H_n$:

$$O = 1.2O - 0.001OH$$

$$H = 1.3H - 0.002OH.$$

For $(O, H) = (0, 0)$, substitution yields

$$O_{n+1} = 1.2(0) - 0.001(0)(0) = 0$$

$$H_{n+1} = 1.3(0) - 0.002(0)(0) = 0.$$

For $(O, H) = (150, 200)$, substitution yields

$$O_{n+1} = 1.2(150) - 0.001(150)(200) = 150$$

$$H_{n+1} = 1.3(200) - 0.002(150)(200) = 200.$$

b) Consider the following initial populations near the equilibrium value (**Figures 10.12–10.14**):

	Owls	Hawks
Case A	151	200
Case B	150	201

FIGURE 10.12.

Case A

Years	Owls	Hawks
0	151	200
1	151	199.6
2	151.0604	199.2008
3	151.1811	198.7783
4	151.3658	198.3088
5	151.6218	197.7671
6	151.9604	197.1256
7	152.3972	196.3527
8	152.953	195.4113
9	153.6549	194.2572
10	154.5373	192.8372
11	155.6442	191.0873
12	157.0314	188.9303
13	158.7697	186.2734
14	160.9491	183.0063
15	163.6842	178.9988
16	167.1218	174.0999
17	171.4502	168.1381
18	176.913	160.9249
19	183.8259	152.2629
20	192.6012	141.9621
21	203.7793	129.8666
22	218.0711	115.8983
23	236.4112	100.1197
24	260.0241	82.81673
25	290.4945	64.59306
26	329.8295	46.44312
27	380.4771	29.73943
28	445.2574	16.03092
29	527.1709	6.564424
30	629.1446	1.612604
31	753.9589	0.067263

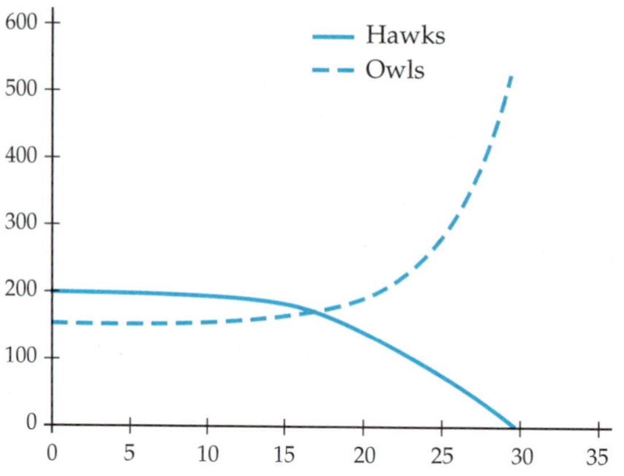

FIGURE 10.13.
Behaviour Case A.

Corbis

Case B

Years	Owls	Hawks
0	150	201
1	149.85	201
2	149.7002	201.0603
3	149.5414	201.1809
4	149.3648	201.3654
5	149.1609	201.6212
6	148.9191	201.9596
7	148.6273	202.3962
8	148.2711	202.9518
9	147.8335	203.6536
10	147.2933	204.536
11	146.6252	205.6433
12	145.7978	207.0313
13	144.7726	208.7713
14	143.5028	210.9539
15	141.9308	213.6952
16	139.9871	217.1438
17	137.5872	221.4923
18	134.6301	226.991
19	130.9963	233.9687
20	126.5465	242.8612
21	121.1226	254.2531
22	114.5513	268.9375
23	106.6544	288.0044
24	97.26837	312.9718
25	86.27978	345.9789
26	73.68476	390.0706
27	59.67945	449.6072
28	44.78303	530.8248
29	29.96769	642.5283
30	16.70614	796.7766
31	6.736306	1009.188
32	1.285371	1298.347
33	−0.12641	1684.514

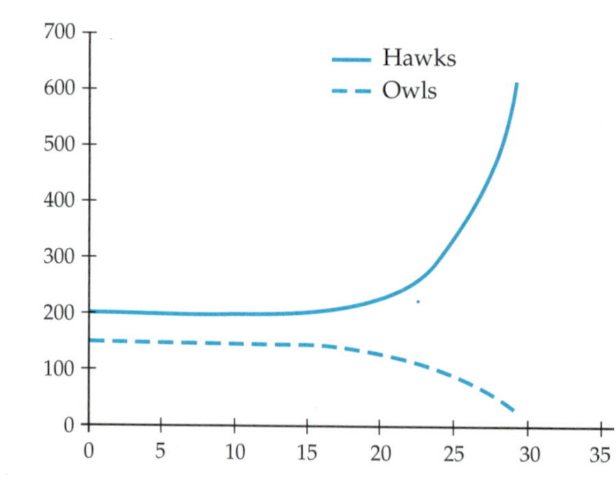

FIGURE 10.14.
Behavior in Case B.

Behaviors for initial values very near the equilibrium value differ wildly. If the initial population is 150 owls and 200 hawks, the model predicts the populations will remain there forever. However, starting with only one more owl, the model predicts the owls win; starting with one more hawk, the model predicts the hawks win. Therefore, the equilibrium value $(O, H) = (150, 200)$ is unstable.

The model in Example 8 is obviously not precise. It does not capture many of the complexities of the habitat containing the owls or hawks (weather, disease, growth of the food source, etc.), nor would one expect the interaction between the owls and hawks to be as simple as assumed in the model. Even if the model were a precise law governing the growth of the species, it would be difficult to determine exact values for the constants (parameters) a, b, c, and d. However, ecologists do use such models for gaining qualitative insight into the coexistence of the species. The analysis of the model indicates that coexistence of these two species is highly unlikely, a principle known as Gause's *Principle of Mutual Competitive Exclusion.*

Exercises 10.4

1. In Example 8, investigate the stability of the equilibrium value at $(O, H) = (0, 0)$.

2. Refer back to Example 8.

   a) Interpret the constants c and d. What is the significance of raising the value of c?

   b) Consider the interaction term cO_nH_n. How does the detrimental effect of the interaction change as either O_n or H_n increases? If O_n is very small, what kind of growth would you expect the hawks to demonstrate?

3. Consider a three-party system with Republicans, Democrats, and Independents. Assume that in the next election, 75% of those who voted Republican in the last election will vote Republican again; 5% will vote Democrat; and 20% will vote Independent. Of those who voted Democrat in the last election, 60% will vote Democrat again; 20% will vote Republican; and 20% will vote Independent. Of those who voted Independent in the last election, 40% will vote Independent again; 20% will vote Democrat; and 40% will vote Republican. Assume these tendencies continue from election to election, and that no additional voters enter or leave the system.

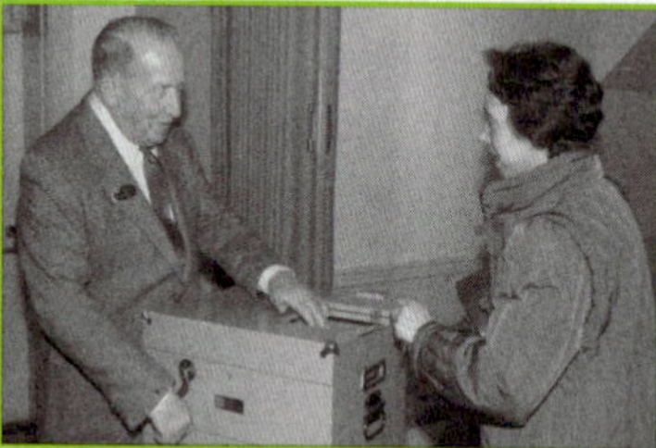

Case	Republicans	Democrats	Independents
1	222,221	77,777	100,000
2	399,998	0	0
3	0	399,998	0
4	0	0	399,998

TABLE 10.28.
Initial values.

   a) Write difference equations to model this situation.

   b) Predict the long-term behavior for each initial number of voters shown in **Table 10.28**.

4. Suppose the spotted owls' primary food source is a single prey: mice. An ecologist wishes to predict the population levels of spotted owls and mice in a wildlife sanctuary. Letting M_n represent the mouse population after n years, and O_n the predator owl population, the ecologist has suggested the model:

$$M_{n+1} = 1.2M_n - 0.001O_nM_n$$

$$O_{n+1} = 0.7O_n + 0.002O_nM_n.$$

The ecologist wants to know if the two species can coexist in the habitat, and if the outcome is sensitive to the starting populations.

Exercises 10.4

a) Compare the signs of the coefficients of the above model with the signs of the coefficients of the Owls-Hawks model in Example 8. Explain the sign of each of the four coefficients 1.2, –0.001, 0.7, and 0.002 in terms of the predator-prey relationship being modeled.

b) Test the initial populations in **Table 10.29**, and predict the long-term outcome.

	Owls	Hawks
Case A	150	200
Case B	150	300
Case C	100	200
Case D	10	20

TABLE 10.29.
Initial populations.

5. An economist is interested in the variation of the price of a single product. It is observed that a high price for the product in the market attracts more suppliers. However, increasing the quantity of the product supplied tends to drive the price down. Over time, there is an interaction between price and supply. The economist has proposed the following model, where P_n represents the price of the product at year n, and Q_n the quantity.

$$P_{n+1} = P_n - 0.1(Q_n - 500)$$

$$Q_{n+1} = Q_n + 0.2(P_n - 100)$$

a) Does the model make sense intuitively? What is the significance of the constants 100 and 500? Explain the significance of the sign of the constants –0.1 and 0.2.

b) Test the initial conditions in **Table 10.30** and predict the long-term behavior.

	Price	Quantity
Case A	100	500
Case B	200	500
Case C	100	600
Case D	100	400

TABLE 10.30.
Initial values.

Exercises 10.4

6. Recall Activity 10.1. In the Battle of Trafalgar in 1805, a combined French and Spanish naval force under Napoleon fought a British naval force under Admiral Nelson. Initially, the French-Spanish force had 33 ships, and the British had 27 ships. Modeling assumptions in Activity 10.1 were:

- A naval battle takes place in stages, or encounters.
- During an encounter, each side suffers a loss equal to 10% of the number of ships in the opposing force.
- Fractional values are meaningful and indicate that one or more ships are not at full capacity.

Using these assumptions, you saw that if the two forces engaged, the British lost the battle and about 26 ships, while the French-Spanish force lost about 15 ships. You also saw that Lord Nelson could overcome a superior force by employing a divide-and-conquer strategy.

An alternative strategy for overcoming a superior force is to increase the technology employed by the inferior force. Suppose that the British ships were equipped with superior weaponry, and that the French-Spanish suffered losses equal to 15% of the number of ships in the opposing force, while the British continued to suffer casualties equal to 10% of the opposing force.

a) Formulate a system of difference equations to model the number of ships possessed by each force. Assume the French-Spanish start with 33 ships and the British start with 27 ships.

b) Build a numerical solution to determine who wins under the new assumption.

CHAPTER 10

MODELING CHANGE WITH DISCRETE DYNAMICAL SYSTEMS

Chapter 10 Review

1. Summarize the important mathematical ideas of this chapter.

2. Compute the first five terms of the sequence $a_{n+1} = 3a_n + n^2$, $a_0 = 2$.

3. Write out the first three algebraic equations represented by $a_{n+1} = 5a_n + 2$, where the sequence starts with the term a_0. Express a_3 in terms of a_0.

4. You presently use a credit card that charges 1.5% interest each month. You make monthly payments of $40. Determine when you will pay off the charge card if you make no new purchases and your current debt is

   a) $1500

   b) $2800

5. The following data represent the velocity of an object. Examine the change in velocity. Build a difference equation that models the velocity.

Time t (seconds)	0	1	2	3	4	5	6	7
Velocity v (ft/sec)	0	16	48	80	112	144	176	208

6. Use a CBL to measure a falling body. Using your data, repeat Exercise 5.

7. Find the equilibrium value, if one exists, for each difference equation.

   a) $a_{n+1} = -1.1a_n$

   b) $a_{n+1} = -1.1a_n + 20$

   c) $a_{n+1} = 1.3a_n$

   d) $a_{n+1} = 1.3a_n - 30$

8. You presently use a credit card that charges 1.5% interest each month and you make monthly payments of $40. (See Exercise 4.) Build a difference equation model and find the equilibrium value. Interpret the equilibrium value in terms of the monthly interest and the monthly payment. Is the equilibrium value stable or unstable?

9. Consider the interaction of two species whose populations after n years are represented by the numbers x_n and y_n. Consider the following model.

$$x_{n+1} = 0.8x_n + 0.002x_n y_n$$

$$y_{n+1} = 0.9y_n + 0.001x_n y_n$$

$$x_0 = 50$$

$$y_0 = 100$$

   a) Interpret the magnitude of the coefficients 0.8 and 0.9. What happens to either species in the absence of the second species?

   b) Interpret the algebraic sign of the coefficients 0.001 and 0.002. Is the effect of the presence of the second species harmful?

   c) Iterate the system. Can the two species coexist?

   d) Is there an equilibrium value? Explain the significance of the equilibrium value in this context. Experiment to find out what happens in the vicinity of the equilibrium value.

10. Name several behaviors not mentioned in the text that can be modeled by a difference equation.

11. Name several behaviors not mentioned in the text that can be modeled by a system of difference equations.

Appendix A: The Laws of Exponents

Exponents are shorthand notations used to represent many factors multiplied together. All of the rules for manipulating exponents may be deduced from the laws of multiplication and division that you are already familiar with.

EXPONENTIAL NOTATION

Repeated multiplication is represented using exponential notation, for example:

$3 \times 3 \times 3 \times 3 = 3^4.$

There are four factors in the product, each of which is a 3. In the mathematical expression 3^4, 4 is called the exponent and 3 is usually called the base.

For a real number a and a positive integer n, exponentials are defined by:

$a^n = a \times a \times a \ldots a \times a,$
a total of n factors, each of which is an a.

A special case of exponential notation to note:

$a^1 = a.$

This makes sense, since a^1 should have only one factor of a.

RULES FOR COMBINING EXPONENTS

Suppose that a and b are real numbers and that m and n are positive integers.

1. $a^n \cdot a^m = a^{n+m}.$
2. $\dfrac{a^n}{a^m} = a^{n-m}.$
3. $(a^n)^m = a^{n \cdot m}.$
4. $(a \cdot b)^n = a^n \cdot b^n.$
5. $\left(\dfrac{a}{b}\right)^n = \dfrac{a^n}{b^n}.$

Justifications of these laws are based on the laws of multiplication and division that you are familiar with. Rules 1, 3, and 4 are justified below, and the ideas of the justifications for Rules 2 and 4 described. (The justifications of Rules 2 and 4 are exercises for this appendix.)

For the first exponential rule,

$a^n \cdot a^m = (a \times a \times a \ldots a \times a) \cdot (a \times a \times a \ldots a \times a)$
total of n factors total of m factors

$= a \times a \times a \ldots a \times a \times a \times a \times a \ldots a \times a$
total of $n + m$ factors
$= a^{n+m}.$

For the third exponential rule,

$(a^n)^m = (a \times a \times a \ldots a \times a) \cdot (a \times a \times a \ldots a \times a).$
$\ldots \cdot (a \times a \times a \ldots a \times a)$
total of m sets of brackets
each bracket has n copies of a
$= a \times a \times a \ldots a \times a \cdot a \times a \times a \ldots$
$a \times a \cdot \ldots \cdot a \times a \times a \ldots a \times a$

the total number of copies of a is $m \cdot n$
$= a^{mn}.$

For the fourth exponential rule,

$(a \cdot b)^n = (a \cdot b) \times (a \cdot b) \times (a \cdot b) \times \ldots \times (a \cdot b)$
n copies of $(a \cdot b)$ so there are n copies of a and n copies of b

$= (a \times a \times a \times \ldots \times a) \cdot (b \times b \times b \times \ldots \times b)$
n copies of a n copies of b

$= a^n \cdot b^n.$

The idea for justifying the second exponential rule is this: There are n copies of a in the numerator of the fraction, and m copies of a in the denominator of the fraction. Some of these copies of a will cancel out.

Lastly, the idea for the justification of the fifth exponential rule is this: There are n copies of the fraction $\frac{a}{b}$. Thus, the numerator will consist of n copies of a multiplied together, so the numerator is a^n. The denominator will consist of n copies of b multiplied together, so the denominator is b^n. (To review concepts such as numerator, denominator, and multiplication of fractions see Appendix E.)

Many of these rules also work when m and n are not positive integers—for example, if m and n are fractions. Some special considerations apply when m and n are not positive integers in order to make sure that the expression makes good mathematical sense. For example, while it is possible to write down a collection of symbols like:

$$(-2)^{1/2} = \sqrt{-2}.$$

The collection of symbols $(-2)^{1/2}$ should represent the number that is the square root of negative two. However, since squaring any real number gives a result that is greater than or equal to zero, there is no real number that, when squared, gives –2. While it is possible to write down the collection of symbols: $(-2)^{1/2}$, there is no real number that equals this collection of symbols.

EXAMPLE A.1

Use the laws of exponents to simplify the following algebraic expressions when possible.

a) $4b \cdot (3b)^5$.

b) $(2x^3)^2 \cdot \left(\frac{3}{x^2}\right)^4$.

SOLUTION:

a) $4b \cdot (3b)^5 = 4b \cdot 3^5 \cdot b^5 = 4 \cdot 243 \cdot b \cdot b^5 = 972 \cdot b^6$.

b) $(2x^3)^2 \cdot \left(\frac{3}{x^2}\right)^4 = 2^2 \cdot (x^3)^2 \cdot \left(\frac{3^4}{(x^2)^4}\right) =$

$4 \cdot 81 \cdot x^6 \cdot \frac{1}{x^8} = 324 \cdot x^{6-8} = 324 \cdot x^{-2}$.

Part (b) illustrates an interesting point—the application of the exponent rules can generate algebraic expressions such as x^{-2}, which we have no way of interpreting at this point.

INTERPRETING AND SIMPLIFYING FRACTIONAL AND NEGATIVE EXPONENTS

The way that we defined exponential notation as a shorthand for repeated multiplication makes complete sense for exponents that are positive integers. However, exactly what an expression like x^{-2} should mean is not so clear.

Let a be a real number, and let m and n be positive integers.

$a^0 = 1$ (provided that $a \neq 0$)

$a^{-n} = \frac{1}{a^n}$ (provided that $a \neq 0$)

$a^{1/n} = \sqrt[n]{a}$ (if n is an even integer, then this only makes sense when $a \geq 0$)

$a^{m/n} = \sqrt[n]{a^m} = \left(\sqrt[n]{a}\right)^m$ (if n is an even integer, then this only makes sense when $a \geq 0$)

EXAMPLE A.2

Using the laws of exponents, simplify the following expressions as much as possible.

a) $\left(\frac{3^{-1} \cdot L}{L^{-2}}\right)^{-3}$.

b) $(2w^2y^3)^{-1} \cdot \frac{1}{y^{-8}}$.

c) $(a^2 + r^2)^2$.

SOLUTION:

a) $\left(\frac{3^{-1} \cdot L}{L^{-2}}\right)^{-3} = \frac{(3^{-1} \cdot L)^{-3}}{(L^{-2})^{-3}} = \frac{(3^{-1})^{-3} \cdot L^{-3}}{L^6}$

$3^3 \cdot L^{-3-6} = 27 \cdot L^{-9}$.

b) $(2w^2y^3)^{-1} \cdot \frac{1}{y^{-8}} = 2^{-1} \cdot (w^2)^{-1} \cdot (y^3)^{-1} \cdot \frac{1}{y^{-8}} =$

$\frac{1}{2} \cdot w^{-2} \cdot y^{-3} \cdot \frac{1}{y^{-8}} = \frac{1}{2} \cdot w^{-2} \cdot y^{-3-(-8)} = \frac{1}{2} \cdot w^{-2} \cdot y^5$

c) $(a^2 + r^2) = (a^2 + r^2) \cdot (a^2 + r^2) = a^4 + 2a^2r^2 + r^4$.

EXAMPLE A.3

a) $\sqrt[3]{8w^{18}}$.

b) $\sqrt{\frac{64k^2}{9T^9}}$.

c) $\sqrt[n]{a^n}$.

SOLUTION:

a) $\sqrt[3]{8w^{18}} = \sqrt[3]{8} \cdot \sqrt[3]{w^{18}} = 2 \cdot w^6$.

b) $\sqrt{\frac{64k^2}{9T^9}} = \frac{\sqrt{64k^2}}{\sqrt{9T^9}} = \frac{\sqrt{64} \cdot \sqrt{k^2}}{\sqrt{9} \cdot \sqrt{T^9}} = \frac{8k}{3T^{9/2}}$.

c) The simplification of $\sqrt[n]{a^n}$ is not completely straightforward. If a is a positive number or zero, then

$$\sqrt[n]{a^n} = (a^n)^{1/n} = a^1 = a$$

as you might expect. If a is a negative number, then the situation is more complicated. If n is an odd number, then $\sqrt[n]{a^n} = a$. To illustrate this consider the case where $n = 3$ and $a = -2$,

$$\sqrt[3]{(-2)^3} = \sqrt[3]{-8} = -2.$$

If n is an even number, then $\sqrt[n]{a^n} = |a|$. To illustrate this consider the case where $n = 2$ and $a = -2$,

$$\sqrt{(-2)^2} = \sqrt{4} = 2.$$

Exercises for Appendix A

For Problems 1–10, evaluate the quantity (if possible) without using a calculator.

1. 7^2.
2. 3^0.
3. $\frac{9^2}{9^3}$.
4. $\sqrt{9^9}$.
5. $(-1)^4$.
6. $(-1)^5$.
7. $\frac{(12)^3}{(-12)^3}$.
8. $\sqrt[3]{8}$.
9. $\sqrt[3]{-8}$.
10. $\sqrt{(-8)^2}$.

For Problems 11–20, simplify the expression as much as possible.

11. $e \cdot e^{2t} \cdot e^t \cdot 2^e$.
12. $\frac{x^{2n+1} \cdot (2y)^{n+1}}{(x^2 \cdot y)^n}$.
13. $(a-b)^{5/2} \cdot \sqrt{a+b}$.
14. $(T^2 \cdot w^4)^{1/2}$.
15. $e \cdot \sqrt[3]{e^{3r}}$.
16. $\left(\frac{2w^2 \cdot y^{3/2}}{\sqrt{w+y}}\right)^2$.
17. $(C \cdot e^{wt})^2$.
18. $(5xy)^{-1} \cdot (xy^2)^2$.
19. $\sqrt{36u^2v}$.
20. $\frac{8P^{-2}}{4P^{-4}}$.

For Problems 21–25, decide whether each of the statements are true or false.

21. $(a+b)^2 = a^2 + b^2$.
22. $10q^{-2} = \frac{1}{10q^2}$.
23. $\frac{1}{a+b} = \frac{1}{a} + \frac{1}{b}$.
24. $x^n \cdot x^m = x^{n+m}$.
25. $z^2 + z^2 = z^4$.

In Problems 26 and 27 you will justify the last two exponential rules.

26. Suppose that a is a real number, and that n and m are positive integers. Explain why:
$$\frac{a^n}{a^m} = a^{n-m}.$$
27. Suppose that a and b are real numbers, and that n is a positive integer. Explain why:
$$\left(\frac{a}{b}\right)^n = \frac{a^n}{b^n}.$$

Appendix B: Expanding Algebraic Expressions

Algebraic expressions often involve parentheses (i.e. brackets). For example,

$x(1 + x)$ or $(x + 3)(x + 7)$.

This appendix concentrates on multiplying out algebraic expressions that involve parentheses. This operation is often described as expanding the brackets.

THE DISTRIBUTIVE PROPERTY

The distributive property of real numbers can be expressed through a pair of equations:

$$a \cdot (b + c) = a \cdot b + a \cdot c$$

$$(a + b) \cdot c = a \cdot c + b \cdot c,$$

where a, b ,and c are real numbers. The distributive property tells you how to multiply out algebraic expressions involving parentheses.

EXAMPLE B.1

Expand the following expressions and simplify as much as you can. (Use the Laws of Exponents to simplify.)

a) $x \cdot (1 + x)$.

b) $(2^x + x) \cdot x^2$.

c) $\sqrt{x} \cdot (x^2 + 1)$.

SOLUTION:

a) $x \cdot (1 + x) = (x \cdot 1) + (x \cdot x) = x + x^2$.

b) $(2^x + x) \cdot x^2 = (2^x \cdot x^2) + (x \cdot x^2) = 2^x \cdot x^2 + x^3$.

c) $\sqrt{x} \cdot (x^2 + 1) = \sqrt{x} \cdot x^2 + \sqrt{x} \cdot 1 =$
$x^{1/2} \cdot x^2 + \sqrt{x} = x^{5/2} + \sqrt{x}$.

(See Appendix E for adding fractions.)

EXAMPLE B.2

Expand the following expressions and simplify as much as you can.

a) $(x + 3)^2$.

b) $(x - 3)^2$.

c) $(x + 3) \cdot (x - 3)$.

SOLUTION:

a) $(x + 3)^2 = (x + 3)(x + 3) =$
$x \cdot (x + 3) + 3 \cdot (x + 3) =$
$(x \cdot x) + (x \cdot 3) + (3 \cdot x) + (3 \cdot 3) =$
$x^2 + 3x + 3x + 9 = x^2 + 6x + 9$.

b) $(x - 3)^2 = (x - 3)(x - 3) =$
$x \cdot (x - 3) - 3 \cdot (x - 3) =$
$(x \cdot x) - (x \cdot 3) - (3 \cdot x) + (3 \cdot 3) =$
$x^2 - 3x - 3x + 9 = x^2 - 6x + 9$.

c) $(x + 3)(x - 3) = x \cdot (x - 3) + 3 \cdot (x - 3) =$
$(x \cdot x) - (x \cdot 3) + (3 \cdot x) - (3 \cdot 3) =$
$x^2 - 3x + 3x - 9 = x^2 - 9$.

The three algebraic expressions from Example B.2 illustrate three important special cases of expanding brackets. These three special cases will be used to help with factoring algebraic expressions in Appendix C. In general, for real numbers a and b:

$(a + b)^2 = a^2 + 2ab + b^2$.
A "perfect square."

$(a - b)^2 = a^2 - 2ab + b^2$.
A "perfect square."

$(a + b)(a - b) = a^2 - b^2$.
The "difference of two squares."

When each factor includes two terms, the product normally contains four terms:

$$(a + b)(c + d) = ac + ad + bc + bd,$$

although as Example B.2 illustrates, sometimes "like" terms can be combined.

EXAMPLE B.3

Expand the following expressions and simplify as much as you can. If you can combine like terms, indicate which terms you can combine.

a) $(x + 3)(x + 7)$.

b) $(3^x + 7)(\frac{1}{3^x} + 2)$.

c) $(4 - 3x)^2$.

SOLUTION:

a) $(x + 3)(x + 7) = x^2 + 3x + 7x + 21 = x^2 + 10x + 21$. The two terms "$3x$" and "$7x$" have the same power of x and may be combined.

b) $(3^x + 7)(\frac{1}{3^x} + 2) = \frac{3^x}{3^x} + \frac{7}{3^x} + 3^x \cdot 2 + 14 =$
$1 + \frac{7}{3^x} + 3^x \cdot 2 + 14 = \frac{7}{3^x} + 3^x \cdot 2 + 15$.
The two constant terms may be combined. (See Appendix E for more information on multiplying fractions.)

c) $(4 - 3x)^2 = (4 - 3x)(4 - 3x) = 16 - 12x - 12x + 9x^2 = 16 - 24x + 9x^2$.
The two terms "$-12x$" have like powers of x and may be combined.

Exercises for Appendix B

For Problems 1–5, simplify the quantity as much as possible.

1. $-(t + 2) + 3(1 - 2t)$.
2. $(y + 9) - 2(2 - y)$.
3. $7(h + 1) + (7h + 1)$.
4. $(x + 3)6 + 3(x + 6)$.
5. $-s + 1 + -1(s + 1)$.

For Problems 6–15, expand each of the following expressions as much as possible.

6. $x(x^2 + 1)$.
7. $9y^2(2y + 3)$.
8. $4(r + 4)^2 + 1$.
9. $(2w - 1)(4p + 1)$.
10. $(2w - 1)(2w - 1)$.
11. $6y(x - y) + 3(y + 1)^2$.
12. $9w(w - w^2) - 9w^2$.
13. $(z + 2)(y + 3)$.
14. $-6(x + 1)^2$.
15. $(5u - 1)(5u + 1)$.

For Problems 16–20, simplify by expanding and collecting like terms.

16. $(t^2 + 1)(t + 1) - t^3$.
17. $(r + 1)\left(\frac{r^2}{2} + 1\right)$.
18. $(x^2 - y^2)(x + y)$.
19. $(x^2 - y^2)^2$.
20. $\left(\frac{e^u + e^{-u}}{2}\right)^2 - \left(\frac{e^u - e^{-u}}{2}\right)^2$.

Appendix C: Factoring Algebraic Expressions

"Factoring" algebraic equations is the reverse of "expanding" algebraic expressions discussed in Appendix B. Factoring algebraic equations can be a great help when trying to find solutions of equations.

EXAMPLE C.1

Find all of the real numbers x that satisfy the algebraic equation:

$$x^2 - 2x + 1 = 0.$$

SOLUTION:

One possible approach is just to guess a value of x and plug this guess into the equation to check. For example, if you guess $x = 2$, plugging in gives:

$$2^2 - 2 \cdot 2 + 1 = 4 - 4 + 1 = 1.$$

This is not equal to zero, so $x = 2$ is not a solution of the algebraic equation. The next step would be to guess another value for x and then check this.

Here is an alternative to find the value(s) of x that satisfy the algebraic expression. This method is based on realizing that $x^2 - 2x + 1$ is a "perfect square:"

$$x^2 - 2x + 1 = (x - 1)^2.$$

So, the algebraic expression may be written as:

$$(x - 1)^2 = 0.$$

Taking the square root of each side of this:

$$x - 1 = 0,$$

so $x = 1$. (See Appendix F for more information on solving linear equations.)

The crucial step in the second approach of Example C.1 was recognizing that $x^2 - 2x + 1$ was the same as the product of factors: $(x - 1)(x - 1)$. The operation of converting the "expanded" expression, $x^2 - 2x + 1$, into a product of two factors is called "factoring" the algebraic expression.

Not every algebraic expression can be factored, and factoring is not always a straightforward process. In the remainder of this appendix, we will outline four strategies that can help you to factor algebraic expressions.

STRATEGY 1: COMMON FACTORS

Often, all terms in an expression will have a common factor. A useful simplification can be to extract this factor from each term in the expression. This operation can be thought of as the distributive law in reverse.

EXAMPLE C.2

Factor each of the following expressions. If you are able to find a common factor, say what that factor is.

a) $x^{3/2} + x + \sqrt{x}$.

b) $e^{2t} + e^t + e^{\sin(t)+1}$.

c) $y^3 + 2y^2 + y$.

SOLUTION:

a) $x^{3/2} + x + \sqrt{x} = \sqrt{x} \cdot (x + \sqrt{x} + 1)$.
The common factor is $\sqrt{x}$.

b) $e^{2t} + e^t + e^{\sin(t)+t} = e^t \cdot (e^t + 1 + e^{\sin(t)})$.
The common factor is e^t.

c) $y^3 + 2y^2 + y = y \cdot (y^2 + 2y + 1)$.
The common factor is y.

STRATEGY 2: GROUPING LIKE TERMS

Many algebraic expressions do not have a common factor that is shared by all terms. However, some of the terms may have a common factor. It can be useful to group these "like" terms and extract the common factor from them.

EXAMPLE C.3

Factor each of the following expressions.

a) $e^{2x} + x^2 + x \cdot e^x$.

b) $e^{2x} + x^2 + 2 \cdot x \cdot e^x$.

c) $A(1 + e^x) + A(1 + e^x) \cdot e^x$.

SOLUTION:

a) $e^{2x} + x^2 + x \cdot e^x =$
$e^x \cdot (e^x + x) + x^2 =$
$e^{2x} + x \cdot (x + e^x)$.

b) $e^{2x} + x^2 + 2 \cdot x \cdot e^x =$
$e^{2x} + x \cdot e^x + x^2 + x \cdot e^x =$
$e^x \cdot (e^x + x) + x \cdot (x + e^x) =$
$(e^x + x) \cdot (e^x + x)$.

c) $A(1 + e^x) + A(1 + e^x) \cdot e^x =$
$A(1 + e^x) \cdot (1 + e^x)$.

STRATEGY 3: PERFECT SQUARES

In Appendix B, several special cases of multiplying brackets were noted. Two of these are the "perfect squares:"

$$(a + b)^2 = a^2 + 2ab + b^2$$

$$(a - b)^2 = a^2 - 2ab + b^2.$$

EXAMPLE C.4

Factor each of the following expressions.

a) $r^2 - 4r + 4$.

b) $x + 2\sqrt{x} + 1$.

c) $2^{2x} + 2^{x+1} + 1$.

Solution:

a) $r^2 - 4r + 4 = (r + 2)^2$.

b) $x + 2\sqrt{x} + 1 = (\sqrt{x} + 1)^2$.

c) $2^{2x} + 2^{x+1} + 1 = (2^x)^2 + 2 \cdot 2^x + 1 = (2^x + 1)^2$.

STRATEGY 4: DIFFERENCES OF SQUARES

The last special case of multiplying brackets from Appendix B was the "difference of two squares:"

$$(a + b)(a - b) = a^2 - b^2.$$

EXAMPLE C.5

Factor each of the following expressions.

a) $t^2 - 81$.

b) $t^4 - 16$.

c) $\sin^4(x) - \cos^4(x)$.

SOLUTION:

a) $t^2 - 81 = (t - 9)(t + 9)$.

b) $t^4 - 16 = (t^2 - 4)(t^2 + 4) = (t - 2)(t + 2)(t^2 + 4)$.

c) $\sin^4(x) - \cos^4(x) =$
$(\sin^2(x) - \cos^2(x))(\sin^2(x) + \cos^2(x)) =$
$(\sin(x) - \cos(x))(\sin(x) + \cos(x))$.

Note that in part (c), the trigonometric identity, $\cos^2(x) + \sin^2(x) = 1$ was used to simplify the expression.

FACTORING QUADRATIC EXPRESSIONS

As indicated by Example C.1, being able to factor quadratic expressions can be a very useful tool for solving equations. In Appendix H you will use inequalities to determine the sign of an algebraic expression. In such a situation, factoring the algebraic expression can also help with the mathematical analysis (see Example H.4).

EXAMPLE C.6

Factor: $r^2 + 12r + 32$.

SOLUTION:

We are trying to find two numbers, say a and b, so that:

$$(r + a)(r + b)$$

will multiply out to give: $r^2 + 12r + 32$. From Appendix B, $(r + a)(r + b)$ multiplies out to give:

$$(r + a)(r + b) = r^2 + ar + br + ab = r^2 + (a + b)r = ab.$$

Comparing this algebraic expression with $r^2 + 12r + 32$, you are looking for a and b so that:

$$a + b = 12, \text{ and } ab = 32.$$

Two numbers that do this are: $a = 8$ and $b = 4$. Thus:

$$r^2 + 12r + 32 = (r + 8)(r + 4).$$

Example C.7

Factor: $e^{2t} + 3e^t + 2$.

Solution:

The laws of exponents from Appendix A give that: $e^{2t} = (e^t)^2$. Using this, the algebraic expression that you have to factor begins to resemble a quadratic expression:

$$e^{2t} + 3e^t + 2 = (e^t)^2 + 3e^t + 2.$$

If you write r instead of e^t, this expression looks just like a quadratic expression, which can be factored just as in Example C.6:

$$e^{2t} + 3e^t + 2 = (e^t)^2 + 3e^t + 2 = r^2 + 3r + 2 = (r + 1)(r + 2).$$

Converting back by substituting e^t for r gives:

$$e^{2t} + 3e^t + 2 = (e^t + 1)(e^t + 2).$$

Exercises for Appendix C

For Problems 1–15, factor the quantity as much as possible.

1. $x - x^2 + 1$.
2. $x^2 - 3x + 2$.
3. $x \cdot \ln(x) - x$.
4. $x \cdot \sin(x) + x \cdot \cos(x)$.
5. $c^2x + d^2y$.
6. $2x + 4y^2$.
7. $a^2 - 4$.
8. $A \cdot e^{2t} + A \cdot t \cdot e^{2t} + A \cdot t^2 \cdot e^{2t}$.
9. $p(p + q) - q(p + q)$.
10. $(1 + 2t)^4 - w^2$.
11. $a^2 + 2ac + 2c^2$.
12. $4x + 16$.
13. $1 - \cos^2(u)$.
14. $Ar^2(1 + r) + 2A(1 + r)r + A(1 + r)$.
15. $x^2 + h^2 + 2hx - h^2$.

For Problems 16–20, factor the quadratic expression.

16. $x^2 + 6x + 8$.
17. $3x^2 + 9x + 6$.
18. $e^{4t} + 2e^{2t} + 1$.
19. $y^2 + 7y + 12$.
20. $t^2 + t + 2$.

Appendix D: Completing the Square and the Quadratic Formula

Factoring quadratic expressions such as:

$x^2 + 6x + 8$

was one of the topics introduced in Appendix C. Factoring quadratic expressions is a useful skill that can help you to find the solutions of equations. However, quadratics are not always easy to factor. Sometimes quadratics cannot be factored completely. Two procedures that can help you to factor a quadratic are "completing the square" and the quadratic formula.

COMPLETING THE SQUARE

In Appendix A, two special cases of expanding brackets were considered:

$(a + b)^2 = a^2 + 2ab + b^2$

$(a - b)^2 = a^2 - 2ab + b^2.$

These were called "perfect squares." In Appendix C, two strategies suggested for factoring algebraic expressions were to look for perfect squares and to look for differences of squares.

EXAMPLE D.1

Factor the quadratic expression:

$x^2 + 6x + 8$

by looking for perfect squares and differences of squares.

SOLUTION:

The given algebraic expression is very close to the perfect square:

$(x + 3)^2 = x^2 + 6x + 9.$

So: $x^2 + 6x + 8 = (x^2 + 6x + 9) - 1 = (x + 3)^2 - 1.$

Since $1 = 1^2$, the last expression is a difference of two squares, and so:

$x^2 + 6x + 8 = (x + 3)^2 - 1^2 =$
$(x + 3 + 1)(x + 3 - 1) = (x + 4)(x - 2).$

Example D.1 is a very elaborate and counter-intuitive way to factor a straightforward algebraic expression like $x^2 + 6x + 8$. The process outlined in Example D.1 has one great advantage over quicker, more intuitive methods: the method of Example D.1 will work when the numbers involved are more complicated and it is harder to factor the quadratic expression.

EXAMPLE D.2

Factor the quadratic expression:

$x^2 + 66x + 8$

by looking for perfect squares and differences of squares.

SOLUTION:

The perfect square that this quadratic expression is related to is:

$(x + 33) = x^2 + 66x + 1089.$

The quadratic expression that we have to factor is related to this perfect square:

$x^2 + 66x + 8 = (x^2 + 66x + 1089) - 1081 =$
$(x + 33)^2 - 1081.$

This last algebraic expression is a difference of two squares, so:

$x^2 + 66x + 8 =$
$(x + 33 + 32.879)(x + 33 - 32.879) =$
$(x + 65.879)(x + 0.121).$

The number 32.879 appears because the square root of 1081 is approximately 32.879.

The process of factoring quadratic equations illustrated in Examples D.1 and D.2 is called "completing the square." As indicated, not all quadratics can be completely factored. The process of completing the square can tell you when this is the case.

Example D.3

Try to factor the quadratic expression:

$x^2 + 6x + 10$

by looking for perfect squares and differences of squares.

Solution:

Following the pattern set out in Examples D.1 and D.2, you could start by finding a perfect square that is closely related to $x^2 + 6x + 10$. The perfect square:

$(x + 3)^2 = x^2 + 6x + 9$

is very closely related to $x^2 + 6x + 10$. In fact,

$x^2 + 6x + 10 = (x^2 + 6x + 9) + 1 = (x + 3)^2 + 1.$

This is where the pattern established in Examples D.1 and D.2 breaks down, because the expression that we have is not a difference of two squares. The expression $(x + 3)^2 + 1$ cannot be factored any further because it is the sum rather than the difference of two squares.

THE QUADRATIC FORMULA

The process of factoring a quadratic expression by completing the square can be summarized as an algebraic formula. The formula is this:

If you are trying to factor the quadratic expression:

$ax^2 + bx + c,$

then the factors are:

$$a\left(x + \frac{b + \sqrt{b^2 - 4ac}}{2a}\right)\left(x + \frac{b - \sqrt{b^2 - 4ac}}{2a}\right).$$

There is one important caveat when using this method to factor a polynomial: because you have to take a square root, the quantity $b^2 - 4ac$ must be greater than or equal to zero. If $b^2 - 4ac$ is negative, then it is not possible to factor the quadratic expression $ax^2 + bx + c$.

Example D.4

Use the quadratic formula to factor the following quadratic expressions. If it is not possible to factor any of the quadratic expressions, indicate why you think this to be the case.

a) $6x^2 + 9x + 2.$

b) $x^2 + x + 1.$

c) $x^2 + 2x + 1.$

Solution:

a) To factor $6x^2 + 9x + 2$, note the similarities between this expression and $ax^2 + bx + c$. The correspondence is that $a = 6$, $b = 9$, and $c = 2$.

Before plugging into the formula, it is wise to make sure that the quantity $b^2 - 4ac$ is greater than or equal to zero. (If the quantity $b^2 - 4ac$ is negative, then the quadratic expression cannot be factored, so there will be no sense in trying.)

$b^2 - 4ac = 9^2 - 4 \cdot 6 \cdot 2 = 81 - 48 = 33.$

The number 33 is greater than or equal to zero, so the quadratic will factor. Plugging $a = 6$, $b = 9$, and $c = 2$ into the formula gives:

$6x^2 + 9x + 2 = 6(x + 1.2287)(x + 0.2713).$

b) To factor $x^2 + x + 1$, again note the similarities between this expression and $ax^2 + bx + c$. In this case, the correspondence is that $a = 1$, $b = 1$, and $c = 1$.

To check whether or not the expression $x^2 + x + 1$, can be factored, you can check the sign of $b^2 - 4ac$. In this case,

$b^2 - 4ac = 1 - 4 \cdot 1 \cdot 1 = 1 - 4 = -3.$

Since $b^2 - 4ac$ is negative, the quadratic expression $x^2 + x + 1$ cannot be factored.

c) Here the analysis is just like the previous two cases, except that $a = 1$, $b = 2$, and $c = 1$.

Checking the quantity $b^2 - 4ac$ gives

$$b^2 - 4ac = 4 - 4 \cdot 1 \cdot 1 = 0.$$

As this is not negative, the quadratic expression $x^2 + 2x + 1$ will factor. Plugging $a = 1$, $b = 2$, and $c = 1$ into the formula gives:

$$(x + 1)(x + 1) = (x + 1)^2.$$

An alternative (and equally valid) way to factor this particular quadratic expression would have been to realize that $x^2 + 2x + 1$ was a perfect square, so that you can determine that $x^2 + 2x + 1 = (x + 1)^2$ without having to use the formula.

USING COMPLETING THE SQUARE TO OBTAIN THE QUADRATIC FORMULA

You might wonder how the formula for the factors of the quadratic expression $ax^2 + bx + c$ was obtained. The working shown below indicates how the process of completing the square may be used to obtain the formula. The working presented below is quite formidable because it features a lot of symbols, rather than just concrete numbers. Use the explanatory notes for each step to follow what is going on.

$$ax^2 + bx + c$$

$$= a\left(x^2 + \frac{b}{a}x + \frac{c}{a}\right)$$

Factor out the a.

$$= a\left(x^2 + 2\left(\frac{b}{2a}\right)x + \left(\frac{b}{2a}\right)^2 - \left(\frac{b}{2a}\right)^2 + \frac{c}{a}\right)$$

Make the expression as much like the perfect square: $\left[x + \frac{b}{2a}\right]^2$ as possible.

$$= a\left(\left[x + \frac{b}{2a}\right]^2 - \left[\left(\frac{b}{2a}\right)^2 - \frac{c}{a}\right]\right)$$

Factor the perfect square and group the leftover terms together.

$$= a\left(\left[x + \frac{b}{2a}\right]^2 - \left[\frac{b^2}{4a^2} - \frac{c}{a}\right]\right)$$

Simplify the term: $\left(\frac{b}{2a}\right)^2 = \frac{b^2}{4a^2}$.

$$= a\left(\left[x + \frac{b}{2a}\right]^2 - \frac{b^2 - 4ac}{4a^2}\right)$$

Put the two terms over a common denominator. (See Appendix E.) Realize that what you have is a difference of two squares.

$$= a\left(x + \frac{b}{2a} + \sqrt{\frac{b^2 - 4ac}{4a^2}}\right)\left(x + \frac{b}{2a} - \sqrt{\frac{b^2 - 4ac}{4a^2}}\right)$$

Use the difference of squares to factor.

$$= a\left(x + \frac{b}{2a} + \frac{\sqrt{b^2 - 4ac}}{2a}\right)\left(x + \frac{b}{2a} - \frac{\sqrt{b^2 - 4ac}}{2a}\right)$$

Take the square roots of the numerators and denominators.

$$= a\left(x + \frac{b + \sqrt{b^2 - 4ac}}{2a}\right)\left(x + \frac{b - \sqrt{b^2 - 4ac}}{2a}\right)$$

Combine like terms.

Exercises for Appendix D

For Problems 1–10, complete the square for the given quadratic expressions.

1. $y^2 + 2y$.
2. $2a^2 + 8a + 5$.
3. $u^2 - 14u$.
4. $x^4 + 4x^2 + 1$.
5. $3r^2 + 12r + 16$.
6. $-x^2 + 10x + 1$.
7. $t^2 - 7t - 8$.
8. x^2.
9. $3r^2 - 24r + 14$.
10. $w^2 + 3w$.

For Problems 11–15, factor the quadratic expressions (if possible).

11. $r^2 + 7r + 12$.
12. $2y^2 + 5y + 3$.
13. $e^{2x+1} + 4e^{x+1} + 4e$.
14. $3t^2 + 13t + 9$.
15. $-u^2 + 4u - 5$.

Appendix E: Manipulating Fractions

The rules for manipulating fractions that involve algebraic expressions are exactly the same as the rules for manipulating fractions that involve numbers.

The fundamental rules for combining and manipulating fractions are listed below. The uses of these rules are illustrated more completely later in this appendix.

a) Adding fractions

$$\frac{a}{c}+\frac{b}{d}=\frac{a\cdot d+b\cdot c}{c\cdot d}$$

b) Subtracting fractions

$$\frac{a}{c}-\frac{b}{d}=\frac{a\cdot d-b\cdot c}{c\cdot d}$$

c) Multiplying fractions

$$\frac{a}{c}\cdot\frac{b}{d}=\frac{a\cdot b}{c\cdot d}$$

d) Dividing fractions

$$\frac{a/c}{b/d}=\frac{a}{c}\cdot\frac{d}{b}=\frac{a\cdot d}{b\cdot c}$$ "Invert and multiply."

NUMERATOR AND DENOMINATOR

Fractions express a ratio of two quantities. For example, the fraction $\frac{a}{b}$ expresses the ratio of quantity a to quantity b. The quantity that appears on the top of the fraction is called the numerator. In this case, the numerator is a. The quantity that appears on the bottom of the fraction is called the denominator. In this case, the denominator is b.

FINDING A COMMON DENOMINATOR

Adding and subtracting fractions usually requires a common denominator, that is, all of the fractions involved have the same denominator.

EXAMPLE E.1

Find common denominators for the following collections of fractions. Express the fractions using this common denominator.

a) $\frac{1}{2}, \frac{1}{x}$.

b) $\frac{1}{x^2-x}, \frac{1}{x}$.

c) $\frac{1}{x^2+2x+1}, \frac{1}{x+1}$.

SOLUTION:

a) The common denominator is $2x$. The two fractions can be expressed by multiplying both numerator and denominator (top and bottom) by whatever factor is needed to convert the denominator to the common denominator.

$$\frac{1}{2}=\frac{1}{2}\cdot\frac{x}{x}=\frac{x}{2x}$$

$$\frac{1}{x}=\frac{1}{x}\cdot\frac{2}{2}=\frac{2}{2x}.$$

b) A common denominator can always be obtained by just multiplying each of the denominators together. A possible common denominator is

$x\cdot(x^2-x)=x^3-x^2$.

Expressing the two fractions using this common denominator:

$$\frac{1}{x^2-x}=\frac{1}{x^2-x}\cdot\frac{x}{x}=\frac{x}{x^3-x^2}$$

$$\frac{1}{x}=\frac{1}{x}\cdot\frac{x^2-x}{x^2-x}=\frac{x^2-x}{x^3-x^2}.$$

Note, however, that $x^2-x=x\cdot(x-1)$. The significance of this observation is that since x^2-x already has the denominators of both $\frac{1}{x^2-x}$ and $\frac{1}{x}$ as factors. So, x^2-x is already a common denominator for both of these fractions and could be used instead of x^3-x^2.

c) One way to obtain a common denominator is to just multiply the denominators of the two fractions $\frac{1}{x^2+2x+1}, \frac{1}{x+1}$ together. However, $x^2+2x+1=(x+1)^2$. Since $x+1$ is a factor of x^2+2x+1, x^2+2x+1 can be used as a common denominator. Expressing the two fractions using the common denominator x^2+2x+1:

$$\frac{1}{x^2+2x+1}$$

$$\frac{1}{x+1} = \frac{1}{x+1} \cdot \frac{x+1}{x+1} = \frac{x+1}{x^2+2x+1}.$$

The advantage of using $x^2 + 2x + 1$ as a common denominator (rather than, say, the more obvious $(x + 1) \cdot (x^2 + 2x + 1) = x^3 + 3x^2 + 3x + 1$) is that the fractions that you obtain using the common denominator are usually the simplest possible.

ADDING AND SUBTRACTING FRACTIONS

Fractions can only be added and subtracted when the two fractions have the same denominator. If the two fractions do not have the same denominator, then a common denominator must be found before the fractions can be added or subtracted. The necessity of finding a common denominator is why the product of the two denominators (i.e. $b \cdot d$) appears in the rules expressed below. Expressed using algebraic symbols, the rules for adding and subtracting functions are:

$$\frac{a}{c} + \frac{b}{d} = \frac{a}{c} \cdot \frac{d}{d} + \frac{b}{d} \cdot \frac{c}{c} = \frac{a \cdot d + b \cdot c}{c \cdot d}$$

$$\frac{a}{c} - \frac{b}{d} = \frac{a}{c} \cdot \frac{d}{d} + \frac{b}{d} \cdot \frac{c}{c} = \frac{a \cdot d - b \cdot c}{c \cdot d}.$$

EXAMPLE E.2

Evaluate and simplify each of the algebraic expressions.

a) $\frac{1}{x+h} - \frac{1}{x}$.

b) $\frac{(x+h)^2}{h} - \frac{x}{h}$.

c) $\frac{\sqrt{x}}{x+1} + \frac{x+1}{\sqrt{x}}$.

d) $\frac{r+1}{r^2+5r+6} + \frac{r+2}{r^2+4r+3}$.

SOLUTION:

a) $\frac{1}{x+h} - \frac{1}{x} = \frac{x-(x+h)}{x(x+h)} = \frac{-h}{x^2+xh}$.

b) $\frac{(x+h)^2}{h} - \frac{x}{h} = \frac{x^2+2hx+h^2-x}{h} = \frac{x^2+(2h-1)x+h^2}{h}$.

(The two fractions already have the same denominator.)

c) $\frac{\sqrt{x}}{x+1} + \frac{x+1}{\sqrt{x}} = \frac{(\sqrt{x})\cdot(\sqrt{x})+(x+1)\cdot(x+1)}{(x+1)\cdot\sqrt{x}} = \frac{x+x^2+2x+1}{(x+1)\cdot\sqrt{x}} = \frac{x^2+3x+1}{(x+1)\cdot\sqrt{x}}$.

d) $\frac{r+1}{r^2+5r+6} + \frac{r+2}{r^2+4r+3} = \frac{(r+1)}{(r+2)(r+3)} + \frac{(r+2)}{(r+1)(r+3)} = \frac{(r+1)^2+(r+2)^2}{(r+1)(r+2)(r+3)}$.

Some further simplifications are possible in part (d). For example, you could multiply out all of the brackets. The important point in part (d), however, is that by using the simplest (or "least") common denominator, the algebraic form of the result will be as simple as possible. In the case of part (d), the simplest common denominator is $(r + 1)(r + 2)(r + 3)$, rather than the more obvious common denominator $(r + 1)(r + 2)(r + 2)(r + 3)$.

MULTIPLYING FRACTIONS

Multiplying two fractions is perhaps the most straightforward of all operations. You simply multiply the numerators and multiply the denominators. Expressed using algebraic symbols, this rule is:

$$\frac{a}{c} \cdot \frac{b}{d} = \frac{a \cdot b}{c \cdot d}.$$

EXAMPLE E.3

Evaluate and simplify the following fractions as much as possible.

a) $\frac{2y}{y+1} \cdot \frac{\sqrt{y}}{y+2}$.

b) $\frac{\sin(\theta)+\cos(\theta)}{1+\theta} \cdot \frac{\sin(\theta)+\cos(\theta)}{1-\theta}$.

SOLUTION:

a) $\frac{2y}{y+1} \cdot \frac{\sqrt{y}}{y+2} = \frac{2y \cdot \sqrt{y}}{(y+1)\cdot(y+2)} = \frac{2y^{3/2}}{y^2+3y+2}$.

b) $\frac{\sin(\theta)+\cos(\theta)}{1+\theta} \cdot \frac{\sin(\theta)+\cos(\theta)}{1-\theta} = \frac{\sin^2(\theta)+2\sin(\theta)\cos(\theta)+\cos^2(\theta)}{1-\theta^2}$.

DIVIDING FRACTIONS

Many people remember the rule for dividing one fraction by another by remembering that you must "invert the denominator and multiply." Expressed as algebraic symbols, the rule is:

$$\frac{a/c}{b/d} = \frac{a}{c} \cdot \frac{d}{b} = \frac{a \cdot d}{b \cdot c}.$$

EXAMPLE E.4

Evaluate or simplify each of the fractions given below.

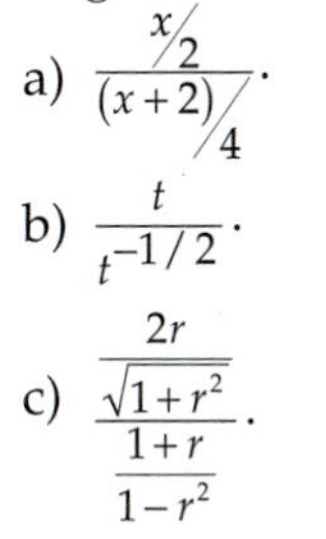

a) $\dfrac{x/2}{(x+2)/4}$.

b) $\dfrac{t}{t^{-1/2}}$.

c) $\dfrac{\frac{2r}{\sqrt{1+r^2}}}{\frac{1+r}{1-r^2}}$.

SOLUTION:

a) $\dfrac{x/2}{(x+2)/4} = \dfrac{x}{2} \cdot \dfrac{4}{x+2} = \dfrac{2x}{x+2}$.

b) $\dfrac{t}{t^{-1/2}} = \dfrac{t}{\frac{1}{\sqrt{t}}} = t \cdot \dfrac{\sqrt{t}}{1} = t^{3/2}$.

c) $\dfrac{\frac{2r}{\sqrt{1+r^2}}}{\frac{1+r}{1-r^2}} = \dfrac{2r}{\sqrt{1+r^2}} \cdot \dfrac{1-r^2}{1+r} = \dfrac{2r}{\sqrt{1+r^2}} \cdot \dfrac{1-r}{1} = \dfrac{2r-r^2}{\sqrt{1+r^2}}$.

SIMPLIFYING COMPLICATED FRACTIONS

Sometimes, fractions will have other fractions in their numerator or denominator (or both). To simplify and evaluate complicated fractions, evaluate the numerator and denominator separately, and then divide the two.

EXAMPLE E.5

Simplify each of the complicated fractions as much as possible.

a) $\dfrac{\frac{1}{x+h} - \frac{1}{x}}{h}$.

b) $\dfrac{(x+y)^{-1/2} + \sqrt{x+y}}{x+y}$.

c) $\dfrac{\frac{r}{r+1} + \frac{1}{r}}{\frac{r^2}{9-r} + \frac{r}{9-r}}$.

SOLUTION:

a) $\dfrac{\frac{1}{x+h} - \frac{1}{x}}{h} = \dfrac{\frac{x-(x+h)}{x \cdot (x+h)}}{h} = \dfrac{\frac{-h}{x \cdot (x+h)}}{h} = \dfrac{-h}{h \cdot x \cdot (x+h)} = \dfrac{-1}{x \cdot (x+h)}$.

b) $\dfrac{(x+y)^{-1/2} + \sqrt{x+y}}{x+y} = \dfrac{\frac{1}{\sqrt{x+y}} + \sqrt{x+y}}{x+y} =$

$\dfrac{\frac{1+\left(\sqrt{x+y}\right)\left(\sqrt{x+y}\right)}{\sqrt{x+y}}}{x+y} = \dfrac{1+x+y}{(x+y) \cdot \sqrt{x+y}} = \dfrac{1+x+y}{(x+y)^{3/2}}$.

c) $\dfrac{\frac{r}{r+1} + \frac{1}{r}}{\frac{r^2}{9-r} + \frac{r}{9-r}} = \dfrac{\frac{r^2+1}{r \cdot (r+1)}}{\frac{r^2+r}{9-r}} = \dfrac{r^2+1}{r \cdot (r+1)} \cdot \dfrac{9-r}{r^2+r} =$

$\dfrac{(r^2+1) \cdot (9-r)}{r \cdot (r+1) \cdot (r^2+r)} = \dfrac{-r^3+9r^2-r+9}{r^4+2r^3+r^2}$.

Exercises for Appendix E

For Problems 1–10, evaluate the quantity without using a calculator.

1. $\dfrac{2}{7} + \dfrac{3}{9}$.
2. $\dfrac{1}{2} - \dfrac{1}{3}$.
3. $\dfrac{-1}{4} + \dfrac{1}{2}$.
4. $\dfrac{2}{3y} + \dfrac{y}{8}$.
5. $\dfrac{2a}{x} - \dfrac{3}{7x}$.
6. $\dfrac{2}{7} \cdot \dfrac{a}{9b}$.
7. $\dfrac{\frac{7}{x}}{x^2}$.
8. $\dfrac{\frac{9}{b^2}}{\frac{27}{b^3}}$.

9. $\dfrac{3z^2}{\frac{2z}{11}}$.

10. $\dfrac{3+y}{3-y}+\dfrac{3}{y-3}$.

For Problems 11–20, perform the operation indicated. Simplify your answers as much as possible.

11. $L + w + \dfrac{L}{L+w}$.

12. $\dfrac{a-\dfrac{b}{a-\frac{b}{a}}}{}$

13. $\dfrac{(x+h)^2-(x-h)^2}{2h}$.

14. $\dfrac{2+b^{-1}}{7+b^{-2}}$.

15. $\dfrac{q^2-1}{q^2+2q+1}$.

16. $\dfrac{2}{7}+\dfrac{x}{x-4}$.

17. $\dfrac{1}{\sqrt{t}}+\dfrac{t^{3/2}}{\sqrt{t}-\frac{1}{\sqrt{t}}}$.

18. $\dfrac{\dfrac{x^2+2xy+y^2}{x}}{x^2+2xy+y^2}$.

19. $\dfrac{a^{-2}+a^{-3}}{1+a^{-1}}$.

20. $\dfrac{1+2\cdot\sin(\theta)\cdot\cos(\theta)}{\sin(\theta)+\cos(\theta)}$.

For Problems 21–25, decide whether each of the statements is true or false.

21. $\dfrac{1}{x}+\dfrac{1}{y}=\dfrac{2}{x+y}$.

22. $\dfrac{x+y}{y}=1+\dfrac{x}{y}$.

23. $\dfrac{y}{x+y}=1+\dfrac{y}{x}$.

24. $\dfrac{\frac{2}{x}}{\frac{x}{2}}$ … $\dfrac{\frac{2}{x}}{x^2}=2x$.

25. $\dfrac{2a^2+2ab+b^2}{a^2+2ab+b^2}=1+a^2$.

Appendix F: Solving Equations

THE GOAL OF SOLVING EQUATIONS

When you are trying to solve an equation like $x^2 = 4$, you are trying to determine all of the numerical values of x that you could plug into that equation. In this case, the numerical values that you could plug in are $x = 2$ and $x = -2$.

The numerical values that you find are called the solutions of the equation.

Some people find it helpful to interpret algebraic equations as verbal sentences in order to remind themselves of what they are trying to accomplish. Some examples are given in **Table F.1**.

Algebraic equation	Equivalent sentence	Solution(s)
$x^2 = 4$	"What numbers, when you square them, give you 4?"	$x = 2, x = -2$.
$x + 2 = 8$	"What number, when you add 2 to it, gives you 8?"	$x = 6$.
$2x = 16$	"What power of 2 gives you 16?"	$x = 4$.
$\frac{(x-1)(x+2)}{x^3+17} = 0$	"What numbers, when you plug them into the numerator, give you zero?"	$x = 1, x = -2$.

Table F.1.
Algebraic equations, equivalent sentences, and solutions.

SOLVING LINEAR EQUATIONS

A linear equation is one that only involves x—there are no powers of x, no radicals involving x, no fractions with x in the denominator.

Solving linear equations involves expanding any brackets, grouping like terms and simplifying as much as possible.

Linear equations always have exactly one solution.

EXAMPLE F.1

Solve the following equations to find x.

a) $3x + 1 = 10$.

b) $\frac{x+16}{4} = 9$.

c) $y = m \cdot x + b$.

SOLUTION:

a) $3x = 10 - 1 = 9$.

$x = 9/3 = 2$.

b) $x + 16 = 4 \cdot 9 = 36$.

$x = 36 - 20$.

c) $m \cdot x = y - b$.

$x = \frac{y-b}{m}$.

The working in part (c) involves the important assumption that m is not equal to zero.

SOLVING QUADRATIC EQUATIONS

Quadratic equations are equations that involve only x^2, x and constants. Quadratic equations may have one, two, or no solutions.

Quadratic equations can be solved by factoring.

EXAMPLE F.2

Find all solutions of the following quadratic equations. If you find less than two solutions, explain how you know that you have found all of the possible solutions of the quadratic equation.

a) $x^2 + 4x + 4 = 0$.

b) $x^2 + 9x = -18$.

c) $3x^2 + x = -2$.

SOLUTION:

a) The lefthand side of this quadratic expression is a perfect square,

$(x + 2)^2 = (x + 2)(x + 2) = 0$.

To find the values of x that can be plugged in, ask yourself, "What values of x should I plug in to make each factor equal to zero?" In this case, there is only one possible value of x, namely $x = -2$.

b) The first step here is to get all of the non-zero terms on the lefthand side of the equation:

$x^2 + 9x + 18 = 0.$

The lefthand side can now be factored and analyzed in the same was as (a). Factoring the lefthand side of the quadratic equation:

$(x + 6)(x + 3) = 0.$

The values of x that can be plugged in to make the factors equal to zero are $x = -6$ and $x = -3$. This quadratic equation has two solutions.

c) Proceeding in the same fashion as (b):

$3x^2 + x + 2 = 0.$

This quadratic cannot be factored. You can verify this by calculating the quantity $b^2 - 4ac$:

$b^2 - 4ac = 1 - 4 \cdot 3 \cdot 2 = -23.$

Since this is negative, the quadratic expression $3x^2 + x + 2$ cannot be factored. Because there are no factors, the quadratic equation:

$3x^2 + x + 2 = 0$

has no real number solutions.

SOLVING POWER AND RADICAL EQUATIONS

Power equations are equations involving only a single power of x and constants. For example,

$$x^2 = 4 + 0.5x^2$$

is a power equation, because it involves only a single power of x (in this case, the power is x^2) and constants.

An equation like

$$x^2 + 2x = 9$$

is not a power equation because it involves more than one power of x (this equation involves $x = x^1$ and x^2).

Radical equations are like power equations, except that in a radical equation, the power of x can be a fraction.

Solving power and radical equations is an application of the laws of exponents (see Appendix A):

$(a^n)^{1/n} = a$

$(a^{m/n})^{n/m} = a.$

EXAMPLE F.3

Find solutions of the following equations. If you are unable to find any solutions explain why it is mathematically impossible to find real numbers that satisfy the given equation.

a) $x^{18} = 5.$

b) $x^{1/2} = -1.$

c) $x^2 = -1.$

SOLUTION:

a) Applying the laws of exponents to both sides of the equation:

$(x^{18})^{1/18} = (5)^{1/18}$

$x = (5)^{1/18} = 1.09353243.$

b) Applying the laws of exponents to both sides of the equation, and noting that $\frac{1}{½} = 2$ gives:

$(x^{1/2})^2 = (-1)^2$

$x = 1.$

c) This equation has no real numbers x that can be plugged in to it. One way to see this is to convert the algebraic statement of the equation into a verbal sentence. In this case, the sentence might be something like, "What number, when squared, gives negative one?" Since squares are always greater than or equal to zero, it is impossible to find a real number that is negative when you square it.

SOLVING EQUATIONS INVOLVING FRACTIONS

The usual strategy for solving involving fractions is to get rid of the fraction by finding a least common denominator and then multiplying all terms by this common denominator. This will convert the equation involving fractions to one of the other types of equations.

Not every equation involving fractions will have a solution.

EXAMPLE F.4

Find the solutions of the following equations.

a) $\frac{1}{x}+\frac{1}{x+2}=\frac{4}{3}$.

b) $\frac{2x}{x-1}+\frac{1}{x^2-x}=5$.

c) $\frac{1}{x-1}=0$.

SOLUTION:

a) There are three fractions in this equation. The strategy is to find a common denominator, and then multiply all of the terms by this common denominator. In order to find the common denominator it is only strictly necessary to take notice of the fractions with x in the denominator. However, you can find a common denominator for all of the fractions in the equation and still solve the equation correctly. Both ways of working this problem are shown below.

Using the common denominator of $\frac{1}{x}$ and $\frac{1}{x+2}$:

The common denominator here is $x(x + 2)$. Multiplying all of the terms in this equation by this common denominator gives:

$$x(x+2)\cdot\left[\frac{1}{x}+\frac{1}{x+2}\right]=x(x+2)\cdot\left[\frac{4}{3}\right]$$

Multiply by the common denominator.

$$(x+2)+x=\frac{4}{3}x^2+\frac{8}{3}x$$

Expand the brackets.

$$\frac{4}{3}x^2+\frac{2}{3}x-2=0$$

Collect like terms.

$$\frac{4}{3}(x-1)(x+1.5)=0$$

Solve the quadratic by factoring.

$x = 1$, $x = -1.5$

The solutions of $\frac{1}{x}+\frac{1}{x+2}=\frac{4}{3}$.

Using the common denominator of all three fractions:

The common denominator here is $3x(x + 2)$. Multiplying all of the terms in this equation by the common denominator gives:

$$3x(x+2)\cdot\left[\frac{1}{x}+\frac{1}{x+2}\right]=3x(x+2)\cdot\left[\frac{4}{3}\right]$$

Multiply by the common denominator.

$$3(x + 2) + 3x = 4x(x + 2)$$

Expand the brackets.

$$4x^2 + 2x - 6 = 0$$

Collect like terms.

$$4(x - 1)(x + 1.5) = 0$$

Solve the quadratic by factoring.

$x = 1$, $x = -1.5$

The solutions of $\frac{1}{x}+\frac{1}{x+2}=\frac{4}{3}$.

b) There are several choices for the common denominator here. One of these choices is $(x^2 - x)$. This choice is the "least common denominator."

$$(x^2-x)\cdot\left[\frac{2x}{x-1}+\frac{1}{x^2-x}\right]=5\cdot(x^2-x)$$

Multiply by the common denominator.

$$2x^2 + 1 = 5x^2 - 5x$$

Expand the brackets.

$$3x^2 - 5x - 1 = 0$$

Collect like terms.

$$3(x + 0.18046)(x - 1.847) = 0$$

Solve the quadratic by factoring.

$x = -0.18046$, $x = 1.847$.

The solutions of $\frac{2x}{x-1}+\frac{1}{x^2-x}=5$.

You do not have to use the least common denominator when solving equations that involve fractions—any common denominator will do. However, the least common denominator will often lead to the simplest equation to solve after you have expanded the brackets and collected like terms.

c) The common denominator here is $(x - 1)$. Multiplying both sides of the equation by this common denominator gives:

$$(x-1)\cdot\left(\frac{1}{x-1}\right)=(x-1)\cdot 0$$

Multiply by the common denominator.

$1 = 0$

Expand the brackets.

The last equation is puzzling, because one and zero are not equal. The significance of this result can be understood by carefully examining the logic embodied in the working here. We usually start the working by assuming that the equation has at least one solution, and then proceed on this assumption with a series of algebraic manipulations designed to find the numerical values of the solution(s). If we are able to find sensible numerical values, then this confirms that the initial assumption (that there is at least one solution) is correct. If, on the other hand, the algebraic manipulations lead to nonsense (like $1 = 0$), then the initial assumption (that there is at least one solution) must be flawed. That is, if the algebraic manipulations lead to nonsense, then the equation does not have any solutions.

SOLVING EXPONENTIAL EQUATIONS

Solving an exponential equation usually involves finding the value of an exponent (x in the equation below) that will satisfy a given equation:

$$5^x = 10.$$

It is important to remember that exponentiation and multiplication are not the same thing. The solution to the equation $5^x = 10$ is $x = 1.43067$, not $x = 2$.

Solving exponential equations involves the use of the logarithm function. The logarithm rule most useful for solving exponential equations is:

$$\log(a^p) = p \cdot \log(a)$$

where a is a positive real number and p is any real number.

Natural logarithms (normally written "ln" instead of "log") can also be used to solve exponential equations. The working involved is basically identical, except that you write "ln" instead of "log."

EXAMPLE F.5

Use logarithms to solve the exponential equation:

$$5^x = 10.$$

SOLUTION:

Applying logarithms to both sides of the equation:

$$\log(5^x) = \log(10)$$

and using the law of logarithms to expand the left side:

$$x \cdot \log(5) = \log(10).$$

All that remains is to make x the subject of the equation and evaluate the logarithms numerically using a calculator:

$$x = \frac{\log(10)}{\log(5)} = \frac{1}{0.6989700043} = 1.430676558.$$

EXAMPLE F.6

Carbon-14 is a radioactive isotope of carbon with a half-life of 5730 years. The radioactive decay of carbon-14 is used in biology and archaeology to establish the dates of ancient relics. The mass, M, of carbon-14 remaining in a relic after T years is given by the formula:

$$M = A \cdot (0.9998790392)^T,$$

where A is the mass of carbon-14 in the relic when it was brand new.

The "mega-tooth" shark (*Charcharodon megalodon*) is an extinct, giant shark that scientists think was about the size of a Greyhound bus. Much of what we know about this shark comes from fossilized teeth that have been found in coastal regions of Virginia and North Carolina. One fossil tooth found had about 0.0000001% of the carbon-14 remaining. How old is the tooth?

Solution:

The objective of this problem is to find a value for T. Since T appears in the exponent, this will involve solving an exponential equation.

We are not told exactly how much carbon-14 is in the tooth in the beginning—that is, we are not told A in this problem. However, we are told the relationship between M and A. In symbols, this is:

$$M = 0.000000001 \cdot A.$$

Substituting this expression for M into the formula describing the decay of carbon-14 gives:

$$0.000000001 \cdot A = A \cdot (0.9998790392)^T.$$

This equation has two unknowns, A and T. We want to find the value of T. Luckily, A appears as a factor on both sides of the equation and can be canceled out:

$$0.000000001 = (0.9998790392)^T.$$

Applying logarithms to both sides of this equation, and rearranging to make T the subject of the equation:

$$\log(0.000000001) = \log((0.9998790392)^T)$$

$$\log(0.000000001) = T \cdot \log(0.9998790392)$$

$$T = \frac{\log(0.000000001)}{\log(0.9998790392)} = \frac{-9}{-0.0000525357854} = 171311.8 \cdot$$

From this data, the "mega-tooth" appears to be about 171,000 years old.

SOLVING EQUATIONS APPROXIMATELY USING A GRAPHING CALCULATOR

It is not easy to solve some straightforward equations. For example, the equation:

$$2^x - 2x = 0$$

has two solutions: $x = 1$ and $x = 2$. However, none of the techniques for solving different types of equations discussed in this appendix will help you to easily find these solutions. In such a situation, the graphing calculator can be a useful tool for finding approximate solutions to equations.

Example F.7

Find solutions for the equation:

$$2^x - 2x = 0.$$

Solution:

In this situation, you can use a graphing calculator to try to find the solution(s) to this equation by graphing the curve:

$$y = 2^x - 2x.$$

(See **Figure F.1** below.) In trying to find the places where $2^x - 2x = 0$, you are trying to find the places where $y = 2^x - 2x = 0$. The places where $y = 0$ are the x-intercepts of the curve $y = 2^x - 2x$. From Figure F.1, the x-intercepts of $y = 2^x - 2x$ are located at $x = 1$ and $x = 2$.

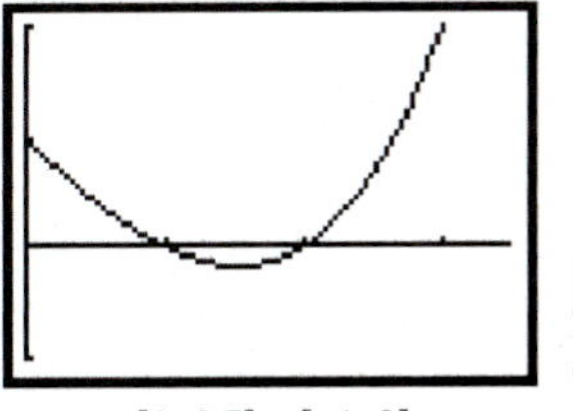

[0, 3.5] x [−1, 2]

Figure F.1. Finding solutions of $2^x - 2x = 0$ using a graphing calculator.

Example F.8

Find approximate solutions for the equation:

$$2^x - 2x = 1.$$

SOLUTION:

The working here follows a similar pattern to Example F.7. First, you use a graphing calculator to plot a graph of $y = 2^x - 2x$. This time, however, you are looking for the places where $y = 2^x - 2x = 1$, that is where the graph of $y = 2^x - 2x$ has height equal to one. **Figure F.2** shows a plot of $y = 2^x - 2x$ along with a plot of a horizontal line of height 1.

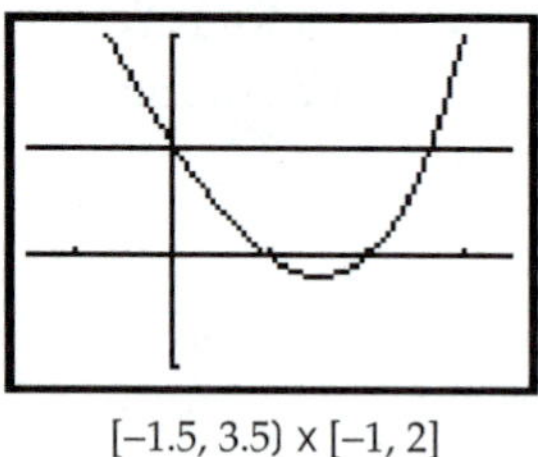

[–1.5, 3.5) x [–1, 2]

Figure F.2. Finding solutions of $2^x - 2x = 1$ using a graphing calculator.

The x-values where the curve and the horizontal line intersect are the solutions of the equation $2^x - 2x = 1$. Using the TRACE or INTERSECTION functions of a graphing calculator allows you to determine where these x-values are: $x = 0$ and $x \approx 2.659$.

Note that the second solution ($x \approx 2.659$) is not an exact solution of the equation. If you plug 2.659 into the equation, you get:

$2^{2.659} - 2 \cdot 2.659 = 0.99795$,

which is not precisely 1. The "solution" $x \approx 2.659$ is a close approximation to the true value of the solution.

Exercises for Appendix F

For Problems 1–10, find all of the solutions of the given equation.

1. $x^2 + 2x + 1 = 1$.
2. $3y + 4 = 9$.
3. $4 \cdot 5^t = 100$.
4. $5^{2t} + 2 \cdot 5^t + 1 = 4$.
5. $\sqrt{2w+1} = 11$.
6. $s^2 + 6s = -8$.
7. $\dfrac{y+2}{y} = 4$.
8. $18 - 3x = 3(12 + 2x)$.
9. $x = \sqrt{x+2}$.
10. $3 \cdot e^u = 12$.

For Problems 11–20, solve the given equation for the variable indicated.

11. $PV = nRT$. Solve for T.
12. $I = NeVA$. Solve for e.
13. $\sqrt{t^2 + a} = \dfrac{1}{a}$. Solve for t.
14. $\dfrac{z-1}{z+1} = Q$. Solve for z.
15. $e^{kt} = \dfrac{P}{A+P}$. Solve for P.
16. $x^2 + 2xy + y^2 = 9$ Solve for y.
17. $V = \frac{1}{3}\pi r^2 h$. Solve for h.
18. $\left(\dfrac{1}{3}\right)^x = e^{cx}$. Solve for c.
19. $s = ut + \frac{1}{2}at^2$. Solve for u.
20. $\dfrac{ax+b}{cx+d} = a$. Solve for x.

For Problems 21–25, use a calculator to find approximate solutions to the given equations. How many solutions does each equation have?

21. $2^x - 4x = 0$.
22. $x^2 + 2x = 0$.
23. $(10^x)^2 - 2 \cdot 10^x = -1$.
24. $2^{\sqrt{x}} = 4$.
25. $e^x + x = 0$.

Appendix G: Solving Systems of Equations

Often, you will be able to solve a problem that you are interested in by finding a solution for a single equation. More complicated problems may require you to solve more than one equation.

EXAMPLE G.1

A shoe company manufactures running shoes and walking shoes. Each day, the company orders in enough materials to make 80 pairs of shoes, and has 100 person-hours of labor available. Running shoes and walking shoes take the same amount of materials each. Each pair of walking shoes takes 1 hour to assemble, and each pair of running shoes takes 2 hours to assemble.

a) Represent this information using one or more equations.

b) If the company wants to use all of the materials and all of the labor each day, how many pairs of running and walking shoes should they manufacture?

SOLUTION:

a) Let R represent the number of pairs of running shoes that the company makes each day and W represent the number of pairs of walking shoes that the company makes each day.

The company has enough materials to make 80 pairs of shoes, or in other words the number of pairs of running shoes plus the number of pairs of walking shoes should add up to ninety. Expressing this as an equation:

$R + W = 80.$

Each pair of running shoes takes 2 hours to assemble, and each pair of walking shoes takes 1 hour to assemble. Each day, there are 100 hours of labor available, so:

$2R + W = 100.$

b) In order to find the number of pairs of each type of shoe that the company should manufacture, we have to find a numerical value of R and a numerical value of W. We want to find numerical values that satisfy both of the equations:

$$R + W = 80 \qquad (1)$$
$$2R + W = 100 \qquad (2)$$

One way to proceed is to make W the subject of Equation (1),

$W = 80 - R$
and to substitute this for W in Equation (2):
$2R + (80 - R) = 100.$

This gives an equation that only involves R, and allows the numerical value for R to be determined:

$2R + (80 - R) = 100.$
$R = 100 - 80.$
$R = 20.$

Knowing the numerical value for R, you can substitute this back into either Equation (1) or Equation (2) to find the numerical value of W:

$20 + W = 80.$
$W = 80 - 20.$
$W = 60.$

The final conclusion to this problem is that the company should manufacture 20 pairs of running shoes and 60 pairs of walking shoes.

The collection of equations that you wish to solve is called a system of equations. In order to be certain that you will obtain a solution, you need to have the same number of equations as variables. In Example G.1, there were two variables (R and W) and two equations (one for materials and one for labor). In Example G.1, all of the equations were linear equations. The equations in a system of equations will often be linear, but they do not have to be.

EXAMPLE G.2

Strontium-90 is a radioactive substance. Strontium-90 is sometimes released during nuclear accidents and is regarded as a very dangerous substance because it becomes incorporated into bone tissue. The mass, M, of strontium-90 remaining in a person's bones after T years is described by an exponential decay formula:

$$M = M_0 \cdot B^T,$$

where M_0 is the mass of strontium-90 that was absorbed into a person's bones when exposed to radioactivity, and B is a number.

Bone samples indicated that 12 years after exposure, the person had 0.75 grams of strontium-90 in his/her body, and 30 years after exposure, this person had 0.488 grams of strontium-90 in his/her body.

a) What is the value of the constant B ?

b) How much strontium-90 did the person have in his/her body just after being exposed to the radiation?

SOLUTION:

Before attempting to answer questions (a) and (b), it can be useful to identify the important information in the problem, and put it all together.

We are told that when $T = 12$, $M = 0.75$. Putting this together with the equation gives:

$$0.75 = M_0 \cdot B^{12} \qquad (1)$$

Likewise, we are told that when $T = 30$, $M = 0.488$, so that:

$$0.488 = M_0 \cdot B^{30} \qquad (2)$$

There are two equations (Equations (1) and (2)), and two unknowns, M_0 and B.

a) The strategy in Example G.1 was to try to combine the two equations in order to eliminate one of the variables. This can be accomplished here by division to cancel out the M_0's:

$$\frac{0.488}{0.75} = \frac{M_0 \cdot B^{30}}{M_0 \cdot B^{12}} = \frac{B^{30}}{B^{12}} = B^{30-12} = B^{18}.$$

(The Laws for Exponents were used to combine the powers of B—see Appendix A.) We are left with an equation that includes only one of the unknown quantities, B:

$$B^{18} = 0.65$$

$$B = (0.65)^{1/18} = 0.9764.$$

b) In Example G.1, once the numerical value of one of the quantities was known, the numerical value of the other quantity could be found by plugging back into one of the original equations. To find the numerical value of M_0, we can plug $B = 0.9764$ into Equation (1) and solve for M_0:

$$0.75 = M_0 \cdot (0.9764)^{12}$$

$$\frac{0.75}{(0.9764)^{12}} = M_0$$

$$M_0 = 0.9989.$$

So, just after exposure to the radiation, the person had 0.9989 grams of strontium-90 in their bones.

Exercises for Appendix G

For Problems 1–5, solve the system of linear equations.

1. $x + y = 2.$
   $x - y = -2.$
2. $3p + 2q = 10.$
   $p - q = 0.$
3. $r - 2s = 1.$
   $s + r = 4.$
4. $x = 8y + 9.$
   $x + 2y = 29.$
5. $t = 4w + 9.$
   $t = 37.$

For Problems 6–10, solve the system of equations.

6. $x^2 + y^2 = 1.$
   $x - y = 0.$
7. $A \cdot e^{2k} = 10.$
   $A \cdot e^{4k} = 20.$
8. $r = 4 - s^2.$
   $s + r = 1.$
9. $y = 2^x.$
   $y = 2x.$
10. $p = \dfrac{1}{q+1}.$
    $p - q = 1.$

Appendix H: Interpreting and Working with Inequalities

THE GOAL OF INTERPRETING INEQUALITIES

The goal of solving an equation was usually to find one or more numbers that could be plugged into the equation.

The goal of interpreting an inequality is to determine a range of numerical values that work in the inequality.

EXAMPLE H.1

Use words to describe the range of numerical values that work in each of the following inequalities.

a) $-2 \leq x \leq 7$.

b) $-2 < x < 7$.

c) $100 > r \geq 4$.

d) $-\infty < y < 9$.

SOLUTION:

a) The range of numerical values is all real numbers between –2 and 7, including both –2 and 7.

b) The range of numerical values is all real numbers between –2 and 7, but not including either –2 or 7.

c) The range of numerical values is all real numbers between 4 and 100, including 4, but not including 100.

d) The range of numerical values is all real numbers less than 9, not including 9.

MANIPULATING INEQUALITIES

Inequalities may be manipulated in many of the same ways as equations can be manipulated. The main difference between manipulation of inequalities and manipulation of equations lies in the effects of the distributive law.

Suppose that a, b and c are all real numbers, and that $a < b$. Then:

$a + c < b + c$

$a - c < b - c$

$c \cdot a < c \cdot b$,
provided that c is a positive number.

$c \cdot a > c \cdot b$,
provided that c is a negative number.

The last two rules indicate how manipulating inequalities differs from manipulating equations. If you multiply (or divide) an inequality by a negative number, then this reverses the direction of the inequality.

WORKING WITH LINEAR INEQUALITIES

Linear inequalities are like linear equations, except that the equality (=) is replaced by an inequality (<, >, ≤, or ≥). Linear inequalities may be manipulated in exactly the same way as linear equations, except that multiplying (or dividing) by a negative number reverses the direction of the inequality.

EXAMPLE H.2

Use words to describe the range of numerical values that work in each of the following inequalities.

a) $-2x + 9 \leq x - 7$.

b) $\frac{x+1}{-2} \geq 4$.

c) $2 < 3x + 1 < 11$.

SOLUTION:

a) If you were solving the linear equation

$-2x + 9 = x - 7$,

a reasonable strategy would be to group like terms, and then make x the subject of the equation. The working to implement this strategy would look something like this:

$-2x - x = -7 - 9$
Group like terms.

$-3x = -16$
Simplify by adding the like terms together.

$x = \frac{-16}{-3} = \frac{16}{3}$.
Make x the subject by dividing both sides by -3.

The working to interpret the inequality is similar, except that you have to remember that multiplying (or dividing) by a negative number reverses the direction of the inequality.

$-2x - x \leq -7 - 9$
Group like terms.

$-3x \leq -16$
Simplify by adding the like terms together.

$x \geq \frac{-16}{-3} = \frac{16}{3}$.
Make x the subject by dividing both sides by -3.

The direction of the inequality is reversed.

The range of numerical values that satisfy the inequality is the set of numbers that are greater than or equal to 16/3.

b) Multiplying both sides of the inequality by -2 will get rid of the fraction, but it will also reverse the direction of the inequality.

$x + 1 \leq -8$.

Subtracting one to each side of this inequality will make x the subject of the inequality:

$x \leq -9$.

The range of numerical values that satisfy the inequality is the set of numbers that are less than or equal to -9.

c) When manipulating the inequality,

$2 < 3x + 1 < 11$,

you need to be careful to perform the manipulations on the left side, the middle, and the right side of the inequality. Like the previous cases, the objective is to isolate x.

Subtracting one:

$2 - 1 < 3x < 11 - 1$

$1 < 3x < 10$.

Dividing by 3:

$\frac{1}{3} < x < \frac{10}{3}$.

The range of numerical values that satisfy the inequality are the set of numbers that are greater than 1/3, but less than 10/3, including neither 1/3 nor 10/3.

MORE COMPLICATED INEQUALITIES

Manipulating inequalities may involve raising both sides of an inequality to a power, or manipulating fractions that have x in the denominator. Manipulating these more complicated inequalities is possible with the following rules.

Suppose that a and b are positive real numbers with $a < b$. Then:

$a^2 < b^2$

$\frac{1}{a} > \frac{1}{b}$.

One way to remember the second rule is that when you "flip" fractions, you reverse the direction of the inequality, i.e. if

$\frac{a}{1} < \frac{b}{1}$ then $\frac{1}{a} > \frac{1}{b}$.

EXAMPLE H.3

Use words to describe the range of numerical values that work in each of the following inequalities.

a) $0 < \frac{1}{x+3} \leq 4$.

b) $1 < \sqrt{3-x} < 2$.

c) $\frac{x-3}{\sqrt{x}} > 1.$

SOLUTION:

a) There are two things to be careful of here. Firstly, in order to isolate x, it will be necessary to flip the fraction,

$\frac{1}{x+3}$,

and you will need to remember that flipping a fraction reverses the direction of the inequality. Secondly, the left side of the inequality is zero. This cannot be flipped, as it is impossible to have zero in the denominator of a fraction.

To get around this, you can break this into two inequalities and consider them separately.

To ensure that $0 < \frac{1}{x+3}$, it is enough to ensure that $0 < x + 3$, and this is accomplished when $x > -3$.

To ensure that $\frac{1}{x+3} \leq 4$, flip both sides of this inequality and reverse the direction of the inequality:

$x + 3 \geq 1/4$

$x \geq 13/4.$

So, the range of numerical values that work in the inequality $0 < \frac{1}{x+3} \leq 4$ are all numbers that are greater than or equal to 13/4.

b) Since all of the numbers and quantities involved in the inequality are positive, squaring the numbers and quantities will not alter the directions of the inequalities. Squaring all of the parts of the inequality:

$1 < 3 - x < 4$

$-2 < -x < 1$

$2 > x > -1.$

(Note that multiplying through by -1 has reversed the directions of the inequalities.)

So, all numerical values between -1 and 2 will work in the inequality.

c) As a first manipulation designed to get rid of the fraction, you could multiply both sides of the inequality by $\sqrt{x}$:

$x - 3 > \sqrt{x}.$

This is valid, so long as the square root is positive. The next manipulation would be to square both sides of the inequality to eliminate the square root:

$(x - 3)^2 > x.$

Next, multiplying out the brackets, collecting like terms gives:

$x^2 - 6x + 9 > x$

$x^2 - 7x + 9 > 0.$

The x-values where $x^2 - 7x + 9 > 0$ can be determined by factoring and studying each factor (see later in this appendix), or by graphing

$y = x^2 - 7x + 9$

on a computer or calculator (see **Figure H.1**). The graph shows that $x^2 - 7x + 9 > 0$ when $x < 1.697$ and when $x > 5.30277$. The original inequality,

$\frac{x-3}{\sqrt{x}} > 1$

is satisfied when $x > 5.30277$.

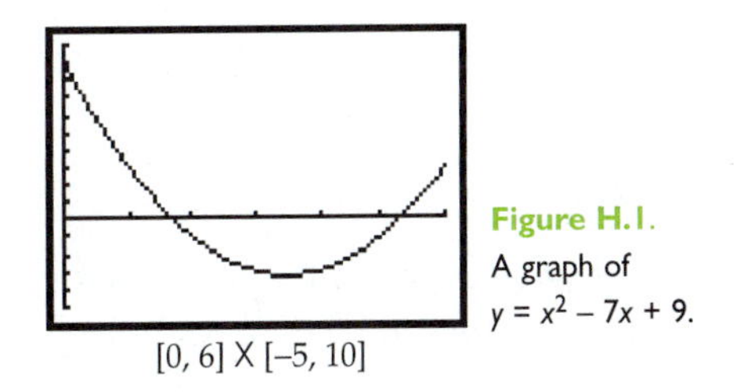

Figure H.1. A graph of $y = x^2 - 7x + 9$.

DETERMINING THE SIGN OF AN ALGEBRAIC EXPRESSION

One of the most important applications of calculus is optimization. Optimization is the science of determining how a process can be most efficiently, profitably, or cost-effectively run. Part of a typical optimization problem from calculus involves deciding where an algebraic expression is positive and where it is negative.

This process can be accomplished by factoring the algebraic expression as much as you possibly can, determining where each factor is positive and negative, and finally combining this information using the familiar rules for multiplying positive and negative numbers:

(positive) X *(positive)* = *(positive)*

(positive) X *(negative)* = *(negative)*

(negative) X *(negative)* = *(positive)*

EXAMPLE H.4

Determine where the algebraic expression,

$x^3 + 3x^2 + 2x$

is positive, and where it is negative.

SOLUTION:

We begin by factoring $x^3 + 3x^2 + 2x$ as much as is possible. Each term in this expression has a common factor of x, so you can factor that out:

$x^3 + 3x^2 + 2x = x(x^2 + 3x + 2).$

The brackets contain a quadratic expression that can be factored in the usual fashion (see Appendix B and Appendix D) to:

$x^2 + 3x + 2 = (x + 1)(x + 2).$

Thus,

$x^3 + 3x^2 + 2x = x(x + 1)(x + 2).$

Next, the places where each of the three factors is positive and negative can be determined.

The first factor, x, is positive when $x > 0$, and negative when $x < 0$. The second factor, $x + 1$, is positive when $x > -1$ and negative when $x < -1$. Lastly, the third factor, $x + 2$, is positive when $x > -2$, and negative when $x < -2$.

Number lines can be a useful device for organizing the information about the three factors (see **Figure H.2**). The information about the expression $x^3 + 3x^2 + 2x$ can be deduced by reading down the number lines, and applying the rules for multiplying positive and negative numbers.

	$x < -2$	$-2 < x < -1$	$-1 < x < 0$	$x > 0$
x	– – –	– – –	– – –	+ + +
$x + 1$	– – –	– – –	+ + +	+ + +
$x + 2$	– – –	+ + +	+ + +	+ + +
Product of the three	– – –	+ + +	– – –	+ + +

(Boundaries at –2, –1, 0.)

Figure H.2.
Number lines for determining the sign of $x^3 + 3x^2 + 2x$.

So, the conclusion is that the algebraic expression $x^3 + 3x^2 + 2x$ is:

- positive when $-2 < x < -1$ and when $x > 0$, and,
- negative when $x < -2$ and when $-1 < x < 0$.

You can check these conclusions by graphing the algebraic expression $y = x^3 + 3x^2 + 2x$ on a computer or calculator (see **Figure H.3**).

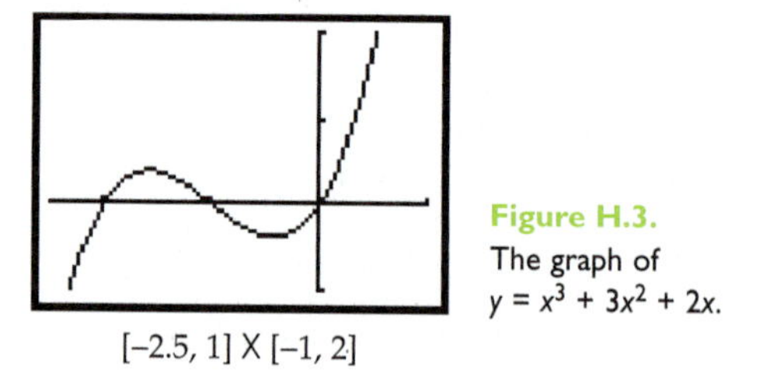

[–2.5, 1] X [–1, 2]

Figure H.3.
The graph of $y = x^3 + 3x^2 + 2x$.

In today's world where you have ready access to graphing calculators and computers, the ability to deduce where an algebraic expression has positive numerical values and where it has negative numerical values is probably not as important as it used to be. After all, you can just graph a formula and see for yourself where the graph is above the x-axis (this is where the formula is positive) and where the graph is below the x-axis (this is where the formula is negative). The kind of analysis presented here is still important for three reasons.

1. If you don't have a calculator available to you for some reason, then you can still determine the sign of an algebraic expression.
2. The skills you develop here can help you to build an intuitive sense of how algebraic expressions behave. This intuition can help to guide your efforts to analyze mathematical expressions and problems more effectively.
3. You can double-check the answers you get using a calculator or computer—especially if the algebraic expression is one that a calculator is not able to display very well.

INTERVAL NOTATION

An easy way to write ranges of numerical values is to use to use interval notation. A closed interval $[a, b]$ indicates all real numbers x for which $a \le x \le b$. Closed intervals include their endpoints. An open interval (a, b) indicates all real numbers x for which $a < x < b$. Open intervals exclude their endpoints. Half-open or half-closed intervals are denoted by $(a, b]$ or $[a, b)$. These notations indicate all real numbers x for which $a < x \le b$ (that is, open at a and closed at b) and $a \le x < b$ (open at b and closed at a) respectively.

Caution: The notation (3, 4) can be interpreted as the point (3, 4) or as the open interval consisting of all numbers strictly between 3 and 4. Context determines the meaning.

Exercises for Appendix H

For Problems 1–10, use words to describe the range of numerical values that will satisfy the given inequality.

1. $x + 3 > 0$.
2. $2y - 1 < 2$.
3. $\frac{1}{3x+1} < 2$.
4. $\sqrt{u-1} \ge 3$.
5. $|x| < 9$.
6. $\frac{1}{x^2} > 4$.
7. $\frac{1-x}{x^2+7} \le 0$.
8. $w^2 + 2w + 1 < 0$.
9. $-3(p + 7) > 0$.
10. $2 + \frac{1}{x} > 0$.

For Problems 11–20, use words to describe the range of numerical values that will satisfy the given inequality.

11. $0 < \frac{x}{\sqrt{1+x^2}} < 1$.
12. $-1 < 4x + 3 < 1$.
13. $1 < \sqrt{1+x^2} < 2$.
14. $-2 < \frac{1-2u^2}{1+2u^2} < 3$.
15. $0 < \frac{1}{x+1} - 2 < 1$.
16. $0 < t^2 + t - 2$.
17. $0 < \sqrt{\frac{L}{32}} < 1$.
18. $-6 < |y + 5| < 10$.
19. $\frac{1}{x^2} > \frac{1}{4x-4}$.
20. $\sqrt{x^2+1} < \sqrt{1-x}$.

For Problems 21–25, decide where each of the following algebraic expressions is positive and where each of the following algebraic expressions is negative.

21. $w^2 - 4$.
22. $u^3 - u$.
23. $e^t \cdot (t + 1)$.
24. $\frac{r+2}{r-1}$.
25. $\frac{y^2+3y+2}{y+3}$.

Solutions for Appendices Exercises

APPENDIX A

1. 49.
2. 1.
3. 1/9.
4. 81.
5. 1.
6. –1.
7. –1.
8. 2.
9. –2.
10. 8.
11. $2^e \cdot e^{3t+1}$.
12. $2^n xy$.
13. $(a + b)^3$.
14. $|T| \cdot w^2$.
15. e^{r+1}.
16. $(4w^4y^3)/(w + y)$.
17. $C^2 \cdot e^{2wt}$.
18. $(1/5)xy^3$.
19. $6 \cdot |u| \cdot (v)^{1/2}$.
20. $2P^2$.
21. False.
22. False.
23. False.
24. False.
25. False.
26. The basic idea here is cancellation of common factors from the numerator and denominator. There are n factors of a in the numerator and m factors of a in the denominator. In the case where $n > m$, all of the factors of a in the denominator will cancel with factors of a in the numerator, leaving $n - m$ factors of a remaining in the numerator. In the case where $n < m$, the n factors of a in the numerator will cancel with factors in the denominator, leaving $m - n$ factors of a in the denominator. Thus, the simplified expressions will be: $(1/a)^{m-n} = a^{-1(m-n)} = a^{n-m}$.
27. The basic idea here is that you will have n factors, each of which is a/b. Thus, the overall numerator will consist of n copies of a all multiplied together, i.e. the overall numerator is a^n. Similarly, the overall denominator is b^n.

APPENDIX B

1. $-7t + 1$.
2. $5y + 5$.
3. $14h + 8$.
4. $9x + 36$.
5. $-2s$.
6. $x^3 + x$.
7. $18y^3 + 27y^2$.
8. $4r^2 + 32r + 65$.
9. $8wp + 2w - 4p - 1$.
10. $4w^2 - 4w + 1$.
11. $6xy - 3y^2 + 6y + 3$.
12. $-9w^3$.
13. $zy + 3z + 2y + 6$.
14. $-6x^2 - 12x - 6$.
15. $25u^2 + 1$.
16. $t^2 + t + 1$.
17. $(r^3 + r^2)/2 + r + 1$.
18. $x^3 + x^2y - xy^2 - y^3$.
19. $x^4 - 2x^2y^2 + y^4$.
20. 1.

APPENDIX C

1. $x(1 - x) + 1$.
2. $(x - 1)(x - 2)$.
3. $x(\ln(x) - 1)$.
4. $x(\sin(x) + \cos(x))$.
5. Without assuming anything about c and d, there is no way to factor this that provides obvious simplifications.
6. $2(x + 2y^2)$.
7. $(a + 2)(a - 2)$.
8. $A \cdot e^{2t}(1 + t + t^2)$.
9. $(p + q)(p - q)$.
10. $((1 + 2t)^2 + w)((1 + 2t)^2 - w)$.
11. $a(a + 2ac) + 2c^2$, or $a^2 + 2c(a + c)$, or $a(a + c) + c(a + 2c)$.
12. $4(x + 4)$.
13. $(1 + \cos(u))(1 - \cos(u))$.
14. $A(1 + r)^3$.
15. $x(x + 2h)$.
16. $(x + 2)(x + 4)$.
17. $3(x + 1)(x + 2)$.
18. $(e^{2t} + 1)^2$.
19. $(y + 3)(y + 4)$.
20. This does not factor. To see this, $b^2 - 4ac = 1 - 8 = -7$, which is negative.

APPENDIX D

1. $(y + 1)^2 - 1$.
2. $2(a + 2)^2 - 3$.
3. $(u - 7)^2 - 49$.
4. $(x^2 + 2)^2 - 3$.
5. $3(r + 2)^2 + 4$.
6. $-(x - 5)^2 + 24$.
7. $(t - 3.5t) - 20.25$.
8. x^2.
9. $3(r - 4)^2 - 34$.
10. $(w + 3/2)^2 - 2.25$.
11. $(r + 3)(r + 4)$.
12. $2(y + 1)(y + 3/2)$.
13. $e(e^x + 2)^2$.
14. $3(t + 3.468)(t + 0.865)$.
15. This quadratic does not factor. To see this, $b^2 - 4ac = 16 - 4(-1)(-5) = -4$.

APPENDIX E

1. $13/21$.
2. $1/6$.
3. $1/4$.
4. $(16 + 3y^2)/(24y)$.
5. $(14a - 3)/(7x)$.
6. $(2a)/(63b)$.
7. $7/(x^3)$.
8. $b/3$.
9. $(33z)/2$.
10. $y/(3 - y)$.
11. $(L^2 + 2wL + w^2 + L)/(L + w)$.
12. $a(a^2 - 2b)/(a^2 - b)$.
13. $2x$.
14. $b(2b + 1)/(7b^2 + 1)$.
15. $(q - 1)/(q + 1)$.
16. $(9x - 8)/(7x - 28)$.
17. $(t^2 + t^{1/2} + t^{-1/2})/(t - 1)$.
18. $1/x$.
19. $1/a^2$.
20. $\sin(\theta) + \cos(\theta)$.
21. False.
22. True.

23. False.

24. False.

25. False.

APPENDIX F

1. $x = 0$ and $x = -2$.
2. $y = 5/3$.
3. $t = 2$.
4. $t = 0$.
5. $w = 60$.
6. $s = -2$ and $s = -4$.
7. $y = 2/3$.
8. $x = -2$.
9. $x = 2$.
10. $u = \ln(4) \approx 1.38629$.
11. $T = (PV)/(nR)$.
12. $e = I/(NVA)$.
13. $t = \pm(1/a^2 - a)^{1/2}$.
14. $z = (1 + Q)/(1 - Q)$.
15. $P = Ae^{kt}/(1 - e^{kt})$.
16. $y = 3 - x$, or $y = -3 - x$.
17. $h = (3V)/(\pi r^2)$.
18. $c = \ln(1/3) \approx -1.0986122289$.
19. $u = (s - 0.5at^2)/t$.
20. $x = (ad - b)/(a - ac)$.
21. There are two solutions. One is located near $x = -2.454386$, and the other is located near $x = 4.8845$.
22. There are no solutions of this equation.
23. There is only one solution: $x = 0$.
24. There is only one solution: $x = 4$.
25. There is only one solution, located near $x = -0.5671433$.

APPENDIX G

1. $x = 0$ and $y = 2$.
2. $p = 2$ and $q = 2$.
3. $s = 1$ and $r = 3$.
4. $x = 25$ and $y = 2$.
5. $t = 37$ and $w = 7$.
6. There are two solutions: $(x, y) = (1/\sqrt{2}, 1/\sqrt{2})$ and $(x, y) = (-1/\sqrt{2}, -1/\sqrt{2})$.
7. $A = 5$ and $k = \ln(2)/2 \approx 0.34657359$.
8. This system of equations has no solutions.
9. There are two solutions: $(x, y) = (1, 2)$ and $(x, y) = (2, 4)$.
10. There are two solutions to this equation: $(p, q) = (1, 0)$ and $(p, q) = (-1, -2)$.

APPENDIX H

1. All real numbers greater than –3.
2. All real numbers less than 3/2.
3. All real numbers greater than –1/2.
4. All real numbers greater than or equal to 10.
5. All real numbers between –9 and +9 (not including either –9 or +9).
6. All real numbers between –1/2 and +1/2, excluding –1/2, +1/2 and zero.
7. All real numbers greater than or equal to 1.
8. There are no real numbers that satisfy this inequality.
9. All real numbers less than –7.
10. All real numbers greater than zero, and all real numbers less than –1/2.
11. All real numbers greater than 0.
12. All real numbers greater than –1 and less than –1/2.
13. All real numbers greater than 0 and less than 1.
14. All real numbers.

15. All real numbers greater than 1/3 and less than 1/2.
16. All real numbers greater than 1 and all real numbers less than –2.
17. All real numbers greater than 0 and less than 32.
18. All real numbers greater than –15 and less than 5.
19. All real numbers less than 1.
20. All real numbers greater than –1 and less than zero.
21. Positive when $w < -2$ or when $w > 2$. Negative when $-2 < w < 2$.
22. Positive when $-1 < u < 0$, and when $u > 1$. Negative when $u < -1$ and when $0 < u < 1$.
23. Positive when $t > -1$. Negative when $t < -1$.
24. Positive when $r > 1$ and when $r < -2$. Negative when $-2 < r < 1$.
25. Positive when $-3 < y < -2$ and when $y > -1$. Negative when $y < -3$ and when $-2 < y < -1$.

Index

Defined terms and their page numbers are in bold.

Symbols

A

B

C

D

E

H

I

K

L

M

N

O

P

U

V

W

X

Y

Z

Acknowledgements

PROJECT LEADERSHIP

Solomon Garfunkel
COMAP, Inc., Lexington, MA

Nancy Chrisler,
Pattonville School District, St. Ann, MO

Gary Froelich
COMAP, Inc., Lexington, MA

AUTHORS

Nancy Crisler
Pattonville School District, St. Ann, MO

Gary Froelich
COMAP, Inc., Lexington, MA

CONTRIBUTING AUTHORS

Sharon North
St. Louis Community College, St. Louis, MO

Dale Winter,
Harvard University, Cambridge, MA

REVIEWERS

Henry Pollak
Teachers College, Columbia University, NY

Landy Godbold
The Westminster Schools, Atlanta, GA

Donnie Hallstone
Green River Community College, WA

Donna Brouillette
Georgia Perimeter College, Lilburn, GA

Daniel Harned
Michigan State University, East Lansing, MI

Joanne Brunner
Joliet Junior College, Joliet, IL

Kathy Rodgers
University of Southern Indiana, Evansville, IN

Tom Rosseau
Siena College, Loudonville, NY

Paul Cox
Ricks College, Rexburg, ID

Sue Neal
Wichita State University, Wichita, KS

ASSESSMENT

Dédé de Haan, Jan de Lange,
Anton Roodhardt, Henk van der Kooij
The Freudenthal Institute, The Netherlands

COMAP STAFF

Solomon Garfunkel, Laurie Aragón, Jan Beebe Sheila Sconiers, Sue Rasala, Gary Froelich, Roland Cheyney, Sue Martin, Lynn Aro, Nancy Crisler, George Ward, Daiva Kiliulis, Gail Wessell, Pauline Wright, Michele Doherty, Gary Feldman, Clarice Callahan, Rafael Aragón, Kevin Darcy, Frank Giordano

ART

Lianne Dunn
Sandwich, MA

Mary Reilly
Sommerville, MA

Tom Thaves
Solana Beach, CA
Frank and Ernest reprinted with permission

PHOTO RESEARCH

Michele Doherty
COMAP, Inc.

Susan Van Etten
Wenham, MA

COVER ART

(Clockwise from upper left)
map: Corbis; calculator: courtesy of Texas Instruments Incorporated; traffic: Jim Schafer/eStock Photography/Picture Quest; jet: Corbis; seismograph: Roger Ressmeyer/Corbis; temperature sign: Patrick Endres/Alaska Stock; logs: Bill Varie/Corbis

INDEX EDITOR

Seth Maislin
Focus Publishing Services, Arlington, MA